T0251409

FOOD STORAGE STABILITY

Edited by
IRWIN A. TAUB
R. PAUL SINGH

CRC Press
Taylor & Francis Group
Boca Raton London New York

CRC Press is an imprint of the
Taylor & Francis Group, an **informa** business

Acquiring Editor:	Ron Powers
Project Editor:	Albert W. Starkweather, Jr.
Cover design:	Denise Craig

Library of Congress Cataloging-in-Publication Data

Food storage stability / [edited by] Irwin A. Taub and R. Paul Singh.
      p. cm.
   Includes bibliographical references and index.
   ISBN 0-8493-2646-X (alk. paper)
   1. Food—Storage. 1. Food—Quality. I. Taub, Irwin A. II. Singh, R. Paul
TP373.3.F68 1997
631.5′68—dc21   97-41662
                                                                           CIP

The Editors

Dr. Irwin A. Taub is a Senior Research Scientist at the U.S. Army Natick Research, Development, and Engineering Center. He was educated as a chemist and has been involved in diverse research studies in physical chemistry and in food science and technology. These studies include such topics as inorganic complexation; free radical reactions; the radiation chemistry of ice, proteins, and fats; thermal processing of food; the structural stability of foods; time–temperature indicators; chemical heaters; and food ingredients that enhance mental and physical performance.

Dr. Taub received his B.S. in chemistry from Queens College in New York City in 1955 and his Ph.D. in inorganic and physical chemistry from the University of Minnesota in 1961. He did post-doctoral work at Argonne National Laboratory from 1961 to 1963 and then spent six years as a Research Fellow in the Radiation Laboratory at Carnegie-Mellon University.

Dr. Taub came to Natick in 1969 to participate in the Radiation Preservation of Food Program, and focused his efforts on establishing the chemiclearance principle for approving irradiated foods. After this program was transferred in 1980 to the U.S. Department of Agriculture, he served in various capacities at Natick involving food research, including Chief of the Technology Acquisition Division.

Upon appointment in 1991 as a Department of Defense Senior Research Scientist, Dr. Taub began a research program that currently is focused on the relation of food structure to chemical reactions in foods, on predicting the shelf life of stored foods, on validating the microwave and ohmic sterilization of foods, on gaining approval for irradiated foods, and on formulating rations to enhance mental and physical performance. He is the recipient of two Department of Defense Awards for Technical Achievement, one for work on the chemiclearance principle relating to irradiated foods and the other for work on the structure of bread and its relationship to storage stability.

Dr. Taub has lectured extensively, and has published in all of the fields described. He has written several book chapters, edited symposia and conference proceedings, and authored more than 60 papers.

R. Paul Singh, Ph.D., is a Professor of Food Engineering, Department of Biological and Agricultural Engineering, Department of Food Science and Technology, University of California, Davis.

Dr. Singh graduated in 1970 from Punjab Agricultural University, Ludhiana, India, with a degree in Agricultural Engineering. He obtained an M.S. degree from the University of Wisconsin, Madison, and a Ph.D. degree from Michigan State University in 1974. Following a year of teaching at Michigan State University, he moved to the University of California, Davis, in 1975 as an Assistant Professor of Food Engineering. He was promoted to Associate Professor in 1979 and, again, to Professor in 1983.

Dr. Singh is a member of the Institute of Food Technologists, American Society of Agricultural Engineers, and Sigma Xi. He received the Samuel Cate Prescott Award for

Research, Institute of Food Technologies, in 1982, and the A. W. Farrall Young Educator Award, American Society of Agricultural Engineers in 1986. He was a NATO Senior Guest Lecturer in Portugal in 1987 and 1993, and received the IFT International Award, Institute of Food Technologists, 1988, and the Distinguished Alumnus Award from Punjab Agricultural University in 1989, and the DFISA/FPEI Food Engineering Award in 1997.

Dr. Singh has authored and co-authored nine books and more than 160 technical papers. He is a co-editor of the *Journal of Food Process Engineering.* His current research interests are in studying transport phenomena in foods as influenced by structural changes during processing.

Contributors

Ronald E. Barnett
Research and Development
Campell Soup Co.
Camden, NJ

Toralf Boehme
Federal Research Center
 for Nutrition
Karlsruhe, Germany

Larry D. Brown
University of Missouri
 College of Veterinary Medicine
Columbia, MO

Christine M. Bruhn
University of California at Davis
 Center for Consumer Research
Davis, CA

Armand V. Cardello
U.S. Army Natick RD & E Center
Natick, MA

Pavinee Chinachoti
University of Massachusetts
 Chenoweth Laboratory
Amherst, MA

Fergus M. Clydesdale
University of Massachusetts
 Chenoweth Laboratory
Amherst, MA

Eugenia A. Davis (deceased)
University of Minnesota
 Department of Food Science
 and Nutrition
St. Paul, MN

John T. Fruin
Florida Department of Agriculture
 and Consumer Services
Tallahassee, FL

Jack R. Giacin
University of Michigan
 School of Packaging
East Lansing , MI

Joan Gordon, Professor Emeritus
University of Minnesota
 Department of Food Science and Nutrition
St. Paul, MN

Norman F. Haard
University of California at Davis
 Food Science and Techology
Davis, CA

Ruben J. Hernandez
University of Michigan
 School of Packaging
East Lansing , MI

Adel A. Kader
University of California at Davis
 Department of Pomology
Davis, CA

Hie-Joon Kim
College of Natural Sciences
 Department of Chemistry
Seoul, Korea

Jatal D. Mannapperuma
University of California at Davis
 Department of Biological-Agriculture
 Engineering
Davis, CA

K. Ananth Narayan
U.S. Army Natick RD & E Center
Natick, MA

Wasef W. Nawar
University of Massachusetts
 Chenoweth Laboratory
Amherst, MA

David Reid
University of California at Davis
 Department of Food Science
Davis, CA

Edward W. Ross
U.S. Army Natick RD & E Center
Natick, MA

W. E. L. Spiess
Federal Research Center
 for Nutrition
Karlsruhe, Germany

Alina S. Szczesniak
Retired from General Foods Corp.
Mount Vernon, NY

John Henry Wells
Oregon State University
 Food Innovation Center
Corvallis, OR

Walter Wolf
Federal Research Center
 for Nutrition
Karlsruhe, Germany

Richard C. Worfel
Academy of Health Sciences
 Food Technology and Sanitation Branch
Army Medical Department Center
 and School
Fort Sam Houston, TX

Bruce B. Wright
U.S. Army Natick
 RD & E Center
Natick, MA

Tom C. S. Yang
U.S. Army Natick
 RD & E Center
Natick, MA

Preface

The deteriorative processes that occur in foods after harvesting and during storage and distribution are unavoidable. If food is untreated, microbial deterioration becomes the dominant process affecting safety and quality. Even if the food is treated to reduce or eliminate microbial contamination, biochemical and physical deterioration become the dominant processes in determining storage lifetime and in altering product quality. Accordingly, if technological strategies are to be devised to retard such deterioration and to minimize the consequent loss of quality, it is crucial to understand the nature of constituent instabilities and the factors that control component degradation.

Much has already been written about the nature of microbial growth and the biostatic or biocidal approaches that can be taken, respectively, to retard growth or to destroy partially or completely the contaminating pathogenic or spoilage microorganisms, so this book does not address directly such treatments or storage strategies.

Since the ultimate goal in food production and distribution is to deliver safe, nutritious, and high-quality foods to the consumer, this book on storageability starts and ends with consumer perceptions and attitudes. *Cardello's* chapter on sensory perception puts into perspective the basis for the subjective responses to food quality attributes and the methodologies available to measure and predict such responses. *Bruhn's* chapter on consumer trends provides some insight into societal and psychological factors that influence consumer preferences and the changes that are taking place.

Ultimately, delivering quality foods desired by consumers depends on being able either to modify the instabilities of major constituents or to choose storage conditions that minimize the kinetics of the associated biochemical or physical reactions.

With respect to the food matrix and its constituents, *Haard's* chapter on food as cellular systems deals with physiological processes occurring post-harvest or post-slaughter and the consequence of these processes to food quality attributes. The chemical and/or physical reactions associated with the instabilities of constituent proteins, lipids, carbohydrates, and nutrients are addressed, respectively, in chapters by *Barnett and Kim, Nawar, Gordon and Davis,* and *Narayan.* Protein hydrolysis and crosslinking are processes common to many products either during treatment or upon subsequent storage. Lipid hydrolysis and oxidation are also crucial and need to be measured, assessed, and mitigated. Changes in carbohydrates associated with component interactions and structural reorganization are both physicochemical and physical in nature and tend to compromise quality and acceptance. Many of the same biochemical processes that involve proteins and sugars ultimately lead to undesirable changes in quality and in the bioavailability of essential nutrients. Common to all of these chapters are the implications for minimizing deleterious effects.

With respect to physical or physicochemical changes, the chapters by *Clydesdale, Szczesniak,* and *Chinachoti* on color, texture, and moisture provide a basis for understanding the consequences of such changes to these important quality attributes. Texture, which is particularly crucial to acceptance, is intimately tied to structure and mechanical properties, which in turn depend often on moisture migration and distribution. The interdependence of these factors on the driving forces for physical or chemical changes is important and is addressed.

Since availability of moisture and oxygen directly influences some of the physical and chemical reactions in the food, the ability of protective packaging to control or prevent their permeation must be understood. The chapter by *Hernandez and Giacin* addresses this issue from a structural and thermodynamic perspective. New materials and their properties are discussed.

All of these deteriorative processes are very much dependent on residence times at specific temperatures. Four chapters, those by *Ross*, by *Spiess, Boehme, and Wolf*, by *Wells and Singh*, and by *Wright and Taub*, deal with the modeling, monitoring, and managing of storage effects. *Ross* covers the equations that can be used to assess cycling phenomena; *Spiess, Boehme, and Wolf* treat the cumulative effect on quality due to holding for specific times at different temperatures during storage and distribution; *Wright and Taub* cover the use of time–temperature indicators; and *Wells and Singh* discuss a new concept of inventory management that is based on the "shortest remaining shelf life" principle. All contribute to the possibility of predicting or indicating product quality when delivered to the consumer.

The practical exploitation of some of the strategies for maintaining quality throughout storage is covered in chapters dealing with frozen, chilled, controlled atmosphere, and ambient storage. *Reid* describes the influence of freezing and the consequence of prolonged frozen storage on product quality. *Kader, Singh and Mannapperuma*, show how the judicious choice of packaging, oxygen and carbon dioxide levels, and package size could be used to extend the chilled lifetime of fruits and vegetables. *Yang* provides an overview of storage at ambient temperatures, showing the consequences of high temperatures and describing new ways to minimize some of the deteriorative processes.

As a reminder, in general, of the consequences to public health of either using contaminated foods or mishandling foods, the chapter by *Brown, Fruin, and Worfel* addresses toxicological implications. It is not intended to be pejorative, but only to highlight the importance of maintaining the highest possible standards and of pursuing the best possible means of food storage and distribution.

Inherent in this overall treatment of food storage is the intent to integrate basic concepts with practical applications and to communicate the reliance of a successful technology on sound scientific principles. It is also our intention, by presenting the material in this integrated way, to generate further interest in researching the scientific and technological issues of food deterioration that are as yet unresolved, including those associated with insect infestation and physical abuse. Such resolution should be possible if, as the thematic content of this book suggests, food is thought of as a complex chemical reactor whose dynamics can be understood and, within limits, controlled for purposes of ensuring that safe and appealing foods reach the consumer.

Irwin A. Taub
R. Paul Singh

Acknowledgments

The editors are grateful to several individuals who not only assisted us in the preparation of this book, but also did so patiently and good naturedly: Dr. Matthew Herz, as Associate Technical Director for Technology, for suggesting the effort and providing continued support; Dr. Tom C. S. Yang, for his help in organizing the material; Michael Statkus, for his technical editing and formatting skills; Mrs. Brigitta Whalen, for transcribing several handwritten chapters; and Mrs. Anne T. Otis, for making all the editorial corrections on each of the chapters and for serving as Administrative Assistant for the preparation of this book. Their assistance is sincerely appreciated.

Contents

---

* Adapted from *J. Food Biochem.* 19(3), pp. 191–238, Food & Nutrition Press Inc., Trumbull, CT

chapter one

Perception of food quality

Armand V. Cardello

Contents

1

Introduction

Definition of food quality

It is clear from other chapters in this volume that the term "food quality" has a variety of meanings to professionals in the food industry. To nutritionists it is synonomous with the nutritional value of the food, to microbiologists it refers to the food's safety, while to chemists it may equate to the item's stability. Although each of these interpretations of "food quality" has merit, consumers, through their purchase and/or nonpurchase of the product, must certainly be considered the ultimate arbiters of food quality. With this fact in mind, the meaning of food quality to the consumer becomes of critical importance to the discussion of the quality preservation of food.

The notion that food quality must ultimately be defined by consumer perception is embodied in the frequently cited definition of food quality as "the combination of attributes or characteristics of a product that have significance in determining the degree of acceptability of the product to a user."[1] A critical phrase in this definition is "acceptability of the product to a user," because it serves as the basis for determining both the relative importance of factors that contribute to food quality and the appropriate methods for measuring it. Another definition of food quality that places even greater emphasis on the perceptual aspects of food quality derives from a definition of *sensory quality*.[2] It defines food quality as "the acceptance of the perceived characteristics of a product by consumers who are the regular users of the product category or those who comprise the market segment."[3] In this definition, the phrase "perceived characteristics" refers to the perception of *all* characteristics of the food, not simply its sensory attributes. Thus, it also includes the perception of the food's safety, convenience, cost, value, etc.

By defining food quality in terms of the consumer's perception of it, this chapter will necessarily focus on the factors and mechanisms that can influence consumer perceptions of foods and beverages and on the available methods by which consumer perceptions of food can be measured. In addition, the relationship between perceived food quality and the common notion of "consumer acceptance" will be explored.

Intrinsic vs. extrinsic factors affecting food quality

The factors that affect consumer perception of food and food quality are numerous. Many of these factors are intrinsic to the food, i.e., related to its physicochemical characteristics. These include such factors as ingredient, processing, and storage variables. These variables, by their nature, control the sensory characteristics of the product, which, in turn, are the most salient and important variables determining both the acceptability and perceived quality of the item to the user. In fact, it is usually through sensory characteristics and the changes that occur in them over time that consumers develop their opinions about other aspects of food quality, e.g., safety, stability, and even the nutritional value of the

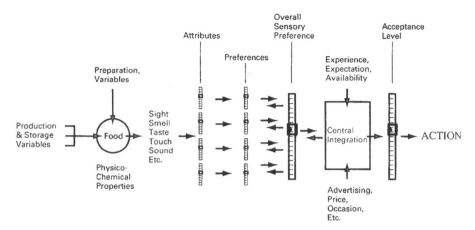

Figure 1 Land's model of the sensory and cognitive factors that contribute to food acceptance action. (From Land, D. G. In *Sensory Quality in Foods and Beverages*. Ellis Horwood: Chichester, U.K., 1983. With permission.)

food. Thus, understanding the relationships between (1) physicochemical characteristics of foodstuffs, (2) the sensory and physiological mechanisms that convert these characteristics into human perceptions of food attributes, and (3) the effects of these perceived attributes on acceptance and/or consumption of the item is critical to an understanding of what constitutes food quality.

Although intrinsic factors are significant determinants of food quality, many factors that are extrinsic to the product also play a role. These factors include variables closely associated with consumers' perceptual/cognitive/emotional systems, e.g., attitudes, expectations, and social or cultural influences. Many of these factors will be addressed in this chapter. Other factors, such as price, convenience, and brand image, i.e., those that researchers have sometimes termed "instrumental" or "expressive"[4,5] are addressed in Chapter 19 of this volume.

McEwan and Thomson[6] reviewed various behavioral models of food acceptance and consumption. Central to all of these models are three major elements first identified by Pilgrim.[7] These are (1) sensation—the role played by the physicochemical characteristics of food as it interacts with the human receptors to produce a sensory experience, (2) attitudes—the role played by human learning and experience, and (3) physiology—the roles played by hunger, thirst, and other internal, biological factors. While all of the models[7-11] address these three components, they differ in the emphasis that they give to each. Land's model,[10] depicted in Figure 1, is the most useful for the present discussion, since it depicts the various influences on the sensory quality of food and the measurement levels involved in the evaluation of perceived food quality.

Land's model starts with "food" on the left side of Figure 1, where are depicted the variables that define the physicochemical or stimulus properties of the food, i.e., preparation, production, and storage variables. Moving to the right in Figure 1, the physicochemical properties can be seen to interact with the human sensory systems to produce modality-specific experiences of the sensory attributes of the food. These sensory attributes of taste, smell, appearance, etc., in turn, lead to sensory "preferences." At this point in the model, central nervous system factors, such as learning and experience, expectations, and cognitive factors arising from price, advertising, etc., may modify overall sensory preference. The integration of all these factors results in the acceptance level for the item. It is this acceptance level that is the precursor to behavior (action), e.g., choice, purchase, and consumption.

In the context of this model, the intrinsic variables that will be discussed in this chapter are those that enter from the left side of Figure 1 and lead to the perception of the sensory attributes of the product. The extrinsic variables are those that enter at the right side and interact with sensory preference to produce behavioral action.

Measurement concepts

As one traces the methods for the measurement of perceived food quality, it is important to keep in mind the distinction between the concepts of "degree of excellence" and "difference from a standard." The former refers to the initial quality of a food item, while the latter refers to changes or deviations from this initial or target quality. Covarying with these measurement concepts is the issue of who performs the judgments. Traditional industry approaches have assumed that product experts, not consumers, are best able to identify good quality. This has led to the establishment of expert grading standards for various commodities to index the degree of excellence of the product, and to the use of expert panels to detect differences from preestablished standards of quality. Although these approaches are useful for classifying products into different categories based on sensory properties, it must be remembered that the final arbiter of food quality is the consumer. Thus, it is essential that consumers be used to the maximum extent possible at every measurement level, both to index the degree of excellence of the product, as well as to provide information about the importance of any deviations from the standard. The former is routinely accomplished under the rubrics of "consumer acceptance testing" and "market research," while the latter is now being accomplished through the integration of descriptive and consumer acceptance techniques using multivariate statistical approaches.

Intrinsic factors

Food as a stimulus

The starting point for our discussion of perceived food quality is the physical matter of food and the interactions that occur between the food's physicochemical properties and human sensory receptor organs. These interactions result in primary sensory experiences of the consumer, the building blocks of more complex perceptions of quality and/or acceptability.

Dimensions of sensory experience

All sensory experience is mediated through specialized receptors associated with each sen;sory modality. It is commonly stated that there are five human senses: vision, hearing, taste, smell, and touch, a classification first proposed by Aristotle. In fact, several other sensory systems must be added to this classification scheme for completeness. These systems are the vestibular sense (balance), the kinesthetic sense (body and limb position), the deep pressure sense (subcutaneous sensation), and the interoceptive sense (receptor systems in internal organs). Although less commonly associated with food perception, several of these latter senses play an important role in the perception of food quality.

For each sensory modality, there exist two basic dimensions of sensory experience, one qualitative and one quantitative. For example, within the sense of taste, one can experience the qualitatively different sensations of "salty," "sweet,""sour," and "bitter." Similarly, within the visual sense one can experience "red," "green," or "blue." These different sensations within each modality are, appropriately, termed sensory *qualities*. However, one can also experience differences in the magnitude or *intensity* of each of these sensations. Thus, saltiness can be of low, intermediate, or high intensity, and lights can vary from dim to bright. Understanding how stimulus objects interact with sensory organs to elicit specific sensations of quality and intensity is the basis for our understanding of how the intrinsic factors of food contribute to the consumer's perception of food quality.

Psychophysics: linking physics and psychology

The study of the human senses and of the physicochemical parameters that elicit experiences of quality and intensity evolved out of mid-19th century Germany and the branch of psychology known as psychophysics. Both the term "psychophysics" and the founding of the field can be traced to the German psychologist, Gustav Fechner, who defined its purpose as establishing the mathematical relationships between the "physical and psychological worlds."[12] In a sense, this functional approach to sensory experience was an early predecessor to later, more generalized psychological models of behavior that are based on a stimulus–response (S-R) paradigm. While the field of psychophysics has grown rapidly since its inception, it is still devoted to the determination of relationships between sensory experience and physicochemical characteristics of the stimulus that elicit the sensation.

Basic sensory processes

It is of no surprise to the reader that the taste, smell, texture, and appearance of a food contributes greatly to the consumer's perception of the quality of that food. But what exactly comprises these characteristics of a food and what are the underlying mechanisms by which they operate?

Taste

Taste is the sensory experience that results from stimulation of chemoreceptors located on the tongue, palate, pharynx, larynx, and other areas of the oral cavity. These chemoreceptors, known as taste buds, are comprised of groups of chemoreceptor cells. The interaction of sapid chemicals with the microvillar surface at the apical end of these taste cells is the physical event that leads to taste transduction.

The most well-studied anatomical aspect of the human taste system is the tongue. Figure 2 shows the dorsal surface of the human tongue and identifies four distinct morphological structures. These structures, known as papillae, include circumvallate, fungiform, filiform, and foliate types. Filiform papillae, widely distributed over the dorsal surface, have no role in taste. However, fungiform papillae, mushroom-shaped structures distributed primarily over the anterior tongue; foliate papillae, folds located on the margins of the tongue; and circumvallate papillae, large, circular structures forming a chevron near the back of the tongue, all house significant numbers of taste buds.

The early view concerning taste qualities, handed down from Aristotle, was that there were only two primary qualities, sweet and bitter, and that other taste sensations were merely subcategories of these. It was not until 1864 that Fick[13] proposed the current view of four primary taste qualities—salty, sweet, sour, and bitter. For each of these taste qualities one can identify a set of stimuli that arouse these sensations. In the case of sourness the adequate stimulus is the hydrogen ion present in Bronsted acids, although the anion and any undisassociated acid can modify its taste.[14] Since stimulation requires access of the compound to the receptor surface, factors that affect such access, e.g., the lipophilicity of the compound, may also affect the perceived taste.[15] The actual mechanism of sour taste transduction occurs through the blockage of potassium conductance in K^+ channels on the taste cell membrane[16] or by direct penetration of the hydrogen ion through sodium channels.[17] Either of these events can result in depolarization of the taste cell and initiation of an action potential to carry taste information more centrally in the nervous system.

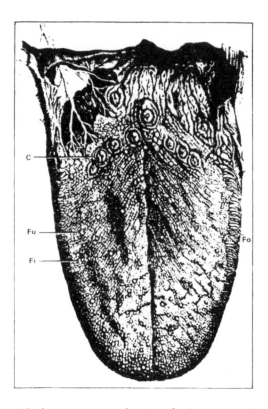

Figure 2   Dorsal surface of the human tongue showing the four types of papillae: C, circumvallate; Fu, fungiform; Fi, filiform; Fo, foliate. (From Geldard, F.A. *The Human Senses*. Wiley and Sons: New York, 1972. With permission.)

For saltiness, the adequate stimulus is the sodium ion (Na^+) and certain cations of other low molecular weight salts. The anions of salts play an inhibitory role. Salt taste transduction occurs as a result of the passive diffusion of sodium ions through amiloxide-sensitive sodium channels on the taste cell membrane.[18,19]

Although sour and salt taste transduction mechanisms are fairly well understood, the same cannot be said for sweetness and bitterness. Sweet taste can be elicited by a variety of organic and inorganic compounds. For many years it has been believed that sweet-tasting substances contain a hydrogen ion bonded to an electronegative atom (AH^+) in close proximity to another electronegative atom (B) so as to allow the creation of a hydrogen bond with the receptor surface.[20] Sweet taste transduction appears to result from conformational changes induced in receptor proteins by these sweet-tasting compounds, resulting in the activation of G-proteins.[21,22] The activiation of G-proteins results in the blockage of potassium channels via second messenger pathways and, ultimately, depo-larization of the taste cell. Evidence now points to as many as six different receptor sites for sugars[23] and more than one type of sweet sensation.[24]

Bitter taste is equally complicated, being aroused by such a diverse set of compounds as the alkaloids, heavy halide salts, and certain amino acids. Moreover, there appears to be a strong genetic component to the perception of bitterness, as evidenced by the fact that compounds such as phenylthiocarbamide and 6-*n*-propylthiouracil are extremely bitter to some people but tasteless to others.[25–28] Current theory concerning bitter taste stimulation parallels that for sweet taste, with some suggestion that an AH–B mechanism is responsible,[29–31] but with a different AH–B distance than that for sweet compounds. As

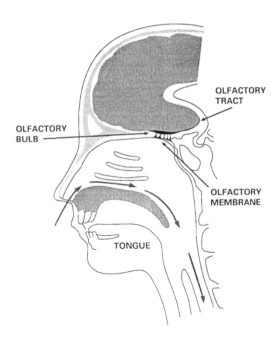

OLFACTORY
TRACT

OLFACTORY
BULB

OLFACTORY
MEMBRANE

TONGUE

Figure 3 Schematic cross section of the head, showing the oral–nasal passages and the olfactory membrane. (From Griffiths, M. *Introduction to Human Physiology.* Macmillan: New York, 1974. With permission.)

with sweetness, it appears that activation of G-proteins is responsible for a bitter taste transduction. This activation leads to the release of calcium ions, which, in turn, activate potassium channels on the taste cell membrane.[32,33] It is also believed that there may be three or more different bitter receptor mechanisms.[34,35]

Although only four taste qualities have been generally recognized through the years, some researchers argue for the existence of a fifth quality. This latter taste quality is called *umami* and is described as delicious or savory.[36,37] Umami taste is most commonly associated with the taste of monosodium glutamate, but can also be elicited by certain L-amino acids and derivatives of 5'-ribonucleotides. Other sensations that are often associated with taste, e.g., "pungency" and "astringency," are really tactile sensations or part of the "common chemical sense." This latter chemosensory system, distinct from taste and smell, mediates sensations of pain and/or irritation caused by chemicals in contact with the skin or mucous membrane.[38]

Smell

Smell is the sensation resulting from stimulation of receptors in the olfactory epithelium of the nose by airborne compounds. These compounds reach the receptor surface through the nares or rostronasally through the mouth. Figure 3 is a schematic of the human oral–nasal region, showing the location of the olfactory receptor membrane, high in the nasal cavity. The arrows indicate the normal passage of inhaled air through the nares. However, chewing of food and other jaw movements will also cause air in the mouth to be passed into the nasal cavity through the nasopharynx. This fact accounts for the common confusion of the odor of a food with its "taste" and may underlie Aristotle's belief that the sensory qualities of smell were the same as those for taste.

During the past century numerous attempts have been made to classify odors according to their quality. Zwaardemaker[39] and Henning[40] were two of the first to do so. Later, Crocker and Henderson[41] suggested the existence of four basic odor qualities: fragrant,

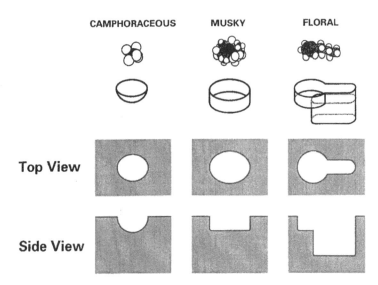

Figure 4 Schematic diagram showing the shapes of odorant molecules and hypothetical shapes of associated receptor sites on the hair cells. (From Amoore, J.E., et al. *Sci. Am.* 1964, 210(2), 42. With permission.)

acid, burnt, and caprylic, and developed a four-digit numeric system for classifying odors. However, these and other classification schemes have had limited success due to the difficulty of describing the wide range of perceptible odors in terms of some small number of sensory qualities. Further complicating the problem of odor classification is the fact that odor thresholds, even in the same subject, vary widely over time.[42] Somewhat greater success has been achieved in odor classification using multidimensional scaling approaches, but even here the success has been limited.[43–45]

Equally difficult has been the identification of the attributes of the stimulus that elicit odor qualities. Over the years, investigators have examined a wide array of chemical and/or physical characteristics of odorant molecules.[46–55] Out of these data, the most widely accepted model of olfactory functioning that has evolved is the stereochemical or "lock and key" theory.[47,56,57] This theory assumes that a specific shape and size of odorant molecule is responsible for each basic smell quality, and that these molecules "fit" into specific receptor sites that have corresponding geometries. Figure 4 depicts this model for three potential primary odors. As can be seen, each odorant molecule fits into a similarly shaped and sized receptor site. The theory accounts for complex odors by proposing that some molecules fit into more than one receptor site. Based on studies of specific anosmia (inability to smell a particular compound), several primary odor qualities have been identified, including sweaty, spermous, fishy, malty, musky, urinous, minty, and camphoraceous.

Perhaps the most important breakthrough in the understanding of olfactory functioning has occurred within the past few years with the isolation of what appear to be the first known odor receptors—genes that code receptor proteins.[58] It is now believed that there may be upwards of 300 to 1000 different olfactory genes, a number so large that it could easily account for the over 10,000 perceptible odors. This large number of receptor types is unique in human sensory systems. It may be that the olfactory sense has evolved so that almost all information processing goes on in the nose and not in the brain. This would be consistent with evolutionary history and the existence of many species with small brains, yet highly evolved senses of smell.

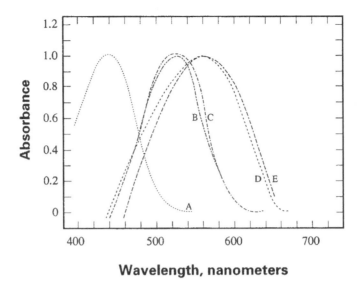

Wavelength, nanometers

Figure 5 Absorption spectra of the visual pigments in five human parafoveal cones: one blue-sensitive(A), two green-sensitive(B,C), and two red-sensitive cones(D,E). (From Wald, G. *Science.* 1968, 162, 230. With permission.)

Vision

Among humans, vision plays an important role in the perception of food quality, because the visual appearance of food often determines whether the item is selected for purchase or consumption. At the biophysical level, visual perception is the result of stimulation of receptors in the retina of the eye by electromagnetic radiation. In the case of color vision there are three distinct receptor types in the human retina. Figure 5 shows the absorption spectra for these three receptor types. Each is differentially sensitive to wavelengths in the red, green, and blue regions of the spectrum, providing evidence that there are, in fact, three primary color qualities in humans—red, green, and blue—and that all other colors are the result of combinations of responses among these three basic receptor types. The three physical properties of light, i.e., its wavelength, intensity, and purity are associated with three psychological dimensions: hue, brightness, and saturation. Their combined effect is what we call "color." Just as with smell, where there are literally thousands of different odors and numerous classification schemes, so too are there thousands of different colors and numerous color classification systems. The most commonly used of these systems are the Munsell, ICI, Lovibond, and Ostwald. Their description, along with those of various instrumental methods of color measurement, can be found in a variety of standard texts on food color measurement.[59,60]

Although color is a critical aspect of the appearance of food, other visual attributes play an important role. Light striking an object may be transmitted, absorbed, or reflected. All foods and beverages absorb some light, and the rest is reflected or transmitted. Light that is reflected in all directions produces a perception of a flat finish on the object. However, light that is reflected in an only a single direction results in the perception of gloss. Gloss (also termed "shine" or "polish") is an important quality characteristic of such foods as apples and pastry glazes. Light that is neither absorbed nor reflected by an object is transmitted through it. The more light that is transmitted, the more "translucent" the object appears. Translucency is an important attribute of such products as beverages, jellies, and sauces. If the transmitted light is scattered by particles contained within the object, as in fruit juices, then the attribute of "turbidity" is perceived in the product.

Size, shape, and surface texture are other important visual attributes of foods. Size and shape contribute to the perception of "wholeness" in a product, an important perceptual aspect of the quality of such foods as nuts, potato chips, and cookies. In foods that come from natural sources these attributes are difficult to control, except through detection and removal of aberrant sizes and shapes. In processed foods, size and shape can be directly manipulated to match consumer and functional requirements. Surface texture can also be important, especially in fibrous products such as meats and fish, in which fiber geometry is observable, or in any product in which a smooth or other characterizing surface texture is desirable.

Hearing

Hearing (audition) is the perceptual experience that results from stimulation of receptors in the cochlea of the ear by sound waves. The latter can be transmitted through air, water, bone, or any other elastic media. Although most lay people do not consider hearing to be related to food quality, the sounds produced during biting and mastication of food have significant effects on quality perception for many foods.

The amplitude, wavelength, and purity of sound waves correspond to three basic psychological dimensions: loudness, pitch, and timbre. Wavelength or its inverse, frequency, corresponds to the primary qualitative dimension—pitch. Multiple wavelengths produce the psychological experience of timbre, in much the same way as multiple wavelengths of light determine perceived saturation. However, unlike vision, there are no true primary qualities in audition. One cannot produce the entire sound spectrum by merely combining some small set of wavelengths.

The study of the psychoacoustic effects of chewing on perceived food quality has had a short but productive history. Although the sound of certain foods, such as crisp vegetables, crackers, and snack food, has long been known to affect their acceptability, only recently have investigators been able to analyze foods and/or their texture by the physical or perceptual sounds emitted during mastication. Early studies were able to analyze chewing sounds for the purpose of classification.[61,62] More recently, analysis of food-crushing sounds has been undertaken to assess both the crispness and crunchiness of foods.[63–67] Figure 6, taken from the work of Vickers and Bourne,[64] shows plots of sound amplitude over time for food items that differ in crispness. The irregular amplitude as a function of time is a characteristic of crisp foods. In addition, the louder the sound or the greater its density, as seen in increasing order from C to B to A, the greater the food's crispness.

Kinesthesis and somesthesis

Kinesthesis and somesthesis are two closely allied senses whose receptor organs are located in the skin or in deep tissue. Kinesthesis is the perception of limb position and limb movement. It is mediated by receptors in muscles, tendons, and joints. Somesthesis is the perception of pressure, pain, and temperature, and is mediated by receptors located in the skin. These two modalities are responsible for all nongustatory oral perceptions, i.e., food texture, temperature, and mouthfeel.

Passive limb movement is known to be mediated by receptors located in the joints, whereas voluntary limb movement and resistance to movement is mediated by receptors in muscle. Since foods provide resistance to active jaw movements (chewing), both kinesthetic joint and muscle receptors are involved in the perception of food texture and quality. The somesthetic sense, on the other hand, is a skin sense; thus, its role in perceived food quality derives from the contribution of skin structures on or near the oral cavity. These structures differ greatly in their sensitivity to simple pressure or touch. The lips and tip

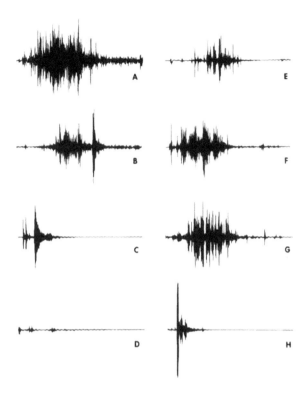

Figure 6 Plots of sound amplitude as a function of time for three food items: (A) water-soaked green pepper, (B) fresh green pepper, and (C) blanched green pepper. Sounds are those produced by a subject biting into each food item. (From Vickers, Z.M.; Bourne, M.C. *J. Food Sci.* 1976, 41, 1158. With permission.)

of the tongue are the most sensitive areas, while posterior parts of the oral cavity have much reduced sensitivity.[68-70] The schematic shown in Figure 7 is known as a sensory homunculus (little man). It is an anthropometric representation of the somatosensory brain area of the post-central gyrus. The size of each body part in the figure reflects the amount of cortical tissue devoted to somesthetic perception from that body area. As can be seen, the facial area, in general, and the lips and tongue, especially, are disproportionately represented.

As everyone knows who has ever bitten their tongue, consumed coffee that is too hot, or bitten into a cayenne pepper, many areas of the mouth and nose contain large numbers of receptors that give rise to painful sensations as a result of intense tactile, thermal, or chemical stimuli. These sensations play an important role in the perceived quality of most spicy foods, e.g., those containing black pepper, chili pepper, ginger root, etc. Depending upon intensity, these sensations can result in such perceptual dimensions as pungency, stinging, biting, chemical cool, and chemical warmth. These sensations, mediated by the trigeminal nerve, belong to what is called the "common chemical sense."[71-73] Even the perception of carbonation in soft drinks is attributable to this generalized skin sense. Yet, in spite of the importance of pain sensations to food perception, many areas of the oral cavity are relatively analgesic. These include the mucous lining of the cheeks, the posterior tongue and mouth, and the lower part of the uvula.

The identification of primary qualities in the somesthetic and kinesthetic senses is still unsettled, due to lack of agreement on differences between modalities and qualities and on whether or not such qualitatively different sensations as pain and tickle may be merely

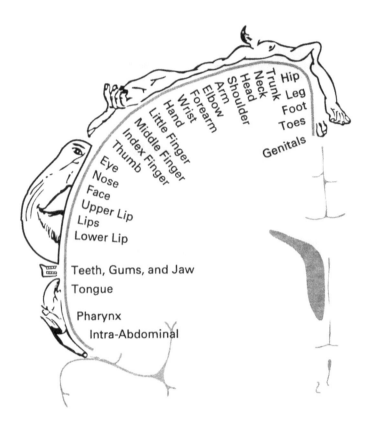

Figure 7   A sensory homunculus. Schematic diagram showing the projection of the human somatic system on the post-central gyrus of the brain. (From Geldard, F.A. *The Human Senses*. Wiley and Sons: New York, 1972. With permission.)

differences in perceived intensity. While these problems have slowed the integration of basic data into applications in the food industry, much progress has been made in the analysis of the combined sensory experience that results from stimulation of these senses by foods in the mouth, i.e., food texture. A discussion of this critical aspect of food quality can be found in the section on applied sensory methodology.

Basic quantitative relationships

There are two basic measurement concepts that must be understood when discussing quantitative aspects of sensory experience. One is the concept of threshold, the other is the concept of intensity scaling. The former refers to the minimal stimulus energy required to elicit a sensation. The latter refers to the measurement of the magnitude of the sensation.

Sensory thresholds

Thresholds are quantitative indices of the minimum stimulus energy required to detect the presence of, or to recognize the nature of ,a given stimulus. They are extremely important to the study of perceived food quality, because they reveal the concentration level above which certain physicochemical stimuli will contribute a taste, odor, etc., to the food. For many noxious or undesirable stimuli, it is essential that such thresholds not be surpassed in food.

There are two ways in which one can conceptualize sensory thresholds. The older, classical view is that there exists an absolute lower energy boundary for sensory systems,

Table 1 Detection Thresholds for Common Tastants and Odorants

Compounds	Taste or Odor	Threshold[a]
Tastants		
Hydrochloric acid	Sour	0.0009 N
Critric acid	Sour	0.0016 N
Lactic acid	Sour	0.0023 N
Sodium chloride	Salty	0.01 M
Potassium chloride	Bitter/salty	0.017 M
Sucrose	Sweet	0.01 M
Glucose	Sweet	0.08 M
Sodium saccharine	Sweet	0.000023 M
Quinine sulfate	Bitter	0.000008 M
Caffeine	Bitter	0.0007 M
Odorants		
Citral	Lemon	0.000003 mg/l
Limonene	Lemon	0.1 mg/l
Butyric acid	Rancid butter	0.009 mg/l
Benzaldehyde	Bitter almond	0.003 mg/l
Ethyl acetate	Fruity	0.0036 mg/l
Methyl salicylate	Winter green	0.1 mg/l
Hydrogen sulfide	Rotten eggs	0.00018 mg/l
Amyl acetate	Banana oil	0.039 mg/l
Safrol	Sassafras	0.005 mg/l
Ethyl mercaptan	Decayed cabbage	0.00000066 mg/l

[a] Threshold values are those reported in a number of standard sources.[267–269]

below which a stimulus signal cannot be detected. This threshold or "limen" may vary due to adaptation or to changes in the physiological state of the organism. However, it is assumed to have a theoretical, absolute value that can be statistically estimated. Within this conceptual framework, a large number of psychophysical methods have been developed to measure either absolute thresholds (the minimum stimulus energy required to elicit any sensation), recognition thresholds (the minimum stimulus energy required to elicit a specific sensation), or difference thresholds (the minimum difference in stimulus energy required to elicit a sensation of a perceptual difference). Some of the methods so developed include the Method of Limits, Method of Constant Stimuli, and Method of Adjustment.[12,74]

Table 1 shows threshold values for a variety of food-related sensory continua, determined via traditional threshold methods. Of interest to the discussion of perceived food quality is the fact that thresholds increase significantly with aging. For example, taste thresholds increase by a factor of 1.5 to 11.5 depending upon the compound and age of the individual, and smell thresholds increase from 2–15 times.[75] Clearly, such dramatic shifts have commensurate effects on preferences and perceived food quality among these populations.

The second model of sensory thresholds holds that thresholds are not absolute. Rather, they vary as a function of the operating characteristics or decision rules of the receiver (observer). The analogy can be made to communication systems, where signals are usually transmitted within a background of noise. For this situation, electrical engineers have established mathematical models to describe the sensitivity of the system. The application of similar models to human perception is called "signal detection theory."[76–78] In this approach to "threshold" measurement, an experimental situation is used in which a stimulus may or may not be presented to an observer. In response to either of these conditions, the observer can say, Yes, I detect it, or No, I do not detect it. This set of

Table 2   Possible Outcomes in a Signal Detection Experiment

		Subject's Reported Observation	
		YES	NO
Stimulus Signal	YES	Correct Detection (Hit)	Miss
	NO	False Alarm	Correct Rejection

outcomes is depicted in Table 2 along with the classification of each possible stimulus-response type. The key to signal detection theory and what makes it different from classical threshold theory is the model's assumption that the observer's responses are not simply a function of the stimulus energy, but also a function of the expected probability that a stimulus will be presented on any given trial and the probable consequences (pay-offs) to the observer for reporting either outcome. Thus, while classical threshold theory assumes that the sensory signal impinging on the observer's sense organ is sufficient to determine the response, signal detection theory assumes that the less likely the occurrence of a sensory signal, the more likely the observer is to say "no," thereby increasing both the "misses" and the "correct rejections." Similarly, if it is very important to the subject to not miss the signal, then he or she will be more likely to say "yes," thereby increasing both "hits" and "false alarms." By systematically varying these biases of the situation and observer, it is possible to generate a "receiver operating curve" that shows how the probability of hits varies with the probability of false alarms. Moreover, a measure of the sensitivity of the observer (known as d') can be determined as an index comparable to that for traditional threshold theory.

Although signal detection theory has gained wide acceptance in areas of applied psychophysics, the food industry has been slow to adopt this approach. However, several studies have applied these techniques to problems in the perception of foods[79,80] and to hedonic testing.[81]

Suprathreshold scaling

Sensory scaling, or the measurement of sensory magnitude, has had a long and contro-versial history within experimental psychology. Much of this controversy has focused on the form of the "psychophysical function," i.e., the mathematical relationship that relates the physical intensity of a stimulus to its perceived magnitude, and on the proper scalar methods by which to measure perceived magnitude.

In 1850 Fechner proposed the first psychophysical "law."[12] He maintained that the perceived magnitude of a stimulus increased as a logarithmic function of the physical intensity of the stimulus: $S = k \log I$. This version of the psychophysical law went unchal-lenged for over a century, although empirical data frequently surfaced that were not well fit by this logarithmic model. It was not until the 1950s that S. S. Stevens proposed an alternative psychophysical law.[82,83] Early in Stevens' career he described a hierarchy of four measurement types (scales): nominal, ordinal, interval, and ratio.[84] Nominal scales simply identify or name objects. They bear no quantitative relationship to one another. Ordinal scales, on the other hand, provide information about the rank of each object along some dimension. Interval scales index the degree of difference between objects on some dimension. However, a critical aspect of interval scales is that there is no true zero point. The lack of a zero point means that the ratio of the numbers assigned to objects has no meaning. For example, the Fahrenheit and Celcius scales of temperature are interval scales. While we know that 50°F reflects more thermal energy than 25°F, we cannot say that it reflects twice as much energy. However, the fourth scale type, the ratio scale, does possess a true zero point, making the ratios between numbers meaningful. The Kelvin scale of temperature is an example of a ratio scale. Here, 50K actually means twice as much thermal energy as 25K.

As a result of his early mathematical training, S. S. Stevens believed that sensory intensity should only be measured using ratio scales. With this assumption, he developed a ratio method in which subjects were allowed to assign their own internal numbers to represent the magnitude of their sensations and named the method "magnitude estimation." This simple yet powerful method has grown in popularity to the point where it is now the most commonly used scaling technique of sensory psychologists.

Stevens' main contention, that sensation magnitude grows as a power function of stimulus intensity, can be expressed mathematically as $S = KI^n$, where S is the magnitude of the sensation, I is the intensity of the stimulus, n is the exponent of the power function, and k is a constant of proportionality. The exponent of the power function (n) is an index of the rate of growth of perceived intensity as a function of physical intensity. It is believed to be an invariant characteristic of the sensory attribute being measured and has been tied to the mechanism of energy transduction at the receptor.[85]

In spite of the mathematical advantages of ratio scales over other types of scales, most research conducted within the food industry utilizes ordinal or interval scales. In the area of food acceptance testing, the classic nine-point hedonic scale[86,87] is an example of such a scale. While this and other labeled category scales are easy to use, they have several problems. One is that the verbal labels often fail to define perceptually equal intervals between the scale points. This has been shown even for the nine-point hedonic scale.[88] Other disadvantages include the fact that many panelists avoid the use of the end-points, and that the fixed upper and lower end-points force panelists to restrict extreme stimuli to an artificially narrow set of scale points.

However, magnitude estimation is not free from problems. For example, it has been demonstrated that magnitude estimation scales are comparable in efficiency to category and line scales when used by a college population, but they are less efficient with a heterogeneous population.[89] Perhaps the most systematic study of scale types was a multilaboratory study conducted by Committee E-18 of the American Society of Testing and Materials.[90] This study compared two types of magnitude estimation scales (unipolar and bipolar) and a nine-point category scale, each used for the scaling of acceptability. Results showed equal reliability, precision, and discrimination among the scales. Thus, the theoretical issues underlying choice of one scale type over another may be less important than their ease of use by subjects and/or other practical concerns of the test design.

Applied sensory methodology

In the previous section, the theoretical bases for the measurement of both the qualitative and quantitative dimensions of sensory experience were outlined. With this in hand, we can now examine the applied methodologies for measuring the perceptual aspects of food and food quality.

The role of experts vs. consumers

As discussed in the Introduction, as well as by several other authors,[2,91-98] food quality must be defined in terms of the consumer's perception of it. Summary judgments of food "quality" by appointed or presumed experts can be misguided, especially when the intent is to make a decision concerning the suitability, marketability, or sensory appeal of the food. The shift toward greater dependence on consumers to define quality is not new, but has grown to the extent that many researchers now view perceived food quality as substantially and conceptually similar to consumer acceptance.[2,96–98] This trend toward consumer definitions of food quality has even influenced the restaurant industry, where many restaurant critics are being replaced by consumer-based alternatives, e.g., Zagat restaurant surveys.[99] Of course, expert judges for quality still have their place. In those commodity areas where expert tasters have historically been used to assess quality, e.g., wine, tea, coffee, etc., the expert is extremely useful, because he/she is typically better able to discriminate differences between products and/or to describe these differences in qualitative terms. Yet, the expert's superior ability for discrimination should not be assumed under all circumstances. For example, standard tea testing procedures require that samples be prepared so that half of the samples are mixed by pouring the milk into the tea and the other half by pouring the tea into the milk. The reason is that tea tasters can detect this apparently subtle difference in preparation. However, in one rather interesting study it was demonstrated that consumers have the same ability to detect the slight sensory difference created by this variation in sample preparation.[100]

Evidence of the problem with using expert opinions of food quality to guide decisions concerning food marketing to the general public abounds. For example, it is well known that the majority of consumers prefer wines with a fruitier flavor over those that are drier in flavor. Yet, many wine experts rate the drier wines as having higher quality. The dependence of quality on individual perception is further reflected in the fact that ratings of wine quality vary significantly from one wine master to another, even for the same wines.[101] Of course, the problem cannot be placed entirely at the feet of the experts. For example, in one report it was concluded that new alternatives to steam sterilization of canned vegetables that produce a fresher tasting product should first be introduced to the institutional food-service market, because home consumers have come to expect a certain level of deterioration in their canned vegetables.[102] To consumers, the product made by the new process, with its novel, fresher tasting quality may well be perceived as a poorer quality product. Clearly, then, consumers simply perceive quality differently than experts. For this reason, expert judgments of quality can be misleading for many consumer applications.

What then is the importance of expert judgments of foods? The most important use of experts is for the identification and description of products with unique sensory character or subtle character differences. This critical descriptive/discriminitive information can then be used *in conjunction* with consumer acceptability data, in order to decide which of several samples possesses attributes that are in keeping with consumer demands. The process of developing food products of high sensory quality (as we have defined it) can be shortened and optimized through the judicious use of trained or expert panelists who can evaluate large numbers of samples in a short period of time and provide accurate descriptions of the attributes of the product and the differences among them. It is these abilities that the consumer lacks.

Descriptive analysis

While "experts" exist in certain older commodity areas, they do not exist for the vast number of consumer products on the market. In response to the need for accurate methods of describing the qualitative sensory aspects of a wide range of food products, various systems of descriptive analysis have been developed. Foremost among these have been methods for descriptive analysis of flavor and texture.

Descriptive flavor analysis. The first widely populatized descriptive method was the flavor profile technique developed by Arthur D. Litte Co.[103,104] This method utilizes a panel of six to eight trained judges, who are selected on the basis of availability, interest, personality factors that lend themselves to positive group interactions, and possession of "normal" taste and smell sensitivity. Each panelist is given training in the basic principles of taste and smell physiology, sensory testing, and flavor vocabulary. The latter is accomplished through the use of established reference standards by which panelists can taste or smell samples that possess specific character notes, e.g., "sweet," "woody," or "rancid." The panel is headed by a leader who schedules meetings, leads profile discussions, obtains the consensus of the panel, and communicates the results of the panel evaluations to the users.

All test sample are evaluated by individual members of the panel. This is followed by group discussion. It is the task of the panel to (1) define the qualitative aroma, taste, flavor, and mouthfeel attributes of the product; (2) indicate the order in which each of these appears; (3) describe any aftertastes; (4) rate the intensity of each attribute; and (5) rate the "amplitude" or overall quality of the product. Intensity is judged by consensus of the panel using the following labeled scale:

0 = not present
)(= threshold
1 = slight
2 = moderate
3 = strong

An improved and broadened approach to descriptive flavor profiling was developed at the Stanford Research Institute.[105] This technique, known as quantitative descriptive analysis (QDA), also uses trained panelists. However, all evaluations are made in individual testing booths to avoid the influence of group dynamics. Also, intensity judgments are made using a linear graphic rating scale, thereby eliminating non-numerical symbols and allowing descriptive and inferential statistics to be performed on the data.

Descriptive texture analysis. The importance of texture to perceived food quality is easily observable among supermarket consumers selecting melons and/or loaves of bread. Such manual tests of texture are used as predictors of subsequent oral texture (and perhaps even of flavor correlates). A descriptive approach patterned after the Arthur D. Little flavor profile was developed for the evaluation of food texture at General Foods Corp.[106,107] This approach attempted to standardize the vocabulary used to describe food texture by using operational definitions of textural attributes. The standardized terminology is predicated on the classification of food texture characteristics into three areas: (1) mechanical—those related to the responses of foods to applied forces; (2) geometrical—those related to the geometrical arrangement of the food matrix, e.g., size, shape, and orientation of particles; and (3) moisture and fat-related—those associated with the water and fat content of food.

As with the ADL flavor profile method and QDA, the General Foods texture profile method uses trained panelists who are selected on the basis of similar criteria. In its original form the method included an enumeration of texture attributes and definitions that could be applied to all foods. For each of these attributes, standard scales (ordered series of food products representing graduated degrees of the attribute) were developed.[108] These scales are used in panel training to familiarize the panel with the attributes in real foods. While all items on the standard scales are numbered to represent approximately equal perceptual intervals, actual scaling of test products is accomplished using the Arthur D. Little scaling system.

In more recent years the texture profile method has been expanded to allow greater flexibility and tailoring of terminology to specific products,[109,110] and investigators now use a variety of techniques to scale intensity. This has also led to the development of a more generalized sensory descriptive method (Spectrum[111]) that can be used to profile all relevant sensory attributes of a product. This latter approach combines the essential aspects of the older descriptive methods with more modern approaches to sensory scaling and the analysis of time–intensity relationships.

Free-choice profiling. Another significant breakthrough in descriptive methodology has come in the past decade with the development of the free-choice profiling method.[112,113] This approach differs from other profiling approaches in its use of *un*trained panelists. Thus, it is more consistent with the *zeitgeist* in sensory quality assessment, i.e., utilizing the consumer in the determination of quality attributes of food. The free-choice profiling approach allows consumers to either create their own vocabulary of terms to describe the attributes of the product or to select terms from a prepared list. The advantage of this approach is that, in addition to having consumers be the judge, the sensory characteristics of the product are described in everyday consumer terminology. Applications of this methodology in commodity areas formerly governed by expert opinion, e.g., wine, scotch, and beer tasting, are now common in the literature.[112,114]

Affective measures

On the far right of Land's model of food quality (Figure 1), just prior to "action," we see the concept of "acceptance level." This represents an emotional state of the consumer elicited by the food item. It is the hedonic state that precedes behavioral approach or avoidance. As such, it has been a primary target of those researchers interested in predicting the behavioral response to food.

The first systematic study of the affective or hedonic response to sensory stimuli was conducted by Beebe-Center.[115] As it applies to food, the study of hedonics (from the Greek word for "pleasure") began to flourish in the years following World War II at the U.S. Army Quartermaster Food and Container Institute in Chicago. It was here that today's most common measure of food acceptance was developed, the nine-point hedonic scale.[86,87] As a direct (self-report) measure of the degree of like/dislike for a food item, this scale has achieved widespread use in the food industry. The popularity of this scale stems from its ease of use and its reliability. In several studies the group reliability coefficient for this scale has been reported at .94 to .99.[116–119] Reports of individual reliability coefficients have also been high, with values of .60[118] and .74[120] reported.

In spite of its widespread use, the nine-point hedonic scale suffers from the same problems noted earlier for other types of interval scales. Other methods that have been used over the years to assess food acceptance include magnitude estimation, linear graphic rating scales, and facial scales (series of faces with smiles/frowns used primarily with children and those with reading difficulties). A simple binary measure of like/dislike has also been proposed as an adjunct for assessing food acceptance.[121–123] This "acceptor set size" index is calculated as the percentage of consumers who like the product (rate it acceptable). The concept behind this approach is that numerical means on a nine-point (or any other) scale do not provide useful data for understanding consumer behavior in response to the food. The accepter set size, on the other hand, tells the product developer what percentage of people like the item and, therefore, will be likely to select, buy, or consume the item. This procedure sacrifices details in individual data in order to obtain more useful data to predict group behavior.

Consumption measures

In some sense, all measures of perceived food quality are attempts to index the behavioral "action" that serves as the end-point in Land's model, where "action" means selection or consumption of the food item. Unfortunately, consumption is a difficult, time-consuming, and costly measure. Direct measures of consumption involve weighing each food item that is served to an individual and then weighing the uneaten portions. Other less complicated methods, such as visual estimation of food intake or aggregate measures of consumption, are less accurate and/or sacrifice information about individual intake. Moreover, unless such variables as hunger, thirst, and sensory-specific satiety (satiety to a specific flavor/food induced by prior consumption) can be controlled, measures of consumption may not accurately reflect ingredient or processing-related aspects of the food item. For the interested reader, several comprehensive reviews of the literature on measurement of food consumption have been published.[117-120]

Extrinsic factors

The preceding sections have dealt with conceptual and measurement issues related to the intrinsic factors that affect food acceptance and perceived food quality. These factors have had a long history of study and a large scientific database has been developed. The study of extrinsic factors, on the other hand, has a much more recent history. Consistent with this shorter history, many fewer facts are known. However, the challenges and opportunities are enormous and new data are rapidly accumulating.

Environmental factors

One obvious set of extrinsic factors that could play a role in the perceived quality, selection, or consumption of food by humans is the ambient physical environment and/or the body's internal environment as affected by heat, cold, or physical exercise. The fact that food and water intake is related to thermoregulation is belied by the control of all three by the hypothalamus. Moreover, phenomenologically, it certainly seems that the perception and acceptance of certain foods, such as ice-cold lemonade, differ, depending upon whether they are consumed while performing physical work on a hot day or while sedentary on a cold day. Yet, the evidence for the effects of both physical environment and bodily exercise on perceived food quality, acceptance, and consumption is limited and, in some cases, contradictory. For example, some investigators have reported that human food consumption increases in cold environments and decreases in hot environments;[124,125] while others have concluded that the effects are minimal.[126,127] The difference appears to be due to the ability of the organism to maintain normal body temperature, e.g., through the choice of clothing. Thus, while cold will increase consumption of food when body temperature is actually reduced,[128,129] it has no effect on well-clothed individuals living in cold climates. Similarly, consumption will not decrease in tropic environments if air-conditioning is available, but consumption will decline in individuals who must maintain thermal balance by reducing their activity in the heat. In the case of physical exercise, both human and animal studies have shown a reduction in food intake after short-term exercise.[130,131] However, sustained exercise has been shown to increase intake among athletes and lean individuals. No effect has been found with untrained individuals.[132]

Just as variable as the effects on overall food intake are the effects of these environmental factors on acceptability and perceived quality. Although some human data show environmentally controlled shifts in preference, such as a greater preference for fruits than meats in the heat[133] and some animal data showing a shift toward greater intake of carbohydrates in the cold,[134] it must be concluded that physical factors play less of a role in food acceptability in most situations than do cultural, cognitive, or social factors.

How then do we reconcile these facts with the striking and common experience noted at the beginning of this section that cold beverages on a hot day (or hot beverages on a cold day) seem to produce a much-enhanced level of acceptability? The answer is that such effects are not due directly to differences in ambient or body temperature, but to the

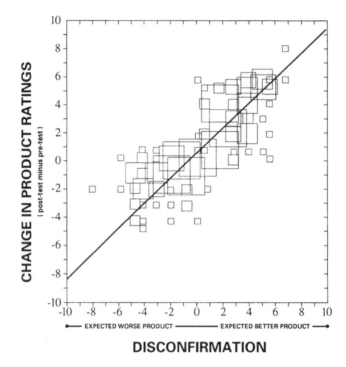

DISCONFIRMATION

Figure 8   Results of a study examining the effect of disconfirmed consumer expectations on product acceptance. Plot shows the change in hedonic rating from baseline as a function of whether subjects expected a better or worse product than they actually received.

for the food item. Moreover, when the expected acceptability of the item is raised or lowered through information provided about the product, perceived acceptability increases or decreases in the direction of the expected acceptability.[203,205–207,209]

Figure 8 shows data taken from one such study in which consumers were led to expect that they would receive one of several cola beverages for which baseline acceptance had been determined in a pilot test.[205] Subjects were subsequently served cola beverages that either confirmed their expectation or disconfirmed it (in the latter case, this meant a cola that they liked better or worse than the one they expected). When the changes in individual subject's product ratings from their baseline (ordinate) were compared to the difference between the expected and actual acceptability of the products they received (abscissa), a high positive correlation was observed. Consumers who expected a "low preference" cola rated the cola less acceptable than they had in the baseline test, while those who expected a "high preference" cola rated the cola more acceptable than they had in the baseline. The implications of such data are (1) that product information, packaging, or advertising appeal that increases the consumer's expectations for the product will increase its perceived acceptability and perceived quality, and (2) that any negative associations with the product that lower expectations will result in decreased acceptance, regardless of "true" product quality.

Social and cultural influences

Cultural factors are, perhaps, the most powerful of the extrinsic factors that can control food acceptance and perceived quality. In fact, one set of authors has concluded, "If one were interested in determining as much as possible about an adult's food preferences and could only ask one question, the question should undoubtedly be: 'What is your culture or ethnic group?'"[210] Cultural factors determine both what is preferred and what is rejected, and many of these food habits reflect perceptions of the self and philosophies of one's role in the universe.[211] In fact, as early as 1000 B.C. in China, the Yellow Emperor's Book of Internal Medicine detailed the relationships between health, food, and medicines within the context of the Chinese philosophical dichotomy of yin and yang. The powerful influence of culture in determining group perceptions of product quality is easily seen in the strict dietary avoidance of pork by Jews and Muslims, the avoidance of beef in India, and the avoidance of eggs, a symbol of fertility, by certain African tribes.[212] Closer to home, one can appreciate the powerful cultural factors that have made horse, cat, and dog consumption a rare and sometimes illegal practice in North America. Yet, in other parts of the world these practices are common. In almost all of these cases, the relationship to perceived food quality is obvious. The individuals who avoid the food items commonly perceive them as "unclean" or "disgusting"; yet those who are not part of the culture perceive the foods to be of excellent quality for consumption.

Although religious taboos provide the most dramatic examples of the effect of culture on perceived food quality, other cultural and societal factors play a role. For example, there are numerous regional preferences/aversions for indigenous/exogenous foods, such as the preference for brown eggs over white in the Northeastern United States. Other social factors, such as status in society, income level, and ethnicity can also affect food purchases, as reflected in the gourmet and health food preferences of the "yuppie culture," the often-reported consumption of pet food by the homeless, and the breadth of exotic vegetables and meats to be found in traditional Chinese, Italian, and other ethnic markets.

Perhaps some of the most interesting effects on food quality perception are those that result from social influences, i.e., effects of parents, peers, leaders, and heroes. For example, a close association has been demonstrated between parental and child food preferences, including gender influences, i.e., father–son and mother–daughter preferences are more similar than father–daughter or mother–son preferences.[213] Although the effects of gender have not been consistently supported, other social imitation effects have been reported with children in studies in which elders, heroes, teachers, and peers have been used to increase preferences for certain foods.[214–218] This strong effect of role models on perceived food quality, consumer choice, and consumption is not lost on commercial food companies, many of whom spend millions of dollars for sponsorship of their products by athletes, film stars, and rock idols.

In adults, several studies have shown a strong social facilitation effect on consumption.[219–222] That is, the total amount eaten in a meal increases when the meal is eaten with others as opposed to when it is eaten alone. Demonstration in adults of role model effects similar to those found in children has been more difficult. However, in a recent study conducted with military personnel, it was shown that positive or negative attitudes expressed by unit leaders toward military food had a significant effect on both food acceptance and consumption by subordinate personnel within the unit.[223] If the expressed attitude (through verbal comments and behavioral action) were positive toward the food, acceptance and consumption increased. If the expressed attitude was negative, acceptance and consumption decreased.

Psychological and emotional factors

Psychological factors that have an effect on food acceptance, selection, and consumption have been postulated for many years. Perhaps the oldest such relationship relates to the "psychosomatic" theory of obesity. Simply stated, this theory holds that food consumption is a coping mechanism against fear or anxiety and that consumption of food is anxiety-reducing. This theory held sway for many years, and addressing the psychological causes of eating was long seen as a prerequisite for modifying dietary behavior.[224,225] However, during the past quarter century, increasing amounts of data have been collected that question the psychosomatic theory. Much of these data come from the work of Schacter,[226-229] who proposed that eating behavior is, in large part, controlled by external cues in the environment, and that obese individuals are merely more sensitive to these external cues than are normal-weight individuals. Schacter's theory suggests that the sensory properties of food—its taste, aroma, appearance—are more salient to obese individuals, and thus, the appearance or odor of food is more likely to motivate obese individuals to consume foods to which they have been only casually exposed. A similar, personality-based approach to account for individual differences in flavor perception and food acceptance has recently been proposed.[230] This theory also focuses on whether individuals respond more to internal (self-produced) cues or external (situational) cues, with evidence of effects on both perceived flavor intensity (saltiness) and hedonic ratings.[230]

The notion that food intake may be controlled by psychological factors that make certain individuals more susceptible to the sensory properties of food has led investigators to examine the relationship between the personality construct of "sensation seeking"[231] and food preferences. In several studies it has been shown that sensation seeking, the desire to seek out new experiences, is correlated with specific food preferences. For example, it has been shown that individuals who score high on a sensation seeking index prefer spicy, sour, and crunchy foods over bland, sweet or soft foods.[232] Similarly, other investigators have shown that sensation seekers prefer both spicy foods and those that have a potential to result in illness, e.g., seafood.[233]

Independently of the role of personality, it is clear that food can take on very strong psychological and emotional significance. This emotional significance is cross-cultural, as evidenced by the common practice of feasting to celebrate socio-religious events. In addition, food is still considered by psychologists to be the primary motivator of living organisms. In fact, almost all that we know today about learning and conditioning comes from studies in which food (or the withholding of food) has been used to motivate organisms to behave in desired ways.

Not surprisingly, psychological and emotional factors can also play a prepotent role in the *failure* to consume food. One particularly interesting and complex problem is the development of aversions to foods based upon their association with offensive or disgusting social, situational, or ideational objects. It has been shown, for example, that contact with a safe but disgusting object will cause an otherwise acceptable food to be perceived as offensive and inedible.[234-236] This "contamination" effect has been interpreted and compared to the basic principles operating in magic and similar primitive belief systems.[237] As examples, it has been shown that drinks that have been stirred with a sterilized dead cockroach will not be consumed, that normally acceptable foods (e.g., fudge) that have been shaped into disgusting objects, such as dog feces, will be rejected, and that even preferred items (soups) stirred with an unused fly swatter will become unacceptable. In each of these cases a food that is otherwise safe, clean, and of high quality is perceived as contaminated and unacceptable. This occurs in spite of the individual's ability to intellectualize and understand that what has been done to the food is perfectly safe and does not change its sensory or other properties. Clearly, such effects on the perceived quality of a food are the result of deep-rooted psychological, cultural, and emotional factors that are, as yet, poorly understood.

In summary, a wide variety of factors that are unrelated to the physical, chemical, nutritional, microbiological, and sensory properties of food can play a significant role in controlling consumers' perception of food quality, acceptance, and consumption of food.

These extrinsic factors must be taken into account when developing products that are intended to have high perceived quality. Although current approaches to food quality assessment do this to some degree, far greater attention to these important factors is warranted.

Perceived food quality: current and future directions

By understanding the nature of both the intrinsic and extrinsic factors affecting the perception of food and its perceived quality, we can begin to appreciate the complexity of food quality perception. However, in day to day practice in the food industry we must be able to both measure and predict food quality. In view of the changes that have occurred in the conceptual approach to defining food quality, i.e., a shift toward greater dependence on the consumer and his/her perceptions and behavior, greater emphasis should be put on consumer-oriented approaches to quality measurement and prediction. Thus, it may be useful to trace through the typical phases in the R&D cycle for a food product and to identify what food quality means and what future trends in perceived quality assessment are on the horizon.

Designing the product

If food quality is in the "eye of the consumer," and the eye of the consumer is subject to such factors as past experiences, expectations, and social and cultural factors, then optimizing food quality requires designing product characteristics to meet consumer needs, and desires as molded by these factors. This means conducting in-depth market research studies in order to understand consumer perceptions, needs and desires, and then following this with strategic marketing initiatives that tailor to the consumer's needs not only the intrinsic attributes of the product, but also the packaging, labeling, product name, and product image.

Qualitative and quantitative market research is the first step. Quantitative techniques, as their name implies, consists of surveys, questionnaires, and structured interviews that seek out information that can be quantified, e.g., frequency of product purchases, estimates of perceived value, number of satisfied buyers, ratings of package aesthetics, etc. Traditional demographic data also fall into this category, as do psychographic data, i.e., information about consumer lifestyles, preferences, behavior patterns, values, and interests. With such data, market segmentation can be accomplished in order to tailor products and advertising to specific subgroups of the consumer population, e.g., the current "nostalgia" themes in advertising used to target the baby boomer audience. Segmentation can also be accomplished on the basis of sensory preferences,[238,239] so that "likers" vs. "dislikers" of certain products can be identified, or those that prefer one or another specific sensory characteristics in their products.

In addition to quantitative market research techniques, a far greater use of qualitative techniques, such as focus groups, has evolved.[240,241] The purpose of qualitative research is to gain insight into factors that control consumer perceptions and behavior. It is focused on soliciting ideas and/or determining the reasons behind certain attitudes, beliefs, behaviors, etc. Focus group interviews involve group discussions among 8–10 homogeneous consumers of the product, led by a trained moderator. Sessions are usually videotaped or tape-recorded so that specific comments of the group can be replayed for full context analysis. Goldman and MacDonald[241] provide an excellent example of the difference between asking a seemingly simple question on a quantitative survey, such as: "Which do you prefer, apple pie or chocolate cake?", and asking the same question in a focus group, where you may get the following exchange:

Q: Which do you prefer, apple pie or chocolate cake?
A: I don't know, I like them both.
Q: Well, if you had to give up one or the other, which would it be, do you think?
A: Actually, it depends. If we're talking about my mother's apple pie then I prefer that to any chocolate cake. But if we're just talking about anyone's apple pie, then I don't know. It depends, I guess, on a bunch of things.
Q: On what other things does it depend?
A: Well, it depends on what I've had for dinner, for example. If I've had a heavy dinner, I think I'd probably go for the pie. And on Thanksgiving, its pie for sure. That's a big tradition in my family. But, I don't know, after something light like fish, chocolate cake is great. That is, if it has frosting. I don't care for plain chocolate cake.

From this simple example, it is clear that focus groups and other qualitative techniques provide a wealth of important information about the characteristics and quality of a product or product concept that survey and questionnaire approaches simply cannot uncover. Only by fully understanding the consumer and the market category to which the intended food product belongs can such products be designed with a high potential for perceived quality and success in the marketplace.

The concept of "market category" is critical to the success of any product, because it defines the arena in which the product will be competing. Examples of market categories include frozen entrees, premium ice creams, diet beverages, etc. Before product design and development can begin, a comprehensive strategy for analysis of the market for the product is needed. Such strategies of analysis fall under the rubric of "category appraisal" research.[238] This research approach consists of a set of procedures by which an entire category of foods, e.g., potato chips, are analyzed in terms of the different consumer segments that exist (e.g., thin and crisp lovers vs. thick and crunchy lovers), the critical sensory factors controlling acceptance, the effect of brand identity or image on acceptance, and even the effect of promotional and advertising programs. Applying correlation and regression techniques to these data provides the product developer with insightful information concerning the roles of each of these factors in controlling consumer purchase and marketshare.[238] Consumer purchase and market share constitute the "action" variables in Land's model (Figure 1) with which food producers typically are most concerned.

It is of no surprise to the reader that literally hundreds of different variables can be examined in the course of market research for a product. Thus, leaders in food product development in the future will depend to a much greater extent on multivariate statistical analysis of these data in order to design products with high perceived quality. Many of these multivariate techniques have been applied by psychologists and sensory researchers for many years, e.g., factor and cluster analysis, multidimensional scaling, principal component analysis and Procrustes analysis. However, other techniques have evolved mostly within the market research arena and are only now seeing broader application among food scientists. Typical of such techniques are the repertory grid method[242-244] and the method of conjoint analysis.[245,246] Conjoint analysis is a method that utilizes consumer choice behavior in order to infer the importance or value of various product attributes to the consumer. Using this method, consumers make choices between product attributes that are "jointly" presented, e.g., "do you prefer crunchy corn chips with low cheese flavor or crispy corn chips with high cheese flavor?" Two or more attributes or levels of each attribute can be presented and rank preferences assigned. Through a decompositional approach, the worth or "utility" of the individual attributes is mathematically determined so that the total utility for any combination of attributes/levels is as monotonic as possible with the original preference judgments. In this way, the product developer can better estimate the likely combination of attributes that will produce a product with high perceived quality/acceptance. These and other advanced statistical techniques will become the norm for future data analytic approaches to food quality perception.

Product formulation

Based on the results of market research to define characteristics of the product required to meet consumers' needs/desires, the food technologists begin their arduous task of formulating a product that has these quality characteristics. In some cases, this development work can be guided by existing products that possess one or more of the desired characteristics. In other cases, an entirely new product is formulated. It is here that traditional sensory techniques of descriptive testing come into play to characterize the product profile and to compare it to existing or ideal target profiles. However, as mentioned previously, even here the consumer is starting to play a greater role through the use of the free-choice method of profiling. Development of this technique is indicative of a growing trend to focus more attention on consumer language in food product development and marketing.[247] However, a cautionary note must be sounded to insure that with certain of these methods, the sensory concepts and associated verbal labels are used consistently among panelists.[248–250]

This stage of product development will also see much greater use of multivariate techniques to develop "perceptual maps" of the way consumers perceive product characteristics and the relationships among product prototypes and existing commercial products in the category. Many contemporary approaches like the repertory grid method and the combination of free-choice profiling with Procrustes analysis will also see wider application. In all cases, the emphasis is, and will continue to be, on the greater use of consumers in making sensory and perceptual judgments of products under formulation and in guiding the direction of further formulations to improve perceived quality.

Product optimization

Although optimum sensory and marketing variables can be identified through many of the techniques discussed above, development of a product that has these optimal characteristics can be difficult, especially when there are scores of ingredient, processing, and marketing variables that can be manipulated. To aid in this process, a number of techniques that fall under the general rubric of "optimization" methodology have been developed and will see much greater use. Such optimization procedures derive from standard multiple regression techniques, enhanced by the speed and power of computer calculations and graphic analyses. The simplest of these optimization approaches is based on first-order or linear regression, in which a dependent variable (y) can be related to a series of independent variables ($x_1 \ldots x_n$) by the function $y = a + b_i x_i \ldots b_n x_n$ Although such simple linear models can be applied to many food optimization problems, their use in optimizing perceived food quality and consumer acceptance is severely restricted by the fact that acceptance and other hedonic judgments are nonlinear functions of ingredient and sensory variables. As such, linear programming approaches are generally inappropriate for the optimization of consumer acceptance and perceived food quality.

The alternative to a linear approach is the use of second- and third-order polynomials that utilize quadratic and cubic terms and associated cross-products. The quadratic version of such an equation is $y = a + b_i x_i + b_2 x_j + b_3 x_i^2 + b_4 x_j^2 + b_5 x_i x_j$. Using this approach, the dependent variable, y, can be consumer acceptance or any desired sensory or market attribute of the product, while the dependent variables, $x_1 \ldots x_n$, can be sensory, instru-

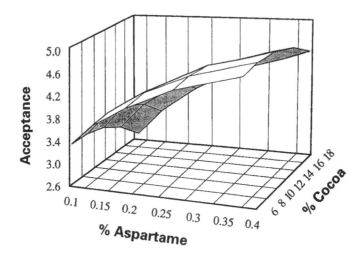

Figure 9   A response surface generated as part of a developmental program to optimize the levels of cocoa and aspartame in a cocoa beverage bar. (From Siegel, S.F. *Activities of the R&D Associates.* 1987, 39(1), 183. With permission.)

mental, ingredient, processing, or other product-related variables. Once the independent and dependent variables and the mathematical model have been decided, a strategy of data collection is employed by using one of a variety of factorial and composite experimental designs that are available for this purpose.[251,252] Samples can then be prepared and data collected on all relevant variables. A least-squares method is then utilized to generate the factor coefficients. The final regression equation is used to plot the dependent variable as a function of the independent variables. The resultant "response surface" may be curvilinear in one or more dimensions.

Figure 9 is a response surface generated during the optimization of the consumer acceptability of a cocoa beverage bar.[253] Two ingredient variables were manipulated: cocoa and aspartame. As can be seen, the response surface that was generated showed highest acceptability at the intermediate cocoa level (12%) for all levels of aspartame. However, peak acceptance was found at the highest aspartame level that was tested. As such, one might conclude that the optimal combination of ingredients for this product is 12% cocoa and 0.4% aspartame (assuming that 0.4% aspartame is an upper limit based on other criteria, e.g., cost or functionality). However, the response surface in Figure 9 is a very simple one. Frequently, more complicated response surfaces are generated that can contain multiple local maxima and minima, making it difficult to identify a single combination of ingredient levels that produces optimum acceptance or quality.[254] A variety of search algorithms have been developed to locate such maxima and minima, and these search strategies are integral aspects of the optimization software systems that have been developed (see Saguy et al.[255] and Floros and Chinan[256] for reviews).

The last set of factors to be considered in product optimization are the "constraints." For example, while one can optimize the level of ingredients and processes to produce a high-quality bread product, the cost of those ingredients and/or their levels may be too prohibitive to produce with any profit. As a result, the manufacturer must perform "constrained optimization." In this process, at each step in the search for an optimal set of independent variables, the cost of this combination of variables is calculated. If the calculated value exceeds the cost (or any other) constraint, the search strategy is prevented from moving in that direction, and another path to the optimal solution is attempted. Ultimately, solutions are identified that optimize the dependent variable within the specified constraints. The reader interested in an in-depth discussion of optimization of food acceptance/quality is referred to texts by Moskowitz.[238,257]

Production, storage, and distribution

This phase in the life cycle of a food product is the one that is most closely associated with the word "quality," because it is here that extensive "quality assurance" procedures are implemented. However, by this time in development, all of the quality that will ever appear in the product has already been built into the product as part of previous development phases. We have effectively switched over from the conceptual view of quality as "degree of excellence" and are now dealing with it as "deviation from a standard." At this point, food quality means eliminating defects and monitoring deterioration.

As it relates to sensory properties, quality assurance begins with the inspection of raw materials and ingredients and continues through to the evaluation of the final packaged product. A wide variety of traditional approaches are used, the majority of which are designed around the use of trained, in-house panels who use discriminative methods, descriptive methods, or direct judgments of overall quality using specified standards. These standards can be actual control samples that are fresh or stored under ideal conditions, written standards that may be in the form of sensory profiles or specifications documents, photographic standards that may be used to assure certain appearance characteristics, or cognitive standards that are based upon extensive prior experience with the product. In all cases, special care must be taken to ensure that many of the common errors found with older grading standard approaches, e.g., combining multiple sensory attributes into a single quality index, are avoided.

The sensory methods most commonly used for quality control purposes have been detailed in numerous texts and symposia, and the interested reader can consult these resources for details on these methodologies and other issues related to sensory food quality assurance.[111,258–261]

Many times, as part of quality assurance procedures or even prior to full-scale product production, storage stability studies are undertaken to ensure prolonged shelf-life of the product under various temperature/humidity conditions. As in the assessment of product defects, standard discriminative and descriptive sensory procedures are applied to the stored samples as well as to control samples, using any one of a variety of sampling designs.[262,263] Of course, in these studies, as well as those noted above, a critical issue is the magnitude and importance of the defect or quality degradation. Although an in-house panel may find a significant sensory difference between a production sample and a control, or between a sample stored for 1 year at 35°C and a control stored at 5°C, this does not necessarily mean that the consumer would notice the sensory change and/or associated decrement in quality. If he/she does not, the producer/store manager may not want to incur the cost of destroying a production run or store-shelf lots of the product. For this reason, here too, consumer tests are commonly conducted to assess the acceptability of stored samples.

Summary

Quality vs. acceptability

The association between perceived food quality and consumer acceptability that has been a central theme of this chapter is not new. As early as 1958, David Peryam wrote about quality scoring that "I'm not sure that it requires any special definition, or that everybody cares to make the distinction we use it for. . . . It is distinguished from preference more in the way the result is arrived at, and who does the judging, than in the actual results."[264] And concerning the scales used to judge quality, Pilgrim and Peryam wrote,

> The main difference from the hedonic scale is in the way the dimension is expressed; however, there is evidence to show that the difference is only superficial and that the two scales (quality scoring vs. hedonic) are just different ways of representing the same basic dimension of preference. The rationale of such a scale is based on the assumption that it is possible for someone to develop a concept of general excellence of quality, completely or partially independent of his own personal preferences, and that when he uses the criterion to evaluate food, his personal biases do not affect the ratings. Experimental evidence indicates that this assumption is unwarranted, even with trained subjects; results on the quality scale correlate as highly with preference results as with themselves. It is therefore evident that they are measuring the same thing.[265]

The truth in Peryam's discussion of the topic can be seen 30 years later in a paper on food quality.[266] Figure 10, taken from this report, shows consumer acceptance (overall liking) and quality ratings for products at varying points in their shelf lives. The correlation coefficient between acceptance and quality is greater than .85. In a still more recent study,[98] the correlations between consumer ratings of food quality and acceptability ranged from .92 to .94, showing clearly that consumers equate quality with liking. Hopefully, 30 years from now we will no longer have to present such data to make the case that food quality should be defined by the consumer's perception of it, and that their perception is intimately linked with what they like.

Retrospective

This chapter has presented an overview of several important issues in food quality assessment. First is the fact that "food quality" means many things to many people. However, the final judge of the quality of any food item is the consumer. Thus, perceived food quality and how we measure it are central to any discussion of food quality. Second, the factors affecting perceived food quality are many and varied, and can be categorized as being intrinsic, i.e., related to the food and its physicochemical features, or extrinsic, i.e., related more to the consumer and his/her cognitive, experiential, or sociocultural environment. In the case of intrinsic factors, the facts are well known and the methods for assessing the effects of these variables are well established for use by both consumers and trained or expert judges. In the case of extrinsic factors, the facts are much less clear, but the effects on perceived food quality can be equally dramatic. Although the methods of measurement of the independent effects of these extrinsic factors are rooted in the complex measurement processes used in the experimental laboratories of social scientists, their combined effect on food quality perception can be assessed through consumer marketing and advanced consumer analytic techniques. In the future, the measurement of perceived food quality will depend even more heavily than it does today on consumer evaluation. In addition, there will be greater reliance on consumer marketing, advanced multivariate statistical techniques, and mathematical optimization methods in order to integrate the myriad of ingredient, processing, sensory, and market variables that contribute to perceived food quality. The challenges to the food industry are great, but the payoff will be in terms of products that have significantly improved food quality for the consumer.

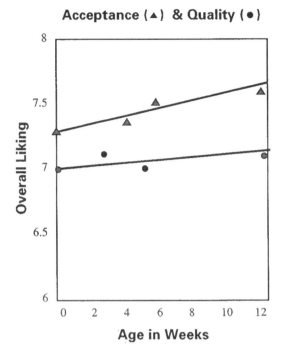

Figure 10 Mean ratings of overall liking and quality for products at 0, 3, 6, and 12 weeks of age. Data are from 125 consumers using nine-point scales. (From Stone, H., et al. *Food. Technol.* 1991, 45(6), 88. With permission.)

References

1. U.S. Department of Agriculture Marketing Workshop Report, 1951. In W.A. Gould, *Food Quality Assurance*. Westport, CT: AVI Publishing, 1977.
2. Galvez, F.C.F., Resurreccion, A.V.A. *J. Sensory Stud.* 1992, 7(4), 315.
3. Cardello, A.V. *Food Qual. & Pref.* 1995, 6, 163.
4. Wierenga, B. *J. Food Qual.* 1983, 6, 119.
5. Steenkamp, J. *J. Food Qual.* 1987, 9, 373.
6. McEwan, J.A.; Thomson, D.M.H. *Food Qual. & Pref.* 1988, 1(1), 3.
7. Pilgrim, F.J. *Am. J. Clin. Nutrit.* 1957, 171.
8. Olsen, J.C. In *Criteria of Food Acceptance*; Solms, J; Hall, R.L., Eds.; Forster-Verlag: Zurich, 1981; p 69.
9. Harper, R. In *Criteria of Food Acceptance*; Solms, J; Hall, R.L., Eds.; Forster-Verlag: Zurich, 1981; p 220.
10. Land, D.G. In *Sensory Quality in Foods and Beverages*; Williams, A.A.; Atkins, R.K., Eds.; Ellis Horwood: Chichester, U.K., 1983; p 15.
11. Williams, A.A. *Chem. & Ind.* 1983, 19, 740.
12. Fechner, G.T. *Elements der Psychophysik*; Breitkopf and Harterl: Leipzig, 1860. (English translation by H.E. Adler; Holt, Reinhart & Winston: New York, 1966).
13. Fick, A. In *Lehrbuch der Anatomie und Physiologie der Sinnesorgane*; M. Schauenberg und Company: Lahr, 1864, p 67.
14. Ganzevles, P.G.J.; Kroeze, J.H.A. *Chem. Senses.* 1987, 12, 536.
15. Gardner, R.J. *Chem. Senses & Flavor.* 1980, 5, 185.
16. Kinnamon, S.C.; Roper, S.D. *J. Gen. Physiol.,* 1988, 91, 351.
17. Gilbertson, T.A.; Avenet, P., Kinnamon, S.C.; Roper, S.D. *J. Gen. Physiol.* 1992, 100, 803.
18. DeSimone, J.A.; Ferrell, F. *Am. J. Physiol.* 1985, 249, R52.
19. Schiffman, S.S.; Simon, S.A.; Gill, J.M.; Beeker, T.G. *Physiol. & Behav.* 1986, 36, 1129.

20. Shallenberger, R.S.; Acree, T. *Nature.* 1967, 216, 480.
21. Tonosaki, K.; Funakoshi, M. *Nature.* 1988, 331, 354.
22. Striem, B. J.; Pace, U.; Zehavi, U.; Naim, M.; Lancet, D. *Biochem. J.* 1989, 260, 121.
23. Jackinovich, W., Jr.; Sugarman, D. *Chem. Senses.* 1988, 13, 13.
24. Boudreau, J.C. *J. Sensory Stud.* 1986, 1, 185.
25. Fox, A.L. *Proc. Nat. Acad. Sci. USA.* 1932, 18, 115.
26. Kalmus, H. In *Handbook of Sensory Physiology, IV: Chemical Senses, 2: Taste*; Springer-Verlag: New York, 1971; p 165.
27. Bartoshuk, L.M. *Science.* 1979, 205, 934.
28. Bartoshuk, L.M.; Fast, K.; Karrer, T.A.; Marino, S.; Price, R.A.; Reed, D.R. *Chem. Senses.* 1992, 17, 594.
29. Kubota, T.; Kubo, I. *Nature.* 1969, 223, 97.
30. Temussi, P.A.; Lelj, F.; Tancredi, T.J. *Med. Chem.* 1978, 21, 1154.
31. Tancredi, T.; Lelj, F.; Temussi, P.A. *Chem. Senses & Flavor.* 1979, 4, 259.
32. Akabas, M.H.; Dodd, J.; Al-Awgati, Q. *Science,* 1988, 242, 1047.
33. Hwang, P. M.; Verma, A.; Bredt, D. S.; Snyder, S. *Proc. Nat. Acad. Sci. USA.* 1990, 87, 7395.
34. McBurney, D.H.; Smith, D.V.; Schick, T.R. *Percep. & Psychophys.* 1972, 11, 228.
35. Herness, M.S.; Pfaffmann, C. *Chem. Senses.* 1986, 11(3), 347.
36. Yamaguchi, S. In *Food Taste Chemistry*; Boudreau, J.C., Ed.; American Chemical Society: Washington D.C., 1979; p 33.
37. Kawamura, Y.; Kare, M. *Umami: Physiology of Its Taste*; M. Dekker: New York, 1986.
38. Green, B.G.; Mason, J.R.; Kare, M.R., Eds. *Chemical Senses: Volume 2, Irritation*; M. Dekker: New York, 1990.
39. Zwaardemaker, H. *Die Physiologie des Geruchs*; Engelmann: Leipzig, 1895.
40. Henning, H., *Der Geruch*; Barth: Leipzig, 1916.
41. Crocker, E.C.; Henderson, L.F. *Am. Perfum. Essent. Oil Rev.* 1927, 22, 325.
42. Stevens, J.C.; Cain, W.S.; Burke, R.J. *Chem. Senses.* 1988, 13, 643.
43. Woskow, M.M. In *Theories of Odors and Odor Measurements*; Tanyloc, N., Ed.; Robert College Research Center: Bebek, 1968; p 147.
44. Davis, R.G. *Chem. Senses.* 1979, 4, 191.
45. Lawless, H.T. *Chem. Senses.* 1989, 14(3), 349.
46. Amoore, J.E. *Perf. Essent. Oil Record.* 1952, 43, 321.
47. Amoore, J.E. *Molecular Basis of Odor*; Thomas: Springfield, 1970.
48. Wright, R.H. *J. Appl. Chem.* 1954, 4, 611.
49. Wright, R.H.; Reid, C.; Evans, G. *Chem. & Ind.* 1956, 37, 973.
50. Wright, R.H. *Annals NY Acad. Science.* 1974, 237, 129.
51. Beets, M. *Amer. Perfum.* 1961, 76, 54.
52. Davies, H.T. *J. Theoret. Biol.* 1965, 8, 11.
53. Mozell, M. In *Olfaction and Taste I*; Pfaffmann, C., Ed.; Rockefeller Univ. Press: New York, 1969.
54. Mozell, M.; Jagodowicz, M. *Annals NY Acad. Science.* 1974, 237, 76.
55. Dravnieks, A.; Laffort, P. In *Olfaction and Taste IV*; Schneider, D., Ed.; Wissench. Verlagsgesellsch: Stuttgart, 1972.
56. Amoore, J.E.; Johnston, J.W., Jr.; Rubin, M. *Scientific Am.* 1964, 210, 42.
57. Amoore, J.E. *Cold Spring Flavour Symposia in Quantitative Biology.* 1965, 30, 623.
58. Buck, L.; Axel, R. *Cell.* 1991, 65, 175.
59. Hunter, R.S. *The Measurement of Appearance*; Wiley: New York, 1975.
60. Francis, F.J.; Clydesdale, F.M. *Food Colorimetry: Theory and Applications*; AVI: Westport, CT, 1975.
61. Drake, B.K. *J. Food Sci.* 1965, 30, 556.
62. Drake, B.K. *Biorheol.* 1965, 3, 21.
63. Drake, B.K.; Halliden, L. *Rheol. Acta.* 1974, 13, 608.
64. Vickers, Z.M.; Bourne, M.C. *J. Food Sci.* 1976, 41, 1158.
65. Vickers, Z.M. In *Food Texture and Rheology*; Sherman, P., Ed.; Academic Press: London, 1979.
66. Vickers, Z.M. *J. Texture Stud.* 1980, 11, 291.
67. Christensen, C.M.; Vickers, Z.M. *J. Food Sci.* 1981, 46, 574.

68. Henkin, R.L.; Banks, V. In *Second Symposium on Oral Sensation and Perception*; Bosma, J.T., Ed.; Thomas: Springfield, 1967; p 182.
69. Ringel, R.L. In *Second Symposium on Oral Sensation and Perception*; Bosma, J.T., Ed.; Thomas: Springfield, 1967; p 323.
70. Ringel, R.L. In *Second Symposium on Oral Sensation and Perception*; Bosma, J.T., Ed.; Thomas: Springfield, 1970; p 309.
71. Jones, M.H. *Am. J. Psychol.* 1954, 67, 696.
72. Cain, W.S. *Sensory Processes.* 1976, 1, 57.
73. Gouindarajan, V.S. In *Food Taste Chemistry*; Boudreau, J.C., Ed.; American Chemical Society: Washington D.C., 1979; p 53.
74. Engen, T. In *Experimental Psychology*, 3rd Edition; Holt, Rinehart and Winston, Inc.: New York, 1971; p 11.
75. Schiffman, S.S. *Contemp. Nutrit.* 1991, 16, 1.
76. Tanner Jr, W.P.; Swets, J.A. *Psvchol. Rev.* 1954, 61, 401.
77. Green, D.M.; Swets, J.A. *Signal Detection Theory and Psychographics*; Wiley: New York, 1966.
78. Swets, J.A. *Science.* 1961, 134, 168.
79. O'Mahony, M. *J. Food Sci.* 1979, 44, 302.
80. O'Mahony, M. In *Sensory Quality in Foods and Beverages. Definition. Measurements and Control*; Williams, A.A.; Atkins, R.K., Eds.; Ellis Horwood: Chichester, U.K., 1983; p 69.
81. Vie, A.; Gulli, D.; O'Mahony, M. *J. Food Sci.* 1991, 56(1), 1.
82. Stevens, S.S. *Psychol. Rev.* 1957, 64, 153.
83. Stevens, S.S. *Science.* 1961, 133, 80.
84. Stevens, S.S. In *Handbook of Experimental Psychology*; Stevens, S.S., Ed.; Wiley: New York, 1951; p 1.
85. Stevens, S.S. *Science.* 1970, 170, 1043.
86. Peryam, D.R.; Giradot, N.F. *Food Eng.* 1952, 24, 58.
87. Peryam, D.R.; Pilgrim, F.J. *Food Technol.* 1957, 11, 9.
88. Jones, L.V.; Peryam, D.R.; Thurstone, L.L. *Food Res.* 1955, 20, 512.
89. Lawless, H.T.; Malone, G.J. *J. Sensory Stud.* 1986, 1(2), 155.
90. Pearce, J.H.; Korth, B.; Warren, C.B. *J. Sensory Stud.* 1986, 1(1), 27.
91. Schutz, H.G.; Judge, D.S. In *Research in Food Science and Nutrition, Vol 4: Food Science and Human Welfare*; McLaughlin, I.V.; McKenna, B.M., Eds.; Boole Press: Dublin, 1984; p 229.
92. Steenkamp, J.B.E.M. *J. Food Qual.* 1986, 9, 373.
93. Cardello, A.V.; Maller, O. In *Objective Methods in Food Quality Assessment*; Kapsalis, J., Ed.; CRC Press: Boca Raton, FL, 1986, p 61.
94. McNutt, K. *Food Technol.* 1988, 42, 97.
95. Civille, G.V. *J. Food Qual.* 1991, 14(1), 1.
96. Fishken, D. *J. Sensory Stud.* 1990, 5, 203.
97. O'Mahoney, M. *J. Food Qual.* 1991, 14, 9.
98. Cardello, A.V.; Bell, R.; Kramer, F.M. *Food Qual. & Pref.* 1996, 7(1), 7.
99. Ryan, N.R. *Rest. & Institut.* 1988, 44.
100. Powers, J.J. *J. Sensory Stud.* 1988, 3(2), 151.
101. Langron, S.P.; Noble, A.C.; Williams, A.A. In *Sensory Quality In Foods and Beverages: Definition. Measurement and Control*; Atkin, R.K.; Williams, A.A., Eds.; Ellis Horwood: Chichester, U.K., 1983; p 325.
102. Anonymous. *Food Process.* 1988, 49(5), 190.
103. Cairncross, S.E.; Sjostrom, L.B. *Food Technol.* 1950, 4, 308.
104. Caul, J.F. In *Advances in Food Research*, Vol. 7; Mrak, E.; Stewart, G.F., Eds.; Academic Press: New York, 1957; p 5.
105. Stone, H.; Sidel, J.L.; Oliver, S.; Woolsey, A.; Singleton, R.C. *Food Technol.* 1974, 24, 28.
106. Szczesniak, A.S. *J. Food Sci.* 1963, 28, 385.
107. Brandt, M.A.; Skinner, E.Z.; Coleman, J.A. *J. Food Sci.* 1963, 28, 404.
108. Szczesniak, A.S.; Brandt, M.A.; Friedman, H.H. *J. Food Sci.* 1963, 28, 397.
109. Civille, G.V.; Szczesniak, A.S. *J. Texture Stud.* 1973, 4, 204.
110. Civille, G.V.; Liska, I.H. *J. Texture Stud.* 1975, 6, 19.

111. Meilgaard, M.D.; Civille, G.V.; Carr, T.B. *Sensory Evaluation Techniques;* Vol. I and II. CRC Press: Boca Raton, FL, 1987.

112. Williams, A.A.; Langron, S.P. *J. Sci. Food Agri.* 1984, 35, 558-68.

113. Steenkamp, J.-B.E.M.; Van Trijp, H.C.M. In *Food Acceptability;* Thomson, D.M.H., Ed.; Elsevier: New York, 1988; p 363.

114. Gains, N.; Thomson, D.M.H. *Food Qual. & Pref.* 1990, 2, 39-47.

115. Beebe-Center, J.G. *The Psychology of Pleasantness and Unpleasantness;* Von Nostrand: Princeton, NJ, 1932.

116. Schutz, H.G. *Am. Psychol.* 1957, 12, 380.

117. Peryam, D.R.; Polemis, B.W.; Kamen, J.M.; Eindhoven, J.; Pilgrim, F.J. *Food Preferences of Men in the Armed Forces;* Quartermaster Food and Container Institute: Chicago, 1960.

118. Waterman, D.; Meiselman, H.L.; Branch, L.; Taylor, M. Technical Report #75-25-FSL, U.S. Army Natick Laboratories, Natick, MA; 1974.

119. Wyant, K.W.; Meiselman, H.L.; Waterman, D. Technical Report #79-041, U.S. Army Natick Research and Development Command, Natick, MA; 1979.

120. Smutz, E.R.; Jacobs, H.L.; Waterman, D.; and Caldwell, M. Technical Report #75-52-FSL, U.S. Army Natick Laboratories, Natick, MA; 1974.

121. Gordon, N.M.; Norback, J.P. *Food Technol.* 1985, 39(11), 96.

122. Lagrange, V.; Norback, J.P. *J. Sensory Stud.* 1987, 2, 119.

123. Jezior, B.; Popper, R.; Lesher, L.; Greene, C.; Ince, V. Interpreting Rating Scale Results: What does a Mean Mean? Paper presented at 32nd Annual Conference of the Military Testing Association, Orange Beach, AL, 5-9 Nov, 1990.

124. Brobeck, J.R. *J. Biol. and Med.* 1948, 20, 545.

125. Logue, A.W. *The Psychology of Eating and Drinking;* W.H. Freedom: New York, 1986.

126. Henschel, A. In *Man Living in the Arctic;* Fisher, F.R., Ed.; Quartermaster Research and Engineering Command: Natick, MA, 1960; p 68.

127. LeBlanc, J. *Man in the Cold;* Thomas: Illinois, 1975.

128. Kissileff, H.R. In *Proceedings of the 3rd International Conference on the Regulation of Food and Water Intake;* Haverford, PA.

129. Portet, R. *Comp. Biochem. Physiol. B.* 1981, 70, 679.

130. Nikoletesas, M.M. *Neurosci. Biobeh. Rev.* 1980, 4, 265.

131. Applegate, E.A.; Upton, D.E.; Stern, J.S. *Physiol. Behav.* 1982, 28, 917.

132. Titchenal, C.A. *Sport Med.* 1988, 6, 135.

133. McCoy, J.L. AMXFC Report No 3-63, U.S. Army Quartermaster Research and Engineering Center, Chicago, IL; 1963.

134. Andik, I.; Bank, J. *Acad. Sci. Hung.* 1954. 6, 37.

135. Cabanac, M. *Science.* 1971, 173, 1103.

136. Moskowitz, H.R.; Kumraih, V.; Sharma, K.N.; Jacobs, H.L.; Sharma, S.D. *Physiol. & Behav.* 1976, 16, 471.

137. Cabanac, M.; LaFrance, L. *Physiol. & Behav.* 1990, 47, 539.

138. Jensen, K. *Genet. Psychol. Mono.* 1932, 12, 361.

139. Nisbett, R.; Gurwitz, S. *J. Comp. Physiol. Psychol.* 1970, 73, 245.

140. Nowlis, G.H. In *Fourth Symposium on Oral Sensation and Perception: Development of the Fetus and Infant;* Bosma, J.F., Ed.; U.S. Government Printing Office: Washington D.C., 1973; p 292.

141. Desor, J.A.; Maller, O.; Turner, R.E. *J. Comp. Physiol. Psychol.* 1973, 84, 496.

142. Desor, J.A.; Maller, O.; Andrews, K. *J. Comp. Physiol. Psychol.* 1975, 89, 966.

143. Lipsitt, L.P. In *Taste and Development: The Genesis of Sweet Preference;* Weiffenbach, J.M., Ed.; U.S. Government Printing Office: Washington D.C., 1977; p 124.

144. Crook, C.K. *Infant Behav. Dev.* 1978, 1, 52.

145. Steiner, J.E. *Adv. Child Dev.* 1979, 13, 257.

146. Rosenstein, D.; Oster, H. *Child Dev.* 1988, 59, 1555.

147. Garcia, J.; Ervin, F.R.; Koelling, R.A. *Psychonom. Sci.* 1966, 5, 121.

148. Garcia, J.; Koelling, R.A. *Psychonom. Sci.* 1966, 4, 123.

149. Garb, J.L.; Stunkard, A. *Am. J. Psychiat.* 1974, 131, 1204.

150. Bernstein, I.L.; Webster, M.M. *Physiol.& Behav.* 1980, 25, 363.

151. Logue, A.W.; Ophir, I.; Strauss, K.E. *Behav. Res. & Ther.* 1981, 19, 319.
152. Pelchat, M.L.; Rozin, P. *Appetite.* 1982, 3, 341.
153. Bartoshuk, L.M.; Wolfe, J.M. Conditioned "Taste" Aversions in Humans: Are They Olfactory Aversions? Paper presented at the meeting of the Association for Chemoreception Sciences, Sarasota, FL, April, 1990.
154. Rozin, P.; Kalat, J.W. *Psychol. Rev.* 1971, 78, 459.
155. Zahorik, D. In *Preference Behavior and Chemoreception*; Kroeze, J.H.A., Ed.; Information Retrieval: London, 1979; p 233.
156. Torrance, E.P. *J. Appl. Psychol.* 1958, 42, 63.
157. Capretta, P.J.; Rawls, L.H. *J. Comp. & Physiol. Psychol.* 1974, 86, 670.
158. Balogh, R.D.; Porter, R.H. *Infant Behav. & Devel.* 1986, 9, 395.
159. Davis, L.B.; Porter, R. H. *Chem. Senses*, 1991, 16(2), 169.
160. Domjan, M. *J. Exp. Psychol.* 1976, 2(1), 17.
161. Kuo, Z.Y. *The Dynamics of Behavior Development: An Epigenetic View*; Random House: New York, 1967.
162. Capretta, P.; Petersik, J.T.; Stewart, D.J. *Nature.* 1975, 254, 689.
163. Birch, L.L.; Marlin, D.W. *Appetite.* 1982, 3, 353.
164. Birch, L.L.; McPhee, L.; Pirok; Steinberg, L. *Appetite.* 1987, 9, 171.
165. Rozin, P.; Schiller, D. *Motiv. & Emot.* 1980, 4, 77.
166. Kamen, J.; Peryam, D. *Food Technol.* 1961, 15(4), 173.
167. Siegel, P.; Pilgrim, F.J. *Am. J. Psychol.* 1958, 71, 756.
168. Zellmer, G. *Hospitals.* 1970, 44, 75.
169. Berlyne, D.E. *Percept. & Psychophys.* 1970, 8, 279.
170. Stang, D. *J. Pers. Soc. Psychol.* 1975, 51, 7.
171. Booth, D.A. *J. Comp. Physiol. Psychol.* 1972, 81, 457.
172. Booth, D.A. *Br. Med. Bull.* 1981, 37, 135.
173. Booth, D.A. In *The Determination of Behavior by Chemical Stimuli: ECRO Symposium*; Steiner, J.; Ganchrow, J., Eds.; Information Retrieval: London, 1982; p 233.
174. Booth, D.A. In *The Hedonics of Taste*; Bones, R.C., Ed.; L. Erlbaum Associates: Hillsdale, 1990.
175. Tordoff, M.G.; Tepper, B.J.; Friedman, M.I. *Physiol. & Behav.* 1987, 41, 481.
176. Mehiel, R.; Bones, R.C. *Anim. Learn. & Behav.* 1988, 16, 383.
177. Birch, L.L.; McPhee, L.; Steinberg, L.; Sullivan, S. *Physiol. & Behav.* 1990, 47, 501.
178. Holman, E. *Learning & Motiv.* 1975, 6, 91.
179. Fanselow, M.; Birk, J. *Anim. Learn. Behav.* 1982, 10, 223.
180. Zellner, D.A.; Rozin, P.; Aron, M.; Kulish, C. *Learn. & Motiv.* 1983, 14, 338.
181. Breslin, P.A.S.; Davidson, T.L.; Grill, H.J. *Physiol. & Behav.* 1990, 47, 535.
182. Tolman, E.C. *Psychol. Rev.* 1955, 62, 315.
183. Hebb, D.O. *The Organization of Behavior*; Wiley: New York, 1949.
184. Helson, H. *Adaptation-Level Theory*; Harper & Row: New York, 1964.
185. Lewin, K.; Dembo, T.; Festinger, L.; Sears, P.S. In *Personality and the Behavior Disorders*, Vol. 1; Hunt, J.M., Ed.; Ronald: New York, 1944.
186. Festinger, L. *A Theory of Cognitive Dissonance*; Row and Peterson: Evanston, IL, 1957.
187. Moskowitz, H.R. In *Preference Behavior and Chemoreception*; Kroeze, J.H.A., Ed.; Information Retrieval: London, 1979; p 131.
188. Schutz, H.G. In *Food Acceptability*; Thomas, D.M.S., Ed.; Elsevier: New York, 1988; p 115.
189. Szczesniak, A.S.; Kahn, E.L. *J. Texture Stud.* 1971, 2, 280.
190. Vickers, Z.M. *J. Food Qual.* 1991, 14(1), 87.
191. Zellner, D.A.; Stewart, W.F.; Rozin, P.; Brown, J.M. *Physiol. & Behav.* 1988, 44, 61.
192. Riskey, D.R.; Parducci, A.; Beauchamp, G.K. *Percep. & Psychophys.* 1979, 26, 171.
193. Lawless, H. *J. Test. & Eval.* 1983, 11, 346.
194. Mattes, R.D.; Lawless, H.T. *Appetite.* 1985, 6, 103.
195. McBride, R.L. *Appetite.* 1985, 6, 125.
196. Conner, M.T.; Land, D.G.; Booth, D.A. *Br. J. Psychol.* 1987, 78, 357.
197. Carlsmith, J.M.; Aronson, E. *J. Abnorm. Soc. Psychol.* 1963, 66(2) 161.
198. Insko, C.A. *Theories of Attitude Change*; Appleton-Century-Crofts: New York, 1967.

199. Oliver, R.L. *J. Appl. Psychol.* 1977, 62(4), 480.
200. Oliver, R.L. In *Theoretical Developments in Marketing*; Lamb, C.W.; Dunne, P.M., Eds.; American Marketing Association: Chicago, 1980; p 206.
201. Latour, S.A.; Peat, S.A. In *Advances in Consumer Research*; Wilkie, W.L., Ed.; Association for Consumer Research: 1979; p 6, 431.
202. Oliver, R.L.; DeSarbo, W.S. *J. Consumer Res.* 1988, 14, 495.
203. Cardello, A.V. In *Measurement of Food Preferences*, Thomson, D.; MacFie, H., Eds.; Blackie Academic: London, 1994; p 253.
204. Cardello, A.V.; Maller, O.; Masor, H.B.; Dubose, C.; Edelman, B. *J. Food Sci.* 1985, 50(6), 1707.
205. Cardello, A.V.; Sawyer, F.M. *J. Sensory Stud.* 1992, 7, 253.
206. Tuorilla, H.; Cardello, A.V.; Lesher, L. *Appetite.* 1994, 23, 247.
207. Helleman, V.; Aaron, J. J.; Evans, R.; Mela, D. J. In *Proceedings of the Food Preservation 2000 Conference*, 1995, In Press.
208. Cardello, A.V.; Melnick, I.M.; Rowan, P.A. In *Proceedings of the Food Preservation 2000 Conference*, Vol 1; Taub, I. and Bell, R., Eds.; Science and Technology Corp, Hampton, VA, p. 295, 1997.
209. Cardello, A.V.; In *Not Eating Enough: Proceedings of the NAS/NRC Committee Meeting on Underconsumption of Military Rations*; Marriott, B., Ed.; National Academy Press: Washington, D.C., 1995.
210. Rozin, P.; Vollmecke, T.A. *Ann. Rev. Nutr.* 1986, 6, 433.
211. Levi-Strauss, C. *Partisan Rev.* 1966, 3, 587.
212. Simoons, F.J. *Eat Not This Flesh*; The University of Wisconsin Press: Madison, 1961.
213. Pliner, E. *J. Nutr. Educ.* 1983, 15, 137.
214. Duncker, K. *J. Abnorm. Soc. Psychol.* 1938, 33, 489.
215. Marinho, H. *J. Abnorm. Soc. Psychol.* 1942, 37, 448.
216. Birch, L.L. *Child Dev.* 1980, 51, 489.
217. Birch, L.L. Zimmerman, S.I.; Hind, H. *Child Dev.* 1980, 51, 856.
218. Harper, L.V.; Sanders, K.M. *J. Exp. Child Psychol.* 1975, 20, 206.
219. Klesges, R.C.; Bartsch, D.; Norwood, J.D.; Kautzman, D.; Haugrud, S. *Int. J. Eating Disord.* 1984, 3, 35.
220. Berry, S.L.; Beatty, W.L.; Klesges, R.C. *Appetite.* 1985, 6, 41.
221. de Castro, J.M.; de Castro, E.S. *Am. J. Clin. Nutrit.* 1989, 50, 237.
222. Brewer, M.E.; De Castro, J.M.; Elmore, D.K.; Orozco, S. *Appetite.* 5, 89.
223. Engell, D.; Kramer, F.M.; Luther, S.; Adams, S. Paper presented at Society for Nutrition Education Annual Meeting, Anaheim, CA, 1990.
224. Babcock, C.G. *J. Amer. Diet. Assoc.* 1948, 24, 390.
225. Fantino, M.; Goillot, E. *Cahiers le Nutrition et de Dieteti.* 1986, 21, 51.
226. Schachter, S.; Goldman, R.; Gordon, A. *J. Pers. & Soc. Psychol.* 1968, 10, 91.
227. Schachter, S. *Am. Psychol.* 1971, 26, 129.
228. Schachter, S. *Emotion. Obesity and Crime*; Academic: New York, 1971.
229. Schachter, S.; Robin, J. *Obese Humans and Rats*; Erlbaum: Potomac, MD, 1974.
230. Stevens, D.A.; Dooley, D.A.; Laird, J.D. In *Food Acceptability*; Thomson, D.M.H., Ed.; Elsevier: New York, 1988; p 173.
231. Zuckerman, M. *Sensation Seeking: Beyond the Optimal Level of Arousal*; Erlbaum: Hillsdale, 1979.
232. Kish, G.B.; Donnenwerth, G.V. *J. Consult. & Clin. Psychol.* 1972, 38, 42.
233. Logue, A.W.; Smith, M.E. *Appetite.* 1986, 7, 109.
234. Rozin, P.; Fallon, A.E. *Appetite.* 1980, 1, 193.
235. Fallon, A.E.; Rozin, P. *Child Devel.* 1984, 55, 566.
236. Augustoni-Ziskind, M.; Fallon, A.; Rozin, P. *Devel. Psychol.* 1985, 21(6), 1075.
237. Millman, L.; Nemeroff, C.; Rozin, P. *J. Pers. & Soc. Psychol.* 1986, 50(4), 703.
238. Moskowitz, H.R. *New Directions for Product Testing and Sensory Analysis of Foods*; Food and Nutrition Press: Westport, CT, 1985.
239. Moskowitz, H.R. In *Food Acceptability*; Thomson, D.M.H., Ed.; Elsevier: New York, 1988; p 311.
240. Kalick, J. *Licensing Product Times.* 1990, Fall, 6.
241. Goldman, E.; McDonald, S. *The Group Depth Interview: Principle and Practice*; Prentice-Hall: Englewood Cliffs, NJ, 1987.

242. Kelly, G.A. *The Psychology of Personal Constructs: A Theory of Personality*; Norton: New York, 1955.
243. McEwan, J.A.; Thomson D.M.H. In *Food Acceptability*; Thomson, D.M.H., Ed.; Elsevier: New York, 1988; p 347.
244. McEwan, J.A.; Thomson, D.M.H. *Food Qual. & Pref.* 1989, 1(2), 59.
245. Green, P.E.; Rao, V.R. "Nonmetric Approaches to Multivariate Analysis in Marketing," Working Paper, Wharton School, University of Pennsylvania, 1969.
246. Green, P.E.; Srinivasan, V. *J. Consumer Res.* 1978, 5, 103.
247. Bone, B. *Food Technol.* 1987, 86, 58.
248. Ishii, R.; O'Mahony, M. *Chem. Senses.* 1987, 12, 37.
249. Ishii, R.; O'Mahony, M. *J. Sensory Stud.* 1990, 4, 215.
250. O'Mahony, M.; Rothman, L.; Ellison, T. *J. Sensory Stud.* 1990, 5(2), 71.
251. Box, G.E.P.; Hunter, W.G.; Hunter, J.S. *Statistics for Experimenters*; John Wiley & Sons: New York, 1978.
252. Mullen, K.; Ennis, D. *Food Technol.* 1985, 39(5), 90.
253. Siegel, S.F. *Activities of the R&D Associates.* 1987, 39(1), 183.
254. Clementi, S.; Cruciani, G.; Giulietti, G.; Bertuccioli, M.; Rosi, I. *Food Qual. & Pref.* 1990, 2, 1.
255. Saguy, I.; Mishkin, M.A.; Karel, M. *CRC Crit. Rev. Food Sci. & Nutrit.* 1984, 20(4), 275.
256. Floros, J.D.; Chinnan, M.S. *Food Technol.* 1988, 2, 72.
257. Moskowitz, H.R. *Product Testing and Sensory Evaluations of Foods*; Food & Nutrition Press: Westport, CT, 1983.
258. Symons, H.W.; Wren, J.J., Eds. *Sensory Quality Control: Practical Approaches in Food and Drink Production*; Society of Chemical Industry: London, 1977.
259. Tanno, L.; O'Mahoney, M. *Food Technol.* 1979, 9, 36.
260. Williams, A.A.; Atkin, R.K., Eds. *Sensory Quality in Foods and Beverages*; Ellis Horwood Limited: Chichester, U.K., 1983.
261. Sanna, M.; Bianco, L. *Food Technol.* 1988, 12, 92.
262. Gacula, M.C., Jr. *J. Food Sci.* 1975, 40, 399.
263. Labuza, T.P.; Schmidl, M.K. *Cereal Foods World.* 1988, 33, 734.
264. Peryam, D.R. Food Acceptability Techniques. Course Outline: FE-504, Food Engineering Dept., Illinois Institute of Technology, 1958; p 8.
265. Pilgrim, F.J.; Peryam, D.R. *Sensory Testing Methods: A Manual*, QMFCI Report #25-58, Quartermaster Food and Container Institute for the Armed Forces: Chicago, 1958.
266. Stone, H.; McDermott, B.J.; Sidel, J.L. *Food Technol.* 1991, 45(6), 88.
267. Pfaffmann, C.; Bartoshuk, L.M.; McBurney, D.H. In *Handbook of Sensory Physiology*; Beidler, L.M., Ed.; Springer-Verlag: Berlin, 1971; p 75.
268. Murphy, C. In *Neurobiology of Taste and Smell*; Finger, T.E.; Silver, W.L., Eds.; John Wiley and Sons: New York, 1987; p 251.
269. Amerine, M.; Pangborn, R.M.; Roessler, E. *Principles of Sensory Evaluation of Food*; Academic Press: New York, 1965; p 185.
270. Geldard, F.A. *The Human Senses*; Wiley and Sons: New York, 1972.
271. Griffiths, M. *Introduction to Human Physiology*; Macmillan: New York, 1974.
272. Amoore, J.E.; Johnston, J.W.; Rubin, M. *Scientific Am.* 1964, 210 (2), 42.
273. Wald, G. *Science.* 1968, 162, 230.

chapter two

Foods as cellular systems: impact on quality and preservation

Norman F. Haard

Contents

Introduction

Classification of foods

Foods as living entities

All foods have a finite storage life that is determined by a mix of biological, chemical, and physical forces. Biological processes are "of, or pertaining to, life or living organisms." The loss of food quality may be influenced by foreign organisms (e.g., microorganisms, parasites, insects) as well as the intrinsic physiology of the food itself. We do not usually think of food as a living organism because it normally does not appear to require external nourishment, exhibit growth, or reproduce. Rather, food is considered as a requirement (nutriment) for human life. However, we can consider as Thomas Aquinas did in, " . . . a thing is said to live insofar as it operates of itself and not as moved by another...." Postharvest asparagus may continue to grow in your refrigerator, a side of beef may continue to utilize glycogen and synthesize adenosine triphosphate (ATP) for cell maintanence, a harvested potato may sprout and reproduce itself after many months of storage, and a grain of wheat or a tomato will undergo death (the correlative of life) if it is deprived of oxygen. Some foods, like canned tomato soup, do not share any of the characteristics of living matter and are not considered in this chapter. Foods can be classified as follows: (1) intact tissue systems, e.g., fruits and vegetables, cereals, pulses, eggs, meats; (2) disrupted tissue systems, e.g., ground meat, purees, flour, tomato catsup; (3) noncellular or disrupted cellular systems, e.g., milk, fruit juices, honey, oils; or (4) combinations thereof.[1] Given that food may contain intact cells, i.e., the structural and functional unit of all living organisms, consisting of a nucleus, cytoplasm, and a limiting membrane, the objective of this chapter is to discuss the importance of postharvest and postslaughter physiology to quality preservation in food storage and distribution.

Food quality attributes

Tissues employed as food or as raw material for processed foodstuffs are normally defined by their nutriment, color and appearance, flavor, texture, safety, and suitability for use by the processor and/or consumer. In theory, for any given food that has viable cells, it is possible that each of these quality attributes will be directly or indirectly influenced by physiological reactions. However, a detailed knowledge of the relationship between quality preservation and physiology is limited and based mostly on research to understand a relatively few attributes that are economically or otherwise important in specific foods. Examples of well-studied relationships include rigor mortis with the tenderness of beef and starch–sugar metabolism with Maillard browning in processed potato products.

Taxonomic diversity of foods

According to the Food and Agriculture Organization of the United Nations, an estimated 4.3 billion metric tons (1986) of plant crops are produced in the world. There are more than 250,000 plant species now recognized by the taxonomist, thousands of which enter the human diet. This genetic diversity at the species level is further complicated by the fact that most agricultural crops include dozens, hundreds, or even thousands (e.g., wheat includes about 30,000) of genetically distinct cultivars or agricultural varieties. Cultivars are a group of plants within a particular specie distinquished by one or more characteristics. Two characteristics that are particularly important to the food scientist are chemical composition (e.g., high-solids tomatoes for paste production), and postharvest physiology (e.g., starch synthesis from sugar in sweet corn).

Although there are several million taxonomic species of animals, more than 95% of which are invertebrates, a relatively small percentage of these enter the human diet. Important sources of animal food products are vertebrate organisms (phylum Chordata), including bony fish (class Osteichthyes); beef, lamb, and pork (class Mammalia); and poultry (class Aves); as well as invertebrates including phylum Arthropoda (e.g., shrimp and lobster) and phylum Mollusca (e.g., clam and snail). The greatest taxonomic diversity of animal foods comes from the bony fish, which includes an estimated 40,000 species, several thousand of which are used by humans for food. One type of fish known to the consumer as Pacific rockfish is represented by as many as 90 species of the genus *Sebastes*. Rockfish meat from different species differs in postmortem biochemistry and associated flesh softening.[2] Even within the same specie, breeds of livestock or stocks of fish may also exhibit differences in chemical composition, quality characteristics, and postslaughter metabolism. Important meat characteristics of different breeds, which relate to postslaughter physiology and have fairly high heritability, include fast-glycolyzing muscle in hogs and tenderness in beef.[3]

Source of biochemical diversity

Intrinsic and extrinsic factors

In addition to the contribution of heritable characteristics, other intrinsic factors that influence quality characteristics are the age of the animal or the maturity of the plant, the sex of the animal, and the location of the tissue in the organism. Extrinsic or intraspecific factors may also influence the quantity and properties of enzymes from a given specie. Examples of external factors that may influence postharvest physiology include diet and exercise of animals, trace minerals in the fertilizer applied to the growing plant, and the growing temperature of poikilothermic fish or plants. For example, "water core" is the result of a postharvest physiological response in cold-stored apples that can be minimized by proper calcium nutrition of the tree.[4] Some examples of intraspecific factors influencing muscle enzymes that may be important in postslaughter physiology are summarized in Table 1 and discussed in more detail elsewhere.[5] The relationship between changes in enzyme composition caused by preharvest factors and postharvest quality deterioration requires further study.

Cell structures and metabolic machinery

Food tissue composed of intact cell systems is normally heterogenous, being composed of mixtures of cell types as well as an extracellular matrix. There is remarkable similarity in the structure and basic metabolism of plant and animal cells.[1,6] Understanding post-harvest and postslaughter physiology is simplified, because components common to more than one cell type are strikingly alike with respect to biochemical pathway (e.g., protein synthesis) or structure (e.g., membrane structure). At the same time, common cell components may differ in size, numbers, functional details, and composition. In addition, some basic biochemical functions (e.g., photosynthesis) and many secondary biochemical pathways important to food quality are only found in certain food groups (e.g., trimethylamine oxide metabolism in marine fish) or are unique to one specie (e.g., S-(1-propenyl)-L-cysteine sulfoxide metabolism in onion). The reader is referred to other sources for a discussion of cell types in fruits and vegetables[7] and myosystems.[5,8]

Biological stress and strain

Successful food handling requires a recognition of the principle of biological stress and strain. Stress imposed by harvest, handling, or a preservative process may be inadvertently accompanied by an unexpected biological response. Recognizing the importance of stress–strain relations in food science, Schwimmer has defined food processing and technology as "... the art and science of the promotion, control and/or prevention of cellular disruption and its metabolic consequences at the right time and at the right place in the food processing chain."[9]

Biological strain

Postharvest and postslaughter physiology may be influenced by stress resulting from the harvesting, handling, and processing of foods. Biochemical responses (strain) may result from chemical (e.g., CO_2 in modified atmosphere oralkaline conditions in lye peeling), biotic (e.g., insects or microorganisms), as well as physical (e.g., cut injury or temperature change) sources of stress. Strain (response to stress) may be "plastic" (irreversible) or "elastic" (reversible) and may be "direct" or "indirect." For example, a strain resulting from a chilling temperature (stress) might be loss of membrane integrity caused by a phase change in the phospholipid bilayer (direct) or may be loss in membrane integrity caused indirectly (e.g., by activating an enzyme which, in turn, catalyzes membrane autolysis).

Stress hormones

Harvesting, handling, and processing raw materials for food may be accompanied by the release of hormones. During slaughter of animals, "stress" hormones include epinephrine and norepinephrine from the adrenal medulla, adrenal steroids from the adrenal cortex, and thyroid hormones from the thyroid gland. In general, stress-susceptible animals have unusually high body temperatures and rapid glycolysis at the time of slaughter and as a result may have poor-quality meat because of the influence of these conditions on postmortem physiology.[10]

In plants, ethylene is a stress hormone that can trigger the biosynthesis of enzyme cascades involved with important events such as ripening of fruit, lignification of vegetables, seed sprouting, discoloration of lettuce, bitterness in carrots, yellowing of broccoli, and other physiological effects. Production of stress ethylene in wounded plants is very rapid (i.e., <15 min) and appears to result from the translation of preexisting RNAs.[11] In addition to being a stress hormone, ethylene is also synthesized at specific stages of the normal ontogeny of plants, notably just prior to the ripening of fruit.

Table 1   Factors that Influence the Content and Properties of Enzymes in Meat

Factor	Reference	Observation/Comment
Exercise	Siebenaller et al.[91]	Compared to the demersal mode, muscle from the pelagic mode (high locomotory) of the fish *Sebastolobus altivelis* contains greater amounts of enzymes involved with aerobic energy metabolism (e.g., malic dehydrogenase) and a lesser amount of enzymes involved with anaerobic energy metabolism (e.g., lactic dehydrogenase). All other factors, equal, we would expect a greater rate of pH decline (associated with anaerobic energy metabolism) in the postmortem muscle of demersal fish.
Age	Kaletha and Nowak[92]	the adult isoform of mammalian AMP deaminase exhibits a different pH optimum and kinetic properties from the juvenile isoform. All other factors equal, we would expect a more rapid conversion of AMP to IMP in adult muscle with the same enzyme concentration as juvenile muscle.
Diet	Ji et al.[93]	Cells are equipped with antioxidant enzyme systems which serve to minimize injury caused by free radicals. Dietary selenium can result in increased concentration of certain protective enzymes in the muscle as well as postslaughter membrane lipid peroxidation. Such protective mechanisms may be important in postslaughter development of rancidity.
Fasting	Mayer et al.[94]	Fasting of animals prior to slaughter increases the concentration of alkaline protease(s) in muscle. These enzymes appear to be important in the disassembly of the cytoskeletal network in postslaughter fish.
Temperature	Tsuchimoto et al.[95]	The rate constant for thermal denaturation of carp Ca^{2+}-ATPase is about six times greater for fish reared at 10°C compared to carp reared at 30°C. This response may be relevant to the thermal gelation properties of meat products.
Pressure	Hand and Somero[96]	Deep-living fish have up to 4 orders of magnitude less lactic dehydrogenase than shallow-living fish. This response may influence the rate of pH decline and glycolysis in postslaugher fish.

The wounding of plants can also lead to formation of other hormones, such as traumatin and jasmonic acid, which are products of linoleic/linolenic acid cascades (see below). Traumatin and related products appear to elicit the plant to form a battery of phytoalexin ("to ward off") compounds that increase defense against invading insects and microorganisms. Such compounds are referred to as "stress metabolites" when formed as a consequence of abiotic stress in postharvest vegetables.[12] Stress metabolites are of concern to the food scientist because they may have bitter taste and some exhibit toxic properties.

Polyunsaturated fatty acid cascades

An example of biological response to stress that occurs in both plant and animal tissues and may influence food quality in a variety of ways is the polyunsaturated fatty acid (PUFA) cascade.[13] Linolenic/linoleic acid cascades were referred to in the above discussion of wound hormones in plants. Although differences exist in the control and detail of PUFA cascade pathways, the process consists of the basic steps outlined in Figure 1. Disruption of plant tissues can trigger PUFA cascades that result in more or less spon-

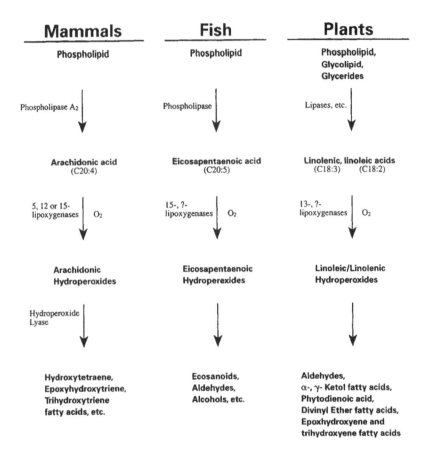

Figure 1   Polyunsaturated fatty acid cascades resulting from wound injury.

taneous formation of characteristic aromas (e.g., following mechanical disruption [chewing] of raw cucumber [Figure 2A]). Grassy/beany flavors are attributed to cascade products such as hexanal and hexenal, whereas compounds such as nonenal and nonadienal have odors described as melon, cucumber, or violet.[14] These pleasant aroma substances may also contribute to off-flavors (e.g., in improperly blanched frozen vegetables) when they accumulate or cause imbalance with respect to other flavor substances. In fish, the release and oxidation of eicosapentaenoic acid leads to the formation of compounds (e.g., octa-1,5c-dien-3-one), that contribute to the delicate, pleasant "seaweed" aroma of very fresh fish (Figure 2B).

Enzymic browning

Stress associated with handling and processing of animal or plant tissues, such as anoxia, chilling, freezing/thawing, ionizing radiation, bruising, cut injury, mincing, electrical stimulation, and modified atmospheres, may cause a variety of other reactions that contribute to deterioration. The "Achilles' heel" of the stressed cell is the biological membrane, because loss of its physical integrity or its selective permeability may lead to a cascade of reactions resulting from decompartmentation of the cell.[15]

    Enzymic browning is an important reaction caused by cell decompartmentation, which occurs in both plant and animal tissues.[16] This reaction causes discoloration on the

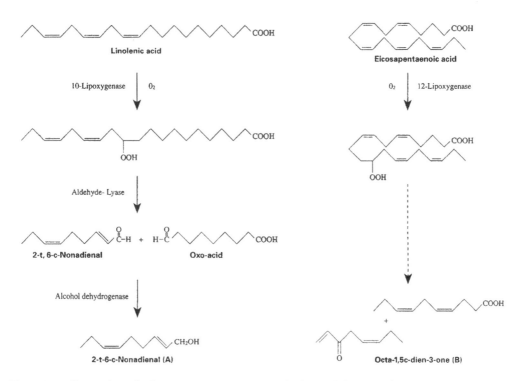

Figure 2 Examples of pleasant aroma compounds formed by specific enzymic oxidation of fatty acids with lipoxygenase: compound contributing to fresh cucumber aroma (A) and fresh marine fish (B).

cut surfaces of some fruits and vegetables (e.g., potato, mushroom, apple, banana) and in the meat adjacent to the exoskeleton of crustacean species (e.g., crab, shrimp, lobster). Enzymic browning also contributes to the character of dried dates, black tea, and to the curing of fresh cacao seeds in chocolate production.

The initial reaction may involve monophenol monooxygenase or tyrosinase (EC 1.14.18.1), and the main reaction is catalyzed by the diphenol oxidases, catechol oxidase (EC 1.10.3.2), or laccase (EC 1.10.3.1). Catechol oxidase is also refered to as polyphenoloxidase or phenolase and appears to be a widespread enzyme in foodstuff. Laccase, and to a lesser extent monophenol monooxygenase, does not appear to be widespread in nature. The basic enzymic reactions of these three activities are illustrated in Figure 3. The benzoquinones formed are very reactive with oxygen nonenzymatically and continue to undergo oxidation resulting in dark colors called melanin. Eskin discusses three groups of phenolic compounds that occur in food material and participate in enzymic browning: simple phenols (e.g., tyrosine, catechol, and gallic acid), cinnamic acid derivatives (e.g., *p*-coumaric acid, caffeic acid, and chlorogenic acid), and flavanoids (e.g., catechin, quercetin, and leucoanthocyanidin).[17]

The biological function of these enzymes is not well understood. In plants, suggested roles include pseudocyclic photophosphorylation, phenol metabolism in senescing tissue, root formation and development, and resistance of the plant to biotic stress through antibiotic action and formation of mechanical barrier. In crustaceans, polyphenol oxidase plays a role in schleritization of the exoskeleton.

A number of methods have been used to prevent enzymic browning including heat treatment, application of sulfites, addition of acidulants, and exclusion of oxygen.

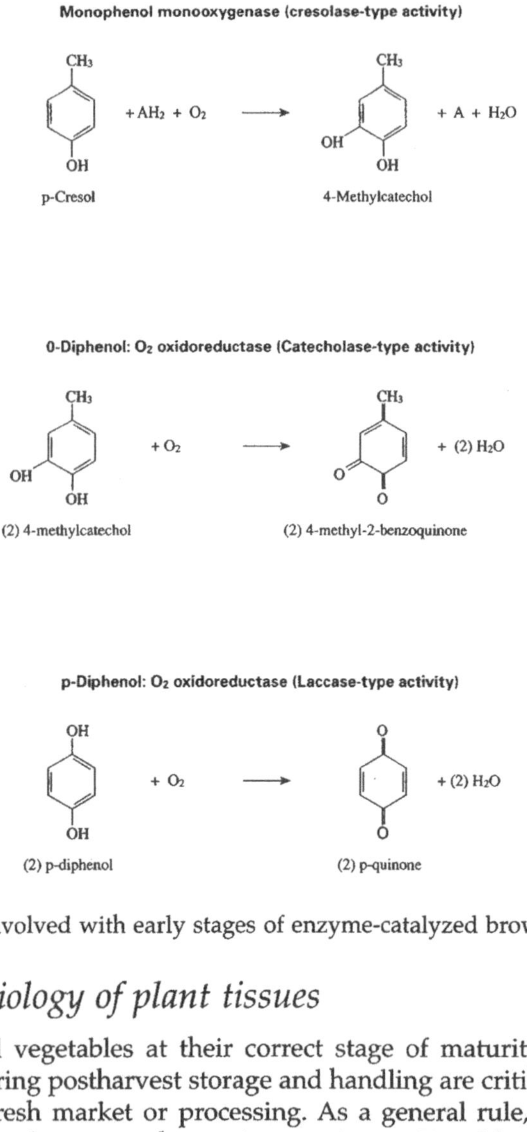

Monophenol monooxygenase (cresolase-type activity)

$$\text{p-Cresol} + AH_2 + O_2 \longrightarrow \text{4-Methylcatechol} + A + H_2O$$

p-Cresol

4-Methylcatechol

0-Diphenol: O_2 oxidoreductase (Catecholase-type activity)

$$\text{(2) 4-methylcatechol} + O_2 \longrightarrow \text{(2) 4-methyl-2-benzoquinone} + (2) H_2O$$

(2) 4-methylcatechol

(2) 4-methyl-2-benzoquinone

p-Diphenol: O_2 oxidoreductase (Laccase-type activity)

$$\text{(2) p-diphenol} + O_2 \longrightarrow \text{(2) p-quinone} + (2) H_2O$$

(2) p-diphenol

(2) p-quinone

Figure 3   Reactions involved with early stages of enzyme-catalyzed browning.

Postharvest physiology of plant tissues

Harvesting fruits and vegetables at their correct stage of maturity and the control of metabolic changes during postharvest storage and handling are critical to obtaining high-quality produce for fresh market or processing. As a general rule, metabolically active tissues lose quality at a faster rate than quiescent tissues. Durable crops, such as cereals and pulses, are characterized by resilient epidermal tissues, low moisture content, low respiration and associated heat production, and a dormant or resting phase of development. Perishable crops, such as ripe raspberries, are characterized by susceptibility to microbial invasion, high moisture content, high respiration and associated heat production, and active metabolism. The metabolic intensity of a given crop is influenced by ontogenic factors, type of plant organ (e.g., stem, flower, fruit), wound injury, and environmental conditions such as temperature.

Ontogeny

Ontogeny refers to the origin and development of an organism from conception through the completion of its life cycle at death (Figure 4). The developmental stage of the plant organ at harvest time varies from one crop to another and may differ for the crop destined for fresh or processing markets. At harvest, some crops are in the growth or

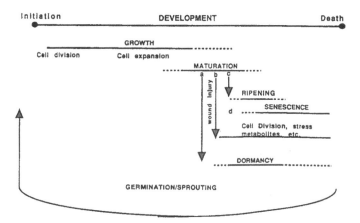

Figure 4   Development stages of plants related to postharvest physiology. (From Haard, N.F., Food as cellular systems: Impact on quality and preservation. A review, *J. Food Biochem.*, 19(3), 191-238, 1995. With permission.)

early maturation stage of development and accordingly may have very high metabolic activity. For such plants, the physiological response to detachment from the parent plant may be continued cell expansion, transpiration, and maturation. Harvested potato tuber (stem tissue), sweet potato (root tissue), or onion bulb (leaf tissue) are physiologically mature tissues primed for dormancy. These crops have a relatively long market life when properly stored, and a key to successful storage is avoiding the breaking of dormancy. Other crops are physiologically primed for a senescence phase (aging) associated with rapid loss in quality attributes and a breakdown in immunity to saprophytic microorganisms. Such crops tend to have a relatively brief market life, although proper storage strategies can slow down metabolism and thereby maintain quality prior to consumption or processing.

Fruits normally undergo a prelude to senescence called "ripening" when the tissue undergoes change from an inedible state to a delicious state (c in Figure 4). Ripening involves a multitude of biochemical reactions that lead to softening, increased sweetness, decreased astringency and tartness, production of pleasant aroma, loss of green hue, and development of yellow, orange, red, and/or blue hues. The word fruit is a derivative of the latin verb "frui" which means to enjoy, delight in, or to have use[18] and in culinary jargon is distinquished from other botanical fruit (e.g., wheat, corn, beans, peas, eggplant, etc.) that tend to have a lower sugar and organic acid content and are eaten with the main part of the meal. Technically, all fruit are the seed-bearing organ developed from the ovary and adjoining plant parts. Fruit that undergo intense metabolic activity during ripening can sometimes be harvested prior to the ripening phase and ripened off the vine, since mature, unripe fruit have much better postharvest storage characteristics than ripe fruit.

Mechanical injury resulting from harvesting or rough handling can lead to formation of stress hormones and lead to characteristic wound response in different tissues (i.e., localized cell division and wound healing, stimulation of ripening and senescence) and other reactions that influence quality.

Respiration

Definition

Physiologically, respiration is the process by which tissues or organisms exchange gases with their environment. In higher plants the pathways of glycolysis, the citric acid cycle, and the mitochondrial electron transport chain are of primary importance in respiratory metabolism. The net reaction of respiration is the conversion of starch or sucrose and oxygen to carbon dioxide and water (Figure 5). Plant tissues contain phosphorylase(s), which catalyzes the formation of glucose-1-phosphate from starch. Phosphorylase appears to be the operative enzyme for depolymerization in postharvest fruits and vegetables (e.g., ripening fruit)[19] and cold-stored potato tuber.[20] Recently, it was shown that ATP required for sucrose synthesis by the enzyme sucrose phosphate synthase is a key step in the regulation of the intense respiration that occurs in ripening banana fruit when almost 20% of the tissue weight is converted to sugar from starch via the phosphorylase reaction.[21]

Amylases are also present in plant tissue, where they may serve to depolymerize starch by hydrolysis rather than by phosphorylation. Amylases have been identified in fruits and vegetables, and there has been some suggestion that they are involved in starch hydrolysis in postharvest commodities.[22] However, the involvement of amylases in fruit and vegetable respiration is not well-substantiated by experimental evidence.

Respiration is an exothermic reaction where the potential energy in carbohydrates is partly conserved in the high-energy phosphate bond of adenosine triphosphate (ATP) and partly released as heat. For conversion of glucose to carbon dioxide and water by glycolysis, citric acid cycle, and the mitochondrial electron transport chain, one would expect about 39% of the potential energy in glucose to be conserved as ATP and 61% to result in the so-called "heat of respiration."

$$6\,O_2 + C_6H_{12}O_6 \rightarrow 6\,CO_2 + 6\,H_2O$$
$$\Delta G = -680 \text{ kcal/mole glucose}$$
$$36\,P_i + 36\,ADP \rightarrow 36\,ATP + 36\,H_2O \tag{1}$$
$$\Delta G = +263 \text{ kcal/mole glucose}$$
$$\text{Heat loss} = 417 \text{ kcal}/680 \text{ kcal} = 61\%$$

The pentose phosphate pathway also occurs in plant tissues, and according to some estimates it can be responsible for up to 30% of the glucose oxidized by plants.[23] The pathway may operate in a cycle (Figure 5) involving decarboxylation of six glucose-6-phosphate to form six ribulose-5-phosphate; conversion of the six ribulose-5-phosphate to four fructose-6-phosphate and two glyceraldehyde-3-phosphate; and conversion of the latter to five glucose-6-phosphate. The net result of the cyclic reaction is the conversion of glucose-6-phosphate (G-6-P) to carbon dioxide and the formation of reduced nicotinamide adenine dinucleotide phosphate (NADPH):

$$G\text{-}6\text{-}P + 12\,NADP^+ + 7H_2O \rightarrow 6\,CO_2 + 12\,NADPH + 12\,H^+ + P_i \quad . \tag{2}$$

Unlike the NADH formed by the citric acid cycle, NADPH does not donate its proton and electron to the mitochondrial electron transport chain and thereby contribute to oxidative phosphorylation. Rather, NADPH serves as a hydrogen donor and electron donor in a variety of reductive biosyntheses. Intermediate reactants in this pathway may also be diverted to other pathways including the glycolytic pathway.

Calorimetry studies indicate that respiration in postharvest fruits and vegetables is less efficient in ATP synthesis than would be predicted by the glycolytic pathway, citric acid cycle, and mitochondrial electron transport chain. That is, in respiring fruits and vegetables, as much as 90% or more of the potential energy of glucose is evolved as heat. The respiration in sensecing flowers and wounded potato tuber is called "thermogenic respiration" because it is not coupled to ATP synthesis in the electron transport chain.[7]

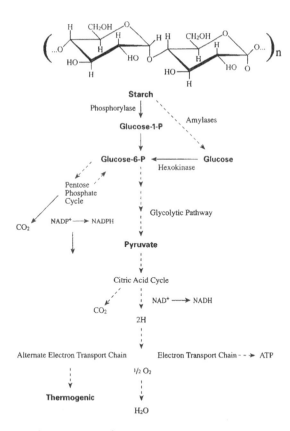

Figure 5   Respiratory pathways in postharvest fruits and vegetables.

Thermogenic or cyanide-insensitive respiration is mediated by an alternate mitochondrial electron transport chain. Evocation of the alternate electron transport pathway may function to decontrol respiration causing (1) rapid catabolism of storage polysaccharides, (2) electron transport without expenditure of reducing power, or (3) generation of heat, and thereby cause volatilization of biologically active molecules or accelerate autolytic reactions. In this regard it is interesting to note that the internal temperature of mango increases by about 10°C during ripening, and this has been attributed to respiration involving the alternate electron transport chain.[24]

Relationship to market life

Respiration plays a major role in the postharvest physiology of fruits and vegetables.[25] The importance of respiration in postharvest crops is best illustrated by comparing rate of respiration with storage life of fruits and vegetables (Figure 6). Commodities that exhibit a very rapid respiration rate (e.g., asparagus, broccoli, pea, sweet corn, mushroom) are generally quite perishable, whereas those with relatively slow respiration rate (e.g., root, tuber, and bulb vegetables) may be stored satisfactorily for longer periods of time. Dry cereal grains have an extremely low respiration rate and show only slight reduction in total sugars and quality even after many years at a cool temperature and low oxygen concentration.[26] Grain respiration is particularly affected by moisture, increasing as the water content of the seed rises above 14%.

Other factors responsible for the wide variation in respiration rate of postharvest crops include the ontogenic stage at harvest, gas exchange (e.g., surface area to volume ratio and the nature of the epidermis), and preharvest factors such as cultural practices.

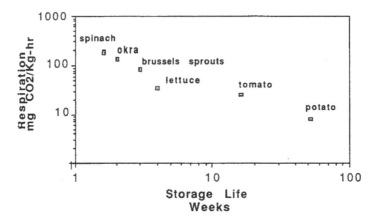

Figure 6   Relationship between storage life and respiration of selected vegetables. (From Haard, N.F., Food as cellular systems: Impact on quality and preservation. A review, *J. Food Biochem.*, 19(3), 191-238, 1995. With permission.)

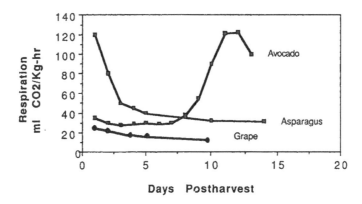

Figure 7   Postharvest respiratory patterns of vegetable climacteric fruit and nonclimacteric fruit. (From Haard, N.F., Food as cellular systems: Impact on quality and preservation. A review, *J. Food Biochem.*, 19(3), 191-238, 1995. With permission.)

Exactly why the spoilage rate of fruits and vegetables is a direct function of respiration rate may involve the following answers: (1) the reducing power and ATP resulting from respiration may be coupled to biosynthetic reactions (e.g., lignin biosynthesis), which can lead to quality deterioration; (2) loss of food reserves in the tissue can lead to loss of weight, sweet taste, etc.; (3) accumulation of CO_2 produced by the commodity can be detrimental depending on the commodities tolerance; and (4) the heat produced by respiration can be a major factor in establishing refrigeration requirements to maintain low temperature.

Control of respiration rate

The respiration rate of most vegetative tissues tends to decline steadily after harvest, more so for stem, root, flower, and immature fruit than for mature fruit.[25] Fruits that undergo a definite ripening phase on or off the vine are different in that they show a marked upsurge in respiration called the "climacteric" (Figure 7). Classification of climacteric and nonclimacteric fruits is shown in Table 2. The magnitude of the climacteric burst varies in different fruits. The climacteric rise in respiration coincides with an increase in endogenous ethylene production, and with changes associated with ripening (i.e., softening,

Table 2   Classification of Fruits According to
Respiratory Patterns

Climacteric	Nonclimacteric
Culinary Fruit	
Apple	Blueberry
Apricot	Cherry
Avocado	Fig
Banana	Grape
Cherimoya	Grapefruit
Feijola	Lemon
Mango	Litchi
Papaya	Melon
Passion Fruit	Orange
Papaw	Pineapple
Peach	Strawberry
Pear	Tamarillo
Fruit Vegetables	
Tomato	Green bean
	Pea
	Cucumber
	Pepper
	Eggplant
	Okra
	Cacao
	Olive

color, increase in sweet taste, and development of aroma). Because ripe fruit has a relatively short market life, the storage of climacteric fruits is best acheived by maintaining them in the preclimacteric state as long as possible.

Research to elucidate the biochemical basis for respiratory control in postharvest fruits and vegetables has focused on the cause of the climacteric rise. Theories to explain the onset of the climacteric rise include (1) membrane leakage and loss of cell compartmentation,[27,28] (2) enhanced ATP turnover,[29] and (3) evocation of the cyanide-insensitive alternate electron transport system.[30] An attractive feature of the latter theory is that the alternate pathway may be involved in the generation of peroxide and free radicals[30a] that act at sensitive subcellular sites and appear to play a pivotal role in maturation and senescence of plant organs.[31] It may also be significant that fruit produce trace amounts of cyanide prior to ripening and ethylene production.[32] More information is needed to fully understand the biochemical basis of respiratory control in senescing plant tissue.

Environmental factors that influence the respiration rate of fruits and vegetables are summarized in Table 3. Temperature reduction is the most important factor in minimizing the rate of respiration and other reactions in postharvest fruits and vegetables. Proper packaging and care in transport is also very important because of the influence of physical stress on respiration (Figure 8). Modified or controlled atmosphere storage of fruits and vegetables (i.e., use of elevated CO_2 and reduced O_2) is discussed in Chapter 16.

Gene expression in the postharvest cell

Definition and importance
Over the past two decades, it has become increasingly clear that plant senescence and wound injury are gene-directed events accompanied by increased transcription of DNA to

Table 3 Some Environmental Factors that Influence the Respiration Rate of Fruits and Vegetables

Environmental Factor	Response
Temperature	Within the physiological temperature range the Q_{10} for respiration is normally 2–3 but may be greater or less than this range for some commodities. The Q_{10} for vegetable respiration normally decreases as the temperature is increased from 0–40°C.[25]
Oxygen concentration	In general, a reduction in O_2 (below 21%) slows respiration. At 2% O_2 and 2.5°C the respiration of broccoli is reduced by about 30%. When oxygen drops below about 2% (depending on commodity, temperature, and duration) anaerobic respiration and production of phytotoxins (e.g., ehtanol, acetaldehyde) will rapidly increase. Alternate chain respiration is affected more than cytochrome oxidase mediated respiration.
Carbon dioxide	In general, an increase in CO_2 (above 0.03%) reduces aerobic respiration. At concentrations above about 20% (depending on commodity, temperature, and duration) anaerobic respiration and damage may result.
Carbon monoxide	CO at 1–10% in air or controlled atmospheres reduces the respiration rate of plant tissue. However, CO may act as an ethylene analog and stimulate the climacteric rise in respiration.
Ethylene	Low concentrations of C_2H_4 trigger the climacteric rise independent of concentration and continued exposure. C_2H_4 induces a climacteric-like rise in respiration of nonclimacteric fruit that is revisible and dependent on concentration.
Stress	Bruising, water stress, ionizing, radiation, fumigants, and biotic agents normally result in an increased rate of plant respiration.

form messenger RNA and its subsequent translation to key enzymes in postharvest metabolism (Figure 9). The importance of gene expression in postharvest systems is illustrated by the following observations: (1) plant senescence, fruit ripening, and wound physiology are each delayed or prevented by applying appropriate inhibitors of gene transcription or mRNA translation to tissue slices; (2) slow-to-ripen or ripening-inhibited mutant varieties of fruit have been isolated by conventional cultivar selection and breeding programs; and (3) many enzymes that catalyze postharvest changes are synthesized *de nova* after harvest. With recent advances in plant cell culture, mutant selection techniques, gene amplification, antisense DNA technology, and the production of transgenic plants, it may become possible to engineer crops for improved storage characteristics.[33] Some proposed genetic engineering solutions to postharvest problems are summarized in Table 4.

Impact on quality

Postharvest metabolism involves catabolic processes (e.g., starch hydrolysis, protein turnover, pigment degradation, pectin hydrolysis) as well as anabolic reactions (e.g., the synthesis of enzymes, other proteins, pigments, polysaccharides, and organic acids). Postharvest physiology is a reflection of developmental processes (e.g., fruit ripening) and response to wound injury as well as physiological processes ongoing in the tissue at the time of harvest. Aldolase, carboxylase, chlorophyllase, phosphorylase, *o*-methyltransferase, peroxidase, phenolase, transaminase, invertase, pectic enzymes, phosphatase, indoleacetic acid oxidase, phenylalanine ammonia lyase, cytochrome P450, and enzymes involved with ethylene synthesis are among the proteins known to increase in ripening fruit.

Genetic-engineered fruits and vegetables can be developed for a number of reasons[34-36] including (1) improved agronomic characteristics (e.g., herbicide resistance, water-use efficiency, or pest resistance); (2) improved chemical composition (e.g., high-solids tomatoes and onions for use in processing, storage proteins in cereals with improved essential

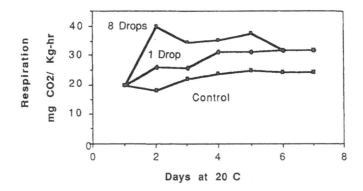

Figure 8 Effect of impact bruising on respiration of tomatoes damaged at mature green stage and ripened at 20°C. (From Haard, N.F., Food as cellular systems: Impact on quality and preservation. A review, *J. Food Biochem.*, 19(3), 191-238, 1995. With permission.)

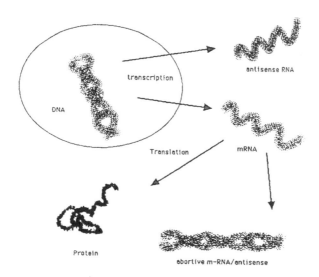

Figure 9 Prevention of protein synthesis by antisense technology. (From Haard, N.F., Food as cellular systems: Impact on quality and preservation. A review, *J. Food Biochem.*, 19(3), 191-238, 1995. With permission.)

amino acid content, augmentation of the level of naturally occuring healthful compounds, and decreased content of naturally occuring toxicants and antinutrients); (3) improved tolerance to external stress (e.g., chilling tolerance); and (4) control of endogenous enzyme levels. The latter strategy appears to be particularly appropriate to postharvest physiology, because benefits can arise from slowing down or eliminating the expression of endogenous enzymes, that contribute to loss in edible quality. Possible benefits of blocking expression of specific genes by antisense technology include extension of market life, improved yield, nutrient retention, textural integrity, color, flavor, and process control.[36]

In eukaryotes multiple gene families occur, thus making deletion of the DNA coding for a target enzyme (i.e., by mutation) more difficult than it is in microorganisms where single genes are common. The use of antisense technology, outlined in Figure 9, is therefore a promising strategy to limit enzyme expression. DNA coding for a target enzyme is synthesized in reverse orientation and cloned. This antisense DNA produces an mRNA transcript that is complementary in sequence to the target mRNA. The antisense mRNA

Table 4 Some Proposed Uses of Genetic Engineering to Improve the
Quality and Storage Characteristics of Plant Crops

Problem	Solution
Chilling sensitive crops cannot be stored below 10–15°C	According to current theory, chilling injury is initiated by a phase transition of phospholipids in cellular membranes. Introduction of gene for fatty acid desaturation from chilling tolerant cyanobacterium may increase the tolerance of the recipient for chilling temperatures.[97]
Vine-ripened tomatoes are damaged by mechanical harvesting and bulk storage.	Transforming plants to contain antisense DNA for key enzyme(s) in the softening during ripening may allow development of flavor and color without excessive softening (see text).
Frozen vegetables become rancid due to the action of lipoxygenase. Blanching inactivates this enzyme but causes a loss in texture and fresh flavor.	Isolate the DNA for lipoxygenase and transform plants to contain antisense DNA for lipoxygenase.
Cultivars of tomato with increased solids content are useful for processed products but have lower field yield.	In the growing plant, the rate of photosynthetic sugar uptake by the fruit is limited by a protein present in tomato fruit cells. The DNA which codes for this protein has been isolated. Increase the number of copies of this DNA so that progeny of the transformed plant produces more of this protein.
Important grains and legume seeds are deficient in essential amino acids, e.g., corn is low in lysine.	Alter the genes coding for key storage proteins to contain more codons for the limiting amino acid.
Broccoli synthesizes ethylene gas which promotes developmental changes of yellowing and senescence.	The gene for 1-aminocyclopropane-1-caroboxylate synthase, a key enzyme in ethylene synthesis, has been isolated. Transform plant to contain antisense ACC synthase.

hybridizes to the target mRNA and results in reduced translation of the target mRNA to the target enzyme.

Prevention of excessive softening during ripening would allow growers of processing tomatoes to fully ripen fruit in the field to maximize quality and at the same time be able to avoid harvest and handling losses due to crushing of fully ripe fruit and culling of unripe fruit. Polygalacturonase is an enzyme that hydrolyzes pectin and thereby was believed to contribute to fruit softening during ripening (Figure 10). Antisense technology has already been used to grow tomatoes engineered to produce low levels of polygalacturonase.[34,37,38] Although polygalacturonase formation during ripening was reduced by as much as 99% in tomatoes transformed by this technique, suprisingly the softening process and the reduction in molecular weight of pectin was not curtailed.[39,40] It is not clear whether the small amount of polygalacturonase present in transgenic fruit is sufficient to cause the same rate of pectin degradation as in normal fruit or whether another unidentified enzyme(s) is involved in the process.

This experience points to the pressing need for food biochemists to identify critical enzymes involved with reactions that lead to quality change in postharvest systems.

Control of biochemical reactions

Plant hormones
Biotechnology aside, the dynamics of chemical change in postharvest fruits and vegetables are subject to naturally occurring biochemical control mechanisms. Phytohormones play

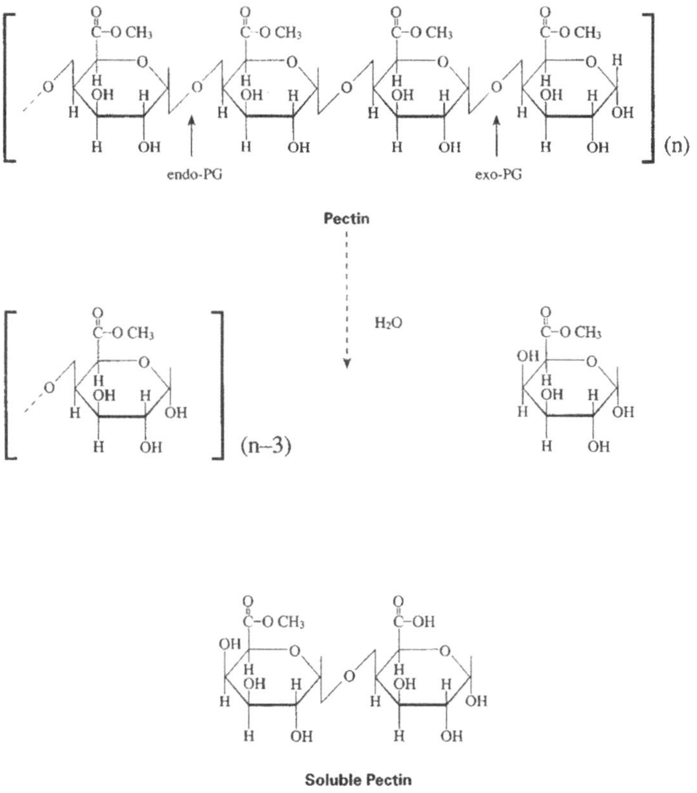

Figure 10 Action of exo- and endo-polygalacturonases. The depolymerization of pectic substances in the middle lamella is associated with texture of softening fruit.

a definite role in the growth and development of plants, although the exact way they function remains obscure with a few exceptions. All the commonly identified plant hormones (i.e., auxins, gibberellins, cytokinins, abscisins, and ethylene) appear to be present in and to influence physiology in harvested fruits and vegetables. Although most attention has been given to the role of ethylene in the postharvest physiology of fruits and vegetables, there is evidence that the balance of several hormones rather than the specific action of one hormone is responsible for phytohormone action.

The ability of ethylene to stimulate fruit ripening can depend on the maturity of the plant organ at the time it is removed from the parent plant; this may reflect changes in concentration of hormones such as auxin and cytokinin. Fruit attached to the plant may display less sensitivity to ethylene than detached fruit; this may reflect transport of hormones between the parent plant and attached organ. Also, certain cultivars of pear are sensitized to ethylene only after the harvested fruit are stored at low, nonfreezing temperatures. Cold conditioning is probably associated with metabolism of hormones other than ethylene. These observations have led to the idea that "juvenility hormones" desensitize fruit to ethylene.[41]

In addition to fruit ripening, phytohormones influence other physiological events important in postharvest commodities including rest, dormancy, regrowth of roots or sprouts, floral induction, and response to biotic and abiotic stress. Some observations relating hormones to deteriorative reactions in postharvest systems are summarized in Table 5. In addition to phytohormones, other endogenous growth factors and inhibitors may also be involved with the control of the physiological change. Although the end result of hormones in developmental changes and stress response in postharvest crops is clear, additional research is needed to understand the mechanism of their action.

Table 5   Changes in Phytohormones and Physiological Changes Related to Postharvest Quality

Commodity	Physiological Change	Phytohormones
Onion	Sprouting	Endogenous abscisic acid (ABA) decreases, and endogenous gibberellins, cytokinins, auxins increase during sprouting; exogenous ABA delays onset of regrowth.
Brussels sprout	Senescence	Yellowing is associated with increased gibberellin and decreased cytokinin and auxin.
Cauliflower	Riciness	Gibberellin appears to control elongation of flower peduncle.
Lettuce	Russet spotting	Caused by endogenous or exogenous ethylene.
Lettuce	Senescence	Yellowing is prevented by exogenous cytokinin.
Celery	Petiole pithiness	Water deprivation and abscisic acid appears to mediate.
Tomato	Ripening	Endogenous or exogenous ethylene or abscisic acid stimulates ripening of mature green fruit; endogenous cytokinins decline during ripening; exogenous auxins, gibberellins, and cytokinins delay ripening.
Carrot	Rooting	Exogenous gibberellins suppress rooting, cytokinins and ethylene promote rooting.

Table 6   Detrimental Effects of Ethylene in Postharvest Fruits and Vegetables

Commodity	Symptoms
Asparagus	Toughness
Beans, snap	Yellowing
Broccoli	Yellowing, floret abscission, off-flavors
Cabbage	Yellowing, leaf abscission
Carrots	Bitterness
Cucumber	Yellowing, softening
Eggplant	Calyx abscission, browning of pulp and seeds, accelerated decay
Lettuce	Russet spotting
Potato	Sprouting
Sweet potato	Flesh browning, off-flavor, failure to soften on cooking
Turnip	Toughness

Ethylene has received much more study by postharvest physiologists than the other phytohormones. In addition to its well-known role as "ripening hormone," other ethylene-mediated processes that lead to quality deterioration in postharvest tissues are summarized in Table 6. Ethylene is synthesized by reactions involving a cycle in which the amino acid methionine participates in the formation of 1-aminocyclopropane carboxylic acid (ACC)[42] (Figure 11). ACC synthase, a key enzyme in this pathway, has been purified.[43] Although a gene coding for ACC oxidase was recently isolated,[44] the enzyme itself has not yet been identified. Various laboratories are investigating the use of antisense technology to halt or slow down the rate of ethylene synthesis in postharvest crops.[34] In crops such as broccoli where ethylene causes yellowing, the goal is to completely shut down ethylene synthesis. In fruits, such as tomato and raspberry, the goal of this research is to slow down or control the rate of ethylene synthesis so that ripening will be delayed. A possible problem with the use of antisense technology to control ethylene biosynthesis is that alteration of ethylene biogenesis may disrupt other developmental processes associated with the growing plant. It now appears that plants contain different ACC synthase genes, one of which is specifically associated with ripening. Therefore, it may be possible to control ethylene biosynthesis without adversely affecting other ethylene-mediated reactions in the growing plant. An antisense gene for ACC oxidase has already been incorporated into transgenic plants.[44]

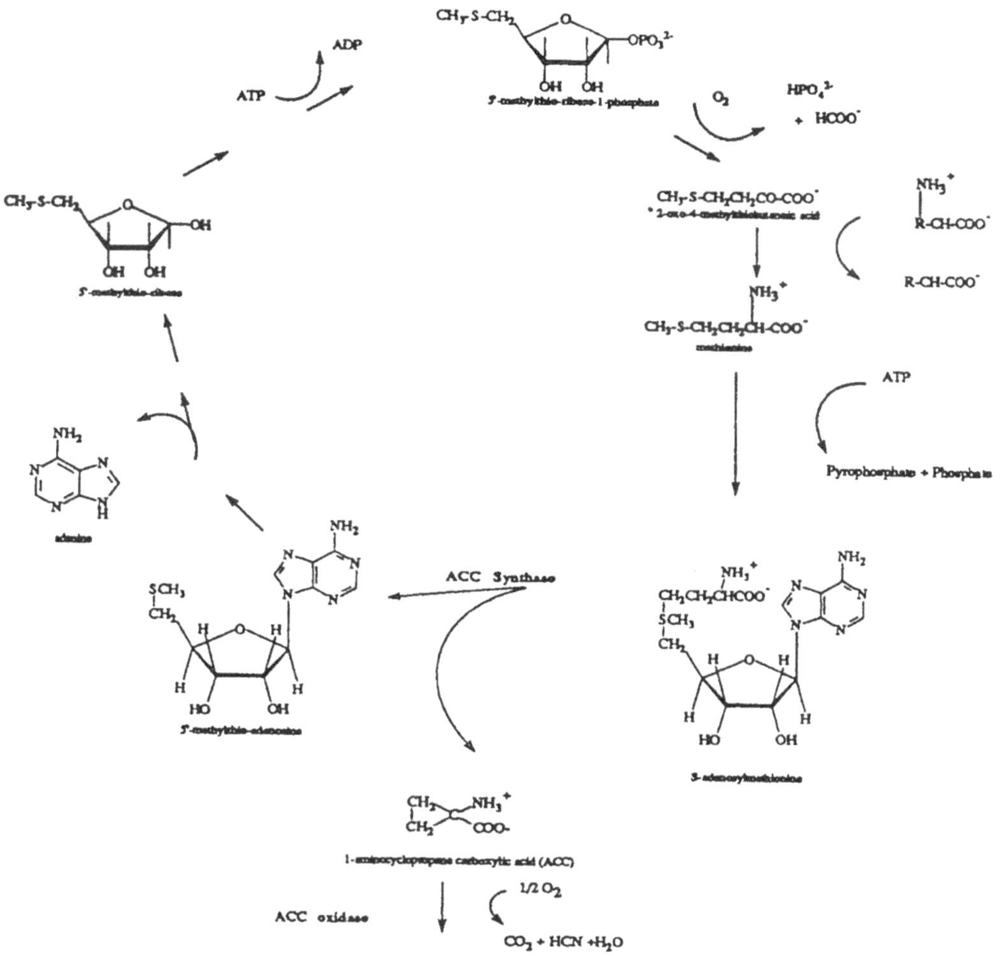

CH$_3$-S-CH$_2$

OPO$_3$$^{2-}$

OH OH

5'-methylthio-ribose-1-phosphate

ADP

ATP

O$_2$

HPO$_4$$^{2-}$
+ HCOO$^-$

CH$_3$-S-CH$_2$

OH

OH OH

5'-methylthio-ribose

CH$_3$-S-CH$_2$CH$_2$CO-COO$^-$
2-oxo-4-methylthiobutanoic acid

NH$_3$$^+$

R-CH-COO$^-$

NH$_3$$^+$

R-CH-COO$^-$

CH$_3$-S-CH$_2$CH$_2$CH-COO$^-$

methionine

ATP

Pyrophosphate + Phosphate

adenine

SCH$_3$
CH$_2$

H O H

H   H

HO   OH

5'-methylthio-adenosine

ACC Synthase

NH$_3$$^+$
CH$_2$CH$_2$CHCOO$^-$
SCH$_3$
CH$_2$

H O H

H   H

HO   OH

S-adenosylmethionine

CH$_2$ NH$_3$$^+$
CH$_2$ COO-

1-aminocyclopropane carboxylic acid (ACC)

ACC oxidase

1/2 O$_2$

CO$_2$ + HCN +H$_2$O

Figure 11 Biosynthesis of the plant hormone ethylene by the methionine cycle. (From Haard, N.F., Food as cellular systems: Impact on quality and preservation. A review, *J. Food Biochem.*, 19(3), 191-238, 1995. With permission.)

Recent advances in biotechnology have also helped researchers to elucidate the mechanism of action of ethylene as ripening hormone. In unripe tomato fruit, the response to ethylene is rapid; mRNAs are formed within 30–120 min, and the types of genes that are expressed depend on ethylene concentration.[45] Ethylene affects transcriptional processes of some genes, and for others both transcription and translation are affected.[46] A DNA binding factor that increases during ripening appears to be involved with gene expression promoted by ethylene.[47] The rapid advance in our understanding of hormone action afforded by genetic engineering techniques bodes well for future prospects to devise new ways to control physiological processes in postharvest fruits and vegetables.

Other control mechanisms

The presence of an enzyme in a tissue does not alone indicate it is active. Catalytic activity may be controlled by the presence of endogenous inhibitors. Some crops undergo "cold sweetening" when stored at a low nonfreezing temperatures. The process involves the conversion of starch to sugars including sucrose, glucose, and fructose. Cold sweetening in potato tuber is problematic because it can cause poor texture after cooking, an undesireable sweet taste, and/or excessive Maillard browning during frying. The mechanism of cold sweetening is complex and involves starch granule composition[48] as well as other factors.[49] However, one contributing factor appears to be a proteinaceous inhibitor of the enzyme invertase. When mature tubers are placed in the cold, there is simultaneous disappearance of invertase inhibitor and *de nova* synthesis of invertase so that sucrose is converted to the reducing sugars glucose and fructose.[50]

Catalytic activity may also be influenced by the amount of substrate or availability of a cofactor present in the tissue. One reason that certain cultivars of fruit (e.g., pear) undergo enzymic browning more rapidly than others is the concentration of phenol substrate in the tissue. The action of an enzyme may also be determined by the specific subcellular site at which it is bound. Calcium concentration in the tissue appears to be responsible for the association–dissociation of peroxidase with cell walls and thereby influences lignification, causing loss of tenderness of crops such as postharvest asparagus.

Brief heating can facilitate enzyme-catalyzed reactions that affect quality of fruits and vegetables. Heat potentiation of enzyme-catalyzed reaction appears to be the result of loss in cell compartmentation, thus allowing enzyme and substrate to interact. Activation of pectin methyl esterase by blanching can prevent sloughing (lack of cohesiveness) of cooked potatoes[51] and green beans[52] in the presence of calcium ions. The firming of vegetables by blanching involves activation of pectin methylesterase, possibly due to leaching of potassium to the extracellular matrix, hydrolysis of methyl esters from pectin, and migration of divalent ions to the pectinic acid to form crossbridges (Figure 12).

Postharvest changes in foods from animals

The most common animal tissues that are used as food include milk, eggs, and myosystems. Physiological processes are most important in myosystems, although enzymes can also contribute to quality deterioration in milk and eggs. Recent biotechnology applications have also included attempts to introduce specific economically important traits into livestock and fish. Transgenic animals with improved growth rate, leaner meat composition, higher content of *x*-3 fatty acids, lower cholesterol content, and altered composition of milk proteins have been developed at the experimental level.[53-55] Future research may be directed to the control of enzyme-catalyzed reactions that influence postharvest changes in quality. As with the application of recombinant DNA technology to control reactions in postharvest fruits and vegetables, our ability to control postmortem reactions through genetic engineering is dependent on our understanding of the biochemical reactions that influence food quality.

Edible fluids of animal origin

Milk

Many enzymes have been identified in milk, including oxidoreductases (e.g., xanthine oxidase, lactoperoxidase, sulfhydryl oxidase), transferases (e.g., UDP-galactosyltransferase, c-glutamyl transferase, ribonuclease), hydrolases (e.g., lysozyme, alkaline protease, lipase) and lyases (e.g., carbonic anhydrase, fructose-1,6-diphosphate D-glyceraldehyde-3-phosphate lyase). Endogenous enzymes may contribute to the flavor and texture stability of milk, particularly during long-term storage after high-temperature–short-time processing.

Hydrolytic rancidity is particularly important in milk and milk products, because milk fat contains substantial quantities of short chain fatty acids, especially butyric, caproic, capric, and caprylic acids. Various types of lipolysis may occur in milk.[15] Induced lipolysis occurs when milk is subject to physical processes such as agitation, foaming, homogeni-

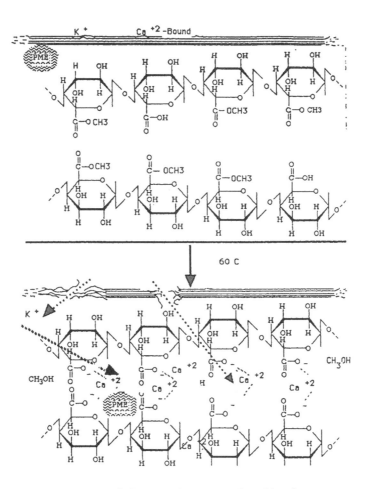

Figure 12 Activation of pectin methylesterase by intermediate blanch temperature. Interaction of calcium ions with carboxylic acid residues promotes cell–cell cohesion. (From Haard, N.F., Food as cellular systems: Impact on quality and preservation. A review, *J. Food Biochem.*, 19(3), 191-238, 1995. With permission.)

zation, and rapid temperature change. A theory to explain induced lipolysis is that physical disruption of the fragile fat globule membrane results in its partial replacement by milk proteins including lipase. Oxidative rancidity may also contribute to flavor defects in milk. Xanthine oxidase, associated with the fat globule membrane, and lactoperoxidase in the serum appear to be an enzymic source of prooxidants for milk lipids.[56] Milk proteases may contribute to the ripening of cheese and to the problem of gelation in high-temperature–short-time processed milk.[57]

Eggs

Relatively little information is available concerning the role of enzymes in the keeping quality of eggs. The enzyme lysozyme is capable of destroying certain bacteria by catalyzing hydrolysis of β-1,4 linkages between glucosamine derivatives in the cell wall. Lysozyme in egg white appears to contribute to resistance to invading microorganisms.[58] The concentration of HCO_3^- and CO_3^{-2} ions in egg are governed by the partial pressure of CO_2 in the external environment. A salient feature of postharvest physiology in shell eggs is the loss of carbon dioxide through the pores in the shell and the resulting increase in pH. During storage of eggs, the pH of albumen increases from about 7.6 to 9.7. Thinning

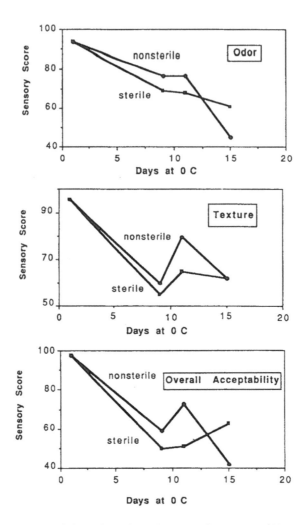

Figure 13   Changes in acceptability of sterile and nonsterile snapper (*Chrysophrys auratus*). (From Haard, N.F., Food as cellular systems: Impact on quality and preservation. A review, *J. Food Biochem.*, 19(3), 191-238, 1995. With permission.)

of egg white in stored eggs is associated with chemical changes in specific protein fractions, namely ovomucin and ovalbumin, and most evidence indicates such changes are nonenzymic reactions promoted by the alkaline pH.[59]

Muscle foods

The term "meat" describes muscle tissue of cattle (beef), hog (pork), sheep (mutton), avian species (poultry), and fish (seafood) that has undergone certain biochemical changes after the death of the animal. Meat quality is influenced by biochemical reactions that take place during and after slaughter. Biochemical reactions are necessary for the normal development of tenderness and juiciness in meat, especially beef. On the other hand, anomolous postmortem physiology can also lead to poor quality meat that has limited market value. In many cases, postmortem disorders are caused by a stressful antemortem environment immediately prior to rendering the animal unconscious and to exsanguination. In other cases, notably fish, biochemical reactions rather than spoilage microorganisms are largely responsible for deterioration in texture and flavor of refrigerated meat. Although postmortem changes contribute to the improved palatability of meats, such as beef, this should not overshadow the equally important role that biochemical reactions may have in the deterioration of quality. This point is well illustrated by studies showing that the market life of sterile and bacterial contaminated fish are similar[60] (Figure 13).

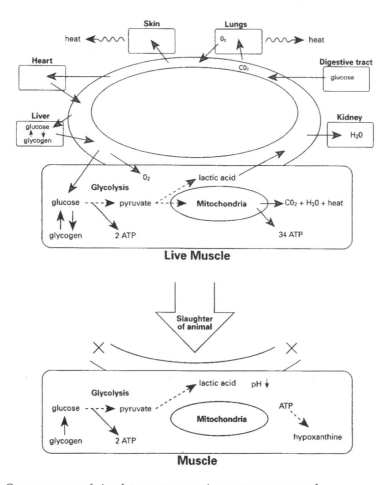

Figure 14   Consequences of circulatory stoppage in postmortem muscle.

The major physiological difference between living muscle and postmortem muscle is stoppage of blood circulation (Figure 14). As a result of severing the circulatory system, the ability of the postmortem muscle cell to maintain the same balanced environment as the living cell is lost. Key changes in the postmortem cell are the inability to maintain temperature, pH, oxygen concentration, and energy supply.

Glycogenolysis and glycolysis
As the circulatory system ceases so does the supply of glucose from the blood. Glucose is supplied to the postmortem muscle by phosphorolysis or by hydrolysis of glycogen. Phosphorlylase is generally believed to be responsible for glycogen degradation in mammalian muscle, while glycogenolytic hydrolases appear to be responsible for glucose formation in fish muscle.[61] The duration of glycolysis in the postmortem cell is influenced by the initial concentration of glycogen in the muscle.[62] Surf clam adductor muscle, which contains almost 40 mg glycogen per gram, continues with postmortem glycolysis for more than 9 days at 4°C.[63]

As the oxygen supply to the muscle becomes depleted after circulatory stoppage, the aerobic oxidation of glucose and fatty acids by the TCA cycle and mitochondrial electron transport chain ceases to function. This stoppage is in contrast to plant tissues that maintain the ability to exchange O_2, H_2O, and CO_2 after harvest. In muscle tissue, the glycolytic pathway continues to convert glucose to pyruvate under anaerobic conditions. In most animals, pyruvate is reduced to lactate with the regeneration of NAD^+ by the enzyme lactate dehydrogenase. However, certain animals such as squid and clam accumulate pyruvate because they lack the enzyme lactate dehydrogenase.[63] The net energy conservation by anaerobic glycolysis is only 2 or 3 mol ATP per mole of glucose compared to 36 or 37 mol ATP per mole of glucose by aerobic respiration. Very low pH (below 5.5), which can develop in some meat as glycolysis proceeds (see below), can be instrumental in cessation of glycolysis. There appears to be a direct effect of pH on critical enzymes including phosphorylase[64] and phosphofructokinase.[8]

Antemortem stress and glycolysis

Stress conditions prior to slaughter, such as exercise, temperature extremes, fasting, or fear, can lead to increased glycogenolysis and rate of glycolysis just prior to or after slaughter. The specific response of an animal to antemortem stress varies with species, breed, and intraspecific factors. As a consequence of different antemortem conditions, the rate and extent of glycolysis can differ after slaughter. Many stress-susceptible animals exhibit very rapid glycolysis while the carcass is still warm and the heat produced by glycolysis reduces the rate of chilling. As a consequence of high carcass temperature and other changes accompanying glycolysis, the result may be poor-quality meat. If stress, such as extensive exercise or fasting of animals, occurs for prolonged periods prior to slaughter, it can lead to depletion of muscle glycogen stores prior to slaughter; therefore, glucose is not available for glycolysis in the postmortem cell.[62]

Postmortem stress and glycolysis

The rate of glycolysis in postmortem muscle appears to be directly related to the concentration of ATP in the tissue and can be faster at 0°C than at higher temperatures[65] (Figure 15). The figure shows that rapid glycolysis, as evidenced by accumulation of lactate in the muscle, does not begin until the ATP concentration of the tissue begins to fall below about 51 mol/g muscle tissue. Likewise, in mammalian muscles an increase in glycolysis may occur at temperatures below 10–12°C. The rapid decline in ATP that occurs in muscle from some animals (e.g., beef or lamb) is associated with a phenomenon called "cold shortening" and tough cooked meat. It is known that ATP exerts feedback control on the glycolytic pathway by inhibiting phosphorylase and phosphofructokinase enzymes. Factors responsible for maintanence and eventual loss of ATP in postmortem muscle are discussed below.

Postmortem glycolysis in muscle is also accelerated by electrical stimulation. Treatment of carcasses with high-voltage pulses can increase glycolytic rate by as much as 150-fold.[66] The initial effect of electrical stimulation is associated with activation of glycogen phosphorylase,[67] and the persistant stimulation of glycolysis following the treatment appears to be associated with rapid ATP hydrolysis, perhaps due to myofibrillar ATPase.[68] Electrical stimulation of caracasses is used commercially because the processor can avoid cold shortening and immediately butcher the carcass without loss in meat quality. Pressurization of muscle is another method that accelerates glycolysis in postmortem muscle and could have similar benefits for the processor as electrical stimulation.[69]

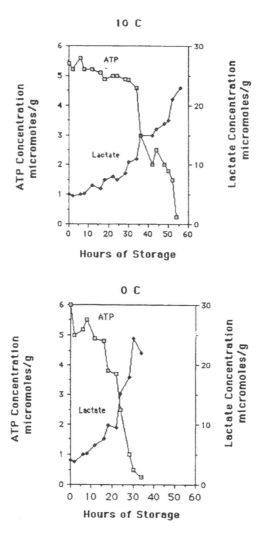

Figure 15   Influence of temperature on the loss of ATP and accumulation of lactic acid in post-mortem fish muscle. (From Haard, N.F., Food as cellular systems: Impact on quality and preservation. A review, *J. Food Biochem.*, 19(3), 191-238, 1995. With permission.)

ATP metabolism in postmortem muscle

ATP is the ultimate source of energy for several steps in muscle contraction, including (1) myosin ATPase, which releases energy for contraction; (2) sarcoplasmic reticulum ATPase for pumping Ca^{2+} ions during relaxation; (3) mitochondrial ATPases for pumping Ca^{2+} ions; and (4) sarcolemma ATPases for pumping Na^+ and K^+ gradients across the plasma membrane. The most immediate source of energy that may be mobilized for ATP synthesis from ADP in the living or postmortem muscle is the Lohmann reaction:

$$\text{creatine kinase}$$

$$ADP + \text{creatine phosphate} \quad \rightarrow \quad ATP + \text{creatine.}$$

(3)

Concentration of creatine phosphate in resting muscle is about twice that of the level of ATP. In the living muscle, rephosphorylation of creatine occurs in the mitochondria. ATP and AMP may also be formed from two molecules of ADP by a reaction catalyzed by adenylate kinase (myokinase).

In postmortem muscle two moles of ADP are rephosphorylated to ATP during anaerobic metabolism of glucose to lactic acid. When the muscle cell is depleted of creatine phosphate and glycogen or an unfavorable low pH is reached, the production of ATP ceases.

The continued activity of ATPases contributes to the depletion of ATP in the muscle. Under some circumstances, it appears the myofibrillar ATPase (myosin) is mainly responsible for the hydrolysis of ATP to ADP in the postmortem cell.[70] This is consistent with the living muscle where an estimated 1000 times more ATP is utilized in the contraction process than in the pumping of Ca^{2+} during relaxation and for maintaining the Na^+ and K^+ gradients across the sarcolemma following a nervous impulse.[10] In the resting muscle the activity of myosin ATPase, although very low, results in the hydrolysis of ATP as follows:

$$ATP + H_2O \rightarrow ADP + PO_4^{3-} + 3\,H^+ + 11.6\ kcal. \tag{4}$$

Very low levels of Ca^{2+} ($<10^{-7}\ M$), necessary to prevent the contraction process, are maintained in the living muscle by transport into the sarcoplasmic reticulum and mitochondria. It appears that Ca^{2+} released from mitochondria under anaerobic conditions cannot be removed rapidly enough by the sarcoplasmic reticulum, especially at low temperatures.[65,71,72] Increased levels of Ca^{2+} in the sarcoplasm may cause stimulation of myofibrillar and other Ca^{2+}-dependent ATPases. Additional research may show that other enzymes that utilize ATP, such as ATP-requiring proteases,[73] are also involved with ATP depletion in postmortem muscle.

In heavily exercised live muscle when glycogen levels become low and in postmortem muscle,[8] adenine nucleotides are deaminated to form inosine monophosphate, and eventually the end products inosine and/or hypoxanthine accumulate (Figure 16). It appears that the rate-limiting reaction in ATP catabolism varies with species. For example, AMP tends to accumulate in postmortem molluscs (e.g., clams) and crustacea (e.g., crab) where it plays an important role in the delicious umami taste of the meat.[74] In most food myosystems, however, the accumulation of IMP from ATP is rapid, and transient accumulation of inosine monophosphate, a flavor enhancer, contributes to the good taste of fresh meat and fish. The subsequent accumulation of inosine and/or hypoxanthine normally occurs at a relatively slower rate. Some species of fish tend to accumulate mainly inosine, others accumulate mainly hypoxanthine, and some accumulate both inosine and hypoxanthine.[75] Hypoxanthine has a bitter flavor and has been reported to cause off-flavor in several meat products including poultry, irradiated beef, and fish. The "K value," based on the quantitative analysis of ATP and its degradation products has been used as a biochemical indicator of fish freshness:

$$K = \frac{[\text{Inosine}] + [\text{Hypoxanthine}] \times 100}{[\text{ATP}] + [\text{ADP}] + [\text{AMP}] + [\text{IMP}] + [\text{Inosine}] + [\text{Hypoxanthine}].} \tag{5}$$

A number of commercial freshness indicators are now available for measuring the K value of fish muscle. Iced fish harvested from cold water lose freshness faster than fish from warm waters,[76] presumably because the enzymes from cold-water fish are less affected by temperature reduction during refrigerated storage. Although nucleotide degradation is generally considered to be the result of endogenous enzymes in muscle, recent studies have shown that spoilage bacteria that contain the enzyme inosine nucleosidase can influence the kinetics of hypoxanthine formation[77] (Figure 17). The "K_o value", which

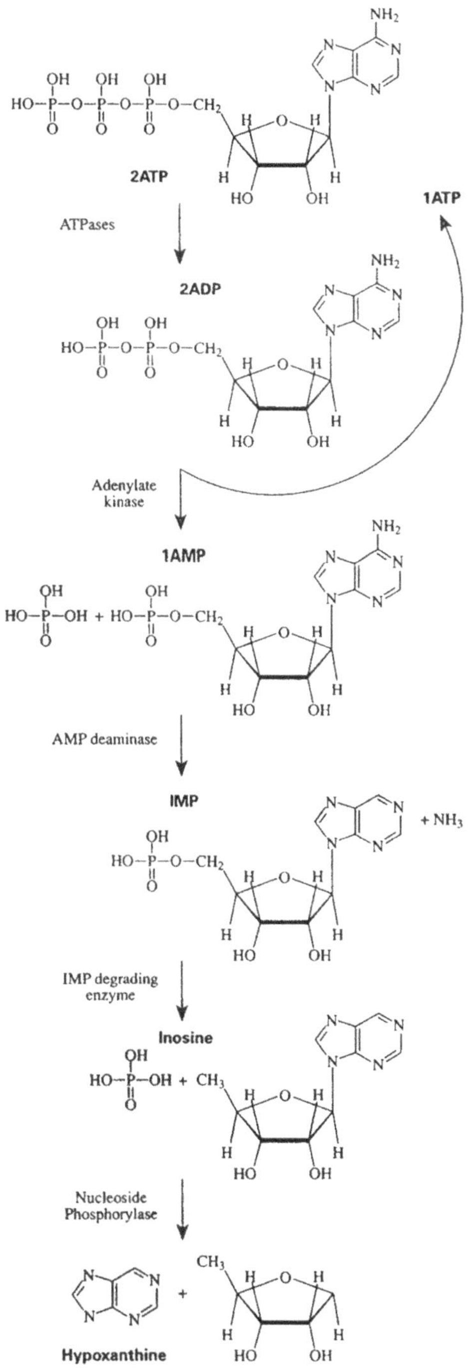

Figure 16   Adenosine triphosphate catabolism in postmortem muscle.

also includes the xanthine concentration, has been used to assess the freshness of beef and rabbit muscles[78] as well as pork and poultry.[79]

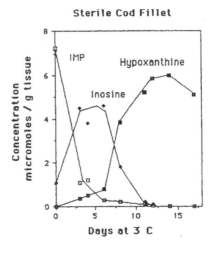

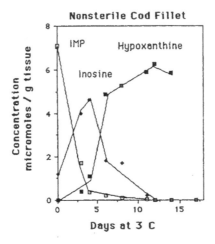

Figure 17   Nucleotide changes in sterile and nonsterile fish muscle. (From Haard, N.F., Food as cellular systems: Impact on quality and preservation. A review, *J. Food Biochem.*, 19(3), 191-238, 1995. With permission.)

Rigor mortis

Definition and importance

A consequence of the decline in concentration of ATP in postmortem muscle is the development of a phenonomenon called "rigor mortis" (Latin for "stiffness of death"). Rigor mortis is due to the formation of irreversible crossbridges between myosin and actin by a similar reaction that forms actomyosin during muscle contraction. The time required for the onset of stiffening may vary from <1 h to several days, depending on factors that influence ATP depletion (i.e., the species and breed of animal, antemortem factors, and temperature). On the one hand, animals that struggle a great deal prior to death, such as trawled fish, may develop rigor mortis very rapidly. On the other hand, animals that are in good nutritional status, well-rested, and undergo little struggle may require 1 or more days prior to developing rigor mortis. Commercially slaughtered chicken normally undergoes rigor in <2 h, while beef normally undergoes rigor in about 24 h.

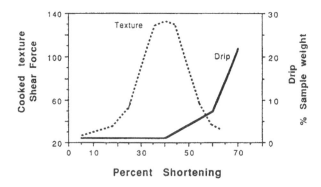

Figure 18 Influence of sarcomere shortening during rigor mortis on the texture and exudative drip of muscle. (From Haard, N.F., Food as cellular systems: Impact on quality and preservation. A review, *J. Food Biochem.*, 19(3), 191-238, 1995. With permission.)

Rigor mortis is important because the muscle shortening that occurs influences texture and other quality characteristics of the meat. Rigor shortening differs from contraction shortening in that there are more bonding sites in the area of overlap between actin and myosin filaments and the process is ireversible because the cross-bridges remain intact, even after the resolution of rigor. The extent of actin–myosin overlap (muscle shortening) occurring during rigor is influenced by various factors, including (1) physical constraint caused by the attachment of the muscle to the carcass and tension on the muscle as the carcass is hung; (2) the temperature of the carcass at the time of rigor onset; and (3) the concentration of Ca^{2+} and ATP in the sarcoplasm at the time of rigor onset.

Cooking ("cook rigor"), freezing ("thaw rigor"), or holding prerigor muscle at just above the freezing point ("cold shortening") can result in excessive shortening, resulting in poor texture[80] and excessive drip[81] in the resulting meat (Figure 18). Extreme shortening can cause physical disruption of the myofibril with a loss of water holding capacity and result in poor-quality meat.

Mechanism
ATP and ADP serve as relaxing factors or plasticizers that prevent the interaction of myosin and actin in the absence of Ca^{2+}. In the prerigor or relaxed state other myofibrillar proteins, troponin and tropomysin, serve to protect the crossbridge sites from interaction in the absence of Ca^{2+}. As the muscle ATP is depleted, cross bridges form between the thick and thin filaments, making the muscle inextensible and causing shortening to the extent that calcium ions and ATP are present in the sarcoplasm and the filaments are physically able to slide.

Resolution of rigor
Because ATP is no longer available in the postmortem cell, the muscle contraction cycle is not able to return to the relaxed state. Nonetheless, rigor mortis is followed by a process called the "resolution of rigor" when the muscle gradually becomes tender and extensible. During this period, key proteins in the myofilaments and cytoskeleton are hydrolyzed by neutral (calcium-activated neutral proteases) and acid (lysosomal cathepsins B and D, etc.) proteases. Although the exact mechanism of the resolution of rigor is not completely understood, considerable research has shown that hydrolysis of key proteins leads tor disruption of the z-bands that link the myofibrils together.[17] Other biochemical changes which appear to be related to the tenderization of beef[82] and the excessive softening of

fish[2] are the increased solubilization of collagen due to a breakdown in the proteoglycan network by glycosidase enzymes (Figure 19). Excessive softening of fish that results from short-term temperature abuse (e.g., 1 h at 30°C) is associated with a complete breakdown of collagen in the extracellular matrix.[83]

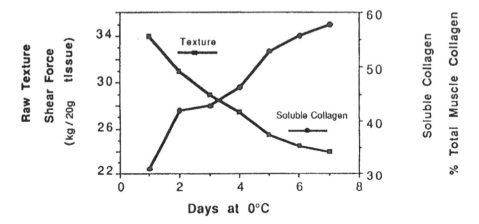

Figure 19   Texture softening and increase in soluble collagen during storage of postrigor Pacific rockfish at 0°C.

pH and meat quality

Cause of pH change

The hydrolysis of ATP in postmortem muscle also causes the pH of muscle to drop from 7.2–7.4 to an acidic value called the "ultimate pH." The ultimate pH of muscle is normally about 5.5 in red meat such as beef, pork, lamb, and tuna; 5.9 in poultry; and 6.2–6.6 in most lean fish. The decline in pH is closely paralleled by the accumulation of lactate, because this end product of anaerobic glycolysis is a measure of ATP turnover in the postmortem cell (Figure 15). Like glycolysis, the rate and extent of pH decline varies with species, type of muscle, antemortem factors, and temperature. White muscles, especially those of fish and shellfish, have a relatively high content of amino acids and peptides that probably serve to buffer changes in pH.

pH and quality

The ultimate pH influences meat quality in a number of ways. Lean muscle contains about 75% water, and an important quality attribute of meat is its "water-holding capacity." The binding of water to muscle proteins by ionic interactions and by entrapment between fibers by capillary action influences the consumer's perception of "juiciness" or "dryness" and the loss of water due to exudation or drip. In the prerigor state, when the pH is near 7, muscle has a high water-holding capacity. When meat with a high ultimate pH is cooked, the meat tends to have a dry consistency. Beef that has been stressed prior to slaughter may have an ultimate pH near 7 and is characterized as dark, firm, and dry (DFD). The high water-holding capacity of DFD beef makes it an excellent raw material for production of processed meat products, although it is not considered suitable for steaks and roasts, etc.

As the pH of the muscle declines to the isoelectric point of myofibrillar proteins, about pH 5.3, the ionic interaction of water with these proteins is reduced, resulting in a lowering of water-holding capacity. A decline in pH also reduces the repulsion of charged muscle fibers and can have a steric effect, causing repulsion of water from between fibers. For beef, an ultimate pH of about 5.8 is considered desirable for optimal eating quality, because the intermediate water-holding capacity is associated with a juicy cooked product that does not exude excessive water.

The rate of pH decline after slaughter can be even more important than the ultimate pH when the carcass temperature has not yet cooled down.[84] Stress-susceptible breeds of hog may develop a disorder called the pale, soft, exudative (PSE) condition, which is of major significance in the meat industry. Rapid pH decline is also associated with poor-quality halibut ("chalky"), tuna ("burnt"), cod ("sloppy"), and beef ("pale, watery"). The combination of high muscle temperature and low pH cause denaturation of muscle proteins.

The water-holding capacity of meat may also be influenced by rapid and extensive shortening during rigor mortis (Figure 18).

In a similar way, the rate and extent of pH decline may influence the appearance and color of meat. Meat with a high ultimate pH may have a dark appearance due to the low reflection of light caused by swollen fibers. Low pH and high carcass temperature may also promote the oxidation of the red meat pigment myoglobin (Fe^{2+}) to the brown pigment metmyoglobin (Fe^{3+}) by denaturing the globin moiety leaving the heme unprotected.[85] It is also possible that the amount of the bright red pigment oxymyoglobin on the surface of meat is lowered because of competition for oxygen by mitochondria[86] and bacteria. Rapid pH decline in fish is associated with advanced onset of an undesirable turbid or opaque appearance of the flesh.[87] Brain spiking of fish at the time of harvest delays the rate of postmortem changes including the pH decline.[88] While this treatment does not influence the eventual ultimate pH, it maintains the translucency of the flesh for longer periods of time.

The texture of meat is also influenced by pH through its effect on water-holding capacity and the denaturation of proteins associated with the above-mentioned physiological disorders linked to rapid pH decline. A low ultimate pH of cod fish meat (pH 6.2) is associated with tough cooked texture.[89] The predisposition of rockfish muscle to become dry and fibrous during frozen storage is also greater for specimens having a low ultimate pH.[90] (Figure 19)

Conclusions

Change in food quality is caused by many different biotic and abiotic reactions. Proper handling and processing serves to minimize or arrest these reactions and thereby extend the useful market life and wholesomeness of food. Physiological processes that occur after harvest of the plant crop or slaughter of the animal may influence food quality attributes in positive as well as in negative ways. The emphasis of this chapter has been on basic physiological changes, notably wound response, fruit ripening, plant senescence, and the conversion of muscle to meat, which are a natural consequence of harvest or slaughter. There are countless biochemical reactions that influence quality, many of which have not been discussed because of space limitation and many of which have yet to be elucidated. Breeding and genetic engineering of crops and livestock for improved quality retention during postharvest storage are promising strategies, but their continued success depends on our basic understanding of inherent physiological processes.

Food preservation methods discussed in detail elsewhere in this book may not arrest all biochemical reactions that affect quality. In some cases, biochemical reactions potentiated by food processing operations may contribute to the desirable character of the product. On the one hand, the desirable flavor, appearance, and texture of traditional sun-dried squid is mediated by biochemical reactions that occur during the several days required to dry the product to a water activity that eventually arrests enzyme-catalyzed reactions. On the other hand, modified atmosphere storage of fish, which is quite effective in retarding microbial growth and thereby delays putrefaction, can actually decrease grade A market life of Pacific rockfish by promoting endogenous biochemical reactions leading to hypoxanthine accumulation. Likewise, modified atmosphere storage or refrigeration of some fruits and vegetables may be limited by physiological disorders caused by abberant metabolism in the presence of excessive concentrations of carbon dioxide or lack of adequate oxygen.

In other ways, biochemical reactions that do not normally occur may be favored by conditions imposed by a food preservation technique. An example is the activation of the muscle enzyme trimethylamine demethylase in gadoid fish during frozen storage in an oxygen-free environment. The enzyme catalyzes the conversion of trimethylamine oxide, a natural constituent of marine fish, to dimethylamine and formaldehyde. The formaldehyde formed as a result of this reaction appears to interact with myosin and thereby contribute to the development of dry, fibrous texture that normally limits the market life of frozen gadoid products.

Another example of a reaction that does not appear to occur in chilled fish is the hydrolysis of collagen by the naturally occuring enzyme collagenase. This enzyme is latent in properly chilled fish, since collagen chains remain intact in sterile fish held on ice for several months. However, short-term exposure of sterile fish to ambient temperatures can initiate an enzyme cascade leading to extensive hydrolysis of collagen and causing loss of fillet integrity after only a few days on ice. However, judicious blanching of fruits and vegetables at intermediate temperatures can potentiate enzyme-catalyzed reactions that result in improved texture, flavor, appearance, and suitability for more rigorous thermo-processing.

References

1. Fennema, O.R., Powrie, W.D., and Marth, E.H. 1973. Low Temperature Preservation of Foods and Living Matter, Marcel Dekker, Inc., New York, p. 79.
2. Kim, K.S. and Haard, N.F. 1992. The degradation of proteoglycans in the skeletal muscle of Pacific rockfish (*Sebastes* sp.) during ice storage. J. Muscle Foods 3:120–121.
3. Price, J.F. and Schweigert, B.S. 1987. The Science of Meat and Meat Products, 3rd edition, Food and Nutrition Press, Westport, CT., p. 349-369.
4. Saks, Y., Sonego, L., and Benarie, R. 1990. Senescent breakdown of Jonathan apples in relation to the water-soluble calcium content of the fruit pulp before and after storage. J. Am. Soc. Hortic. Sci. 115 (4): 615-618.
5. Haard, N.F. 1990. Enzymes from food myosystems. J. Muscle Foods 1: 293-338.
6. Dixon, M. and Webb, E.C. 1979. Enzymes, 3rd edition, Academic Press, New York, p. 633-649.
7. Haard, N.F. 1985. Characteristics of edible plants. In: Food Chemistry (O.R. Fennema, ed.), 2nd edition, Marcel Dekker, New York, p. 869-879.
8. Hultin, H.O. 1985. Characteristics of muscle tissue. In: Food Chemistry (O.R. Fennema, ed.), 2nd edition, Marcel Dekker, New York, p. 725-789.
9. Schwimmer, S. 1972. Cell disruption and its consequences in food processing. J. Food Sci. 37: 530-534.
10. Judge, M., Aberle, E.D., Forrest, J.C., Hedrick, H.B., and Merkel, R.A. 1989. Principles of Meat Science, 2nd edition, Kendall Hunt, Dubuque, IA.
11. Larrigaudiere, C., Latche, A., Pech, J.C., and Triantaphylides. 1990. Short-term effects of g-irradiation on 1-aminocyclopropane-1-carboxylic acid metabolism in early climacteric cherry tomatoes. Plant Physiol. 92: 577-581.
12. Haard, N.F. 1983. Stress metabolites. In: Postharvest Physiology and Crop Protection (M. Lieberman, ed.), Plenum, New York, p. 299-314.
13. Gardner, H.W. Oxidation of lipids in biological tissue and its significance. 1985. In: Chemical Changes in Food During Processing (T. Richardson and J.W. Finley, eds.), AVI Publishing, Westport, CT, p. 177-203.

14. Lindsay, R.C. 1985. Flavors. In: Food Chemistry (O.R. Fennema, ed.), 2nd edition, Marcel Dekker, New York, p. 585-627.
15. Schwimmer, S. 1981. Source Book of Food Enzymology. AVI Publishing Co., Wesport, CT.
16. Mayer, A.M. 1987. Polyphenol oxidases in plants—recent progress. Phytochem. 26: 11.
17. Eskin, N.A.M. 1990. Biochemistry of Foods, 2nd edition, Academic Press, New York, p. 404-406.
18. McGee, H. 1984. On Food and Cooking, Charles Scribner's Sons, New York, p. 161-165.
19. Areas, J.A.C. and Lajolla, F.M. 1981. Determinacao enzimatica espifica de amido, glicose, fructose, e sacarose em bananas preclimactericas e climactericas. An. Farm. Quim. Sao Paulo 20: 307.
20. Isherwood, F.A. 1976. Mechanisms of starch sugar interconversion in *Solanum tuberosum*. Phytochemistry 15:33.
21. Hubbard, N.L., Pharr, D.M., and Huber, S.C. 1990. Role of sucrose phosphate synthase in sucrose biosynthesis in ripening bananas and its relationship to the respiratory climacteric. Plant Physiol. 94: 201-208.
22. Young, R.E., Salimen, S., and Sornrivichai, P. 1974. Enzyme regulation associated with ripening in banana fruit. Colloq. Int. C.N.R.S. 238: 271.
23. ap Rees, T. 1980. Assessment of the contribution of metabolic pathways to plant respiration. In: The Biochemistry of Plants (D.D. Davies, ed.), Vol. 2, Academic Press, New York, p. 1-29.
24. Kumar, S., Patil, B.C., and Sinha, S.K. 1990. Cyanide respiration is involved in temperature rise in ripening mangoes. Biochem. Biophys. Res. Comm. 168 (2): 818-822.
25. Kader, A.A. 1987. Respiration and gas exchange of vegetables. In: Postharvest Physiology of Vegetables (J. Weichmann, ed.), Marcel Dekker, New York, p. 25-43.
26. Pixton, S.W., Warburton, S., and Hill, S.T. 1975. Longterm storage of wheat. III. Some changes in the quality of wheat observed during 16 years of storage. J. Stored Prod. Res. 11: 177.
27. Solomos, T. and Laties, G.G. 1973. Cellular organization and fruit ripening. Nature 245: 390.
28. Guclu, J., Paulin, A., and Soudain, P. 1989. Changes in polar lipids during ripening and senescence of cherry tomato (*Lycopersicon esculentum*)—relationship to climacteric and ethylene increases. Physiologia Plantarum 77(3): 413-419.
29. Biale, J.B. 1969. Metabolism at several levels of organization in the fruit of avocado, *Pesea americana*. Mill. Qual. Plant. Mater. Veg. 19: 141.
30. Theologies, A. and Laties, G.G. 1978. Respiratory contribution of the alternate pathway during various stages of ripening in avocado and banana fruit. Plant Physiol. 69: 122.
31. Leshem, Y.Y. 1988. Plant senescence processes and free radicals. Free Radical Biol. Med. 5(1): 39-49.
32. Lurie, S. and Kleine, J.D. 1990. Cyanide metabolism in relation to ethylene production and cyanide insensitive respiration in climacteric and nonclimacteric fruits. J. Plant Physiol. 135(5): 518-521.
33. Gaser, C.S. and Fraley, R.T. 1989. Genetically engineering plants for crop improvement. Science 244: 1293-1299.
34. Knight, P. 1989. Engineering fruit and vegetable crops. Biotechnology 7: 1233-1237.
35. Harlander, S. 1989. Food biotechnology: yesterday, today and tommorow. Food Technol. September, p. 196-206.
36. Wasserman, B.P. 1990. Expectations and role of biotechnology in improving fruit and vegetable quality. Food Technol. February, p. 68-71.
37. Smith, C.J.S., Watson, C.F., Ray, J., Bird, C.R., Morris, P.C., Schuch, W., and Grierson, D. 1988. Antisense RNA inhibition of polygalacturonase gene expression in transgenic tomatoes. Nature 334: 724.
38. Sheehy, R.E., Kramer, M., and Hiatt, W.R. 1988. Reduction of polygalacturonase activity in tomato fruit by antisense RNA. Proc. Natl. Acad. Sci. USA 85: 8805.
39. Schuch, W., Bird, C.R., Ray, J., Smith, C.J.S., Watson, C.F., Morris, P.C., Gray, J.E., Arnold, C., Seymour, G.B., Tucker, G.A., and Grierson, D. 1989. Control and manipulation of gene expression during tomato fruit ripening. Plant Molec. Biol. 13: 301-311.

40. Giovannoni, J.J., DellaPenna, D., Bennett, A.B., and Fischer, R.L. 1989. Expression of a chimeric polygalacturonase gene in transgenic rin (ripening inhibited) tomato fruit results in polyuronide degradation but not fruit softening. Plant Cell 1:53.
41. Frenkel, C., Dyck, R., and Haard, N.F. 1975. Role of auxin in the regulation of fruit ripening. In: Postharvest Biology and Handling of Fruits and Vegetables (N. F. Haard and D.K. Salunkhe, eds.), AVI Publishing, Westport, CT, p. 19-34.
42. Yang, S.F. and Hoffman, N.E. 1984. Ethylene biosynthesis and its regulation in higher plants. Ann. Rev. Plant Physiol. 35: 155.
43. Yip, W.K., Dong, J.G., Kenny, J.W., Thompson, G.A., and Yang, S.F. 1990. Characterization and sequencing of the active site of 1-1-aminocyclopropane-1-carboxylate synthase. Proc. Nat. Acad. Sci. USA 20: 7930-7934.
44. Hamilton, A.J., Lycett, G.W., and Grierson, D. 1990. Antisense gene inhibits synthesis of the hormone ethylene in transgenic plants. Nature 346(6281): 284-287.
45. Lincoln, J.E., Cordes, S., Read, E., and Fischer, R.L. 1987. Regulation of gene expression by ethylene during *Lycopersicon esculentum* (tomato) fruit development. Proc. Nat. Acad. Sci. USA, 84(9): 2793-2797.
46. Lincoln, J.E. and Fischer, R.L. 1988. Diverse mechanisms for the regulation of ethylene inducible gene expression. Molec. Gen. Genetics 212(1): 71-75.
47. Deikman, J. and Fischer, R.L. 1988. Interaction of a DNA binding factor with the 5'-flanking region of an ethylene-responsive fruit ripening gene from tomato. Embo J. 11: 3315-3320.
48. Barichello, V., Yada, R.Y., Coffin, R.H., and Stanley, D.W. 1990. Low temperature sweetening in susceptible and resistant potatoes: starch structure and composition. J. Food Sci. 55 (4): 1054-1059.
49. ap Rees, T., Dixon, W.L., Pollock, C.J., and Franks, F. 1981. Low temperature sweetening of higher plants. In: Recent Advances in the Biochemistry of Fruits and Vegetables (J. Friend and M.J.C. Rhodes, eds.), Academic Press, New York, p. 41-61.
50. Pressey, R. 1972. Natural enzyme inhibitors in plant tissue. J. Food Sci. 37: 521-523.
51. Bartolme, L.G. and Hoff, J. E. 1972. Firming of potatoes: Biochemical effects of preheating. J. Agric. Fd. Chem. 20: 266-270.
52. Steinbuch, E. 1976. Technical note: Improvement of texture of frozen vegetables by stepwise blanching treatments. J. Food Technol. 11: 313-316.
53. Van Brunt, J. 1988. Molecular farming: transgenic animals as bioreactors. Biotechnology 6(10): 1149-1154.
54. Tave, D. 1988. Genetic engineering. Aquaculture Magazine, March/April, pp. 63-65.
55. Pursell, V.G., Pinkert, C.A., Miller, K.F., Bolt, D.J., Campbell, R.G., Palmiter, R.D., Brinster, R.L., and Hammer, R.E. 1989. Genetic engineering of livestock. Science 244: 1281-1288.
56. Hill, R.D. 1979. Oxidative enzymes and oxidative processes in milk. CSIRO Food Res. 39: 33-37.
57. Fox, P.F. and Morrissey, P.A. 1981. Indigenous enzymes of bovine milk, In: Enzymes and Food Processing (G.G. Birch, N. Blakebrough, and K.J. Parker, eds.), Pergamon Press, Oxford, p. 245-268.
58. Yadav, N.K. and Vadhera, D.V. 1977. Mechanism of egg white resistance to bacterial growth. J. Food Sci. 42: 97-99.
59. Powrie, W.D. and Nakai, S. 1985. Characteristics of edible fluids of animal origin: eggs. In: Food Chemistry (O.R. Fennema, ed.), 2nd edition, Marcel Dekker, New York, p. 829-855.
60. Fletcher, G.C.and Hodgson, J.A. 1988. Shelf-life of sterile snapper (*Chrysophrys auratus*). J. Food Sci. 53 (5): 1327-1332.
61. Burt, J.R. 1966. Glycogenolytic enzymes of cod (*Gadus callarias*) muscle. J. Fish. Res. Board Can. 23: 527.
62. Warriss, P.D., Bevis, E.A., and Ekins, P.J. 1989. The relationships between glycogen stores and muscle ultimate pH in commercially slaughtered pigs. Br. Vet. J. 145(4): 378-383.
63. Lee, Y.B., Rickansrud, D.A., Hagberg, E.C., and Forsythe, R.H. 1978. Postmortem biochemical changes and muscle properties in surf clam (*Spisula solidissima*). J. Food Sci. 43: 35-37, 51.
64. Newbold, R.P. and Lee, C.A. 1965. Post-mortem glycolysis in skeletal muscle. The extent of glycolysis in diluted preparation of mammalian muscle. Biochem. J. 97: 1.

65. Iwamoto, M., Yamanaka, H., Watabe, S., and Hashimoto, K. 1987. Effect of storage temperature on rigor mortis and ATP degradation in plaice *Paralichthys olivaceus* muscle. J. Food Sci. 52(6): 1514-1517.

66. Chrystall, B.B. and Devine, C.E. 1978. Electrical stimulation and lamb tenderness. Electrical stimulation, muscle tension, and glycolysis in bovine sternomandibularis muscle. Meat Sci. 2: 49.

67. Newbold, R.P. and Small, L.M. 1985. Electrical stimulation of post-mortem glycolysis in the semitendinosus muscle of sheep. Meat Sci. 12:1.

68. Horgan, D.J. and Kuypers, R. 1985. Post-mortem glycolysis in rabbit longissimus dorsi muscles following electrical stimulation. Meat Sci. 12: 225.

69. Elkhalifa, E.A., Anglemier, A.F., Kennick, W.H., and Elgasim, E.A. 1984. Effect of prerigor pressurization on post-mortem bovine muscle lactic dehydrogenase activity and glycogen degradation. J. Food Sci. 49: 593.

70. Hamm, R. Dalryymple, R.H., and Honikel, K.O. 1973. Proc. 19th Meet. Eur. Meat Res. Workers 1: 73.

71. Buege, D.R. and Marsh, B.B. 1975. Mitochondrial calcium and postmortem muscle shortening. Biochem. Biophys. Res. Commun. 65: 478-482.

72. Greaser, M.L., Cassens, R.G., Hoekstra, W.G., and Briskey, E.J. 1969. The effect of pH-temperature treatments on the calcium-accumulating ability of purified sarcoplasmic reticulum. J. Food Sci. 34: 633-637.

73. Fagan, J.M. and Waxman, L. 1989. A novel ATP-requiring protease from skeletal muscle that hydrolyzes non-ubiquitinated proteins. J. Biological Chemistry 264 (30): 17868-17872.

74. Konuso, S. and Yamaguchi, K. 1982. The flavor compounds in fish and shellfish. In: Chemistry and Biochemistry of Marine Food Products (R.E. Martin, ed.), AVI Publishing, Westport, CT, pp. 367-404.

75. Ehiro, S. and Uchiyama, H. 1973. Formation of inosine in fish muscle during spoilage in ice. Bull. Tokai Reg. Fish. Lab. 75: 63-73.

76. Tsuchimoto, M., Tanaka, N., Misimi, T., Yada, S., Senta, T., and Yasuda, M. 1988. Resolution characteristics of ATP related compounds in fishes from several waters and the effect of habitat temperature on the characters. Nippon Suisan Gakkaishi 54: 117-124.

77. Surette, M.E., Gill, T.A., and LeBlanc, P.J. 1988. Biochemical basis of postmortem nucleotide catabolism in cod (*Gadus morhua*) and its relationship to spoilage. J. Agric. Food Chem. 36: 19-22.

78. Nakatani, Y., Fujita, T., Sawa, S., Otani, T., Hori, Y., and Takagahara, I. 1986. Changes in ATP related compounds of beef and rabbit muscles and a new index of freshness of muscle. Agric. Biol. Chem. 50: 1751.

79. Fujita, T., Hori, Y., Otani, T., Kunita, Y., Sawa, S., Sakai, S., Tanaka, Y., Takagahara, I., and Nakatani, Y. 1988. Applicability of the K_0 value as an index of freshness for porcine and chicken muscles. Agric. Biol. Chem. 52: 107.

80. Marsh, B.B. and Leet, N.G. 1966. Resistance to shearing of heat denatured muscle in relation to shortening. Nature 221: 635-636.

81. Marsh, B.B. and Leet, N.G. 1966a. Studies in meat tenderness. III. The effects of cold-shortening on tenderness. J. Food Sci. 31: 450-459.

82. Wu, J.J., Dutson, T.P., and Carpenter, Z.L. 1981. Effect of postmortem time and temperature on the release of lysosomal enzymes and their possible effect on bovine connective tissue components of muscle. J. Food Sci. 46: 1132.

83. Cepeda, R., Chou, E., Bracho, G., and Haard, N.F.1990. An immunological method for measuring collagen degradation in the muscle of fish. In: Advances in Fisheries Technology and Biotechnology for Increased Profitability, Technomic Press, Lancaster, PA, pp. 487-506.

84. Fischer, C. and Hamm, R. 1980. Biochemical studies on fast glycolyzing bovine muscles. Meat Sci. 4: 41.

85. Walters, C.L. 1975. Meat colour: The importance of haem chemistry, In: Meat (D.J.A. Cole, and R.A. Lawrie, eds.), AVI Publishing, Westport, CT, pp. 385-401.

86. Ashmore, C.R., W. Parker, and L. Doerr. 1972. Respiration of mitochondria isolated from dark cutting beef: postmortem changes. J. Anim. Sci. 34: 46-48.

87. Cramer, J.L., Nakamura, R.M., Dizon, A.E., and Ikehara, W.N. 1981. Burnt tuna conditions leading to rapid deterioration in the quality of raw tuna. Mar. Fish. Rev. 43(6): 12.
88. Boyd, N.S., Wilson, N.D., Jerrett, A.R., and B.I. Hall. 1984. Effects of brain destruction on postharvest muscle metabolism in the fish kahawai (*Arripis trutta*). J. Food Sci. 49(1): 177-179.
89. Love, M. 1980. The Chemical Biology of Fishes, Vol. 2, Advances 1968-1977, Academic Press, London.
90. Kramer, D.E. and Peters, M.D. 1981. Effect of pH and prefreezing treatment on the texture of yellowtail rockfish (*Sebastes flavidus*) as measured by the Ottawa texture measuring system. J. Food Technol. 16 (5): 493-504.

Bibliography

MacLeod, R.F., Kader, A.A., and Morris, L.L.. 1976. Stimulation of ethylene and CO_2 production of mature green tomatoes by impact bruising. Hort Sci. 11: 604.

Simmons, J.P., McClenaghan, M., and Clark, A.J. 1987. Alteration of the quality of milk by expression of sheep b-lactoglobulin in transgenic mice. Nature xx: 1-3.

Wada, H., Gombos, Z., and Murata, N. 1990. Enhancement of chilling tolerance of a cyanobacterium by genetic manipulation of fatty acid desaturation. Nature 347(6289): 200-203.

chapter three

Protein instability

Ronald E. Barnett and Hie-Joon Kim

Contents

Introduction

Proteins in foods, those that are unprocessed or minimally processed as well as those that are processed to destroy spoilage and pathogenic microorganisms, are susceptible to chemical changes brought about by indigenous or exogenous components and influenced by various conditions of treatment and storage. They can be degraded by hydrolysis of the peptide linkages brought about by enzymes and corresponding to proteolysis, they can undergo aggregative or degradative reactions as well as modifications to substituent moieties that are effectively brought about by interactions with food additives, and they can undergo aggregation and modification due to amine group reactions with indigenous or added components with carbonyl groups such as reducing sugars. Some of the more prevalent and significant reactions occurring during refrigerated or ambient storage are described below, with a focus on proteolysis, additives, and nonenzymatic browning.

Proteolysis

This section covers proteolytic changes occurring in foods that have low but discernible enzymatic activity. Postharvest physiological changes, which is covered in Chapter 2, and proteolysis due to spoilage microorganisms will not be discussed. The main areas of interest are meat tenderization associated with aging and protein changes associated with storage in some dairy products.

Aging of meat

Lean meat is about 22% protein by weight, most of which is myofibrillar protein and can be extracted with 0.3 M potassium chloride. These myofibrilar 1 proteins are responsible for the development of rigor mortis. The other protein fractions are the sarcoplasmic, or water soluble, proteins and the connective tissue, or insoluble, proteins.

The principal proteins in the myofibrils are myosin, actin, troponin, and tropomyosin. Myofibrils contract when the level of Ca^{2+} increases in the muscle cell, with the energy for the contraction coming from the hydrolysis of adenosine triphosphate (ATP). ATP also provides the energy to pump the Ca^{2+} back into the sarcoplasmic reticulum to allow the muscle to relax. In the living organism, the ATP levels are replenished by respiration in the mitochondria of the muscle cell with sugar, glycogen, and fat being oxidized to water and carbon dioxide. With death of the animal, respiration ceases. However, up to 1% of the muscle cell is glycogen, which can provide energy anaerobically by glycolysis to produce ATP. During glycolysis, the glycogen is converted to lactic acid, which can ultimately reach a concentration in the muscle of 0.1 M, or 9 g/kg. The lactic acid causes the pH to fall from near 7.0 to a value between 5.3 and 5.8, depending upon the species, the type of muscle, and the slaughter conditions. Typically, the time required is 6–12 h for pork and 18–40 h for beef.[1]

When the glycogen is exhausted, the ATP concentration falls and rigor mortis occurs, during which the thin and thick filaments of the myofibrils strongly interact to form actomyosin. At this stage, the meat is firm, stiff, and tough. Since the pH is near the isoelectric point of the myofibrillar proteins, protein–protein interactions are strong, with water being excluded from the fibrils. This decrease in water holding capacity causes the meat to drip.

The meat will tenderize with aging, which is presumably due to the action of lysosomal and other proteases.[2] It has also been suggested that catheptic enzymes in muscle tissue phagocytic cells contribute to postmortem tenderization.[3] The time required for aging depends upon the storage temperature, pH, and species. It varies from 1–2 days for poultry to 10–20 days for beef. During aging, protein hydrolysis is occurring, as is evidenced by the appearance of free amino acids.[4] Since no free hydroxyproline is formed, collagen breakdown is not part of the tenderization process.[5] Proteolysis alone may not account for all of the changes occurring during aging. Levels of both salt-extractable protein and water binding increase, which may be related to dissociation of the actomyosin complex.[6] Poultry tenderization normally occurs with 24 h and aging is more effective in ice water than at room temperature, which also implies a more complicated mechanism than simple proteolysis.[7]

Until breakage of the lysosomes occurs, any proteolytic tenderization must be due to free proteases within the muscle cell. Since little degradation of myosin or actin occurs during aging, the substrate specificity of the proteases responsible for tenderization must reflect the actual spectrum of proteins degraded.[8] A possible candidate is the Ca^{2+}-dependent protease, CAF, which is unique in being unable to degrade myosin and actin.

Tenderization of meat can be enhanced by the use of exogenous enzymes such as papain, ficin, and bromelain. This enhancement can be achieved by injecting the enzyme(s) into the live animal prior to slaughter (the Swift process[9]) or topical application of the enzyme(s) as a solution or powder. Care must be taken to avoid excess localized proteolysis, which will produce mushy tissue. Unlike the meat aging process, the exogenous enzymes attack all the muscle proteins including connective tissue.[10–12] The shelf life of the meat may also be shortened because of continued proteolytic activity. Microbial growth will also be enhanced because of the greater availability of low molecular weight substrates.

Meat that is frozen and then thawed will normally deteriorate rapidly even at refrigeration temperatures. This deterioration may be due to the breakage of any intact lysosomes, thereby releasing a wide spectrum of active proteases. Since membranes throughout the tissue will be broken, other compartmentalized proteases may also be released. They will cause the meat to become mushy because of extensive proteolysis, and the resultant low molecular weight substrates will facilitate microbiological growth.

Fish quality deteriorates after rigor mortis, with proteolytic activity being a causal factor. If ground herring meat is treated with a potato extract, which contains protease inhibitors for a broad spectrum of proteases, the production of histamine, putrescine, tyramine, and cadaverine is reduced, as is total volatile nitrogen.[13] The primary source of these compounds is from arginine, tyrosine, and lysine, which are produced by proteolytic action. The bacterial count is also reduced, presumably due to a decrease in the availability of substrates for growth. Proteolytic deterioration of fish meat continues even in the frozen state if the storage temperature is not low enough, with some proteins having a half-life of as little as 5 days.[14]

Dairy products

Cheese manufacture begins with the use of proteolytic enzymes such as rennet to cause milk to coagulate to produce cheese curd. Coagulation is the result of proteolysis of the milk casein. "Casein" is a generic term for a family of related proteins. In milk, the caseins form aggregates of 0.02 to 0.3 μm in diameter called micelles, which are stabilized by β-casein.[15] Rennet causes fragmentation of the β-casein, which leads to destabilization of the micelles. In the presence of Ca^{2+}, the caseins aggregate to form the curd. The cheese curd is then treated in various ways appropriate to the variety of cheese being produced. During aging, continued proteolysis contributes to flavor and texture development. For some cheese types residual protease activity can have an adverse effect on quality. For example, the tensile strength of mozzarella cheese decreases logarithmically with storage time due to protease activity.[16]

UHT-processed milk will eventually gel upon storage and develop off-flavors even though there is no microbial growth. One of the proteases responsible for is plasmin, which probably enters the milk from blood in the form of its precursor, plasminogen.[17] In fresh milk, most of the enzyme is present as the precursor. Increased plasmin activity is observed after UHT treatment, and plasminogen levels decrease while plasmin activity increases upon storage.[18]

Interaction with additives

Polyphosphates

Polyphosphate salts are polyanions that can participate in polyelectrolyte interactions with proteins and gums in food systems. They are good chelators of metal ions[19,20] and can also increase the amount of water binding in a food product.[21] The polyphosphates tend to hydrolyze in food systems, the rate of hydrolysis being dependent upon the pH, temperature, the presence of phosphatases, and the types of metal ions bound.

The addition of sodium hexametaphosphate, $(NaPO_3)_n$, to milk converts it to a translucent, yellow-green fluid. The effect is not a consequence of calcium ion chelation, since ethylene diamine tetraacetic acid (EDTA) does not produce the same result. It appears that a large stable complex is formed between casein and polyphosphate.[22,23] If an excess of the polyphosphate is added to the milk, the casein micelles become completely dispersed and there is an accompanying loss of opacity. Sodium hexametaphosphate also stabilizes milk against gelation[24]; since the hexametaphosphate rapidly hydrolyzes to pyrophosphate, the pyrophosphate may be the active agent that causes protein unfolding and dispersion of the casein. The polyphosphate treated milk may also be more digestible.[25,26]

Polyphosphates can also be used as cheese emulsifying salts.[27] They form complexes with the calcium ions and the cheese proteins to stabilize the protein on the surface of the fat globules to prevent coalescence. The bound phosphate also increases the water-binding capacity of the protein.

Polyphosphates can have dramatic effects on the tenderness and flavor of meat.[28-33] They cause actomyosin to dissociate to actin and myosin and increase the water-binding capacity of the proteins. As is the case in milk stabilization by polyphosphates, hydrolysis to pyrophosphate may be necessary. Calcium ion chelation is probably not the cause of the effects of polyphosphates on meats,[34-37] but it is more likely the result of pH alteration and associated protein unfolding and of protein–protein dissociation due to phosphate binding.

Reducing agents

The most common reducing agents in food systems are ascorbic acid (vitamin C) and erythorbic acid, which is an isomer of ascorbic acid. A major use of these is as color preservatives in cured meats (see Nitrites below). Ascorbic acid is also used as a dough conditioner in the baking industry where the primary effects are oxidative,[38] catalytically resulting in disulfide bond formation in the gluten proteins. The overall mechanism appears to be the following[39]:

$$\text{ascorbic acid} + 1/2\, O_2 \quad \rightarrow \quad \text{dehydroascorbic acid}$$

$$\text{dehydroascorbic acid} + 2\,RSH \quad \rightarrow \quad \text{ascorbic acid} + R \pm S \pm S \pm R$$

Ascorbic acid or dehydroascorbic acid can also react with the amino groups of proteins and lead to protein crosslinking.[40-42]

Sulfites

Sulfites are widely used as antimicrobials and to preserve color in a variety of food systems. In the U.S., sulfites cannot be used in meats, although its use is allowed in some foreign countries. The U.S. prohibition is a result of the effect of sulfites on hemoglobin and myoglobin, in which an S-heme complex is formed that has the same bright red color as oxyhemoglobin and oxymyoglobin. Consequently, sulfite-treated meat would have the appearance of fresh meat regardless of its age, and so discolored meat could be adulterated to appear to be fresh. Sulfites also destroy any thiamine in the product. Most of the chemistry of the sulfites depends upon the multiple equilibria among the various forms of sulfite and the nucleophilic properties of the bisulfite ion:

$$SO_2 + H_2O \rightleftharpoons H_2SO_3$$

$$H_2SO_3 \rightleftharpoons H^+ + HSO_3^- \qquad pK = 1.86^{43}$$

$$HSO_3^- \rightleftharpoons H^+ + SO_3^{2-} \qquad pK = 7.18^{44}$$

$$2HSO_3^- \rightleftharpoons S_2O_5^{2-} + H_2O \qquad K = 0.076^{45}$$

With the exception of the reactivity of SO_2 toward heme proteins, the reactive species is bisulfite (HSO_3^-), which is a potent nucleophile. Bisulfite reacts reversibly with aldehydes and some ketones to form hydroxysulfonates and irreversibly with α, β-unsaturated carbonyl compounds. This is the mechanism by which Maillard browning is inhibited. The primary effect on proteins is to react with disulfide bonds[46]:

$$R \pm S \pm S - R + HSO_3^\pm \rightarrow R - S - SO_3^\pm + HS \pm R$$

Nitrites

The main uses of nitrites in food systems are as antimicrobials and for curing of processed meats. Of interest here is the chemistry of the curing process and any other chemistry that involves food proteins. The nitrite ion is relatively stable and unreactive, but nitrous acid is highly reactive and unstable:

$$HONO \rightarrow NO_2^\pm + H^+ \qquad pK = 3.23^{47}$$

Nitrous acid can react with itself and halides to give powerful nitrosating agents[48,49]:

$$2HONO \rightarrow N_2O_3 + H_2O$$

$$2HONO + HX \rightarrow NOX + H_2O$$

These nitrosating agents can react at a number of sites in meat tissue, and can decompose to nitric oxide, nitrogen dioxide, and nitric acid. In a typical cured meat, the nitrite nitrogen ends up distributed as follows: myoglobin (5–15%), sulfhydryl groups (5–15%), lipid (1–5%), protein (20–30%), nitrate (1–10%), nitrite (5–20%), and gaseous products (1–5%).[50]

The characteristic color of cured meats comes from reactions between myoglobin and nitrite.[51-54] Myoglobin (red) is oxidized with assistance from nitrite to metmyoglobin (brown). Metmyoglobin forms a complex with nitrite that is then reduced to nitric oxide myoglobin (red). Ascorbate and erythorbate accelerate this step. Heating denatures the nitric oxide myoglobin to form the stable pink pigment nitrosylhemochrome. The pigment contains two nitric oxide ligands.[55]

As pointed out above in the case of sulfite reaction with hemoglobin and myoglobin, excessive use of nitrite could lead to fresh-looking but microbiologically unsafe meat products. Recently, higher total plate counts were found in intensely red-colored Chinese-style sausage sold in traditional markets and containing higher residual nitrite levels than in similar but less intensely colored products sold in supermarkets and containing less nitrite.[56]

Nitrosation of the protein itself also occurs.[57-59] After enzymatic hydrolysis the following amino acids have been identified: 6-hydroxynorleucine, 2-nitrotyrosine, 3,4-dihydroxyphenylalanine, and S-nitrosocysteine. The 6-hydroxynorleucine came from nitrosation of lysine, and the two aromatic amino acids resulted from nitrosation of tyrosine. Up to 25% of the nitrite added to cured meats can end up as S-nitrosocysteine.

N-nitroso derivatives of secondary amines are carcinogenic, as shown in animal feeding studies.[60] The levels of nitrite that are approved for curing meats produce no more than trace levels, which probably represent no health risk. Furthermore, ascorbate or erythorbate, which are usually included in curing formulas, inhibits the formation of N-nitroso compounds. N-nitrosation of the peptide bond occurs, but the products are unstable and would not persist in a cured meat product.[61]

Hydrolyzed vegetable proteins

Vegetable proteins are hydrolyzed in 10–20% HCl at atmospheric or elevated pressures and then neutralized with NaOH to pH 6.[62] After filtration and charcoal treatment, the hydrolyzed vegetable protein (HVP) is sold as is or spray-dried. HVP is high in sodium chloride and glutamate, and several amino acids are modified or destroyed during hydrolysis, namely, tryptophan, serine, threonine, cysteine, cystine, and methionine. HVP imparts meat-like flavor to products, and is sometimes used to enhance flavor in cured meats and sausages.[63] Because of the high ionic strength, buffering capacity, and pH of HVP, it will increase water binding capacity of the meat and dissociate protein–protein aggregates, increasing tenderness and storage stability.

Nonenzymatic browning

Significance

In this section the physicochemical changes taking place in proteins during long-term storage of foods will be discussed. Chemical changes in proteins often lead to changes in color and texture. The changes can also lead to decreased digestibility and the loss of reactive amino acids such as lysine. Therefore, one key question is whether the deteriorative change in the sensory attribute or the decreased nutritive value of the foods is the limiting factor in the shelf-life of the protein-rich foods.

Common physical and chemical changes in proteins during cooking and storage have been extensively reviewed.[64-69] Initially, denaturation without chemical changes takes place upon heating. For example, myofibrillar proteins in meat upon mild heating become unextractable with salt solution, but they can be extracted with guanidine-HCl and show the same protein profile by sodium docecylsulfate–polyacrylamide gel electrophoresis (SDS-PAGE) as the unheated meat. Chemical changes in food proteins take place upon further heating. Chemical changes under alkaline conditions are irrelevant during thermal processing or storage, because most foods are at neutral or slightly acidic pH. The extent of racemization and similar reactions would be relatively small under typical processing and storage conditions.

One of the most common reactions in foods is the nonenzymatic browning reaction, the Maillard reaction. It can involve reducing sugars and the amino groups on the proteins. Consequent changes in color and texture of the proteins can occur under abusive storage conditions, because its Q_{10}, the increase in the rate constant as temperature is raised by 10°C, is 3–4. It can also lead to decrease in nutritive values with respect to reduced digestibility and the loss of lysine, due to the crosslinking of the proteins as demonstrated in freeze-dried meat.[70] The relative significance of these deteriorative changes during

Table 1 The Effect of Temperature on Browning, Texture Score, and Extractable Protein of Chicken-a-la-King after 3-Year Storage

Storage Temperature (°C)	Browning[a] (A_{420})	Texture[b] Score	Extractable Protein[c] (mg/mg dry sample)	
			w/ DTT	w/o DTT
Control[d]	0.11	6.53	0.91	0.91
4	0.47	5.45	0.36	0.21
21	0.68	5.18	0.27	0.14
30	1.02	4.36	0.14	0.09

a Portions (50 g) of pulverized freeze-dried chicken meat were digested with 2% pronase in 2 ml 0.1 *M* ammonium bicarbonate buffer (pH 8.0) overnight at 37°C. The digest was centrifuged and the lipid in the supernatant was extracted with 1 ml chloroform. Absorbance of the clear aqueous phase was determined at 420 nm. Average of duplicate determinations.

b From M. Klicka, Ration Design and Evaluation Branch, Food Engineering Directorate, U.S. Army Natick RD&E Center. A 1–9 hedonic scale was used.

c Portions (10 mg) of freeze-dried chicken meat were stirred with 5 ml 8 *M* guanidine-HCl solution with or without 20 m*M* DTT. The extract was filtered with nitrocellulose membrane filter (Schleicher & Schuell) and the amount of protein in the filtrate was determined by a dye-binding method with boving serum albumin as a standard.[71] Average of duplicate determinations.

d Without storage.

storage has not been thoroughly investigated in the past. As will be demonstrated here, using the chicken-a-la-king as representative, consumer acceptance based on sensory attributes such as color and texture determines the shelf-life of protein-rich foods. It will also be demonstrated that the decrease in nutritive value of the proteins is not significant even when the sensory value of the protein food decreased significantly and the product became unacceptable.

Case study

An earlier version of chicken-a-la-king from the military Meal, Ready-to-Eat (MRE) showed substantial browning and decreased texture score after a 3-year storage at either 21°C or 30°C. The color measurements, texture score, and the amount of protein extractable with 8 *M* guanidine-HCl solution with or without added dithiothreitol (DTT), a reducing agent, are summarized in Table 1. After 3 years at 30°C the chicken meat became significantly dark to the eye, which was also reflected in the increase in absorbance at 420 nm of the protein digest. A significant drop in the texture score by the technical panel, from 6.53 to 4.36, accompanied the browning reaction. In the 1–9 hedonic scale, a score below 5 is considered unacceptable. Toughness and dryness of the meat were noted for samples stored for 3 years at either 21°C or 30°C. A significant decrease in extractable protein was also observed with or without DTT. The decreased extractability with guanidine-HCl reflects covalent crosslinking of protein molecules as observed before in freeze-dried meat.[70] The difference in the extractability with and without DTT represents the amount of proteins that became unextractable by disulfide crosslinking. The low extractability of proteins in the samples stored at 4°C probably reflects crosslinking during the retort processing of the MRE pouch. It appears that further crosslinking, color, and textural changes take place during storage at a high temperature. As expected, SDS-PAGE revealed little protein extractable with 6 *M* urea–2% SDS–10 mM DTT from chicken meat stored at all three temperatures.

Since Maillard-type browning has been shown to lead to protein crosslinking with accompanying textural changes in freeze-dried meat, we suspected that a similar reaction

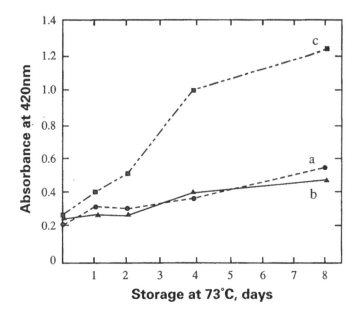

Figure 1 Comparison of browning of chicken meat heated with nonfat dry milk (c) and lactose-free nonfat milk (b). Control was unheated (a).

involving lactose from nonfat dry milk in the chicken-a-la-king might be responsible for the reaction. To demonstrate the involvement of lactose, the following experiment was performed. In a beaker containing 40 g precooked chicken meat, 159 ml water, and 0.93 g table salt, 4.2 g nonfat dry milk (Carnation Co., 53% carbohydrate by weight) or 4.2 g lactose-free nonfat milk (prepared by dialysis of the nonfat dry milk and containing 0.24% of original lactose as measured by the phenol-sulfuric acid method) was added and thoroughly mixed. The mixture was divided into six aliquots and one was analyzed without heating. The remaining five aliquots were heated in a sealed can for 7 min at 120°C and further heated at 73°C for 0, 1, 2, 4, and 8 days.

Both control samples, treated with water only or with lactose-free nonfat milk, showed a slight increase in browning. A significant increase in browning was observed from chicken meat treated with nonfat dry milk containing lactose (Figure 1). In a separate experiment, it was shown that addition of nonreducing sucrose does not promote browning as the addition of lactose, a reducing sugar. These results clearly indicate that lactose is responsible for the browning of the chicken meat through the Maillard-type reaction. Consequently, lactose was removed from the MRE chicken-a-la-king.

Protein bioavailability

The covalent crosslinking of proteins in the meat is expected to reduce the bioavailability of the proteins due to the direct involvement of the lysine residues as well as the reduced digestibility by hydrolytic enzymes. In order to study the effect of protein crosslinking on the bioavailability, we determined, by gel filtration chromatography, the molecular weight distribution of the pronase digest of the chicken meat. Pronase is a nonspecific enzyme and hydrolyzes all peptide bonds except the ones near the crosslinking sites. One can assume that the amount of amino acids released by pronase is a good approximation of it released by a combination of proteolytic enzymes in the body. Rayner and Fox reported a good correlation between *in vitro* available lysine using pronase digestion and "Silcock available" lysine.[72]

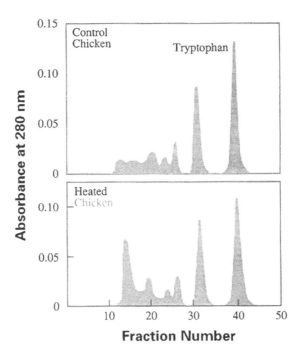

Figure 2 Separation on a Bio-Gel P-2 column of pronase digest from control chicken (upper frame) and from chicken heated with 0.04 *M* lactose at 73°C for 8 days. Brown color associated with the limit peptide was observed in the void volume from the heated chicken, and a 12% decrease in tryptophan content was found.

Since tryptophan has a high extinction coefficient in the UV region around 280 nm, it is useful for monitoring the extent of digestion, as demonstrated by Kim and Haering.[73] Figure 2 shows the separation, using a Bio-Gel P-2 (Bio-Rad, Richmond, CA) column (1.5 × 46 cm, 18 ml/h flow rate, 2.3 ml/fraction), of amino acids and peptides released from the control and browned chicken meat upon pronase digestion (see Table 1 for the digestion condition). For both the control chicken meat (heated at 73°C for 8 days in water) and the browned chicken meat (heated the same way in the presence of 0.04 *M* lactose), two major peaks associated with absorption at 280 nm were observed in the late fractions (around fraction 31 and 39).

The compound corresponding to the large peak around fraction 39 was identified as tryptophan by several methods. First, its UV absorption spectrum was identical to that of authentic tryptophan. Second, authentic tryptophan eluted at fraction 39 from the same gel filtration column. Third, when the material in fraction 39 and authentic tryptophan were dansylated and separated by high-performance liquid chromatography (HPLC), identical chromatograms were obtained. The identity of the other UV-absorbing compound around fraction 31 was not determined. A dipeptide, glycyl-L-tryptophan, was found to elute around fraction 23. Therefore, it is unlikely that the material in fraction 31 is a tryptophan-containing peptide.

The chromatogram in Figure 2 for browned chicken meat shows a decrease in tryptophan (fraction 39) relative to the control chicken. The amount of material in fraction 31 did not change upon heating. Therefore, it could be used as an internal standard. The amount of tryptophan released from the browned chicken meat was approximately 12% less than that from the control chicken. Clearly lactose facilitated crosslinking of the meat proteins. Some tryptophan is either involved directly in the

Table 2   The Effect of Lactose Concentration on the Loss of Reactive Lysine
in Chicken Meat Heated at 73°C for 8 Days

Lactose Conc. (M)	Storage Temp. (°C)	Total Basic (mmol/g)	His + Arg (mmol/g)	Reactive Lys (mmol/g)	%
	4	1.362 (0.017)	0.673 (0.006)	0.689 (0.018)	100
0.04					
	73	1.269 (0.029)	0.670 (0.010)	0.599 (0.031)	86.9
	4	1.367 (0.012)	0.679 (0.018)	0.688 (0.022)	100
0.01					
	73	1.335 (0.013)	0.673 (0.026)	0.662 (0.029)	96.2

Note: Average of four measurements; standard deviation in parentheses.

crosslinking or in close proximity to the crosslinking site and is not susceptible to release by the nonspecific pronase.

From the browned chicken meat, a substantial increase in absorption at both 280 and 420 nm was observed near the void volume of the column. Brown color in the pronase digest observed by the eye was visible only in these early fractions after separation by size exclusion. Clearly these fractions correspond to the limit peptide pigment derived from protein crosslinked through condensed sugar moieties described by Clark and Tannenbaum.[74] The decrease in free tryptophan measured by absorbance at 280 nm was roughly accounted for by the increase of the limit peptide pigment in the void volume.

Reactive lysine loss

Friedman reviewed various aspects of the nutritional value of proteins.[75] Availability of lysine is an important index of protein quality. The extent of the involvement of lysine in the crosslinking can be assessed by the dye-binding method.[76] In this method the dye-binding capacity of basic groups (histidine, arginine, and lysine) in the proteins for acid orange 12 is measured before and after reaction with propionic anhydride to block the lysine groups. The difference in the dye-binding capacity corresponds to the reactive lysine content, which is an estimate of the biologically available lysine. The lysine content in chicken meats heated at 73°C for 8 days in the presence of 0.01 and 0.04 *M* lactose was determined by the dye-binding method. The control samples showed 0.69 mmol reactive lysine per gram of meat (Table 2). When the meat was heated in the presence of 0.01 *M* lactose, approximately 4% loss of lysine was observed. When the lactose concentration was increased to 0.04 *M*, the lysine loss was increased to 13%.

Incidentally, the percent loss of lysine (13%) is about the same as the percent loss of tryptophan (12%) discussed above for the same sample. Due to the three-dimensional nature of the proteins, only a portion of the lysine will be exposed to react with the lactose. An even smaller portion of the lysine residues will be involved in crosslinking with the condensed sugars as a bridge. The probability is small of the tryptophan residue being in the immediate vicinity of the lysine residue modified by lactose and not released by pronase. However, the crosslinking will decrease the digestibility of the proteins by pronase, as reflected by the 12% decrease in released tryptophan. Using the loss of either the lysine or the tryptophan, one could estimate that the nutritive value of the proteins in the chicken meat heated at 73°C for 8 days with a high concentration of reducing sugar is about 13%. If Q_{10} of 4 is assumed for the browning reaction,[77] heating for 8 days at 73°C corresponds to a storage for 22 years at the ambient temperature (23°C), which exceeds the military shelf-life requirement of 3 years at the ambient temperature. As a matter of fact, the browning of the meat in the 30°C, 3-year chicken-a-la-king sample in Table 1 is about the same as that from the chicken meat heated with nonfat dry milk at 73°C for 4 days (Figure 1). Therefore, one can conclude that the loss of the protein nutritive value under typical storage conditions would be much less than 10%.

There is ample data in the literature on the loss of protein nutritive value in dry or low-moisture systems,[69,73–75] but not much in high-moisture systems. It is well known that the Maillard reaction is slower at a high moisture content than at an intermediate moisture content. From an equilibrium consideration, it is not favored at high moisture, because the advanced Maillard reaction, as well as the early formation of the Schiff base, involves removal of water.[79] For example, when skim milk powder (adjusted to 15% moisture content) was stored for 32 days at 37°C, 33% loss of total lysine was observed after acid hydrolysis.[73] When UHT milk was stored at the same temperature, it took 3 years to reach the same level of lysine loss (approximately 31%).[80] The results described above suggest that the loss of the protein nutritive value in chicken meat is less than 10% during a long-term storage under nonabusive temperature conditions. The slower reaction in the meat compared to milk is expected, because the lactose concentration inside the muscle will be lower than that in the milk.

Sensory attributes

A significant deterioration in sensory attributes takes place under conditions similar to those in which browning and nutritive loss occur. Table 1 shows that the texture score drops from 6.5 to about 5 after 3 years at ambient temperature. The technical panel score on appearance for the same MRE chicken-a-la-king samples was 5.6, 5.0, and 3.19 after 3 years at 4°C, 21°C, and 29°C, respectively.[81] Clearly, the foods became unacceptable (below 5 in texture and appearance scores). When the nonfat dry milk was replaced by powdered vegetable shortening in the new MRE chicken-a-la-king, the browning problem was eliminated and the product became more acceptable.

Though a significant decrease in the sensory quality of protein foods can occur during storgage with minimal change in the nutritive value, it can be avoided by a proper selection of ingredients. Conversely, the nutritional value of the proteins should not be a significant problem as long as the food is acceptable to the consumer from a sensory point of view.

Conclusions

Some protein instability, such as proteolysis, is biochemical in nature; others, such as interaction with additives and nonenzymatic browning, are strictly chemical. Some are advantageous to the food; others are deleterious. Deleterious reactions, even though they may not affect the nutritional quality of the proteins significantly, often compromise the sensory quality of the foods and can be minimized by controlling the physicochemical environment of the proteins.

References

1. Honikel, K.O., Kim, C. J., *Fleischwirtschaft*, 1985, 65, 1125.
2. Lawrie, R.A., *Meat Science*; Pergammon: New York, 1985; 3rd ed., p 348.
3. Tappel, A.L., *The Physiology and Biochemistry of Muscle as a Food*; Briskey, E.J., Cassens, R.G., Trantman, J.C., Eds.; Univ. Wisconsin: Madison, 1966; p 237.

4. Reed, C., *Enzymes in Food Processing*; Academic: New York, 1966; pp 348-366.
5. Dahl, O., *Nahrung*, 1962, 6, 492.
6. Wierbicki, E., Kunkle, L.E., Cahill, V.R., Deatherage, F.E., *Food Technol.*, 1954, 8, 506.
7. Jansen, E.F., *The Quality and Stability of Frozen Foods*; Van Arsdel, W.B., Copley, M.J., Olson, R.L., Eds.; Wiley-Interscience; New York 1969, pp. 28-29.
8. Goll, D.E., Otsuka, Y., Nagainis, P.A., Shannon, J.D., Sathe, S.K., Muguruma, M., *J. Food Biochem.* 1983, 7, 137.
9. Beuk, J.F., Savich, A.L., Goeser, P.A., Hogan, J.M., U.S.Patent 2 903 362, 1959.
10. Hinricks, J.R., Whitaker, J.R., *J. Food Sci.*, 1962, 27, 250.
11. Yatco-Manzo, E., Whitaker, J.R., *Arch Biochem. Biophys.*, 1962, 97, 122.
12. El-Gharbawi, M., Whitaker, J.R., *J. Food Sci.*, 1963, 28, 168.
13. Aksnes, A., *Norway J. Sci. Food Agr.* 1989, 49, 225.
14. French, J.S., Kramer, D.E., Kennish, J.M., *J. Food Sci.*, 1988, 53, 1014.
15. Walstra, P., Jenness, R., *Dairy Chemistry and Physics*; Wiley: New York, 1984; Chapter 13, pp 229-253.
16. Kikuch, E., Izutsu, T., Kobayashi, H., Kusakabe, I., Murakami, K., *Jpn. J. Zootech. Sci.*, 1988, 59, 388.
17. Walstra, P., Jenness, R., *Dairy Chemistry and Physics*; Wiley: New York, 1984; Chapter 7, p 133.
18. Manji, B.S., Ph.D. Dissertation, University for Guelph, Guelph, Ont., 1987.
19. Van Wazer, J.R., Campanella, D.A., *J. Amer. Chem. Soc.*, 1950, 72, 655.
20. Van Wazer, J.R., Callis, C.F., *Chem. Rev.*, 1958, 58, 1011.
21. Kutscher, W., *Dtsch. Lebensm.-Rundsch.*, 1961, 57, 140.
22. Odagiri, S., Nickerson, T.A., *J. Dairy Sci.*, 1964, 47, 1306.
23. Odagiri, S., Nickerson, T.A., *J. Dairy Sci.* 1965, 48, 19.
24. Leviton, A., *J. Dairy Sci.*, 1964, 47, 670.
25. Hall, R.E., U.S. Patent 2 064 110, 1936.
26. Schwartz, C., U.S. Patent 2 135 054, 1938.
27. Roesler, H., *Milchwissenschaft*, 1966, 21, 104.
28. Bendall, J.R., *J. Sci. Food Agr.*, 1954, 5, 468.
29. Fukazawa, T., Hashimoto, Y., Yasui, T., *J. Food Sci.*, 1961, 26, 541.
30. Fukazawa, T., Hashimoto, Y., Yasui, T., *J. Food Sci.*,1961, 26, 550.
31. Yasui, T., Sakamishi, M., Hashimoto, Y., *J. Agr. Food Chem.*, 1964, 12, 392.
32. Yasui, T., Fukazawa, T., Takahashi, K., Sakamish, M., Hashimoto, Y., *J. Agr. Food Chem.*, 1964, 12, 399.
33. Klopacka, M., *Przem. Spozywczy*, 1965, 19, 501.
34. Kotter, L., *Zur Wirkung Kondensierter Phosphate und Anderer Salze auf Tierisches Eiweiss*; M. & H. Schaper, 1960.
35. Kotter, L., *Kondensierte Phosphate in Lebensmitteln*; Springer-Verlag: New York, 1958, p 99.
36. Sherman, P., *Food Technol.*, 1961, 15, 79.
37. Swift, C.E., Ellis, M.G., *Food Technol.*, 1956, 10, 546.
38. Matz, S.A., *Bakery Technology and Engineering*; AVI, Westport, CT, 1972; 2nd ed.
39. Tsen, C.C., *Cereal Chem.*, 1965, 42, 86.
40. Bensch, K.G., Fleming, T.E., Lohmann, W., *Proc. Nat. Acad. Sci. U.S.A.*, 1985, 82, 7193.
41. Ortwerth, B.J., Olesen, P.R., *Biochim. Biophys. Acta*, 1988, 956, 10.
42. Larisch, B., Pischetsrieder, M., Severin, T., *J. Agric. Food Chem.*, 1996, 44, 1630.
43. Huss, A., Eckert, C.A., *J. Phys. Chem.*, 1977, 81, 2268.
44. Smith, R.M., Martell, A.E., *Critical Stability Constants*; Plenum: New York, 1976; Vol. 4.
45. Bourne, D.W.A., Higuchi, T., Pitman, I.H., *J. Pharm. Sci.*, 1974, 63, 865.
46. Wedzicha, B.L., *Interactions of Food Components*; Birch, G.G., Lindley, M.G., Eds., Elsevier, New York, 1986, Chapter 7.
47. Streitwieser, A., Heathcock, C.H., *Introduction to Organic Chemistry*; MacMillan: New York, 1976; p 977.
48. Ridd, J.H., *Q. Rev. Chem. Soc.*, 1961, 15, 418.
49. Challis, B.C., *Safety Evaluation of Nitrosatable Drugs and Chemicals*; Gibson, G.G.; Ioannides, C., Eds; Taylor and Francis: London, 1981; pp 16-55.

50. Cassens, R.G., Woolford, G., Lee, S.H., Goutefongeal, R., *Proc. 2nd Int. Symp. on Nitrite in Meat Products;* Tinbergen B.J., Krol, B., Eds., PUDOC, Wageningen, 1977, pp 95-100.
51. Fox, J.B., *J. Agric. Food Chem.,* 1966, 14, 207.
52. Govindarajan, S., *Crit. Rev. Food. Technol.,* 1973, 4, 117.
53. Bard, J., Townsend, W.E., *The Science of Meat and Meat Products;* Price, F.J., Schweigert, B.S., Eds., Freeman, San Francisco, 1971, pp 452-83.
54. Fox, J.B., Ackerman, S.A., *J. Food Sci.,* 1968, 33, 364.
55. Lee, S.H., Cassens, R.G., *J. Food Sci.,* 1976, 41, 969.
56. Chou, C.-K., Lin, K.-J., *J. Chinese Soc. Animal Sci.,* 1995, 24(1), 87.
57. Knowles, M.E., Gilbert, J., McWeeny, D.J., *Proc. 4th Int. Congr. Food Sci. Technol.,* Selegraf, Valencia, 1974; pp 314-19.
58. Olsman, W.J., *Proc. 2nd Int. Symp. on Nitrite in Meat Products;* Tinbergen, B.J., Krol, B., Eds., Taylor and Francis, London, 1977, pp 101-9.
59. Olsman, W.J., van Leenwen, C.M., *Proc., 4th Int. Congr. Food Sci. Technol.;* Selegraf, Valencia, 1974, pp 314-19.
60. Magee, P.M., Barnes, J.M., *Adv. Cancer Res.,* 1967, 10, 163.
61. Challis, B.C., Hopkins, A.R., Milligan, J.R., Mitchell, R.C., Massey, R.C., *Proc 8th Int. Symp. on N-nitroso Compds;* O'Neill, I.K., Von Borstel, R.C., Miller, C.T., Long, J., Bartsch, H., Eds., International Agency for Research on Cancer: Lyon, 1984, pp 64-70.
62. Pomeranz, Y. *Functional Properties of Food Components;* Academic, London, 1985, pp 179-183.
63. Kramlich, W.E., Pearson, A.M., Tauber, F.W., *Processed Meats;* AVI, Westport, CT, 1973, p 46.
64. T.R. Dutson, M.W. Orcutt, J. Chem. Ed., 1984, 61, 303.
65. R.E. Feeney et al., "Chemical Reactions of Proteins" in Chemical Changes in Food During Processing; Richardson and Finley, eds., AVI, Westport, CT, 1985.
66. J.W. Finley, "Environmental Effects on Protein Quality" in Chemical Changes in Food During Processing; Richardson and Finley, eds., AVI, Westport, CT, 1985.
67. J. Davidek, J. Velisek, J. Pokorny, "Proteins, Peptides and Amino Acids" in Chemical Changes During Food Processing; Elsevier, New York, 1990.
68. R.E. Feeney, "Overview on the Chemical Deteriorative Changes of Proteins and Their Consequences" in Chemical Deterioration of Proteins; Whitaker and Fujimaki, eds., ACS Symposium Series 123, American Chemical Society, 1980.
69. J.E. Kinsella, "Relationships between Structure and Functional Properties of Food Proteins" in Food Proteins, Fox and Condon, eds., Appl. Sci. Publ., New York, 1982.
70. Kim, H.-J., Loveridge, V.A., Taub, I.A., *J. Food Sci.,* 1984, 49, 699.
71. Bradford, M.M., *Anal. Biochem.,* 1976, 72, 248.
72. Rayner, C.J., Fox, M., *J. Sci. Food. Agric.,* 1976, 27, 643.
73. Kim, H.-J., Haering, C., *J. Agric. Food Chem.,* 1994, 42, 915.
74. Clark, A.V., Tannenbaum, S.R., *J. Agric. Food Chem.,* 1974, 22, 1089.
75. Friedman, M., *J. Agric. Food Chem.,* 1996, 44, 6.
76. Hurrell, R.F., Lerman, P., Carpenter, K.J., *J. Food Sci.,* 1979, 44, 1221.
77. Saltmarch, M., Vagnini-Ferrari, M., Labuza, T.P., "Theoretical Basis and Application of Kinetics to Browning in Spray-Dried Whey Food Systems" in Progress in Food and Nutrition Science, Vol. 5, Maillard Reactions in Food, Pergamon Press, New York, 1981, p. 331.
78. Obanu, Z.A., Biggin, R.J., Neales, R.J., Ledward, D.A., Lawrie, R.A., *J. Food. Technol.,* 1976, 11, 575.
79. Hodge, J.E., Osman, E.M., "Carbohydrates" in Principles of Food Science, Part I, O.R. Fenema, ed., Marcel Dekker, New York, 1976.
80. A.B. Moller, "Chemical Changes in Ultra Heat Treated Milk During Storage" in Progress in Food and Nutrition Science, Vol. 5, Maillard Reactions in Food, Pergamon Press, New York, 1981, p. 331.
81. M. Klicka, personal communication.

chapter four

Biochemical processes: lipid instability

W. W. Nawar

Contents

Introduction

Lipids are important constituents in practically all foods. An understanding of the chemical and physical changes that lipids can undergo as well as of the mechanisms and consequences of such changes is fundamental to any consideration of food quality.

There is a commonality in such changes among all lipids despite their wide diversity in food with regard to their total amount, chemical composition, physical state, etc. Oxidative or hydrolytic decomposition of lipids can lead to serious problems in bulk oil in a storage tank, in the continuous phase of triacylglycerols in butter, in milk fat globules, in the phospholipid bilayer in cell membranes, in tissue lipids in lean fish, in skin fat in chicken, and in frying oil or fried food.

Extensive reviews are available on the chemistry involved in the decomposition of pure lipid substrates, e.g., the autoxidation of unsaturated fatty acids.[1] Such basic information from studies with model systems is essential to our understanding of oxidative mechanisms. In most foods, however, the situation is vastly different from that of the model system. The presence of many other major and minor constituents in food and the reactions and interactions of these can exert significant influences on the chemical and physical behavior of lipids. Furthermore, the medium in which substrate molecules exist, their molecular orientation, and the physical state of both substrate and medium are important factors affecting oxidation rates.

In this chapter, the basic aspects of lipid hydrolysis and of lipid oxidation are considered only briefly, while more emphasis is placed on the changes that occur in food.

Lipolysis

The majority of natural lipids consists of fatty acids attached to glycerol through carboxylic ester bonds. Hydrolysis of the ester bonds, catalyzed by acid, alkali, heat, moisture, or lipolytic enzymes, results in the liberation of free fatty acids.

$$\begin{array}{ccc}
CH_2OCOR_1 & & CH_2OH \\
| & & | \\
CHOCOR_2 + H_2O & \longrightarrow & CHOCOR_2 + R_1COOH \\
| & & | \\
CH_2OCOR_3 & & CH_2OCOR_3
\end{array} \tag{1}$$

Lipolytic enzymes usually act at oil–water interfaces. Most have specificity for positions on the triacylglycerol molecules. For example, pancreatic lipase is specific for certain positions. Such enzymes may be present naturally in food or in constituents mixed with the food; some could be associated with microbial contamination. Temperature, moisture, and pH are among the factors that control lipase activity.

Potential consequences

A review by Shewfelt covers relevant mechanistic aspects of hypolysis in food.[2] Several undesirable changes may occur. Among the most significant are the development of unacceptable off-flavor (hydrolytic rancidity); changes in functional properties; and increased susceptibility to fatty acid oxidation.

In some cases, desirable flavors can be produced by lipolysis, e.g., in the ripening of some cheeses.

Means of protection

In general, several strategies can be followed to protect lipids from hydrolysis; foremost among these is the selection of high-quality raw materials. Inactivating lipolytic enzymes by heat and avoiding storage conditions that promote growth of contaminating microorganisms of high lipolytic activity are also effective stategies.

Illustrative lipolysis

Off-flavors resulting from hydrolytic rancidity (such as soapy taste in milk or coconut oil) are more likely to occur in fats containing relatively short chain fatty acids (i.e., C4-10).

The potential for lipolysis in milk, however, is minimized due to the structure of the milk emulsion, which limits physical contact between the triacylglycerol substrates residing in fat globules and the lipase enzyme in the skim milk.[3] Agitation during processing, storage, or transportation can disrupt the native milk structure and promote enzyme–substrate interaction. Although heat inactivates the lipolytic microorganisms, the lipases produced by them can survive normal pasteurization temperatures.

Among the cereal crops, oats have high levels of lipolytic activity. Increase in free fatty acids usually occurs during growth and postharvest treatment and results in poor-quality oats and oat products.[4] Lipolysis in improperly stored wheat flour results in progressive deterioration of its baking quality.

In frying potato chips, potatoes with their high moisture content are brought in contact with hot oil (~180°C) and consequently suffer significant hydrolysis. Factors that lead to high levels of free fatty acids and poor shelf-life include long oil turnover periods, temperature, contamination from frying tanks, and poor packaging and storage of the finished product.

Lipase activity in spices and seasonings often results in poor quality of food products, particularly among those containing coconut oil.[5,6]

Many typical cheese flavors are produced where short-chain fatty acids are hydrolyzed by natural milk or microbial lipolytic enzymes.

Oxidation

Lipid oxidation is of paramount importance to food quality. It may lead to the development of rancid off-flavors, cause changes in color or texture, reduce shelf life, and/or impair nutritional quality. However, a limited degree of lipid oxidation is sometimes desirable, as in the formation of typical flavors and aromas that are associated with cheeses and fried food.

Autoxidation

Although saturated fatty acids can react with oxygen, the susceptibility of lipids to oxidation is significantly greater with higher unsaturation in the fatty acid chains. Fatty acids containing one or more of the nonconjugated pentadiene systems, $-CH=CH-CH_2-CH=CH-$, are especially sensitive.

In general, lipid oxidation proceeds via a typical self-propagating free radical mechanism where oxygen attack occurs mainly at positions adjacent to the double bonds. The chain reaction is preceded by an initiation step in which free radicals are produced by some catalytic means, (e.g., metal catalysis, exposure to light or high energy radiation, singlet oxygen, or decomposition of added peroxides). The propagation reaction can be summarized as follows:

$$R \cdot + O_2 \longrightarrow ROO \cdot$$
$$ROO \cdot + RH \longrightarrow ROOH + R \cdot$$

(2)

where RH is the substrate fatty acid, H is an α-methylenic hydrogen atom.

Reaction of the R· radical with molecular oxygen yields peroxy radicals that may abstract hydrogen from α-methylenic groups of other substrate molecules, RH, to form hydroperoxides, ROOH, and R· radicals, which in turn combine with oxygen to form

ROO· and so on. Thus absorption of oxygen continues and hydroperoxides, which are the primary intermediates of lipid oxidation, continue to accumulate. Heat or trace metals cause the lipid hydroperoxides to decompose according to a variety of sequential, simultaneous, and/or competitive reaction pathways, giving rise to a myriad of breakdown products including aldehydes, hydrocarbons, ketones, alcohols, cyclic compounds, dimers, and polymers. Of course, the oxidation products themselves can further oxidize and form more breakdown products.

According to the above scheme, linoleates, the major substrate for lipid oxidation in food, undergo hydrogen abstraction at position 11 producing a pentadienyl radical intermediate, which upon reaction with molecular oxygen produces an equal mixture of conjugated 9- and 13-diene hydroperoxides:

$$
\begin{array}{ccc}
13 & & 9 \\[4pt]
\underset{\displaystyle \overset{|}{\underset{\displaystyle O}{\overset{|}{O}}} H}{-C}-C{=}C{-}C{=}C{-} & + & -C{-}C{=}C{-}C{=}\underset{\displaystyle \overset{|}{\underset{\displaystyle O}{\overset{|}{O}}} H}{C}-
\end{array}
$$

$$\tag{3}$$

Scission at the oxygen–oxygen bond of each hydroperoxide group gives rise to alkoxy and hydroxy radicals:

$$
\underset{\displaystyle \overset{|}{\underset{\displaystyle O}{\overset{|}{O}}} H}{R-CH-R'} \longrightarrow \underset{\displaystyle \overset{|}{O\cdot}}{R-CH-R'} \; + \; \cdot OH
$$

$$\tag{4}$$

Cleavage on the carboxyl (or ester) side results in formation of an aldehyde and an acid (or ester); while scission on the hydrocarbon (or methyl) side produces a hydrocarbon and an oxoacid (or oxoester). If, however, a vinylic radical results from such cleavage, an aldehydic functional group is formed. Thus, the 9-hydroperoxide would produce nonenal, methyl-9-oxononaoate, 2,4-decadienal, decenal and methyl octanoate (Figure 1), and so on.

In addition to the homolytic cleavage described above, heterolytic cleavage may occur under certain conditions between the hydroperoxide group and the allylic double bond, giving rise to hexanal, nonanal, methyl-9-oxononanoate, and methyl-12-oxododecenoate.[7,8]

The alkyl, alkoxy, and peroxy radicals of linoleate can also combine to form a variety of dimeric and polymeric products with carbon–oxygen–carbon or carbon–oxygen–oxygen-carbon crosslinks (Figure 2).

These radicals can also undergo intermolecular cyclization. Among the products formed are the cyclic monomers, which have been the subject of concern because of their toxicity. However, it is generally recognized that, while toxic compounds can be generated in fat by abusive heating and/or oxidation, moderate ingestion of foods fried in high-quality oils using recommended practices is unlikely to pose a significant hazard to health.

Singlet oxygen oxidation

In addition to the ground state of molecular oxygen (triplet oxygen, $^3O_2$), an excited state (singlet oxygen, $^1O_2$) plays an important role in lipid oxidation. $^1O_2$ reacts much faster than $^3O_2$ with moieties of high electron density, such as C=C bonds. The formation of hydroperoxides by $^1O_2$ proceeds via mechanisms different from those of free radical

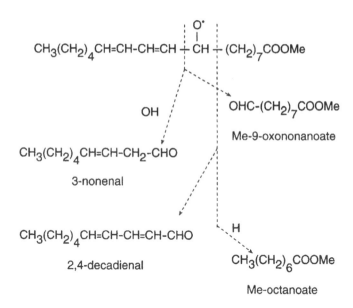

Figure 1 Decomposition of methyl-9-hydroperoxy-10, 12-octadecadienoate.

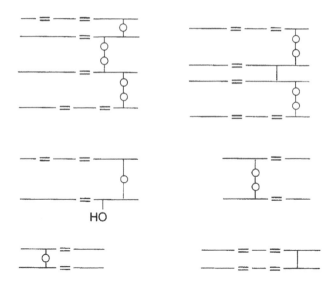

Figure 2 Dimers and polymers of linoleate.

autoxidation. Oxygen is inserted at both ends of the double bond, which then shifts to yield an allylic hydroperoxide in the *trans* configuration. Accordingly, oleate produces the 9- and 10-hydroperoxides, linoleate produces 9-,10-,12-, and 13-hydroperoxides, and lino-lenate produces a mixture of 9-,10-,12-,13-,15-, and 16-hydroperoxides. These hydroper-oxides can then cleave to initiate conventional free radical reactions.

Singlet oxygen can be generated in a variety of ways. In foods, probably the most important is via photosensitization. Several substances commonly found in fat-containing foods can act as photosensitizers to produce $^1O_2$. These include natural pigments, such as chlorophyll-a, pheophytin-a, hematoporphyrin, and myoglobin. The synthetic colorant, erythrosine, also acts as an active photosensitizer. β-Carotene is a most effective $^1O_2$ quencher, while tocopherols are somewhat effective.

Enzymic oxidation

Lipoxygenases are a diverse group of enzymes widely spread throughout both plant and animal kingdoms. They are present in legumes, cereals, fruits, and potato. In animals, they are found in platelets, microsomal membranes, and other specialized tissues. These enzymes act mainly on polyunsaturated fatty acids containing *cis,cis*-1,4-pentadiene systems, and they catalyze the formation of hydroperoxide intermediates similar to those formed by nonenzymic autoxidation. However, lipoxygenases are specific with respect to substrate and to the positional and stereo-isomeric hydroperoxides they produce. The different lipoxygenases present in different plants, or isolipoxygenases in the same plant, vary in their substrate and product specificties, as well as in their optimum parameters (e.g., pH).

It is believed that iron, a constituent of lipoxygenase, plays a key role in the mechanism by which the enzyme catalyzes hydroperoxide formation[9,10]:

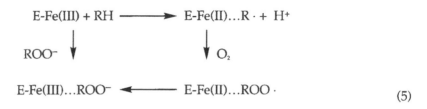

$$\text{E-Fe(III)} + \text{RH} \longrightarrow \text{E-Fe(II)...R} \cdot + \text{H}^+$$

$$\text{ROO}^-$$

$$\text{O}_2$$

$$\text{E-Fe(III)...ROO}^- \longleftarrow \text{E-Fe(II)...ROO} \cdot \tag{5}$$

The breakdown of hydroperoxides leading to the formation of volatile and nonvolatile products may also be catalyzed by enzymes (i.e., hydroperoxidases). Obviously, due to the greater specificity in enzyme-catalyzed formation and decomposition of fatty acid hydroperoxides, fewer and more specific oxidation end products are encountered.

Factors influencing lipid oxidation

In addition to the nature of the substrate, a number of factors influence the rate of oxidation. These include the following:

Free Fatty Acids (FFA). In general, FFA oxidize faster than their glyceryl esters. The presence in commercial oils of relatively large amounts of FFA can enhance the pick up of trace metals from equipment or storage tanks and thereby increase the rate of oxidation.

Oxygen Concentration. At very low oxygen pressure, the rate of oxidation is approximately proportional to oxygen pressure. If the supply of oxygen is unlimited, the rate of oxidation is independent of oxygen pressure. The availability of oxygen clearly plays a critical role in determining competitive oxidative pathways, (e.g., peroxidation vs. homolytic cleavage or polymerization).

Temperature. Rates of reaction increase with increase in temperature. However, as the temperature increases, the increase in rate with increasing oxygen concentration becomes less evident, because oxygen becomes less soluble.

Water Content. In dried food with very low moisture content (aw <0.1), oxidation proceeds very rapidly. Increase in aw to about 0.3 retards lipid oxidation, reportedly by reducing metal catalysis, quenching free radicals, promoting nonenzymic browning, and/or impeding oxygen accessibility. At higher aw (0.55–0.85), the rate of oxidation increases again, presumably due to increased mobilization of the catalysts present.

Physical Condition. The oxidation of an oil occurs at the interface with oxygen. The rate of oxidation increases inproportion to the surface area of the lipid exposed to air. In oil-in-water emulsions, oxygen must gain access to the lipid by diffusion into the aqueous phase and passage through the oil–water interface. The rate of oxidation will depend on the interplay between a number of factors, including type and concentration of emulsifier, size of oil droplets, surface area of interface, viscosity of the aqueous phase, composition and porosity of the aqueous matrix, and pH. Molecular conformation and arrangement of the substrates, and the medium in which they exist, play an important role in lipid oxidation. When the oxidative stability of polyunsaturated fatty acids (PUFA) was studied in an aqueous solution at 37°C with Fe^{2+} and ascorbic acid present as a catalyst, stability increased as the degree of unsaturation increased, contrary to what would be expected. This was attributed to the conformations of the PUFA existing in the aqueous medium.[11] Moreover, the oxidation of ethyl linoleate at 60°C proceeded more rapidly in a monolayer than it did in the bulk phase because of greater access to oxygen afforded by the monolayered linoleate. However, at 180°C, the reverse was true (Figure 3). The molecules of bulk-phase linoleate and associated free radicals were sufficiently mobile to offset the difference in access to oxygen.[8] In tissues and membranes, lipid oxidation is also influenced by molecular arrangement and physical state of the lipid.

Fatty Acid Positions in Triacylglycerols. The relationship between oxidative stability and relative positional distribution of fatty acids in triacylglycerol molecules is controversial. Some investigators observed faster oxidation of some oils after radomization,[12,13] while others found no effect provided antioxidants were removed.[14] Working with synthesized triacylglycerols, Hoffman et al. concluded that oxidative stability is lower if the same fatty acid occupies the 1- and 2-positions, e.g., OOL, than when it occupies the 1- and 3-positions, e.g., OLO.[15]

Pro-oxidants. As discussed above, enzymes can catalytically promote lipid oxidation. Transition metal ions, particularly copper and iron, are also major catalysts for oxidation. At very low concentrations, <0.1 ppm, they can decrease the induction period and increase the rate of oxidation. Such metal ions either in free and bound forms, occur naturally in plant and animal tissues, membranes, and enzymes. They are also introduced into food by contact with metallic equipment used in processing or storage.

Antioxidants. These are substances, occurring naturally or synthesized, that can delay the onset or slow the rate of oxidation. The main antioxidants used in food are mono- or polyhydric phenols with ring substitutions. They either inhibit the formation of free radicals in the initiation step or interrupt the propagation of the free radical chain reaction. For maximum efficiency, primary antioxidants are used in combination with other phenolic antioxidants or with metal-sequestering agents. Excellent reviews are available in the literature on antioxidants and antioxidation.[16–19]

As mentioned above, many of the minor components that occur naturally in food play extremely critical roles in oxidation/antioxidation. Because of the importance of such compounds, they are separately discussed below.

Minor components of major significance

In addition to the prooxidants and antioxidants mentioned above (e.g., trace metals, enzymes, chlorophyll, β-carotene, tocopherol), several other minor components that play important roles in the oxidative/antioxidative balance are widely distributed in oils and fat-containing foods.

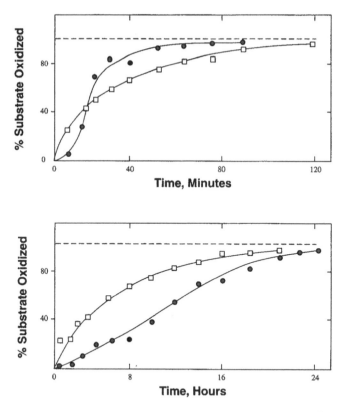

Figure 3   Rate of ethyl linoleate oxidation on silica (□) and in bulk (●) at 60°C (lower frame) and 180°C (upper frame).

In most natural fats and oils, the unsaponifiable fraction ranges from 0.3–2.3% (Table 1) and contains mainly hydrocarbons, sterols, and terpenes. The amounts of total unsaponifiables and their individual components vary among different oils. The major sterols in most oils are β-sitosterol, stigmasterol, and campesterol. Δ^5-Avenasterol and Δ^7-stigmasterol are also found, but in smaller amounts (Table 2). Only trace amounts of brassicasterol can be detected in most oils, except rapeseed oil where it amounts to about 10% of the sterol fraction. Sterols are also important minor components in seafood. While cholesterol is usually the main sterol, brassicasterol, 24-methylene cholesterol, 22-dehydrocholesterol, and other C26 and C29 sterols are commonly encountered.20

Two triterpene alcohols, cycloartenol and 24-methylenecycloartanol, are found in most oils. In addition α- and β-amyrin, euphorbol, buterospermol, and cycloaudenol are occasionally found.[21,22]

The antioxidative properties of various hydrocarbons, triterpene alcohols, sterols, and phenols have been established.[23-25]

Plant tissues, seeds, spices, and the oils extracted from them contain varying amounts of phenolic compounds; they are minor components of the polar fraction. A number of polyphenols and extracts containing the phenolic fraction from natural products have been found to exhibit varying degrees of antioxidant activities. The phenols present in olive oil include tyrosol, hydroxytyrosol, 4-hydroxyphenylacetic acid, syringic acid, 4-hydroxybenzoic acid, vanillic acid, *o*-coumaric acid, *p*-coumaric acid, and 3,4-dihydroxyphenyl acetic acid.

Papadopoulos and Boskou studied the antioxidant activity of these compounds.[25] Although refined olive oil oxidized relatively rapidly, this oil with an added 200 ppm of

Table 1 Terpene and Sterol Fractions from Vegetable Oil Unsaponifiables (% of Oil)

Oil	Neutralized Unsaponifiables (%)	Triterpenes (%)	Sterols (%)
Peanut	0.9	0.14	0.5
Olive	1.2	0.2	0.6
Rice bran	0.8	0.1	0.4
Corn	2.3	0.11	1.38
Coconut	0.3	0.02	0.14
Palm	0.4	0.02	0.17
Rapeseed	0.8	0.03	0.59
Sunflower	0.6	0.22	0.25
Cocoa butter	0.7	0.05	0.3
Soybean	0.7	0.06	0.42

Table 2 The Major Sterols in Some Oils (% of Total Sterols)

Oil	Campesterol (%)	Stigmasterol (%)	Stitosterol (%)
Soybean	20	20	53
Corn	23	6	66
Olive	2	1	91
Peanut	15	9	64
Cottonseed	4	1	93
Safflower	13	9	52
Cocoa butter	9	26	59
Coconut	8	13	58
Palm	14	8	74
Rice bran	28	15	49
Rapeseed	25	Trace	58

a polar fraction of virgin olive oil that contained the phenols showed remarkable stability at 63°C. The tyrosol, which sometimes makes up to 40% of the total phenolic fraction, showed practically no antioxidant activity, but the hydoxytyrosol and caffeic acid were particularly effective (Figure 4). Although the production of virgin olive oil does not involve refining, degumming, or bleaching, a significant amount of the phenols may be lost during the extraction step where the pulp is treated with warm water. The extent of such loss depends on the concentration of phenols in the pulp and the parameters of the extraction process.[21]

With few exceptions (e.g., olive oil), edible oils of plant origin usually undergo various processing treatments to improve their flavor, color, and oxidative stability. The amounts of the minor components remaining in the oil depend on the type of processing and treatment parameters. Table 3 shows the effects of the different processing steps on the minor components in soybean oil.[26] Oxidative stability was found to be in the order crude > deodorized > degummed > refined > bleached.

In addition to the minor components occurring naturally in food, various compounds that form during processing or storage may act as prooxidants or antioxidants. The antioxidative effects of Maillard reaction products have long been recognized and are believed responsible for increased stability in many processed foods. The prooxidative or antioxidative properties of the many decomposition products of food constituents, however, have not been studied in detail.

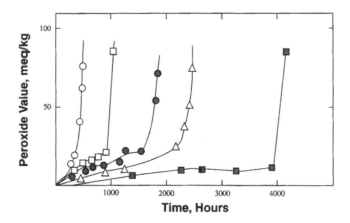

Figure 4   Effect of individual polyphenols (200 ppm) on the autoxidation rate of refined olive oil at 63°C. Symbols: ○, control test; □, protocatechuic acid; ●, BHT; △, caffeic acid; and ■, hydroxytyrosol. From Papadopaulos and Boskou, 1991.[25]

Table 3   Effects of Processing Steps on the Contents of Minor Compounds in Soybean Oil

Oil	Phosphorus (ppm)	Iron (ppm)	Chlorophyll (ppm)	FFA[a] (%)	PV[b] (meq/kg)	Tocopherol (ppm)
Crude	510.0	2.90	0.30	0.74	2.4	1,670
Degummed	120.0	0.78	—	0.36	10.5	1,579
Refined	5.0	0.55	0.23	0.02	8.8	1,546
Bleached	1.4	0.30	0.08	0.03	16.5	1,467
Deodorized	1.0	0.27	0.00	0.02	0.0	1,138

[a] Free fatty acid.

[b] Peroxide value.

From Jung, M.Y., Yoon, S.H., and Min, D.B. *J. Am. Oil Chem. Soc.* 66:118 (1989). With permission.

Interactions of major significance

Not only do the minor components greatly influence the oxidation of lipids, but the interactions of either the oxidizing lipid itself or its oxidation products with other components also decisively influence oxidation and antioxidation. Lipids usually exist as mixtures (e.g., triacylglycerols of different fatty acid constituents, phospholipids, cholesterol). In foods, both lipid and nonlipid components (e.g., protein, carbohydrate, water) often exist in close proximity or in some bound form. The mobility of the food constituents and their breakdown products depend to a great extent on the physical structure of the food.

It is not surprising, therefore, that under oxidative conditions some very complex interactions will take place. Interactions of the free amino groups (e.g,. in amino acids, primary amines, or phosphatidylethanolamine) with aldehydes produced by either oxidation of the lipids or decomposition of the carbohydrate or protein constituents have long been recognized as important (Figure 5). Their reaction products can affect color, texture, and flavor, and some may have prooxidative or antioxidative properties. Carbohydrates in food can also have significant influence on lipid oxidation.[27,28]

Much less clear, however, are the effects of different oxidizable components in a mixture on the oxidation of each other. Work in this author's laboratory with various mixtures of triacylglycerol, phospholipid and cholesterol substrates, and cell membranes indicates that an oxidizable substrate can inhibit and/or accelerate the oxidation of another

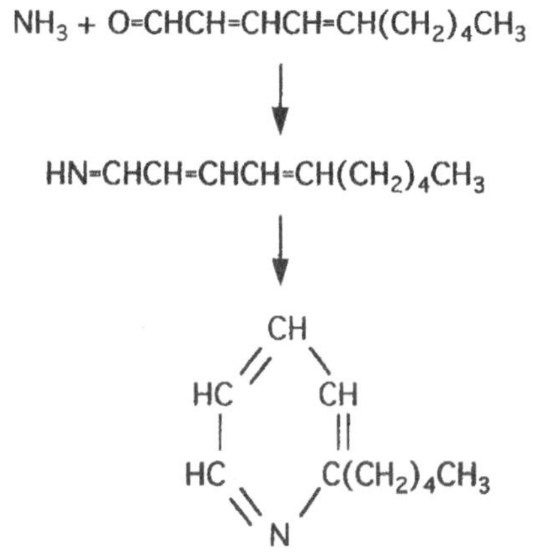

$$NH_3 + O=CHCH=CHCH=CH(CH_2)_4CH_3$$

$$\downarrow$$

$$HN=CHCH=CHCH=CH(CH_2)_4CH_3$$

$$\downarrow$$

2-Pentylpyridine

Figure 5   Interaction of a linoleate oxidation product (2,4-decadienal) with free amino groups.

component of the mixture depending on the interplay of a number of parameters (e.g., nature of the substrates and their concentrations, temperature, pH, and physical state).[29-31] Such interactions are exceedingly complicated but, in view of their importance to food quality, much more research in this area is critically needed.

Detection and measurement

It is clear from the above discussion that lipid oxidation is a complex process involving numerous reactions that give rise to numerous chemical and physical changes. These reactions may occur stepwise, simultaneously, and/or competitively. The rates of these reactions, and the nature and fate of the products they form, are influenced by many variables. Consequently, no single test can possibly measure all oxidative events, and no single test can be universally applicable to all foods, all conditions of processing, or all stages of the oxidative process. Obviously, a more reliable evaluation can be obtained by using a combination of tests. Some of the commonly used methods are discussed below.

Sensory Evaluation. Many of the lipid oxidation products have distinctive flavor characteristics that contribute to the overall perception of objectionable changes in an oxidized food. However, depending on the type of food and the stage and conditions of oxidation, such off-flavors vary not only in their intensity but also in their character. Terms such as rancid, stale, beany, metallic, cardboard, green, fishy, etc., are often used to describe the state of lipid oxidation in foods. The testing of oxidative off-flavors can be conducted by trained taste panels using special score forms.[32]

Peroxide Value (PV). Peroxides are primary intermediates of lipid autoxidation. Standard methods for their measurement are based on their ability to liberate iodine from potassium iodide or to oxidize ferrous to ferric ions. Peroxide values are usually expressed in terms of milliequivalents of oxygen per kilogram of fat (Cd 8-53, Off-Methods, AOCS). A comparison of techniques for the measurement of PV can be found in a report by Barthel and Grosch.[33] Although PV is useful for monitoring oxidation at the early stages, it can be misleading because its accuracy may vary with the procedure used and the treatment history of the food product. As the oxidation progresses, PV reaches a peak and then declines. If the food is subjected to elevated temperatures (e.g., as in frying), peroxides will decompose and PV will decline. Accordingly, a low PV does not always mean a low level of oxidation.

Analysis of Specific Volatiles by Gas Chromatography. Linoleic acid is the most abundant polyunsaturated substrate in vegetable fats. As shown here, hexanal and pentane are major decomposition products arising from breakdown of the 13-hydroperoxide of linoleate:

$$CH_3(CH_2)_4-|-CH-|-C=C-C=C-(CH_2)_7COOH$$

pentane

hexanal   (6)

Both compounds are common to the oxidation of all omega-6 unsaturated fatty acids, and they can easily be measured by gas chromatography. The corresponding products from the omega-3 acids (e.g., EPA and DHA, which are abundant in fish oils) are ethane and propanal.

Thiobarbituric Acid (TBA).[34-36] Oxidation products of unsaturated systems produce a color reaction with TBA that can be measured spectrophotometrically. It is believed that the chromagen arises from condensation of 2 mol of TBA with 1 mol of malondialdehyde. The test can be conducted directly on the food. However, malondialdehyde itself is not always present in oxidized samples, and many other compounds have been reported to produce the characteristic pigment upon reaction with TBA. Such interference may arise not only from other products of lipid oxidation, but also from nonlipid food components. The amount of TBA-reactive substances (TBARS) varies depending on the balance between the rates of their formation and the rates of their decomposition. Consequently, although this test may be applicable for comparison of a single material at different stages of oxidation or of samples subjected to different treatments under controlled conditions, caution is necessary when evaluating foods of unknown history.

Ultraviolet Spectrophotometry. Double bond shifts in the course of lipid oxidation result in diene and triene conjugation that show typical absorption bands at 234 nm and 268 nm, respectively. Correlation with the degree of oxidation is not always satisfactory.

Fluorescence. The interaction of carbonyl compounds arising from lipid oxidation with cellular constituents possessing free amino groups produces certain fluorescent compounds. Tests based on distinctive flourescence emissions are more sensitive than those based on TBA.[37]

Total Volatile Carbonyls. Decomposition products are recovered by distillation under atmospheric or reduced pressure and the associated carbonyl compounds are analyzed by gas chromatography or by reaction with appropriate reagents.

Other Tests. Relatively less specific tests include the following: anisidine value, based on measurement of a yellowish color formed by the reaction of aldehydes with *p*-anisidine in the presence of acetic acid; totox or oxidation value, which is equivalent to 2 PV + anisidine value; Kreis test, based on the measurement of a red color believed to result from the reaction of epihydrin aldehyde or other oxidation products with phloroglucinol; and the oxirane test, based on the addition of halides to the oxirane group. The epoxide group reacts with hydrogen bromide in the presence of acetic acid and crystal violet, giving rise to a bluish green end point. The reaction of picric acid with the oxirane group is reported to provide more sensitivity.

Predictive Tests. The following accelerated tests are also available to predict the resistance of an oil to oxidative rancidity: The Schaal oven test, in which the samples are stored at about 65°C and periodically tested organoleptically or by measuring PV; the active oxygen method (AOM), which involves maintaining the sample at 97.8°C while air is continuously bubbled through it at a constant rate and then determining the time required to obtain a specific PV; and oxygen absorption, where the amount of oxygen absorbed by the sample is determined by the time required to produce a specific pressure decline in a closed chamber under specific oxidizing conditions.

Quick Tests. Since dielectric constant increases with increased polarity, changes in the dielectric constant of oil as measured in an instrument known as the Food Oil Sensor can be correlated with lipid decomposition.[38] Other diagnostic kits for assessing oil deterioration involve addition of a reagent to the sample and matching the developed color with a color standard. One is the Fritest, based on the presence of carbonyl compounds, and another is Oxifrit, based on the presence of oxidized compounds; both are sold by Merck (Dramstadt, Germany). Verify-FAA.500, based on free fatty acids, and Verify-TAM.150, based on "alkaline material," are available from Libra Labs (Piscataway, NJ).

Illustrative examples

Instant potato powders have a relatively short shelf-life, despite their low fat content (<1% of the dry weight). This susceptibility to oxidation is a consequence of fat in potato being relatively high in unsaturation and the surface area of the powder being large.

The oxidative stability of lipids in meats varies significantly among different species and among different parts within the same species. Poultry lipids are more prone to oxidation than red meats, due to their higher degree of unsaturation. Table 4 demonstrates the remarkable variation in lipids and lipid classes among the different parts of chicken broilers. Thighs contain the highest and breasts the lowest amount of total lipids. The skin is very high in total lipids, which consist mainly of triacylglycerols. Only the breasts contain higher amounts of phospholipids than triacylglycerols.

Pikul and Kummerow found higher oxidation levels in raw chicken parts containing higher levels of phospholipids.[39] Lipid oxidation, as measured by TBARS and lipid oxidation fluorescent products (LOFP), increased upon roasting and refrigerated storage. These authors also reported that pieces roasted separately had lower concentration of LOFP and better sensory evaluations than halves and quarters; they attributed this difference to shorter roasting times and lower interior temperatures in case of the smaller pieces. The higher oxidation in the skin than in the muscle was explained on the basis of greater exposure of skin to high oven temperatures during cooking and more oxygen penetration during refrigerated storage.

Table 4 Total Lipid, Triacylglycerol (TG), and Phospholipid (PL)
Contents in Chicken Parts

	Muscle			Skin		
	Lipid (%)	TG (% of lipids)	PL	Lipids (%)	TG (% of lipids)	PL
Breasts	1.2	36	57	22	97	2.3
Thighs	3.2	67	27	32	98	1.2
Drumsticks	2.4	55	39	22	97	1.9
Wings	2.7	59	32	26	97	1.5

From Pikul, J. and Kummerow, F.A. *J. Am. Oil Chem. Soc.* 55:30 (1990). With permission.

Raw nuts usually contain high amounts of fat relatively high in unsaturation. Handling prior to processing (e.g., shelling, blanching, and shipping) can activate lipoxygenase in raw nuts. If fried, nuts absorb frying oil, and the oil in the nut and the absorbed oil will undergo faster oxidation if the temperature and time of processing or the conditions of storage are poorly controlled. If roasted, it is advisable to roast before chopping. Slicing or chopping before roasting results in cell breakdown, increase in surface area, oil release, and greater exposure to the hot air. If nuts of low resistance to oxidation are used as ingredients, they will contribute significantly to faster development of rancidity in the whole food in which they are used.

Milk contains prooxidants (e.g., xanthine oxidase, lactoperoxidase) and antioxidants (e.g., ascorbic acid, tocopherols, metal-binding proteins). Although the milk fat globule membrane (MFGM) itself contains metallo-enzymes, the triacylglycerols inside are protected as long as the membrane remains intact. Processes that cause disruption of the MFGM facilitate contact between substrate and catalyst and thus enhance lipid oxidation.[3] In butter, metal ion catalysis is a major cause of rancid and "fishy" off-flavors, particularly if the butter is poorly wrapped or exposed to fluorescent light as in supermarkets. Caution must be exercised if trace metals are added for nutritional purposes to milk or to dairy products. These supplements can, for example, be added in complexed or encapsulated forms.

Control measures

A number of precautionary measures, based mainly on the various aspects considered above, can be recommended for prolonging shelf-life that is limited by oxidation and for minimizing undesirable changes in the quality of edible oils and fat-containing foods:

1. Select high-quality raw material (e.g., seeds with minimum damage, oils with low FFA content and high resistance to oxidation).
2. Use high-quality food ingredients (e.g., milk, nuts, spices).
3. Use techniques that reduce substrate-catalyst interaction (i.e., avoid cell disruption, contact with enzymes).
4. Minimize contact with oxygen, light, and/or trace metals.
5. Minimize exposure to elevated temperatures.
6. Use packaging that provides a reasonable gas barrier during storage and distribution.
7. Minimize surface area in contact with air.
8. Design and maintain proper storage tanks and pipelines (e.g., stainless steel, if possible; glass lining; free of copper, copper alloys, and dead corners; frequent cleaning; minimal headspace; lowest practical temperatures; protection from contamination with microorganisms; and regular inspection).
9. Use appropriate antioxidants.

Conclusion

It is clear from the above that while the oxidation of pure fatty acids is fairly well understood, new approaches to the examination and control of oxidative stability in the more complex food systems are urgently needed, with a *focus on the role of physical state* in lipid oxidative/antioxidative reactions.

References

1. Frankel, E.N. *Prog. Lipid Res.* 22:1 (1982).

2. Shewfeld, R.L. *J. Food Biochem.* 5:79 (1982).
3. Allen, J.C. Rancidity in Dairy Products. In *Rancidity in Foods.* J.C. Allen and R.J. Hamilton, Eds., Applied Science Publishers, New York, pp. 169-178 (1983).
4. Galliard, T. Rancidity in Cereal Products. In *Rancidity in Foods.* J.C. Allen and R.J. Hamilton, Eds., Applied Science Publishers, New York, p. 109, (1983).
5. Gross, A.F. 1969 IFT Meeting Abstracts paper no. 114 (1969).
6. Halbert, E. and Weeden, D.G. *Nature* 212:1603 (1966).
7. Gardner, H.W. and Plattner, R.D. *Lipids* 19:294 (1984).
8. Hau, L.B. and Nawar, W.W. *J. Am. Oil Chem. Soc.* 65:1307 (1988).
9. de Groot, J.J.M.C., Veldink, G.A., Viliegenbart, J.F.G., Boldingh, J., Wever, R. and van Gelder, B.F. *Biochem. Biophys. Acta* 377 (1975).
10. Vliegenthart, J.F.G. and Veldink, G.A. Lipoxygenase-Catalyzed Oxidation of Linoleic Acid. In *Autoxidations in Foods and Biological Systems.* M.G. Simic and M. Karel, Eds., Plenum Press, New York, p. 529 (1980).
11. Miyashita, K., Nara, E. and Ota, T. *Biosci. Biotech. Biochem.* 56:1638 (1993).
12. Raghuveer, K.G. and Hammond, E.G. *J. Am. Oil Chem. Soc.* 44:239 (1967).
13. Tautorus, C.L. and McCurdy, A.R. *J. Am. Oil Chem. Soc.* 67:525 (1990).
14. Zalewski, S. and Gaddis, A.M. *J. Am. Oil Chem. Soc.* 44:576 (1967).
15. Hoffmann, G., Stroink, J.B.A., Polman, R.G. and van Ooster, C.W. The Oxidative Stability of Some Diacyl-Triglycerides. Proc. Internat. Symp. on Deterioration of Lipids. H. Niewiadomski, Ed., Panstwow Wydawnictwo Naukowe, Warsaw, pp. 93-97 (1973).
16. Dugan, L.R. Natural Antioxidants. In *Autoxidantions in Foods and Biological Systems.* M.G. Simic and M. Karel, Eds., Plenum Press, New York, p. 261 (1980).
17. Porter, W.L. Recent Trends in Food Applications of Antioxidants. In *Antioxidations in Foods and Biological Systems.* M.G. Simic and M. Karel, Eds., Plenum Press, New York, pp. 295-365 (1980).
18. Simic, M.G. and Hunter, P.L. Antioxidants. In *Chemical Changes in Food During Processing.* T. Richardson and J.W. Finley, Eds., Avi Publishing, Westport, CT (1985).
19. Shahidi, F., Janitha, P.K. and Wanasunadara, P.D. *Crit. Rev. Food Sci. Nutr.* 32:67 (1992).
20. Kritchevsky, D., Tepper, S.A., DiTullo, N.W. and Holmes, W.L. *J. Food Sci.* 32:64 (1967).
21. Fedeli, E. *Rev. Franc. Corps Gras* 30(2):51 (1983).
22. Fedeli, E., Lanzani, A., Capella, P. and Jacini, G. *J. Am. Oil Chem. Soc.* 43:254 (1966).
23. Boskou, D. and Katsikas, H. *Acta Alimentaria* 8:317 (1979).
24. Gordon, M.H. and Magos, P. *Food Chem.* 10:141 (1983).
25. Papadopoulos, G. and Boskou, D. *J. Am. Oil Chem.* Soc. 68:669 (1991).
26. Jung, M.Y., Yoon, S.H. and Min, D.B. *J. Am. Oil Chem. Soc.* 66:118 (1989).
27. Sims, R. Inform 5:1020 (1994).
28. Shimada, Y., Roos, Y. and Karel, M. *J. Agr. Food Chem.* 39:637 (1991).
29. Kim, S.K. and Nawar, W.W. Oxidative Interactions of Cholesterol with Triacylglycerols. *J. Am. Oil Chem. Soc.* In Press (1991).
30. Nawar, W.W., Kim, S.K. and Vajdi, M. *J. Am. Oil Chem. Soc.* 68:496 (1991).
31. Chen, Z.Y. and Nawar, W.W. *J. Food Sci.* 56:398 (1991).
32. Jackson, H.W. *J. Am. Oil Chem. Soc.* 58:227 (1981).
33. Barthel, G. and Grosch, W. *J. Am. Oil Chem. Soc.* 51:540 (1974).
34. Dahle, L.K., Hill, E.G. and Holman, R.T. *Arch. Biochem. Biophys.* 98:253 (1962).
35. Gray, J.I. *J. Am. Oil Chem. Soc.* 55:539 (1978).
36. Yu, T.C. and Sinnhuber, R.O. *J. Am. Oil Chem. Soc.* 44:256 (1967).
37. Dillard, C.J. and Tappel, A.L. *Lipids* 6:715 (1971).
38. Fritsch, C.W., Egberg, D.C., and Magnuson, J.S. *J. Am. Oil Chem. Soc.* 56:746 (1979).
39. Pikul, J. and Kummerow, F.A. *J. Am. Oil Chem. Soc.* 55:30 (1990).

chapter five

Biochemical processes: carbohydrate instability

*Joan Gordon and Eugenia A. Davis**

Contents

Introduction

Stability of carbohydrates in food systems can be viewed from the standpoint of changes that often reflect the postharvest physiology of specific fruits, vegetables, or cereal grains. These changes are usually mediated by endogeneous enzymes, but without the control of the actively synthesizing unit. Furthermore, additional mediating factors are introduced

---

* Eugenia A. Davis is deceased.

0-8493-2646-X/97/$0.00+$.50
© 1998 by CRC Press LLC

by environmental conditions associated with storage, transport, processing, and preparation conditions.

Stability of carbohydrates in food systems can also be viewed from the standpoint of their contributions as components of formulated foods. Although the chemistry of individual carbohydrates has been studied extensively, component interactions in a formulated food could change the reaction pathways and overall kinetics, and ultimately the sensory response. The primary emphasis of this chapter is on carbohydrates as components of formulated foods.

Understanding food component interactions has become increasingly important as the food industry has shifted its emphasis to the development of formulated, partially or totally processed foods. It is basic for guiding formulation and manufacturing processes so as to meet consumer demands for quality, economy, and nutritive value as well as to satisfy industry requirements for efficiency and conservation of resources.

An example of the importance of understanding component interactions can be found in the numerous problems accompanying the expanded use of microwave heating. Formulations and practices based on conventional heating could not be easily extrapolated to microwave heating. As a result, it became important to understand, at a fundamental level, how component intreractions are modified by each heating method, and how these interactions, in turn, control heat and mass transport properties during heating by each method.

Understanding component interactions is also important as a technology associated with a commodity area advances, eventually leading to the availability of new components that can be utilized in developing new product formulations. For example, fractionation technology in the dairy industry has led to a range of new components that are primarily protein, lipids, sugars, or minerals. The functionality of these new components needs to be assessed at the component interaction level as a basis for their use as ingredients in new product applications. Chemical modifications of starch and nonstarch polysaccharides have been introduced to provide certain functionalities for specific formulations. Various carbohydrates have also been used in developing low-fat formulations of traditional foods and so-called fat mimetics.

Food microstructure

In recent years, attention has been focused on the key role of food microstructure as a determinant of quality parameters. Study of food microstructure has been advanced by the application to food systems of increasingly sophisticated experimental techniques that make possible the study of component interactions at the molecular and atomic level. Among the techniques that have been utilized are different types of electron microscopy (EM), differential scanning calorimetry (DSC), X-ray scattering (both small-angle [SAXS] and wide-angle [WAXS]), electron spin resonance spectroscopy (ESR), and nuclear magnetic resonance spectroscopy (NMR).[1]

Studies of microstructural development have identified water emission rate during heating as a key indicator of structural development in cereal-based formulations.[2] They have also shown that water emission rates can be modified by changes in formulation, which has led to studies of model systems to investigate specific component interactions in detail. As will be discussed in subsequent sections, water plays a key role in specific carbohydrate transitions during heating and in other changes that take place after heating.

Degradative reactions

In considering stability of carbohydrates, reactions at the covalent bond level are important in monosaccharides. These reactions include browning, hydrolytic reactions, mutarotation, enolization, and dehydration reactions. These reactions result in compounds that are colored or that participate in the series of reactions with amino groups, notably from proteins, that result in brown pigments.

Browning reactions

The sequences of reactions that make up nonoxidative browning, or Maillard, reactions have been extensively studied, and detailed reviews are available, so they will not be discussed in detail in this review.[3-5] Reineccius comments that, in fact, these reactions are "thousands of reactions" and that at least 2500 volatile compounds are formed via Maillard reactions.[6]

Browning reactions contribute to flavor as well as to the color of foods. The flavor development could contribute to the characteristic and expected flavor of specific foods or it could contribute to off-flavors. The absence of browning in microwave heating is considered to be responsible for flavor differences between microwave-heated and conventionally heated foods. Not all browning that is observed comes directly from the Maillard sequence of reactions. Browning could also be associated with thermal degradation, such as carmelization, and the process can result in complex mixtures. Maillard reactions, which also involve amino groups, can result in decreased nutritional quality, especially to the extent that the ε-group of lysine is coupled.

Among the factors that affect browning reactions during storage are temperature, moisture content and water activity, degradative reactions of the reactants (either carbohydrates or proteins), and indirect reactions that could affect pH.

The absence of browning may not be related to the volatilization and loss of water-soluble flavors, according to Wharton and Reineccius.[7] They utilized temperature, pressure, liquid and vapor water profiles derived from theoretical modeling[8] to explain the distribution in cakes of specific flavor compounds formed during microwave heating that may or may not be related to browning reactions. Roos and Himberg have discussed the role of the glass transition of the matrix, maltodextrin, in controlling low temperature browning of lysine–xylose model systems.[9]

Hydrolytic reactions

Hydrolysis of oligosaccharides and polysaccharides of interest in foods may be acid- or enzymatically catalyzed. Among the well-known hydrolytic reactions of disaccharides are the conversion of sucrose to invert sugar and of lactose to glucose and galactose. These reactions are either acid-catalyzed or enzymatically catalyzed by invertase (β-D-fructofuranoside) or lactase (β-D-galactosidase), respectively.

Production of corn syrups from starch utilizes combinations of acid and enzymatic hydrolysis, depending on the specific product desired, especially on the degree of polymerization (dextrose equivalent, DE) and the glucose–fructose ratios. Such combination processes use initial acid hydrolysis steps followed by treatment with α-amylase and glucoamylase (*exo*-1,4-α-D glucosidase) to yield glucose, which then can be treated with isomerase (xylose isomerase) to give glucose and fructose mixtures. High maltose syrups can be produced by using β-amylases. If enzymes are used instead of acid in the initial degradation of starch polymers, a prior starch gelatinization step is required.[3] Technological developments and potential have been summarized.[10,11]

Degradation of starch polymers by α- and β-amylase is an important step in other industrial processes, such as fermentation. Amylolytic enzymes are also added to doughs.

Sometimes the control of a hydrolytic reaction is as important as the reaction. Recently, enzyme-resistant starches were identified.[12–16] Czuchajowska et al.[17] and Lin et al.[18] studied the enzyme-resistant properties of high amylose starches and their use in certain products to modify quality characteristics such as retrogradation and lipid interactions. Understanding "enzyme-resistant" properties of starches requires characterization of amylases as well as the starch substrates. Progress in relating the primary and resulting three-dimensional structures of this diverse group of enzymes to mechanisms of the degradative reactions has been made.[19–21] A model for the active site consisting of 10 contiguous subsites is proposed for barley malt alpha-amylases, for example.[20]

Lactose hydrolysis to glucose and galactose is utilized in production of low lactose whey protein concentrates, which are of interest in controlling lactose crystallization and in meeting special product formulation as well as dietary needs.

Hydrolytic reactions of pectins include hydrolysis of methyl esters of the carboxyl groups at C-6. Hydrolysis of the polymeric chain can occur via β-elimination reactions catalyzed by pectin lyase in addition to hydrolysis of $1 \rightarrow 4$ galacturonic bonds catalyzed by polygalacturonase.

In contrast to the importance of reactions involving covalent bonds in monosaccharides and oligosaccharides, many of the functional properties of polysaccharides are related to associations that determine secondary and tertiary structures. Moreover, amylose and amylopectin are introduced in the food system as starch granules, which then form an important component of the microstructure of the specifically formulated food.

Models of starch granules

Numerous models have been suggested for the structure of starch granules. Most of these focus on the arrangements of amylose and amylopectin that can accommodate the primarily linear structure of amylose, resulting from the $1 \rightarrow 4$ α-D glucose linkages, and the branched structure of amylopectin, resulting from the additional $1 \rightarrow 6$ branching. In addition, models must also account for ordered structures within the granule observed at several different levels of resolution using different probe techniques: wide-angle X-ray scattering, small angle X-ray scattering, transmission and scanning electron microscopy, before and after enzymatic degradation, and light microscopy, and the birefringence observed by polarizing microscopy. Ordered structure is also implied by the endothermic behavior observed using differential scanning calorimetry, by differential responses to enzymatic attack within various regions of the granule, by rheological measurements, as well as by different mechanical damage to the granules in response to various milling processes.

Although most models focus on amylose and amylopectin, consideration must be given also to arrangements that can accommodate naturally occurring surface and internal lipids and proteins, the potential for binding enzymes and lipids on the surface and internally, and the penetration of water. These properties are part of the biological role of the granule in plant biogenesis when storage carbohydrates are utilized by the developing plants, but become the basis of the functional properties of starches as they are utilized in food and industrial applications.

Most models of the starch granule accommodate the linear amylose as a single or double helix and as a lipid–amylose helical inclusion compound. The branched amylopectin is considered, at the present time, to have double helices in the linear components of the structures. Examples of typical models are those of Lineback and Rasper[22] and Blanshard.[23] These models draw from structures proposed earlier by French.[24]

While these models provide a schematic representation of amylose and amylopectin molecules, an additional concern is the assignment of these structures to crystalline and amorphous regions of the granule. Estimates of crystallinity for various starches range from 15–45%.[25] Nevertheless, starches can be classified as A, B, or C types based on X-ray scatter studies. Transformations of native A-type granule structures can be characterized by changes from A-scattering patterns to V- and B-scattering patterns.

A- B- and C-scattering patterns were assigned originally to starches of different botanical origin by Katz.[26] Most cereal starches show A-scattering patterns. Assignment of specific structures to each scattering pattern has proved to be a difficult task. Recent approaches have utilized computer modeling to aid in assigning structures[26–31] and NMR.[32–33] Assignment of structures to A and B polymorphs of amylose presents similar problems. Double helices are assigned, but the pitch and accommodation of water molecules in the helices are different in the two polymorphs. Interconversions between A and B patterns are possible, and most calculations suggest that more water molecules can be accommodated in B structures than in A structures (30–40 vs. 4). This encourages speculation that transformations of A structures that accommodate water result in B structures. In most cereal starches, the transformation is considered to go from A $\rightarrow$ V $\rightarrow$ B, the V form being the single helix form mediated by lipid. As Zobel points out, the reverse transition of B $\rightarrow$ V requires the disruption of the B structure and the involvement of a complexing agent.[26]

These observations are consistent with molecular modeling[31] that suggests a sequence as follows: single chains associate into double helices that then associate in pairs stabilized by hydrogen bonds and van der Waals forces. Further associations lead to either A or B forms.[31]

Potato starch, which is of the B type, can be transformed to A type[26]; however, models based only on amylose do not provide for contributions by phosphate groups to the structure. Muhrback and colleagues showed that crystallinity in potato starches is related to the level and position of phosphorylation.[34–35]

The role of amylopectin in the models of granule structure derived from amylose polymorphs needs to be clarified, because amylopectins contribute to X-ray scattering patterns. Early studies showed that X-ray scattering patterns of wheat starch were retained after extraction of amylose.[36] Subsequent studies showed that the low temperature endotherm of DSC scans is present in waxy starches, which contain almost 100% amylopectin but are diffuse in high amylose starches.[37–41] It has been recently proposed, based on X-ray scattering studies coupled with computer modeling of structures, that double helices in the linear chains of amylopectin contribute to the A and B scattering patterns.[31] A super-helical structure for amylopectin was proposed recently, based on optical tomography and cryo-electron diffraction of potato starch.[42]

These hypotheses tend to focus on crystallinity of the granules. Amorphous regions of the granules may also contribute to the overall properties of the granule.[43] The contribution of amorphous regions of the granule properties must be considered also. Recent applications of polymer science principles to food polymers by Slade and Levine among others have shown the importance of transformations among glassy, rubbery, and crystalline states.[44–50]

Gelatinization and gelation

The disruption, either partial or complete, of the ordered structures in the starch granule is the basis of the phase transitions collectively known as gelatinization and gelation. The changes with time subsequent to gelatinization and gelation, again collectively, are defined as retrogradation.

The disruption of the ordered starch granule structures during gelatinization is shown in the sequential changes that are measured by several analytical methods that evaluate structure. Many definitions have been proposed for "gelatinization," but the concept of sequential changes is best recognized by the definitions proposed by the Atwell Committee:

> Starch gelatinization is the collapse (disruption) of molecular orders within the starch granule manifested in irreversible changes in properties such as granular swelling, native crystallite melting, loss of birefringence, and starch solubilization. The point of initial gelatinization and the range over which it occurs is governed by starch concentration, method of observation, granule type, and heterogenities within the granule population under observation.[51]

Minimal requirements for gelatinization are water content of at least 25 to 30% and heat. The temperatures needed to initiate the transitions depend on the type of starch and, within a specific starch type, on the presence of other components in the system. For food systems, sugars most effectively elevate gelatinization temperatures. Proteins and lipids may cause some increases in gelatinization temperatures, but usually to a lesser extent than the sugars. The role of water content will be discussed more extensively in the next section. Water may be absorbed reversibly below temperatures associated with the phase transitions. Pasting is another relevant process and is defined by the Atwell Committee as follows: "Pasting is the phenomenon following gelatinization in the dissolution of starch. It involves granular swelling, exudation of molecular components from the granule, and eventually, total disruption of the granules."[51]

A series of morphological changes during the swelling of wheat starch granules was developed by Bowler et al., based on scanning electron microscopy.[52] The sequence from swelling to swelling and twisted forms is illustrated.

This definition of pasting recognizes that both granule remnants and exuded molecular components are present in the system. Viewing gelation from a rheological standpoint and applying concepts developed in the study of polymers has led to the model suggested by Morris.[53] In this model, starch gels are considered to consist of a binary "phase-separated network formed by demixing and subsequent gelation of the two polysaccharides." The starch pastes and gels are viewed as a two-component system with the granule remnants containing amylopectin suspended in the amylose dispersion derived from amylose lost from the granule. The model is similar to that proposed by Miller et al.[54] and by Morris and coworkers from NMR studies.[55-56] This model makes it possible to integrate studies of amylose transformations with those of amylopectin associated with the granule, as will be discussed in the following sections.

Retrogradation, as used by the starch chemist, generally refers to the changes that starch systems undergo with time. The Atwell Committee definition defines retrogradation as a process that involves reassociation into ordered structures, but does not require that the newly ordered structures be those of the original structure: "Starch retrogradation is a process which occurs when the molecules comprising gelatinized starch begin to reassociate in an ordered structure. In its initial phases, two or more starch chains may form a simple juncture point which then may develop into more extensively ordered regions. Ultimately, under favorable conditions, a crystalline order appears."[51]

This definition recognizes that the newly developed structure may include junction points as used in Rees' concept of gel formation mechanisms in nonstarch polysaccharides.[57-58] It is also consistent with the Morris model of a two-phase system with starch granule remnants, now considered to contain amylopectin, and exuded material, now associated with amylose, contributing individually and collectively to the structure of the retrograded system.

For bread systems, retrogradation and staling are linked, and it is often difficult to separate the two phenomena and the specific factors involved. Staling is also a time-dependent phenomenon, which is characterized by rheological changes and the sensory sense of dryness, whether to touch or mouthfeel. Studies of staling as retrogradation have contributed to understanding both phenomena, although some researchers attribute an equal, or greater, role of gluten proteins in the control of staling.[59]

Role of amylose

Initially, a role in retrogradation was assigned to amylose because it was assumed that the linear nature of amylose would more easily result in associations than would the branched amylopectin structure. Current research suggests a broader role for amylose as it focuses on amylose–lipid complex formation and the possibility of building suprastructures[60] within the two-phase model of Morris.[53]

X-ray scattering studies show the sequential transformation of the A pattern of cereal starches to the V pattern of amylose–lipid complexes and the B pattern as associated with retrograded starch. The problems of assigning specific molecular transformations have been discussed in the preceding section.

The most usual model for amylose–lipid complexes is that of binding of the hydrophobic portions of the lipid molecule to the hydrophobic interior of amylose helices. This model was suggested from X-ray scattering studies of Mikus et al.[61] Additional evidence for amylose–lipid inclusion complexes that occur naturally in corn, rice, and oat starches was found in ^{13}C cross-polarization magic-angle spinning NMR (^{13}C CP/MAS-NMR). In these starches, the lipids are free fatty acids and lysophospholipids.[62] ESR studies have shown that the binding of lipids to wheat starch granules can occur at the interior as well as at the external surfaces of the granule; however, it is stronger for waxy than for high-amylose starches.[63] Other studies have shown that amylose–lipid complex formation is affected by the length of the carbon chain on the fatty acids, with more complexing occurring for C18 than for C12 saturated fatty acids. Saturated fatty acids show the greatest complexing ability, followed by *trans*-unsaturated fatty acids, and *cis*-unsaturated fatty acids show the least.[64] These differences are also found in the formation of monoglyceride–amylose complexes. Complex formation is a major functional property of emulsifiers in cereal-based food products. For those emulsifiers that form different dispersion states depending on the water content and temperature, amylose–lipid complex formation is also affected by the dispersion state.[65]

Role of amylopectin

The role of the amylopectin in retrogradation is deduced from the role of amylopectin in the structure of the granule. Furthermore, it was recognized early that bread made with waxy starches staled.[66] At the present time, differential scanning calorimetry is used to follow the redevelopment of the low temperature endotherm that is attributed to amylopectin crystallite melting.[67–68, 40–41] This method has been applied to starch pastes, bread, and pasta[67–71] and is illustrated in Figure 1. These thermograms are in contrast to those of cake batter that had high sucrose concentrations and different emulsive systems[72] (Figure 2). Data for the unheated bread dough are difficult to obtain because of expansion and volatilization during heating in the DSC instrument. Thermograms of uncooked and cooked pasta are shown in Colonna et al.[71]

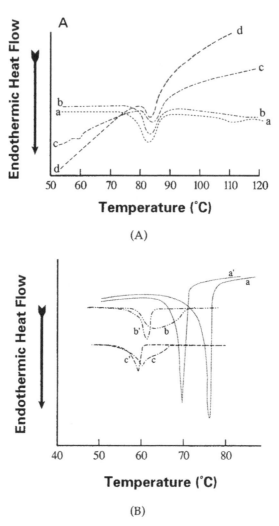

Figure 2 (A) Representative DSC thermograms (a) wheat starch in 42% sucrose; (b) unemulsified cake batter; (c) batter with 10% saturated monoglycerides; (d) batter with 10% unsaturated monoglycerides. (B) Representative DSC thermograms (a) saturated monoglycerides; (a') a reheated; (b) saturated monoglycerides in 42% sucrose; (b') b reheated; (c) saturated monoglycerides in water; (c') c reheated. (From Cloke, J.D.; Gordon, J.; Davis, E.A. *Cereal Chem.* 1983, *60*, 143–146. With permission.)

Yang et al., using small angle X-ray scattering,[73] were also able to show changes in spacings in the 26 to 29.6 nm range during storage of heated model starch–water systems. Radii of gyration decreased from 24.9 ± 1.3 nm to 17.5 ± 2.4 nm. These results were interpreted as supporting the hypothesis that after heating, cooling, and storage, a molecular reorganization occurs that results in a more random structure and smaller repeating units than are found in the native granule. Cameron and Donald used small-angle X-ray scattering to study concomitant water absorption and loss of crystallinity at room temperature and at 51°C.[74]

Water as a participant in gelatinization and retrogradation

Underlying the roles of amylose and amylopectin and the interaction of components such as sugars, lipids, and proteins is the question of whether the major controlling factor is water. Therefore, in this section we examine the roles of water as a participant in gelatinization and retrogradation.

Although changes in viscosity as measured by the Brabender viscosimeter are typically detectable at low starch concentrations (5% starch) and although amylose gels can be formed at concentrations as low as 2% amylose[25], minimal changes in starch granule structure occur when water concentrations are less than 25–30%. At low water contents, low-temperature DSC endotherms are not present; granule swelling is minimal; loss of birefringence is decreased; and the mobility of nonhydrogen-bonding free radical spin probes as measured by ESR is decreased. Minimal water contents are necessary to obtain typical X-ray scattering patterns[26] and to obtain helical structures in amylose.[30-31]

Liu and Lelievre have raised the question of whether these water concentrations represent the interior water concentrations during heating.[75] However, ESR studies indicate relatively free and reversible movement of water into and out of the granule prior to the phase transitions.[76-77]

Behavior in more complex systems, such as starch–sugar–water systems, is sometimes explained on the basis of sugar as a participant in the solvent properties of water.[78] Johnson et al. on the basis of ESR and DSC studies point out that the effects of sugar additions are not the same as decreasing the water content.[77] They examined starch–water systems over a range of 12–50% water. Many formulations that include starch will cover this range of starch–water ratios.

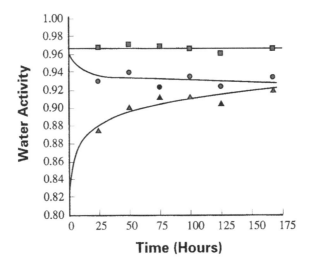

Figure 3 Changes in water activity in crust (▲), center crumb (■), and near crust zone (●) in bread stored for 168 h. (From Czuchajowska, Z.; Pomeranz, Y. *Cereal Chem.* 1989, *66*, 305-309. With permission.)

Associations between water and starch

Associations between starch macromolecules and water, and the relationship between these associations and bulk or free water, have been studied by a variety of techniques. The generalized concept of "bound" water was utilized initially to explain these interactions. Newer analytical techniques allow more detailed descriptions of these associations and supplement earlier studies in which water activity was used as a measure of these associations.[79] These newer techniques are also used to study more complex model systems such as starch–lipid–water, starch–protein–water, and starch–low-weight-carbohydrate–water model systems. The interpretations usually are more difficult when these techniques are applied to the full formulation.

Among these techniques are ESR with stable free radical probes, NMR using ^1H, ^2H, ^{17}O probes,[80-84] nuclear magnetic resonance imaging (MRI),[85-86] and dynamic rheology methods.[87-88] They explore these associations at different levels: starch macromolecule–water mobility,[80] the microenvironment close to the starch macromolecules[76-77, 89-91] and in bulk water. The interpretation of these experiments in terms of free volume theory is discussed by Johnson et al.[77] This interpretation becomes important because proposals by Slade and Levine rely on free volume theory to explain the role of glassy and rubbery transitions of amorphous polymers in starch phase transitions.[78] Free volume theory is also important to the understanding of rheological properties as studied by dynamic rheological methods and to the explanation of this behavior in terms of polymer motion.[87-88, 92-93]

Role of water in starch retrogradation

In high water content gels, release of water, recognized as syneresis, is a characteristic of aging gels. In lower moisture systems, such as bread, staling is associated with the development of firmness in the crumb structure and a mouth feel of dryness. Nevertheless, both water content and water activity of the center crumb, which is in the range of 0.96–0.97 (Figure 3), remain relatively constant. The water activity of the crust, which is initially lower than that of the crumb (0.84 vs. 0.96), tends to equilibrate with that of the crumb near the crust (0.92) after 168 h storage (Figure 3).

Although the issues regarding underlying mechanisms are similar to those discussed for gelatinization, the role of water in this case is in the redevelopment of structure and is the association with the new macromolecular starch structures. Whether its role in the newly formed structures is the same as has been discussed for the amylose and starch crystalline structures continues to be studied. Hydrogen bond formation mechanisms have been suggested by arguments based on the water activity of staled bread crumb,[79] NMR studies,[82] and studies by Martin et al.[59]

Water as a plasticizer

The role of water as a plasticizer in the sense "that it serves as a fluid medium in which motion of the solid particles past one another can take place" was suggested by Bachrach and Briggs in their 1947 study of the role of water in the retrogradation of starch and bread.[79] More recently, a plasticizer role was assigned to water by Slade and Levine in their development of hypotheses based on polymer theory involvement of glassy, rubbery, and crystalline states in the retrogradation process.[44-49] In this view, water concentration is a controlling factor in the temperature dependence of these transitions. The concept of free volume enters into the kinetic analysis of these transformations.

The appropriate methodologies to measure these transitions and to evaluate the state of water continue to be studied. DSC can be used to measure changes in heat capacity associated with these transitions. Thermal mechanical analysis can be used to measure volume changes.[94] Dynamic mechanical thermal analysis (DMTA) is being used to measure structural relaxation over extended temperature ranges.[95-96] Both Johnson et al.[77] in ESR studies of starch–sugar–water systems and Roozen and Hemminga[97] in ESR studies of sugar–water systems point out that ESR spin probe methods can provide information on aqueous microenvironments. These types of studies move closer to explaining more specifically the meaning of "competition" for water between starch and other components of the system.

Sugars

The effects of adding sugars to starch–water systems have long been recognized. Generally, the gelatinization temperatures, whether measured by DSC, microscopy for swelling and birefringence, or rheology, appear to be elevated, but the extent of the changes depends on the specific carbohydrate used.[77, 98–103] As discussed earlier, the mechanisms at the molecular level are less well-understood, despite the range of analytical methods used. Recent studies that utilize both thermal and rheological effects have contributed to separating and defining the contribution of sugars and starch in these systems.[102]

Retrogradation of sugar–starch–water model systems has also been studied by these methods including X-ray scattering.[99–100] The general expectation is that the presence of sugars will retard retrogradation. Isolation of the separate contributions of sugars to controlling retrogradation in full formulations of cereal-based products becomes difficult, however, when the range of water, lipid, and protein levels, the extent of the initial starch transformations, and the contribution of the starch to final product quality are considered. For example, moisture content, sugar concentration, and lipid levels are lower in bread than in cake. Starch contributes to structure in both cases, but the gluten network is more fully developed in bread than in cakes. Sauces and gels, at the other extreme, have high moisture levels and, in some cases, high sugar levels, low protein levels, and lipid levels; the starch contribution to structure is a rheological one.

In these instances, a rheological-based gel model such as those described earlier based on the two-phase system of starch granules remnants in a amylose dispersion becomes appropriate. It also is the basis for the introduction of modified starches and nonstarch polysaccharides such as xanthan gums and carrageenans. Various combinations of these polysaccharides are also added to cake formulations. Extensions of these models have also been made in studies that investigate the effects of sugars on thermal and mechanical properties of starch gels.[101] The role of sugars in inhibiting chain reorganization in the starch gels was demonstrated and attributed to both the amylopectin/amylose ratios of the starches and solute–water relationships of the low molecular weight sugars.

Some types of cookies are examples of high sugar, high lipid, low moisture formulations. In this type of formulation, the role of sugar has been discussed in relation to its state diagram as a guide to changes in glass and rubbery states.[47–49] It should be noted that state diagrams can also be constructed for gluten proteins.[104] Arvanitoyannis et al.[105] utilized Couchman–Karacz and Gordon–Taylor equations to estimate Tg for solutions of mixtures of glucose and fructose. Similar approaches were used for amylopectin–fructose–water systems by Kalichevsky and Blanshard.[88] Dielectric relaxations in liquid and glassy states of glucose were also determined. Sub-Tg relaxations were interpreted on the basis of secondary or β-relaxations. Above Tg, glucose–water mixtures followed patterns shown also by polymers.[106]

The consideration of the contributions of the several components in the context of polymer science will aid in interpretation of experiments such as those of Martin et al.[59] This approach may also aid in interpretation of empirical interactions deduced from response surface methodology in applications such as batter systems as suggested by Box.[107]

Lipids

The contribution of lipids, including those used originally as emulsifiers to the retardation of staling of bread, was recognized early.[108] That cakes containing emulsifiers were structurally less fragile was also recognized soon after the introduction of emulsifiers into cake formulations. The formation of amylose–lipid complexes was also demonstrated.[61] As a result, the starch–lipid binding properties and amylose–lipid complexes have been studied extensively. Because these complexes are viewed as involving hydrophobic interactions between amylose and lipids, it is not surprising that complex formation is related to the degree of unsaturation of the lipid, to whether the bonds are *cis* or *trans*, and to the length of the fatty acid chain.[64, 109] In addition, amylose–lipid complex formation was also found to be related to the initial dispersion state of the lipid, which in turn is related to temperature and water content.[65]

As discussed for sugar additions, the range of cereal-based formulations into which lipids are incorporated is wide, and their contributions to microstructure depends on the formulation as well as on the role of starch in the structure of each product. For example, in breads and cakes, the strengthening of the granule structure is important; in sauces and gels, the amylose–lipid phase and the granule remnant structure are important in their contributions to the rheological properties. In cookies and in pastry with low moisture contents, starch transformation is minimal and the lipid plays a mechanical separation role.

In bread, Krog et al. showed that increasing the level of monoglycerides increases the size of the high temperature amylose–lipid melting endotherm.[69] This increase would be expected if the extent of amylose–lipid complexation increases. However, the development of the amylopectin-melting endotherm during storage was delayed.

In considering the rheological role of the amylose–lipid complexes, Biliaderis and Galloway suggested that crystal type, including the building of extended structures of amylose–lipid complexes, depends on lipid concentration and rate of crystallization.[60] The extent of these extended structures could then affect the rheological properties of the system. Rheological properties may also reflect a behavior of lipid as a plasticizer, supplementing the plasticizer role of water when analyzed in the context of state diagrams.[110] These approaches are basic to developing a rational approach to formulation of low fat versions of traditional food products.

Protein

The contribution of protein to the starch-based systems is also approached differently, depending on the formulation. In systems in which gluten plays a major role in the microstructure, exchange of water between gluten and starch during staling has been considered. In systems that include additional proteins in the formulation, the separate contributions of the proteins are considered. The proteins studied most frequently from this standpoint are the milk proteins.

Transfer of water between starch and gluten was suggested by Willhoft on the basis of rheological studies, the direction being from proteins to starch in bread and additionally to sugars in cakes.[111] The implications of such transfers of water in terms of structure at the molecular level have already been discussed.

Umbach et al., using pulsed gradient NMR, found that the mobility of water decreases when gluten is added to starch in model studies. The mobility of water decreased when gluten was added to starch, but then remained relatively constant with further additions of gluten, suggesting that a balance is reached between macromolecular-associated water and free water.[80]

Incorporation of milk proteins is generally expected to alter the staling and, by implication, starch retrogradation. However, the interactions are quite complex. Consideration must be given to concomitant protein transformations during heating and the consequences for water binding and rheological properties. In full formulations, protein interactions with other components need to be considered as well. ESR studies in which a nonhydrogen-bonding, stable, free radical probe was used showed the effects of milk proteins, individually and in combination, on water microenvironments in starch–water and starch–protein–water systems. In whey protein concentrates (lipid content, 5–7%), the probe was distributed between hydrophobic and hydrophilic regions. If the lipid content was reduced, as in the preparation of whey protein isolates, this distribution was not observed. Mobility of the proteins was studied also by the use of spin-labeled proteins. Some changes in the lipid binding properties of starch in the presence of whey proteins were observed by using spin-labeled lipids.[112–116]

Dielectric properties of starch–water model systems were changed in the presence of milk proteins, and these effects changed during heating. The dielectric properties are important determinants of microwave heating and are strongly, but not completely, related to moisture content at the frequency used for household microwave ovens (2450 MHz). In general, the dielectric properties of mixtures of starch and whey proteins were dominated more by the dielectric properties of starch than those of the whey protein.[116–117] Furthermore, dielectric properties were better indicators of phase transitions than were changes in viscosity.[118]

Studies of mixtures of amylopectin with casein and gluten in which glass transitions were measured by DSC and dynamic mechanical thermal analysis (DMTA) indicated that the polymers appeared to be immiscible and showed two glass transitions as a result.[119]

Additional applications

In the examples discussed to this point, relatively high water activity products such as cakes, bread, and sauces and gels were used as illustrations. In fact, structural development can be monitored in part by water loss rate, especially as the system goes from an emulsion-foam to a solid-foam state, as in cake systems.[2, 120] The complexity of gas incorporation and retention and the characteristics of the resulting solid-foam is well known.[121–122] Cakes made with saturated and unsaturated monoglycerides, which showed differences in water emission rates during initial heating, showed relatively small differences upon reheating.[123] Many starch-based foods, although prepared initially at water contents high enough to facilitate gelatinization, are at lower water contents and activities by the end of processing. This situation is typical of some food products manufactured by extrusion processes. In these cases, V patterns have developed by the end of processing, showing the formation of amylose–lipid complexes. The low final moisture content tends to stabilize the basic starch network. The approaches described here are also used to investigate textural differences in high starch foods such as legumes and rices. Recent approaches include the use of size exclusion high performance liquid chromatography (SE-HPLC) to investigate structural properties amylopectin[124–125] in rice cultivars with different textural properties. Properties of maize starch of known genetic history are also studied by these approaches.[126–127]

Microwave heating of cereal-based foods, both during the initial heating and the reheating, tends to produce regions that are firm. The development of these regions may reflect the temperature profiles that develop during heating. These profiles, in turn, reflect differences in the power distribution within the sample. Theoretical modeling combined with experimental studies demonstrated the importance of sample size in considering the importance of reflections at oil–water interfaces that are found in some cereal-based systems.[128–133] These firm regions may be similar to those that develop during staling. There are differences, however. The regions in microwave heated samples often do not show the reversibility of typical staling that is related to the amylopectin component nor are they as easily controlled by additions affecting starch–lipid interactions. Swelling of starch granules during initial heating for cakes and bagels[120, 134–135] was similar to that during conventional heating, and in bagels, the water mobility as evaluated by ¹H NMR was similar.[135] Goebel et al.[136] and Zylema et al.[137] in earlier studies of model starch–water showed by microscopy that no new structures were formed during microwave heating, but the extent of swelling may be different. Differences in gluten–starch interactions were found, however, in ^{13}C NMR studies of gluten–starch–water model systems.[138] These results, as with other processes, demonstrate the complexity of the starch transformations.

Nonstarch polysaccharides

Nonstarch polysaccharides are characterized by a greater diversity of chemical structures than is present in starch polymers. They are derived from a variety of botanical and microbiological sources (Table 1) and are seldom used commercially in a native, organized structure comparable to that found in starch granules. The extent of purification and extraction methods may vary so that even within a general type, variations may be present that reflect the source or the extraction methods. In some instances (for example, pectins and gellan) modifications of primary structure may be introduced by hydrolysis of the methoxyl groups in pectin and deacylation in gellan gums. Chemical modifications of the type used in starches are not frequently used, although propylene glycol derivatives of alginates are available.

Fiber preparations are an exception in that most fiber preparations are composed essentially of cell wall fragments. They may contain proteins and lignins in addition to nonstarch polysaccharides in organized structures that are typical of the specific source.

The diversity of chemical structures in nonstarch polysaccharides includes the type of unit monosaccharide, the sequence of specific monosaccharides, oxidation and derivatization of unit monosaccharides, D- and L-series. The polymers may differ in types of linkages such as α- and β-glycosidic linkages, and 1→3 and 1→2 linkages in addition to 1→4 and 1→6 linkages found in starch polymers, and a range of molecular weights. Some are characterized by repeating sequences, some by alternating sequences of branched polymers and linear polymers as in galacto- and glucomannans and in pectins. The branched regions are sometimes referred to as "hairy" in contrast to "smooth" or unbranched regions. The interruption of a sequence is sometimes viewed as a "kink." Examples are the 1→2 linkage of rhamnose in the galactouronic acid chain of pectin, and the interruption of the 3, 6 anhydro galactose sulfate chains by galactose units in carrageenan. In purified β-glucans from oats, the "kink" in the chain comes from the interruption of sequences of β(1→4)-D-glucopyranose units by β(1→3)-D-glucopyranose units.[139]

Nevertheless, the mechanisms of gelation and the rheological behavior of nongelling polysaccharides are remarkably similar when viewed from general polymer theory. Junction points are characterized for associations between polymer chains. There may be regions of helical structure or cation-mediated junction points for the uronic-acid-containing polymers. For some polysaccharides, copolymers are required for gelation. Initial clarifications of gelation mechanisms were made by Rees and coworkers[57–58] based initially on studies of dry fibers and confirmed subsequently by techniques such as circular dichroism, optical rotary dispersion, NMR, ESR, and DSC, and more recently by DMTA.[140] Structures are also being characterized by scanning tunneling microscopy.[141]

Morris has summarized four types of models of binary polymer gel networks: a network based on association of one polymer, with the second polymer contributing to the overall viscosity; "independent" gelation of each polymer; intermolecular binding between two polymers; and "phase-separated" networks in which "demixing" of two polymers occurs, followed by gelation of each polymer.[53] The phase-separated model was discussed for the starch system in the previous section. Dea summarized the crosslinking that occurs in the transformation from random coils to junction points for specific polysaccharides, which includes double helices for carrageenans and agarose, calcium-mediated crosslinkages for uronic-acid-containing polysaccharides such as the alginates, and binary associations as in agarose–galactomannan and xanthan–galactomannan gel systems.[142–143]

Table 1 Representative Nonstarch Polysaccharides Used in Food Systems

Name	Source	Unit Monosaccharides	Repeating Structures
Carrageenan	Irish moss *Chrondrus crispus*	3,6-Anhydro D-galactose D-Galactose (sulfate substituted)	Alternating 1,3-linked α-D-galactose and 1,4-linked 3,6-anhydro-α-D-galactose
Agar (agarose)	Red seaweed *Rhodophyceae*	3,6-Anhydro-L-galactose D-Galactose	Alternating 1,3-linked β-D-galactose and 1,4-linked-L-galactose
Locust bean gum	Carob bean	D-Mannose D-Galactose	β(1→4)-D-mannose backbone α(1→6)-D-galactose side chains
Guar gum	*Cyamopsis tetragonoloba* (Endosperm polysaccharide)	D-Mannose D-Galactose	β(1→4)-D-mannose backbone α(1→6)-D-galactose side chain
Konjac mannan	Konjac	D-Mannose D-Galactose	β(1→4)-linked mannose and (1→4)-linked β-D-glucose
Alginate	Brown seaweed *Phaeophyta*	D-Mannuronic acid L-Guluronic acid	Sequences of β(1→4)-D-mannuronic acid α(1→4)-L-guluronic acid alternating sequence also
Xanthan gum	*Xanthomonas campestris*	D-Glucose D-Mannose D-Glucuronic acid	β(1→4)-D-glucose backbone 1→3-linked side chain with α-D-mannose-6-0-acetyl, 1→2-linked β-D-glucuronic acid 1→4-linked β-D-mannose and pyruvic acid
Gellan	*Auromonas elodea*	D-Glucose D-Glucuronic acid L-Rhamnose	→3-β-D-glucose (1→4)-β-D-glucuronic acid-(1→4)-β-D-glucose (1→4)-α-L-rhamnose
Gum arabic	Acacia (exudate)	D-Galactose D-Glucuronic acid L-Rhamnose L-Arabinose	1–3-linked D-galactose
Gum tragacanth	Astragalus (exudate)	D-Galacturonic acid D-Galactose L-Fucose D-Xylose L-Arabinose	Complex mixture of fractions

Later studies have used DSC to demonstrate additive gelation mechanisms in systems such as agarose and κ-carrageenan.[144]

These models are built on ideas of flexible and rod-like polymers.[92] Because of the size of the polymer chains and the potential for interchain interactions, polymer interactions make major contributions to macroscopic behavior, such as viscoelastic and gelation, although there are other properties associated with the segments of the polymers. Stability of a specific polymer system, therefore, depends both on the stability of the associations and the stability of the polymer chain in terms of resistance to degradation to lower molecular weight segments.

The potential for freely rotating chains is related to motion around primary bonds. Differences between the flexibility of polypeptides and polysaccharide chains may contribute to the properties of mixed protein and polysaccharide systems. Mobility of polysaccharide polymer chains has been studied less frequently than that of polypeptide chains, but recent studies have used spin-labeled glucomannan[145] and κ-carrageenan[146] in ESR spectroscopy to examine molecular motion. NMR techniques have also been used to examine molecules as in the study of gluco- and galactomannans using ^{13}C NMR.[147]

Heat reversibility for various polysaccharides was shown during heating and cooling cycles by differential scanning calorimetry. Melting and reforming of the gel structures is faster than that for starch polymers in which reversibility is measured in hours rather than minutes, as for most nonstarch polysaccharide polymers. Some hysteresis was observed in the heating and cooling curves. Hysteresis was found for ι, but not κ-carrageenan.[148] More recently, hysteresis was found for gellan gums if NaCl was present but not if it was absent. Hysteresis was attributed to a two-stage gelation process.[149] The two stages of the association process were detectable in the cooling cycle, but the successive melting points were not detected in the heating cycles.

Disruption of initial structures by heating is now considered an essential step in the formation of the "sandwich" binary system of xanthan gums and galactomannans.[150]

Applications

These general concepts form the basis for the properties of nonstarch polysaccharides in specific applications. Response to shear, including both shear-thinning and shear-thickening, becomes a factor. Response to pH ranges and specific ions also relates to specific rheological properties. Heat- and freeze-stability are additional factors in selection of specific systems.

Some examples of stabilizing functions related to rheological effects, but not necessarily gelation, are the use of carrageenans in evaporated milk and milk drinks. Gum arabic, gum acacia, modified food starches, and gum tragacanth are used as suspending agents in beverages.[151] Relationships between dispersion states and flavor suppression were studied by Baines and Morris.[152-153] Reduction of flavor intensity paralleled the transitions in structure such as random coil → coil overlap and entanglements, weak gels, and strong gels.

The role of nonstarch polysaccharide in emulsion and foam stabilization is variously attributed to viscosity effects via Stokes' law effects and interfacial properties. Those polysaccharides that participate at the interface usually have amphiphilic character, either naturally or through chemical modification.

Concluding remarks

In this chapter, we have discussed the stability of carbohydrates at several levels of structure. We have also discussed the experimental evidence derived from the application of a range of instrumental analysis methods applied to model systems. These studies provide information at the atomic and molecular level about component interactions. This information forms the basis for understanding cereal-based formulated systems at the microstructural level, which in turn contributes toward understanding macroscopic properties of the full formulation.

We have tried to show the range of studies of carbohydrates, and particularly those of the polysaccharides, within this framework. Much progress has been made, but new information continues to become available from which to develop a rational framework to respond to the availability of new ingredients, new technologies, and changing consumer requirements rather than relying on short-term formulation testing.

References

1. Eads, T.M. *Trends Food Sci. Technol.* 1994, *5*, 147-159.
2. Cloke, J.D.; Davis, E.A.; Gordon, J. *Cereal Chem.* 1984, *61*, 363-371.
3. Whistler, R.L.; Daniels, J.R. In *Food Chemistry,* 2nd ed.; Fennema, O., Ed.; Marcel Dekker: NY, 1985; pp 69-137.
4. *The Maillard Reaction in Food and Nutrition;* Waller, G.R.; Feather, M.S., Eds.; ACS Symposium Series 215; American Chemical Society: Washington, D.C., 1983.
5. *The Maillard Reaction in Food Processing, Human Nutrition, and Physiology;* Finot, P.A.; Aeschbacher, H.U.;Hurrell, R.F.; and Liardon, R., Eds.; Birkhauser Verlag: Basel, 1990.
6. Reineccius, G.A. In *The Maillard Reaction in Food Processing, Human Nutrition, and Physiology;* Finot, P.A.; Aeschbacher, H.U.; Hurrell, R.F.; and Liardon, R., Eds.; Birkhauser Verlag: Basel, 1990; pp 157-170.
7. Wharton, C.; Reineccius, G.A. *Cereal Foods World.* 1990, *35*, 553-559.
8. Wei, C.K.; Davis, H.T.; Davis, E.A.; Gordon, J. *AIChE J.* 1985, *31*, 842-848.
9. Roos, Y.H.; Himberg, M.-J. *J. Ag. Food Chem.* 1994, *42*, 893-898.
10. Munro, E.M. *Cereal Foods World.* 1994, *39*, 552-555.
11. Doane, W.M. *Cereal Foods World.* 1994, *39*, 556-563.
12. Englyst, H.N.; Kingman, S,M,; Cummings, J.H. In *Plant Polymeric Carbohydrates;* Meuser, F.; Manners, D.J.; Seibel, W., Eds.; Special Publication 134; Royal Society of Chemistry: Cambridge, 1993; pp 137-146.
13. Chiu, C.-W.; Henly, M.; Altieri, P. U.S. Patent 5 281 276, 1994.
14. Eerlingen, R.C.; Delcour, J.A. *J. Cereal Sci.* 1995, *22*, 129-138.
15. Cairns, P.; Sun, L.; Morris, V.J.; Ring, S.G. *J. Cereal Sci.* 1995, *21*, 37-47.
16. Annison, G.; Topping, D.L. In *Annual Review of Nutrition;* Olson, R.E.; Bier, D.M.; McCormick, D.B., Eds.; Annual Reviews; Palto Alto, CA, 1994; Vol 14, pp 297-319.
17. Czuchajowska, Z.; Sievert, D.; Pomeranz, Y. *Cereal Chem.* 1991, *68*, 537-542.
18. Lin, P.-Y.; Czuchajowska, Z.; Pomeranz, Y. *Cereal Chem.* 1994, *71*, 69-75.
19. Janecek, S.; MacGregor, E.A.; Svensson, B. *Biochem J.* 1995, *305*, 685-688.
20. MacGregor, E.A.; MacGregor, A.W.; Macri, L. J.; Morgan, J.E. *Carbohydrate Res.* 1994, *257*, 249-268.
21. MacGregor, E.A. *Starch.* 1993, *45*, 232-237.
22. Lineback, D.R.; Rasper, V.F. In *Wheat: Chemistry and Technology,* 3rd ed.; Pomeranz, Y., Ed.; American Association of Cereal Chemists: St. Paul, MN, 1988; Vol. 1, pp 277-372.
23. Blanshard, J.M.V. In *Starch: Properties and Potential;* Galliard, T., Ed.; Critical Reports on Applied Chemistry; Society of Chemical Industry; Wiley: Chichester, England, 1987; Vol. 13, pp 16-54.
24. French, D. In *Starch Chemistry and Technology,* 2nd ed.; Whistler, R.L.; BeMiller, J.N.; and Paschall, E.F., Eds.; Academic Press: Orlando, FL, 1984; pp 183-247.
25. Zobel, H.F. *Starch.* 1988, *40*, 44-50.
26. Zobel, H.F. *Starch.* 1988, *40*, 1-7.
27. Sarko, A.; Wu, H.-C.H. *Starch.* 1978, *30*, 73-78.
28. Kim, Y.-H.; Nelson, J.T.; Glynn, A.B. *Cereal Foods World.* 1994, *39*, 8-18.
29. *Computer Modeling of Carbohydrate Molecules;* French, A.D.; Brady, J.W., Eds.; ACS Symposium Series 430; American Chemical Society: Washington, D.C.; 1990.
30. Hinrichs, W.; Buttner, G.; Steifa, M.; Betzel, Ch.; Zabel, V.; Pfannemuller, B.; Saenger, W. *Science.* 1987, *238*, 205-208.
31. Imberty, A.; Buleon, A.; Tran, V.; Perez, S. *Starch.* 1991, *43*, 375-384.
32. Gidley, M.J.; Bociek, S.M. *J. Am. Chem. Soc.* 1985, *107*, 7040-7044.
33. Kasemsuwan, T.; Jane, J. *Cereal Chem.* 1994, *71*, 282-287.
34. Muhrback, P.; Svensson, E.; Eliasson, A.-C. *Starch.* 1991, *43*, 466-468.
35. Muhrback, P.; Eliasson, A.-C. *J. Sci. Food Agric.* 1991, *55*, 13-18.
36. Montgomery, E.M.; Senti, F.R. *J. Polym. Sci.* 1958, *28*, 1-9.
37. Stevens, D.J.; Elton, G.A.H. *Starch.* 1971, *23*, 8-11.
38. Wooton, M.; Bamunurachchi, A. *Starch.* 1979, *31*, 201-204.

39. Kugimiya, M.; Donovan, J.W. *J. Food Sci.* 1981, *46*, 765-770, 777.
40. Russell, P.L. *J. Cereal Sci.* 1987, *6*, 133-145.
41. Russell, P.L. *J. Cereal Sci.* 1987, *6*, 147-158.
42. Oostergetel, G.T.; van Bruggen, E.F.J. *Carbohydrate Polym.* 1993, *21*, 7-12.
43. Biliaderis, C.G. In *Water Relationships in Foods*; Levine, H.; Slade, L., Eds.; Plenum Press: NY, 1991, pp 251-273.
44. Slade, L.; Levine, H. In *Industrial Polysaccharides*; Stivala, S.S.; Crescenzi, V.; Dea, I.C.M., Eds.; Gordon and Breach: NY, 1987; pp 327-430.
45. Slade, L.; Levine, H. *Carbohydrate Polym.* 1988, *8*, 183-208.
46. Slade, L.; Levine, H. In *Water Relationships in Foods*; Levine, H.; Slade, L., Eds.; Plenum Press: NY, 1991; pp 29-101.
47. Slade, L.; Levine, H. *J. Food Eng.* 1994, *22*, 143-188.
48. Slade, L.; Levine, H. In *The Glassy State in Foods*; Blanshard, J.M.V., Ed. Nottingham University Press, Loughborough, England, 1993; pp 35-101.
49. Slade, L.; Levine, H.; Ievolla, J.; Wang, M. *J. Sci. Food Agric.* 1993, *63*, 133-176.
50. *Phase Transitions in Foods*; Roos, Y.H.; Academic Press: San Diego, 1995.
51. Atwell, W.A.; Hood, L.F.; Lineback, D.R.; Varriano-Marston, E.; Zobel, H.F. *Cereal Foods World.* 1988, *33*, 306-311.
52. Bowler, P.; Williams, M.R.; Angold, R.E. *Starch.* 1980, *32*, 186-189.
53. Morris, V.T. In *Food Polymers, Gels, and Colloids*; Dickinson, E.; Special Publication 82, The Royal Society of Chemistry: Cambridge, England, 1991; pp 310-321.
54. Miller, B.S.; Derby, R.I.; Trimbo, H.B. *Cereal Chem.* 1973, *50*, 271-280.
55. Miles, M.J.; Morris, V.J.; Ring, S.G. *Carbohydr. Res.*1985, *135*, 257-269.
56. Miles, M.J.; Morris, V.J.; Orford, P.D.; Ring, S.G. *Carbohydr. Res.* 1985, *135*, 271-281.
57. Rees, D.A. In *Advances in Carbohydrate Chemisrty and Biochemistry*; Wolfrom, W.L.; Tipson, R.S.; Horton, D., Eds.; Academic Press: NY, 1969; Vol. 24, pp 267-332.
58. Rees, D.A. *Biochem. J.* 1972, *126*, 257-273.
59. Martin, M.L.; Zeleznak, K.J.; Hoseney, R.C. *Cereal Chem.* 1991, *68*, 498-503.
60. Biliaderis, C.G.; Galloway, G. *Carbohydrate Res.* 1989, *189*, 31-48.
61. Mikus, F.F.; Hixon, R.M.; Rundle, R.E. *J. Am. Chem. Soc.* 1946, *68*, 1115-1123.
62. Morrison, W.R.; Law, R.V.; Snape, C.E. *J. Cereal Sci.*1993, *18*, 107-109.
63. Pearce, L.E.; Davis, E.A.; Gordon, J.; Miller, W.G. *Cereal Chem.* 1987, *64*, 154-157.
64. Riisom, T.; Krog, N.; Erickson, J. *J. Cereal Sci.* 1984, *2*, 105-118.
65. Krog, N.J. In *Food Emulsions*, 2nd ed.; Larsson, K.; Friberg, S.E., Eds.; Marcel Dekker: NY, 1990; pp 127-180.
66. Noznick, P.P.; Merritt, P.P.; Geddes, W.F. *Cereal Chem.* 1946, *23*, 297-305.
67. Russell, P.L. *J. Cereal Sci.* 1983, *1*, 297-303.
68. Zeleznak, K.J.; Hoseney, R.C. *Cereal Chem.* 1986, *63*, 407-411.
69. Krog, N.; Oleson, S.K.; Toernaes, H.; Joensson, T. *Cereal Foods World.* 1989, *34*, 281-285.
70. Czuchajowska, Z.; Pomeranz, Y. *Cereal Chem.* 1989, *66*, 305-309.
71. Colonna, P.; Barry, J.-L.; Cloarec, D.; Bornet, F.; Gouilloud, S.; Galmiche, J.-P. *J. Cereal Sci.* 1990, *11*, 59-70.
72. Cloke, J.D.; Gordon, J.; Davis, E.A. *Cereal Chem.* 1983, *60*, 143-146.
73. Yang, M.; Grider, J.; Gordon, J.; Davis, E.A. *Food Microstructure.* 1985, *11*, 107-114.
74. Cameron, R.E.; Donald, A.M. *Carbohydrate Res.* 1993, *244*, 225-226.
75. Liu, H.; Lelievre, J. *Carbohydrate Res.* 1991, *219*, 23-32.
76. Pearce, L.E.; Davis, E.A.; Gordon, J.; Miller, W.G. *Food Microstructure.* 1985, *4*, 83-88.
77. Johnson, J.M.; Davis, E.A.; Gordon, J. *Cereal Chem.* 1990, *67*, 286-291.
78. Slade, L.; Levine, H. In *Food Structure—Its Creation and Evaluation*; Blanshard, J.M.V.; Mitchell, J.R., Eds.; Butterworths: London, 1988; pp 115-147.
79. Bachrach, H.L.; Briggs, D.R. *Cereal Chem.* 1947, *24*, 492-506.
80. Umbach, S.L.; Davis, E.A.; Gordon, J.; Callaghan, P.T. *Cereal Chem.* 1992, *69*, 637–642.
81. Callaghan, P.T.; Jolley, K.W.; Lelievre, J.; Wong, R.B.K. *J. Colloid Interface Sci.* 1983, *92*, 332-337.
82. Wynne-Jones, S.; Blanshard, J.M.V. *Carbohydr. Polym.* 1986, *6*, 289-306.

83. Lechert, H.; Maiwald, W.; Kothe, R.; Basler, W.-D. *J. Food Processing Preservation*. 1980, *3*, 275-299.
84. Richardson, S.J. *Cereal Chem*. 1989, *66*, 244-246.
85. Callaghan, P.T. *Principles of Nuclear Magnetic Resonance Microscopy*; Oxford University Press: Oxford, 1993.
86. McCarthy, M.J. *Magnetic Resonance Imaging in Foods*; Chapman and Hall: NY, 1994.
87. Kalichevsky, M.T.; Jaroszkiewicz, E.M.; Ablett, S.; Blanshard, J.M.V.; Lillford, P.J. *Carbohydrate Polymers*. 1992, *18*, 77-88.
88. Kalichevsky, M.T.; Blanshard, J.M.V. *Carbohydrate Polym*. 1993, *20*, 107-113.
89. Windle, J.J. *Starch*. 1985, *37*, 121-125.
90. Nolan, N.L.; Faubion, J.M.; Hoseney, R.C. *Cereal Chem*. 1986, *63*, 287-291.
91. Biliaderis, C.G.; Vaughan, D.J. *Carbohydrate Polym*. 1987, *7*, 51-70.
92. Doi, M.; Edwards, S.F. *The Theory of Polymer Dynamics*; Clarendon: Oxford, 1986.
93. Biliaderis, C.G.; Zawistowski, J. *Cereal Chem*. 1990, *67*, 240-246.
94. Biliaderis, C.G.; Page, C.M.; Maurice, T.J.; Juliano, O. *J. Ag. Food Chem*. 1986, *34*, 6-14.
95. Appelqvist, I.A.M.; Cooke, D.; Gidley, M.J.; Lane, S.J. Carbohydrate Polymers. 1993, *20*, 291-299.
96. Nicholls, R.J.; Appelqvist, I.A.M.; Davies, A.P.; Ingman, S.J.; Lillford, P.J. *J. Cereal Sci*. 1995, *21*, 25-36.
97. Roozen, M.J.G.W.; Hemminga, M.A. *J. Phys. Chem*. 1990, *94*, 7326-7329.
98. Bean, M.L.; Osman, E.M. *Food Res*. 1959, *24*, 665-671.
99. Cairns, P.; Miles, M.J.; Morris, V.J. *Carbohydr. Polym*. 1991, *16*, 355-365.
100. I'Anson, K.J.; Miles, M.J.; Morris, V.J.; Besford, L.S.; Jarvis, D.A.; Marsh, R.A. *J. Cereal Sci*. 1990, *11*, 243-248.
101. Eliasson, A.-C. *Carbohydr. Polym*. 1992, *18*, 131-138.
102. Prokopowich, D.J.; Biliarderis, C.G. *Food Chem*. 1995, *52*, 255-262.
103. Sahagian, M.E.; Goff, H.D. *Food Res. Internat*. 1995, *28*, 1-8.
104. Kokini, J.L.; Cocero, A.M.; Madeka, H.; de Graf, E. *Trends in Food Sci. Technol*. 1994, *5*, 281-288.
105. Arvanitoyannis, I.; Blanshard, J.M.V.; Ablett, S.; Izzard, M.J.; Lillford, P.J. *J. Sci. Food Agric*. 1993, *63*, 177-188.
106. Chan, R.K.; Pathmanathan, K.; Johari, G.P. *J. Phys. Chem*. 1986, *90*, 6358-6362.
107. Box, G.E.P. *Biometrics*. 1954, *10*, 16-60.
109. Bice, C.W.; Geddes, W.F. In *Starch and Its Derivatives*; 3rd ed.; Radley, J.A., Ed.; Chapman and Hill: London, 1953; pp 202-242.
109. Godet, M.C.; Bizot, H.; Buleon, A. *Carbohydrate Polym*. 1995, *27*, 47-52.
110. Zhou, N. Ph.D. Thesis, University of Minnesota, November 1995.
111. Willhoft, E.M.A. *J. Texture Studies*. 1973, *4*, 292-322.
112. LeMeste, M.; Duckworth, R.B. *Int. J. Food Sci. Technol*, 1988, *23*, 457-466.
113. LeMeste, M.; Closs, B.; Courthaudon, J.L.; Colas, B. In *Interactions of Food Proteins*; Parris, N.; Barford, R., Eds.; ACS Symposium Series 454; American Chemical Society: Washington, D.C., 1991; pp 137-147.
114. Schanen, P.A.; Pearce, L.E.; Davis, E.A.; Gordon, J. *Cereal Chem*. 1990, *67*, 124-128.
115. Schanen, P.A.; Pearce, L.E.; Davis, E.A.; Gordon, J. *Cereal Chem*. 1990, *67*, 317-322.
116. Tsoubeli, M.N.; Davis, E.A.; Gordon, J. *Cereal Chem*. 1995, *72*, 64-69.
117. Fleischmann, A.M. M.S. Thesis, University of Minnesota, 1991.
118. Barringer, S.A.; Fleischmann, A.M.; Davis, E.A.; Gordon, J. *Food Hydrocolloids*, 1995, *5*, 343-348.
119. Kalichevsky, M.T.; Blanshard, J.M.V. *Carbohydr. Polym*. 1992, *19*, 271-278.
120. Lambert, L.L.P.; Gordon, J.; Davis, E.A. *Cereal Chem*. 1992, *69*, 303-309.
121. Carlin, G.T. *Cereal Chem*. 1944, *21*, 189-199.
122. Gan, Z.; Ellis, P.R.; Schofield, J.D. *J. Cereal Sci*. 1995, *21*, 215-230.
123. Cloke, J.D.; Davis, E.A.; Gordon, J. *Cereal Chem*. 1984, *61*, 371-374.
124. Ong, M.H.; Blanshard, J.M.V. *J. Cereal Sci*. 1995, *21*, 251-230.
125. Ong. M.H.; Blanshard, J.M.V. *J. Cereal Sci*. 1995, *21*, 261-269.
126. Campbell, M.R.; Pollak, L.M.; White, P.J. *Cereal Chem*. 1995, *72*, 281-286.

127. Kasemsuwan, T.; Jane, J.; Schnable, P.; Stinard, P.; Robertson, D. *Cereal Chem.* 1995, *72*, 457-464.

128. Ayappa, K.G.; Davis, H.T.; Crapiste, G.; Davis, E.A. *J. Chem. Eng. Sci.* 1991, *4*, 1005- 1016.

129. Ayappa, K.G.; Davis, H.T.; Davis, E.A.; Gordon, J. *AIChE. J.* 1991, *37*, 313-322.

130. Ayappa, K.G.; Davis, H.T.; Davis, E.A.; Gordon, J. *AIChE. J.* 1992, *38*, 1577-1591.

131. Barringer, S.A.; Davis, E.A.; Gordon, J.; Ayappa, K.G.; Davis, H.T. *AIChE. J.* 1994, *40*, 1433-1439.

132. Barringer, S.A.; Ayappa, K.G.; Davis, E.A.; Davis, H.T.; Gordon, J. *J. Food Sci.* 1995, *60*, 1132-1136.

133. Barringer, S.A.; Davis, E.A.; Gordon, J.; Ayappa, K.G.; Davis, H.T. *J. Food Sci.* 1995, *60*, 1137-1142.

134. Baker, B.A.; Davis, E.A.; Gordon, J. *Cereal Chem.* 1990, *67*, 451-457.

135. Umbach, S.L.; Davis, E.A.; Gordon, J. *Cereal Chem.* 1990, *67*, 355-360.

136. Goebel, N.K.; Grider, J.; Davis, E.A.; Gordon, J. *Food Microstructure.* 1984, *3*, 73-82.

137. Zylema, B.J.; Grider, J.A.; Gordon, J.; Davis, E.A. *Cereal Chem.* 1985, *62*, 447-453.

138. Umbach, S. Ph.D. Thesis, University of Minnesota, October, 1992.

139. Wood, P.J. *Trends Food Sci. Technol.* 1991, *2*, 311-314.

140. Zhan, D.F.; Ridout, M.J.; Brownsey, G.J.; Morris, V.J. *Carbohydr. Polym.* 1993, *21*, 53-58.

141. Gunning, A.P.; McMaster, T.J.; Morris, E.J. *Carbohydr. Polym.* 1993, *21*, 47-51.

142. Dea, I.C.M. In *Food Carbohydrates*; Lineback, D.R.; Inglett, G.E., Eds.; Avi: Westport, CT, 1982; pp 420-457.

143. Dea, I.C.M. In *Industrial Polysaccharides*; Stivala, S.S.; Crescenzi, V.; Dea, I.C.M., Eds.; Gordon and Breach: NY, 1987; pp 367-389.

144. Zhang, J.; Rochas, C. *Carbohydr. Polym.* 1990, *13*, 257-271.

145. Williams, P.A.; Clegg, S.M.; Day, D.H.; Phillips, G.O.; Nishinori, K. In *Food Polymers, Gels, and Colloids*; Dickinson, E. Ed.; Special Publication 82; Royal Society of Chemistry: Cambridge, 1991; pp 339-348.

146. Day, D.H.; Phillips, G.O.; Williams, P.A. *Food Hydrocolloids.* 1988, *2*, 19-30.

147. Gidley, M.J.; McArthur, A.J.; Underwood, D.R. *Food Hydrocolloids.* 1991, *5*, 129-140.

148. Morris, E.R.; Rees, D.A.; Norton, I.T.; Goodall, D.M. *Carbohydr. Res.* 1980, *80*, 317-323.

149. Robinson, G.; Manning, C.E.; Morris, E.R.; Dea, I.C.M. In *Gums and Stabilizers for the Food Industry 4*; Phillips, G.O.; Williams, P.A.; Wedlock, D.J., Eds.; IRL: Oxford, 1988; pp 173-181.

150. Cairns, P.; Miles, M.J.; Morris, V.J. *Nature.* 1986, *322*, 89-90.

151. Tan, C.-T. In *Food Emulsions*, 2nd ed.; Larsson, K.; Friberg, S.E., Eds.; Marcel Dekker: NY, 1990; pp 445-478.

152. Baines, Z.V.; Morris, E.R. In *Food Colloids*; Bee, R.D.; Richmond, P.; Mingins, J., Eds.; Special Publication No. 75., Royal Society Chemistry: Cambridge, 1989; pp 184-192.

153. Morris, E.R. In *Industrial Polysaccharides*; Stivala, S.S.; Crescenzi, V.; Dea, I.C.M., Eds.: Gordon and Breach: NY, 1987; pp 431-457.

chapter six

Biochemical aspects: nutritional bioavailability

K. Ananth Narayan

Contents

0-8493-2646-X/97/$0.00+$.50
© 1998 by CRC Press LLC

Introduction

Overview of nutrient degradation

It is recognized that micronutrients such as thiamin, ascorbic acid, pyridoxine, folacin, tocopherol, retinol, and carotene are to varying extents vulnerable to external and internal stresses, such as temperature, light, water activity, and pH. There is a need to assess and investigate the retention and biological availabilities of macro- and micronutrients as a result of interactions between nutrients and with other constituents. Among those inter-actions that need further investigation are between proteins and reducing sugars[1-5]; between autoxidized lipids and proteins[6-16]; between dehydroascorbic acid and amino compounds[17]; between lipids and carbohydrates, especially during extrusion processing of foods[18-22]; between minerals and oxalates, phytates, and fibers[23-25]; between cobalamin and ascorbic acid[26]; between riboflavin and ascorbic acid[27]; between niacin and peptides or carbohydrates[28]; between thiamin and carbohydrates[29]; between thiamin and municipal drinking water[30] or brewed tea[31]; among thiamin, nicotinamide, and cobalamin[32]; and between riboflavin and thiamin.[33]

Driving forces affecting nutrient bioavailability

It is almost axiomatic to state that temperature is the major driving force influencing nutrient retention and bioavailability in stored foods. Other prominent factors are water activity (A_w), pH, packaging, and visible and ultraviolet radiation. Another important factor is the presence, abundance, and chemical structure of the interacting species. For example, the maintenance of high protein quality in protein-rich foods is affected by even small concentrations of reducing sugars or oxidizing lipids.

Distinction between nutrient content and bioavailability

The distinction between nutrient content and nutrient bioavailability is in many instances (e.g., vitamins) not clearly evident. Nutrient content represents the sum total of the nutrient as measured by recognized assay procedures, which in many cases is a chemical method; nutritional bioavailability represents only that portion of the nutrient which is digested, absorbed, and metabolized. Because there is a need to minimize animal experimentation, reliance upon *in vitro* enzymatic procedures for nutrient bioavailability has increased. In such instances, nutritional bioavailability may be narrowly redefined as that portion which is available after enzymatic digestion for absorption and metabolism in the body.

Some examples illustrating the difference between nutrient content and nutritional availability are instructive. The iron content of spinach is high, but only a small portion (1.3%) is available for absorption due to chelation of iron by oxalic acid in spinach, even when the ingested dose is small (2 mg). Similarly, substantial levels of fat-soluble vitamins ingested with high levels of synthetic fats, such as sucrose polyesters, may be less available due to interference in absorption by these fats. Food proteins interact with oxidizing polyunsaturated fats (PUF) during storage and might be poorly digested by proteolytic enzymes. Thus, biological availability is compromised, whereas the total content of the proteins as determined by Kjeldahl nitrogen remains largely unaffected.

Scope

This review will be restricted to foods in the pH range of 5 to 7. Nutrient–vitamin and nutrient–mineral interactions and their influence upon the bioavailabilities of vitamins and minerals are not covered. The focus will be on those interactions that affect the protein quality of foods. In general, due to the lack of data, the retention of vitamins rather than their bioavailabilities have been reported in the literature. Baker reported on the poor biological availability of several vitamins in some foods: niacin (10%) in corn; biotin (40%) and pathothenic acid (60%) in wheat and barley; and vitamin B_2 (60%) in corn and soymeal.[34] The numerous interactions that influence the bioavailabilities of vitamins and minerals have been the subject of many reviews and research articles.[35-47] The metabolic interrelationships among vitamins, minerals, and hazardous elements have been the subject of an earlier conference.[39] Accordingly, the thrust will be on protein availability changes during storage of foods, with some discussion on processing effects.

This review will not address the metabolic consequences of either macro- or micronutrients, especially where they are in great excess. For example, it will not address the question of nutritional consequences of ingesting of excess lysine, leucine, protein, fat, or fiber in the diet.

Biological availability of proteins

The bioavailability of food proteins is compromised when they become less digestible by proteolytic enzymes as a result of crosslinking reactions between proteins and carbohydrates, proteins and oxidized lipids, and proteins and other components in food.

Protein–carbohydrate interaction

Nature of the reaction

By far the most damaging reactions are those between proteins (or free amino acids) and carbonyl compounds, particularly reducing sugars. The Maillard reaction[48-50] is a complex multistage reaction and has been the subject of several excellent reviews and investigations.[51–67] An extensive review of the early literature (1915–1950) on heat injury to dietary protein is also available,[68] and one must acknowledge that some of the current considerations of this subject were recognized, investigated, and properly interpreted over 80 years ago. For example, in 1915, McCollum and Davis understood the cause of the loss of nutritive efficiency in heated milk.[69] In 1925, Morgan and King reported retardation of growth in rats fed browned food products, such as toasted crackers, toasted whole wheat, toasted white bread, puffed wheat, puffed rice, or the crust of white bread,[70] which has been confirmed and extended in subsequent investigations.[71,72] Tsen et al. showed that microwave and steam-baked breads were superior in protein quality to conventionally baked bread.[71] Browned albumin following interaction with glucose was demonstrated to be a poor protein source.[72]

In the first phase of the Maillard reaction, the Schiff base formed between the free ϵ-amino groups in lysine of proteins and the carbonyl group of reducing sugars cyclize to the corresponding *N*-substituted glucosylamine and then undergoes an irreversible Amadori rearrangement to form ketose sugar derivatives (*N*-substituted 1-amino-1-deoxy, 2-ketose, commonly referred to as the Amadori compounds [ACs]). The ACs decompose in the subsequent stage of the reaction to give rise to a variety of scission products, carbonyl compounds, heterocyclic flavor compounds, as well as brown melanoidin pigments.[1,51] The ACs, as well as the carbonyl compounds formed upon their degradation, are recognized for their specific reducing properties and afford a simple means of monitoring the reaction. Some of these compounds are powerful lipid antioxidants and, hence, are extremely valuable in extending the shelf life of foods containing polyunsaturated fats.[73] The compounds produced[1, 74] during the second stage of the reaction are responsible for many of the pleasing aromas and flavors that are cherished so much in foods, such as bread, coffee, cocoa, roasted nuts, maple syrup, beer, and wine. They enhance the sensory acceptance of foods and indirectly improve shelf life. Unfortunately, these components also degrade and produce undesirable off-flavors.

From a nutritional standpoint, the compromise to protein quality occurs at the first stage of the Maillard reaction (MR), because the ACs are not digested by mammalian proteolytic enzymes and so they limit the biological availability of lysine (LYS) in foods. Data indicates considerable loss of LYS due to MR in model systems during relatively mild storage conditions.[4,75,76] The loss of LYS in an LYS-GLU-cellulose model (where GLU refers to glucose) at a low A_w of 0.2 as a function of time and temperature is shown in Figure 1. As much as 25% of LYS was lost in this model system in 6 days at 40°C. Further LYS loss is significantly less rapid since the reaction obviously does not follow zero-order kinetics. The brown color developed during the final stages is sometimes used erroneously as a measure of damage to protein quality. Confirming other literature data, the study clearly illustrated that as much as 70% of the LYS was lost in the LYS-GLU-cellulose model system *before* there was a significant increase in color (Figure 2) or in fluorescence (not shown).[4] However, furosine correlated well ($r = 0.98$) with LYS loss.

The reaction between LYS and GLU has generally been considered to be first-order. However, semilog plots of LYS concentration vs. time do not give straight lines (beyond 50–70% LYS loss).[4] It has been argued that the reaction more closely follows second-order kinetics than first-order kinetics.[77] After a period of time, LYS concentration levels off faster in a second-order reaction than in a first-order reaction.

From a practical standpoint, however, protein-rich foods containing reducing sugars and stored for long periods (such as in the case of military rations) may not lose their protein quality as rapidly as would be predicted by first-order kinetics. An alternative is to use curve-fitting approaches to predict shelf life over extended periods. Narayan and Andreotti fitted the data from LYS-GLU-cellulose and *N*-alpha-acetyllysine (NAL)-GLU-cellulose models to double exponential models.[4] These adequately describe LYS loss

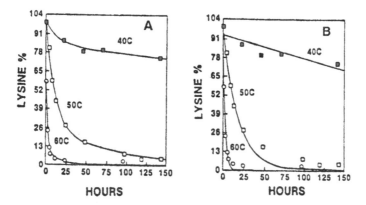

Figure 1   Lysine loss at various temperatures and low A_w in the range of 0.19 to 0.21 with compressed lysine-glucose models. (A) lysine data fit to the double exponential equation, $y = Ae^{Bx} + De^{Fx}$; and (B) lysine data fit to the first order exponential equation, $y = Ae^{Bx}$. (From Narayan, K.A. and Andreotti, R.E. 1989. J. Food Biochem. 13: 105-125. With permission).

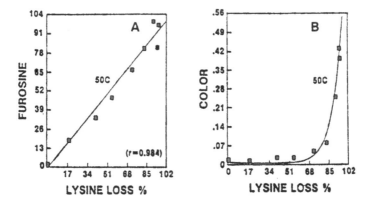

Figure 2   Correlation between the increase in (A) furosine ([10⁷] Data Module 930 peak area units/mg lysine) and (B) color absorbance at 410 nm/mg lysine with the decrease in lysine in the compressed lysine–glucose–cellulose model at 50°C. (From Narayan, K.A. and Andreotti, R.E. 1989. J. Food Biochem. 13: 105-125. With permission).

during storage. An excellent fit ($r > 0.99$) between observed and predicted values is seen for LYS-GLU-cellulose models,[4] as well as for NAL-GLU-cellulose models[77] at 40°C, 50°C, and 60°C. The equation used is

$$C = ae^{bt} + ce^{dt}$$

where C = concentration of lysine (mg/g model) expressed as percentage of initial concentration; t = time; and a, b, c, and d are derived constants.

Amino acid models

Amino acid-reducing sugar models are very useful for determining the effects on the Maillard reaction of temperature, A_w, pH, additives, and blocking agents. The early work of Hannan and Lea, who reacted NAL with GLU at 37°C and 20%, 40%, and 60% RH, showed that as much as 25% of NAL was lost in 3 days at an RH of 20%.[76] Eichner investigated the reactivity of LYS-GLU-cellulose models and observed significant effects

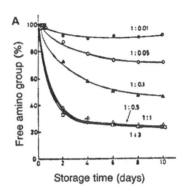

Figure 3 Maillard reaction between ovalbumin and aldohexoses at 50°C and 65% relative humidity. Decrease in free amino group of ovalbumin stored with various levels of glucose and galactose, respectively for 10 days. (From Kato, T., et al. 1986. J. Agric. Food Chem. 34: 351-355. With permission.)

of variables such as A_w, glycerol, and cysteine.[73] From an Arrhenius plot utilizing rate constants for Maillard browning at 80°C, 90°C, and 100°C for glucose–glycine and glucose–aspartame models, Stamp and Labuza have accurately predicted the number of days to reach an optical density (OD) of 0.1 at a low (45°C) temperature, estimating 62 vs. observing 60 days.[78] Amino acid models, such as dextrose and lysine as well as tryptophan and cysteine, have been used in an attempt to predict the shelf life of parenteral nutrition solutions for human use.[75]

Protein–carbohydrate models and food systems

Because of the ease with which optical absorption measurements can be made, Maillard browning rather than Maillard protein quality decrement has been widely investigated. Interesting observations were made by Kato and colleagues in their studies with ovalbumin (OVA)-aldohexose models at 50°C and RH of 65% incubated for 0–26 days.[79,80] The loss in free amino groups was essentially the same with all three sugars (GLU, mannose (MAN), and galactose (GAL)) used (Figure 3).[80] In all cases, the free amino groups decreased with increase in the ratio of sugar to protein. However, GAL produced maximum browning[80] and a high degree of protein insolubilization (Figure 4), which is of concern because this sugar may be present in stored dairy products. In studies comparing OVA-GLU (1:1 w/w) and OVA-lactose (LAC) (1:2 w/w) models (79), it was observed that there was virtually no browning with the LAC model (Figure 5A), but there was substantial loss (70–80%) in free amino groups in both models over a period of 26 days. Once again, this emphasizes that although stored dairy products may be virtually colorless, they could have suffered a significant loss in protein quality. However, such high losses in LYS with a protein–dissacharide model is contrary to normal experience and may be due to the high A_w (65% RH) used.

It is generally accepted that browning is an extension of the first phase of the Maillard reaction[1] in which the ACs first break down to 3-deoxyhexasone, methyl dicarbonyl intermediates, hydroxymethyl furfuraldehyde, and reductones, and subsequently undergo a series of condensation–polymerization reactions to form melanoidin pigments. This work has demonstrated that if the blocked protein is separated from the unreacted sugars, further browning virtually ceases even for the OVA-GLU blocked protein (Figure 5B).[79] Addition of GLU or LAC to the two blocked protein models free of unreacted sugars showed that GLU, but not LAC, was essential for further browning (Figure 5C, 5D).

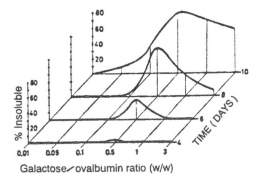

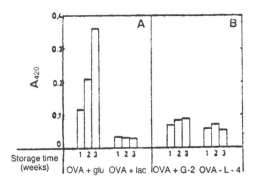

Figure 4 Insolubility of ovalbumin by storage with galactose for 4–10 days at 50°C and 65% RH. (From Kato, T., et al. 1986. J. Agric. Food Chem. 34: 351-355. With permission.)

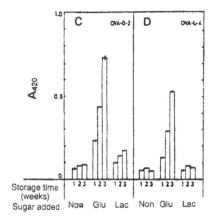

Figure 5 The effect of the type, presence, or absence of reducing sugar on the browning of ovalbumin (OVA). (A) OVA-glucose (GLU) and OVA-lactose (LAC) systems stored for 1–3 weeks at 50°C and 65% RH. (B) OVA-GLU and OVA-LAC systems freed from unreacted sugars after 2 and 4 days of reaction, respectively, and stored for 1–3 weeks. (C and D) OVA-GLU and OVA-LAC systems, freed of sugars, reincubated with or without free sugars (GLU or LAC) for another 3 weeks. "Non" is no sugar added. (From Kato, T., et al. 1988. J. Agric. Food Chem. 36: 806-809. With permission.)

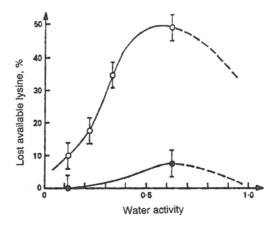

Figure 6 Loss of available lysine as a function of water activity at 1 month's storage. Solid circles, unhydrolyzed milk; open circles, lactose-hydrolyzed milk. The dashed lines are drawn to indicate the probable form of the curves. The bars indicate 95% confidence intervals. (From Burvall, A., et al. 1978. J. Diary Res. 45: 381-389. With permission.)

Hence, undesirable browning and off-flavors may be avoided if the reducing sugars remaining after processing can be encapsulated, neutralized, degraded, or bound.

Stability data at extreme temperatures (30–300°C) are available for various casein-reducing sugar models and have provided limited data on available lysine, protein efficiency ratios (PERs), and protein digestibilities.

There is considerable concern with regard to the stability of dairy products because of their relatively high levels of lactose. In ordinary milk powder, browning is remarkably low during storage.[81] Even at 25°C and an A_w of 0.62, only 25% of the lysine was lost in 9 months. However, with the lactose-hydrolyzed milk (Figure 6), 48% of the lysine was lost in 1 month of storage at 25°C. A loss of 28% and 35% in total lysine and methionine, respectively, has been reported for skim milk powder having $A_w = 0.57$ and stored at 40°C for 1 month.[82] In addition, the pepsin, trypsin, and chymotrypsin digestibilities were reduced. The effect of several processing treatments on lysine availability in milk[2] indicated extensive loss during roller drying (Table 1). The loss, especially in enzymatically available lysine loss, increased with decrease in the speed of the roller. Similar losses during roller drying of cheese whey have been reported.[83] Post-processing, the subsequent storage losses of this product, whether roller dried or spray dried, were similar.

The influence of temperature, A_w and time have been studied with a diafiltered whey protein (2.1% lactose)–lactose system (LYS:LAC, 1:1).[84] At 121°C for 5000 s, irrespective of the A_w of 0.3, 0.5, or 0.7, the losses in enzymatically available lysine exceeded 75% (Figure 7). Even at 100°C and 500 s of heating time, the losses were excessive (40–50%). At an A_w of 0.97, the losses were modest (5–23%).

Besides changes in lysine, color, and digestibility, Hsu and Fennema have investigated changes in functionality (e.g., solubility, foaming properties, and emulsification ability) in a dry whey protein concentrate (52% protein, 36% lactose) during storage for 6 months at temperatures ranging from –40 to 40°C with A_w from 0.15 to 0.41.[85] The functional properties were relatively stable, but browning increased with storage time. Toughening of the texture of meats during storage is attributed, in part, to Maillard reaction. The differences in the two foods may be related to the proportions of fibrous and globular proteins.

Apart from A_w the importance of the crystalline or amorphous (hygroscopic) nature of the reactants involved in Maillard reactions has been underscored by several authors.[86–88] At low A_w the crystalline form has a low capacity for water absorption, while the amor-

Table 1   Influence of Various Treatments on Lysine Values of Standard
Milk (g/16 g N)

Samples	Total Lysine TLV	Furosine FUR	Availability by Enzymic Method ALV_{enz}
Non-treated milks			
Fresh milk	8.22	0	8.37
Freeze-dried	8.22	0	8.08
Spray-dried	8.32	0	8.22
Standard milk (SM)	8.13	0	8.20
Sterilization of SM			
115°C, 10 min	7.97	0.30	7.77
115°C, 20 min	7.96	0.31	7.47
115°C, 30 min	7.80	0.34	6.51
115°C, 40 min	7.70	0.35	5.97
Roller-drying of SM			
6 atu, 11 rpm	5.88	1.24	4.57
6 atu, 5.5 rpm	5.34	1.52	3.04
6 atu, 2 rpm	4.51	1.98	2.15

Note: Methods used: acid hydrolysis, furosine, and enzymatic digestion.

From Finot, P.A., et al. 1981. In: Progress in Food and Nutrition Science. Maillard
Reactions in Food (C. Eriksson, ed.), Pergammon Press, Elmsford, NY, pp 345–355.
With permission.

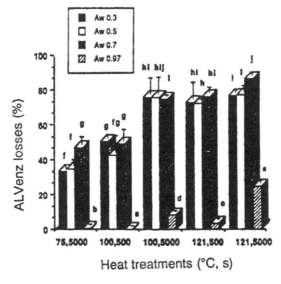

Figure 7   The effect of time, temperature, and A_w on lysine in a diafiltered whey-protein–lactose
system. Results are expressed as enzymatically available lysine losses (%). Each bar is the average
of three determinations (± SD). Means bearing a common letter are not different according to the
LSD test (95% confidence level). (From Desrosiers, T., et al. 1989. J. Agric. Food Chem. 37: 1385-1391.
With permission.)

phous form has high capacity. Nonhygroscopic whey powders have been reported to result
in less browning and lower losses in lysine than hygroscopic whey powders with the same
A_w due to release of water in the hygroscopic powders as the lactose crystallizes.[88] However,
the differences in the two sets of data are significant only at A_w of 0.44 and 0.65 at 45°C,[88]

but not at the lower temperatures (25°C and 35°C). There is some ambiguity in this area, since Huss concluded that there is negligible damage to available lysine in dried skim milk powder when lactose is present in the amorphous form.[87]

Maillard browning of hygroscopic whey powders equilibrated to several A_w is greater in sealed than in open pouches, partly because of the water released during crystallization of amorphous lactose is trapped in sealed pouches resulting in higher A_w than in open pouches.[89] Hence, this factor must be taken into account in the packaging of hygroscopic foods. The role of glass transition in food models has been examined by a number of investigators.[90,91]

The storage stability tests on intermediate moisture food (IMF) models have indicated that significant losses in lysine occur before there is objectionable browning and that a humectant such as glycerol lowers the A_w to the range where the maximum rate for Maillard reaction occurs.[92]

Kramer has assembled a series of tables based on data from the early literature to estimate the maximum storage temperatures for limiting percentage of losses of several nutrients stored for various times.[93] Table 2 is a guide for maximum warehouse temperatures that can be tolerated in order to preserve protein quality of nonfat milk powder. Unfortunately, storage is never at one fixed temperature; since there is always a certain amount of cycling of the temperature during daytime and nighttime; moreover, packaging and other storage conditions can have an effect. Since the prediction of storage life is more complex in these situations, mathematical models making use of data obtained at constant temperatures have been developed and applied to specific foods.[94-96] Nielson et al. have determined the comparative losses of tryptophan (TRY), methionine (MET), and lysine (LYS) in several model systems.[97] During early and advanced stages of the Maillard reaction in whole milk powder stored at 50°C and 60°C for 9 and 4 weeks, respectively, MET and TRY were unchanged, but LYS losses were extensive (21% and 80%, respectively).

Influence of temperature

The reaction between proteins or amino acids and reducing sugars is greatly influenced by temperature, as judged by the Q_{10} values reported in the literature,[4,98] which impact protein bioavailability. This reaction proceeds significantly during normal storage and sometimes occurs even during freezer storage, which is a cause for concern.[99] Additional data are needed during low temperature storage to ensure the nutritional quality of susceptible products during long-term storage.

The loss of 25% of LYS and in a LYS–GLU–cellulose (CEL) model with an A_w of 0.2 stored for 6 days at 40°C was referred to earlier (Figure 1).[4] In subsequent studies with a NAL–GLU (1:1 mol/mol)–cellulose model, 70% of the NAL was lost in about 7 months at 40°C.[77] The estimated activation energy, E_a, of 36 kcal/mol for the NAL–GLU–cellulose model is lower than the value of 57 kcal/mol obtained for the LYS–GLU–cellulose model, which presumably reflects the blocking of the alpha-amino group in lysine by acylation. One can speculate that by acylating the free ε-amino group it might be possible to reduce the temperature sensitivity of the Maillard reaction. Lysine loss curves (Figure 1) during storage of LYS–GLU–cellulose models[4] suggest that a relatively simple way to retard protein quality loss is by reducing storage temperature. However, at freezer temperatures, the water in the food freezes out thereby resulting in concentrated reactants being in a medium with a more favorable A_w in which to react.[99]

The effect of time, temperature, and A_w on a low LAC whey protein (2–4% LAC)–LAC (1:1 LYS:LAC mol/mol) model[84] is shown in Figure 7. Based on the effect of heating samples with common A_w for 5,000 s at 75°C and 100°C, the enzymatically available LYS is almost three times as much at 75°C than at 100°C. Similarly, the available LYS is more

Table 2   Guide for Maximum Storage Temperature (°F) for
Limiting Percentage of Loss of Protein in Dehydrated Products
under Several Storage Conditions

	6	12	18	24
	Months in storage			
Wheat flour				
<10% loss				
in sealed jars	50	38	32	28
in bags	38	26	20	17
<20% loss				
in sealed jars	100+	100+	100	100
in bags	100+	100	70	44
Nonfat milk solids				
<10% loss				
in bags, 60°RH	30	–8	—	—
in bags, 40°RH	52	+6	—	—
in vacuum cans	100	58	38	+5
<25% loss				
in bags, 60°RH	76	30	0	–8
in bags, 40°RH	96	46	20	+8
in vacuum cans	100+	95	68	52
<50% loss				
in bags, 60°RH	80	62	46	38
in bags, 40°RH	100+	92	86	45
in vacuum cans	100+	100+	100	95

[a] Total protein nitrogen.
[b] Protein efficiency ratio.
From Kramer, A. 1974. Food Technol. 28: 50–60.

than twice as much at 100°C than at 121°C after 500 s of heating. In aspartame–glucose and glycine–glucose models with an A_w of 0.80,[78] the rate constants, the time to reach 0.1 absorption units, E_a, and Q_{10} were all estimated. The rate constant in the aspartame–GLU system at 100°C was sixfold greater than at 80°C; in the glycine–GLU system, this ratio was only 3.5, indicating that aspartame added to foods may degrade more rapidly.

The early work of Ben-Gara and Zimmerman provides a lot of information on the stability of several foods (nonfat dry milk, peanut meal, soybean, wheat, rice, etc.) at 20°C, 30°C, and 40°C and RH of 40% and 60%.[100] The decrease in chemically available lysine content of nonfat dry milk after being in 6 months storage at 40% RH and 20–40°C ranged from 18–66%, respectively; at 60% RH, it ranged from 27–84% for the two temperatures. The maximum losses during a 24 month storage at these temperatures were 88% for nonfat dry milk, 34% for peanut meal, 42% for soybeans, and 18% for rice.

Influence of water activity (A_w)
Protein–carbohydrate interactions are profoundly influenced by the A_w of the system. One way to greatly impede this reaction is to make use of dry foods (such as freeze-dried foods) with as low an A_w as possible (i.e., less than 0.2). On the basis of the early studies, such as those of Loncin et al.,[101] it is generally accepted that Maillard browning or lysine loss vs. A_w conforms approximately to a bell-shaped curve with an A_w for maximum damage being in the range of 0.6–0.7 (Figure 8). There is a considerable shift downward to lower A_w for maximum damage when glycerol is included in the reaction medium. Without any additives at low A_w it is likely that the mobility of the reactants is severely

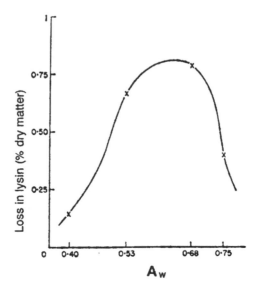

Figure 8 The loss in free (available) lysine of milk powder kept at 40°C for 10 days as a function of A_w. (From Loncin, M., et al. 1968. J. Food Technol. 3: 131-142. With permission.)

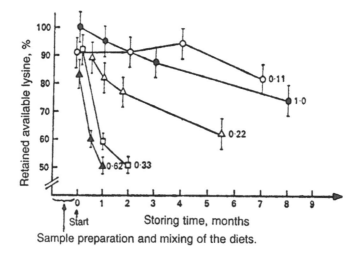

Sample preparation and mixing of the diets.

Figure 9 Storage of lactose-hydrolyzed milk. Retention of available lysine is shown as a function of storage time at 25°C at different A_w (0.11, 1.0, 0.22, 0.33, and 0.62 from top to bottom). The bars indicate 95% confidence intervals. (From Burvall, A., et al. 1978. J. Dairy Res. 45: 381-389. With permission.)

restricted due to low diffusivity, which is responsible for low reactivity. At the other end, at high A_w the reaction rate decreases due to dilution of the reactants.

The effect of several different A_w on lysine availability in lactose-digested milk, LHM, is shown in Figure 9. This product is stable at an A_w of 0.11. It is highly unstable at A_w in the range of 0.33 and 0.62. The rate falls off at an A_w of 0.22. At $A_w = 1.0$, there is a limited, but measurable, loss. Ordinary milk powder is quite stable at $A_w = 0.3$ and A_w approaching 1.0. The greater activity of hygroscopic whey powder as compared with nonhygroscopic whey powder has been discussed above.

The advantages of intermediate moisture foods (IMF) are well recognized: (1) they can be consumed readily without heating (important for many field feeding situations);

(2) they need no refrigeration; (3) they can be conveniently packed in flexible pouches; (4) they can be consumed any time after opening the package, since even open packages are microbiologically safe; and (5) they retain nutrients better than canned products due to limited heat treatment. IMFs are rendered microbiologically safe by reducing their A_w to 0.85. However, the amount of glycerol necessary to reduce the A_w to 0.85 is high (43%). Furthermore, the investigations of Narayan and colleagues on the nutritional effects of glycerol showed that rats fed high levels of glycerol in a low polyunsaturated fat diet produced fatty livers and increased levels of serum lipids, chylomicrons, and high-density lipoproteins.[102–104]

A strategy involving a combination of pH and other adjustments (O_2 tension, antimicrobial agents, and/or mild heat treatments) can be used to control microbial growth in the A_w range of 0.900–0.950.[105,106] Similarly, Fox et al. have also suggested that a combination of mild processing methods and controlled pH and A_w may be used to enhance nutrient quality and acceptance of IMF through minimization of adverse effects caused by humectants and increased processing temperatures.[107] With regard to Maillard browning, the lag time was high at all A_w, so it was not a problem. In order to overcome some of the objections to high levels of humectants (flavor, texture) in present IMFs, it may be possible to reduce the humectants and still stabilize IMF having high A_w foods by reducing pH, by treating mildly with heat, and by adding antimicrobial agents. The combined use of multiple factors such as A_w, temperature, pH, preservatives, redox potential, as well as modified atmosphere packaging, edible coatings, Maillard reaction products, bacteriocin, ultrahigh pressure, etc., result in a series of hurdles which simultaneously disturb the homeostatic mechanisms of microorganisms collectively and synergistically. This results in a safe, nutritious, and highly acceptable product that could not have been possible through the individual action of any one of these hurdles acting separately.[108–110]

Influence of pH

As might be expected, the rate of Maillard reaction is low at low pH and increases progressively as the pH is increased. The effect of pH on amino–carbonyl interaction has been investigated by Wolfrom et al., who showed that in an alanine–glucose system Maillard browning is most intense at pH 7, followed by pH 5 and then by pH 3.[111] Ashoor and Zent have shown the maximum browning occurred at a pH of about 10 for 21 amino acids tested with glucose, fructose, and lactose at 121°C for 10 min.[112]

The effect of pH was studied in solid NAL–GLU–CEL models by Narayan, Cross, and Bibeau (unpublished results). The reactants were dissolved in large volumes of deionized water or various buffers with pHs of 9.0, 7.0, and 4.0, frozen, and then concentrated upon freeze drying. The rates of reaction in samples that were originally at the higher pHs (ca. 9.0 and 4.0) were considerably greater at all times after 4 h than the rates in samples made from pH 7.0 in the powder systems with moisture content 1.7% and an A_w of 0.2 and heated at 60°C. The sample made with deionized water (Milli Q) reacted more rapidly than the samples prepared at pH 7.0 or pH 4.0. It is intriguing that there is a significant effect of pH and buffer salts on reactivity in such dry systems.

Influence of the type of carbohydrate

While only reducing sugars interact with proteins via the Maillard reaction, browning during the storage of model systems containing sucrose and amino compounds has been reported.[113] Under the slightly acidic conditions (pH 5–6) in most foods, the hydrolysis of sucrose could proceed at a slow rate and produce the reactants. If the food contains an acid-like citric acid, the hydrolysis is fairly rapid. In the presence of enzymes, as in the case of lactase in cultured milk or yogurt, the hydrolysis to glucose and galactose is rapid.

Reducing disaccharides and reducing oligosaccharides are less reactive than monosaccharides. The pentoses are, in general, more reactive than the hexoses and the aldoses are more reactive than the ketoses.

Hashiba has listed the order of reactivity with glycine of the several sugars at 120°C: ribose, glycuronic acid, xylose, arabinose, galacturonic acid, galactose, mannose, glucose, and lactose.[114] The large difference in Maillard browning between OVA–GLU and OVA–LAC models (Figure 5) has already been mentioned. Similarly, Burvall et al.[81] and Finot et al.[2] have shown that the reduced lactose milk is far more reactive than ordinary milk powder. As would be expected, the sugar/protein ratio is important; in one study,[115] the rate constant for lysine loss increased with increase in this ratio from 0.5 to 5.0. What is generally not recognized is that in the reaction between free LYS and GLU, as many as four GLU moieties may react with a single free LYS giving rise to tetrafructosyllysine. In principle, this process could lead to a tightly crosslinked polymer through the union of the carbonyl groups in this compound with additional lysine compounds. The bulkiness of the compound formed together with the propensity of ACs formed to degrade easily to carbonyl and amino compounds could hinder such unrestricted polymeric growth. In proteins, the bound LYS can react with two GLU moieties and the two AC carbonyl groups can react with two other LYS moieties, subject to any constraints of steric hindrance.

Influence of antioxidants and metal ions

Experiments have been conducted that appeared to suggest that antioxidants (AOs) are effective in inhibiting the Maillard reaction.[116,117] The data showed that several AOs (ascorbic acid, tocopherol, propyl gallate, butylated hydroxytoluene (BHT), and butylated hydroxyanisole (BHA)) were effective in minimizing the loss in lysine and the formation of furosine after acid hydrolysis in a casein–GLU–AO model incubated at 37°C and 60% RH for 0 to 60 days.[116] However, certain anomalies are apparent. The browning appears to have proceeded at the same rate for 30 days in all systems (with or without AOs), which would be unlikely if the early stages of the Maillard reaction had been arrested. Based on the furosine yield, which correlates linearly with lysine loss (Figure 2), the predicted LYS loss for samples without AOs at the end of 60 days should have been in the range of 35–50%, rather than the 10–18% reported.[116]

It is known that ACs, and especially the carbonyl compounds formed from them, are powerful lipid antioxidants.[1,73,118] Yet it has been observed that Maillard reaction products (MRP) from glutamic acid and either glucose or fructose[119] showed no antioxidant properties. The high stability of fructose–glutamic acid is known. Protein hydrolsates are poor antioxidants, but after reaction with glucose their antioxidative activity improved greatly. Strong antioxidants were obtained from MRP of dipeptides and glucose; the MRP obtained by reacting L-histidylglycine with xylose proved to be a very effective lipid antioxidant.[120] The potential use of such specific products in actual foods containing PUF to inhibit their oxidation remains to be tested.

There is indication that copper, iron, and cobalt salts may enhance amino–carbonyl interactions.[121] Bohart and Carson have reported that even small amounts of manganese (0.4 ppm) inhibits the rate of browning in air.[122] Other metal ions inhibit browning including tin.[121-123] Rendleman and Inglett have extensively investigated the influence of Cu^{2+} in the Maillard reaction and have concluded that cupric ions enhance the rate of Maillard browning in several model systems.[123] The differential effects of metal ions (Na^+, Cu^{2+}, Fe^{3+}, and Fe^{2+}) on this reaction have also been reported by others.[124,125] Melanoidin pigment exhibited a strong affinity for Ca^{2+} and Cu^{2+} ions.[125] Whether this affinity has any significance in the biological availability of calcium and copper remains to be determined.

Protein-oxidized lipid interaction

Nature of the reaction

Polyunsaturated lipids can readily autoxidize during processing or storage in the presence of air, especially when antioxidants are depleted, and can also interact with food proteins to form tightly bound complexes. Following the pioneering investigations by Tappel[6] on the binding of oxidized fatty acids and esters to form stable complexes with proteins, Narayan and colleagues reported on studies relating to the nature of these complexes and the factors influencing their formation.[7,126,127]

These reactions and interactions are of interest in food science as well as in other disciplines. The complex reactions that occur during the oxidation of fats in the presence of proteins are considered to be responsible for the yellowing of bacon and rusting of fish during storage.[128] Physiologically, oxidized fats have been implicated in the formation of brown pigments in the adipose tissue of vitamin E deficient rats as well as ceroid pigments in choline deficient rats.[128,129] Hartroft has reported on the presence of ceroid pigments in the aortas of men exhibiting atheromatous changes; he also observed the formation of ceroid-like pigments upon reacting red cells with cod liver oil in air.[130,131]

Narayan and colleagues investigated the chemical and physical properties of complexes between a food protein (egg albumin) and either oxidized linoleic acid, thermally oxidized (TOCO), or thermally polymerized (TPCO) corn oil.[126,127] With a highly polymerized oxidized corn oil and after 48 h at 60°C, as much as 41% of the lipid was bound and could not be extracted with several solvents by the Soxhlet procedure. With trilaurin, oleic acid, unoxidized linoleic acid, and unoxidized corn oil subjected to a similar treatment, little or no lipid was found bound to the protein. On the basis of further reaction with lauroylchloride, it was concluded that the reactive amino, hydroxy, or sulfhydryl groups in the protein were not involved in the interaction with oxidized lipids. Based on reactions of three aldehydes having differing chain lengths (C2, C9, and C12), it was suggested that such binding is not due to a simple aldehyde–amine interaction. Malonaldehyde, a dialdehyde, was not used, since it is not formed from linoleic acid oxidation. Secondary forces, primarily hydrogen bonding and noncovalent bonding, were deduced to be responsible for binding the oxidized lipid to the aggregated protein.

Many other investigations of the reactions between oxidized lipids and proteins have since been conducted,[8–10,132–139] and it is instructive to consider them briefly because they may have a bearing upon storage stability of foods and could suggest possible strategies for minimizing this interaction. Desai and Tappel[8] proposed a free radical mechanism where the lipid peroxy and lipid alkoxy radicals would interact with the protein and result in lipid–protein copolymers.[8] Nonradical mechanisms involving oxidized lipid scisson products (mainly malonaldehyde) and protein–amino groups have also been proposed.[132] A free radical mechanism has been proposed, in which lipid free radicals lead to the formation of protein free radicals, resulting in protein polymerization with the occlusion or association of the oxidized lipid in some type of lipid–protein complex that does not involve covalent bonds between the two compounds.[9] Roubal later suggested that the denatured protein polymer could effectively trap and stabilize lipid radicals.[134] Gamage et al. separated radical and nonradical products of linoleate oxidation and reacted them with ribonuclease.[137] They reported that the radical product fraction was responsible for protein polymerization, whereas the nonradical product fraction resulted in fluorescence of the reaction product.

Of interest in this connection is an excellent but not often cited review of free radicals in low moisture systems[10] that provides general and specific information on (1) free radical formation in food as a result of ionizing radiation, photolysis, mechanical energy, freezing, and dehydration; and (2) free radical reactions in foods and their effect upon nutrients.

CODE:

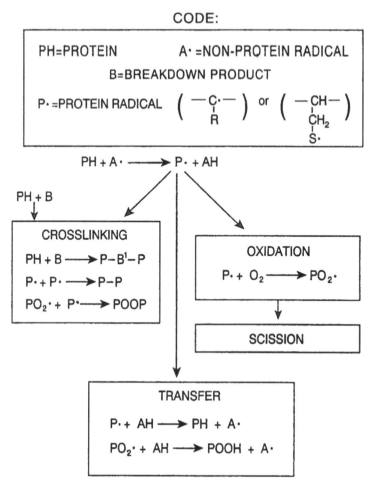

Figure 10 Schematic representation of reactions of proteins with peroxidizing lipids. Redrawn with permission from (139), Copyright, 1975, American Chemical Society, Washington, D.C.

A schematic representation of protein-oxidized lipid interaction as envisaged by Karel et al.[139] is shown in Figure 10. Crosslinking due to interaction of proteins with lipid radicals or lipid breakdown products (dialdehyde), as well as to protein–protein interaction initiated by lipid free radicals, is indicated here. Protein oxidation and scission are also known to occur.

The interaction between thermally oxidized corn oil and food proteins under normal or physiological conditions is important because of the possible formation of such oxidative polymers of vegetable oil in some commercial deep fryers.[140] Rats fed used corn oil from a commercial potato chip fryer were shown to have their growth depressed in comparison with those fed fresh oil.[140,141] The growth depression was greatly overcome by increasing the protein content of the diet from 10% to 20%,[140] which may suggest that the effect is due partly to loss of protein as a consequence of the oxidized lipid–protein interaction and associated diminished protein availability. In addition, there appears to be a synergistic effect of heated fats in carcinogenesis, based on studies with rats fed an extremely low level (0.005%) of *N*-2-fluorenylacetamide (FAA), a hepatocarcinogen.[142] Thermally oxidized lipids and acetaminophen have been implicated in accelerating the induction of liver cirrhosis in rats given 3% ethanol.[143] Oxidized lipids have been identified in human atherosclerotic plaques along with substantial quantities of α-tocopherol and ascorbate.[144] Dietary oxidized lipid has been found in rat very low density lipoproteins as well as in human chylomicrons.[145–146] The role of oxidized low density lipoproteins in atherogenesis has been the subject of a recent symposium.[12–15] It has been suggested that oxidized lipids present in the American diet may be an important risk factor for athero-sclerosis.[16]

Influence of the type of lipid or protein

In studying the reactions of peroxidizing lipids with proteins, linoleic acid, methyl linoleate, linolenic acid, methyl linolenate, methyl arachidonate or ethyl esters of menhaden oil (73% docosahexaenoate and 23% eicosapentaenoate), or oxidized/polymerized corn oil have been generally used. Trilaurin, lauric acid, oleic acid, linoleic acid, or corn oil were ineffective in binding to egg albumin after heating for 2 h at 60°C in the presence of air.[126] Linoleic acid reacted only to the extent of 8% after 24 h, while thermally oxidized corn oil bound to egg albumin to the extent of 4.4% in 2 h.

In the presence of a large amount of water, only egg albumin and lactalbumin were able to form insoluble complexes with oxidized lipids,[126] while amino acids, peptones, gelatin, and sodium caseinate did not yield any similar complexes. However, a number of investigators have reported the complexing of oxidized lipids with gelatin, insulin, lysozyme, cytochrome, and ribonuclease under different experimental conditions.[8,9,134,137,139] Further, Tappel has shown copolymer formation with sodium caseinate, casein hydrolysate, and glycine.[6] These data indicate that food proteins can interact with lipids peroxidizing under both processing and storage conditions and may adversely affect protein quality. In addition, oxidized fat present in foods may interact with digestive enzymes and interfere with the digestion and absorption of foods. At low A_w(< 0.05), protein free radical mobility is restricted, but at high A_w approaching 1.0, their mobility is increased, which it is suggested enhances protein polymerization, decreases solubility, and reduces enzymatic activity.[138]

Influence of temperature

The influence of temperature has not been explored in detail, although it is important in connection with storage of foods containing proteins and fats, especially PUF. The early data of Narayan and Kummerow with egg albumin and thermally oxidized corn oil showed that 60°C was optimum; however, only two other temperatures (30°C and 90°C) were investigated.[126] The rate of the reaction was greater at 30°C than at 90°C.

Influence of pH

Under the conditions that were standardized by Narayan and Kummerow for the reaction between egg albumin and oxidized corn oil,[126] a pH of 7.0 was optimum for maximizing the binding of oxidized lipid (4.3%) to this protein. A pH of 5.0 was found to maximize the formation of polymerized protein-oxidized lipid aggregates, but the amount of lipid bound was only 2.6%. Several investigators have routinely used a pH of 7.0 in their experiments on oxidized lipid–protein interactions.[8,9,136]

Influence of metal ions and antioxidants

The formation of complexes of the type described here between lipids and proteins takes place only when the lipid becomes oxidized.[6,7,126] For example, both unoxidized linoleic acid and unoxidized corn oil reacted very little with egg albumin.[126] The addition of a catalyst such as hematin greatly accelerated the reaction.[6,9] The presence of antioxidants such as BHA and BHT delayed the onset of the characteristic fluorescence by as much as 7 days.[135] Thus, it is possible to retard this reaction by use of metal chelating compounds,

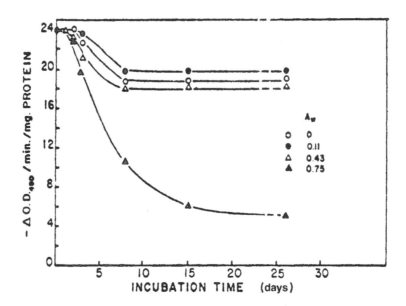

Figure 11 The influence of water activity upon oxidized lipid–protein interactions: Loss of enzymatic activity of lysozyme as a result of reaction with peroxidizing methyl linoleate. Methyl linoleate (400 mg) was emulsified with 2 g of lysozyme in 50 ml of water, freeze dried, equilibriated to various A_w and incubated at 37°C for 0–30 days. (From Kanner, J. and Karel, M. 1976. J. Agric. Food Chem. 24: 468–472. With permission.)

and by including antioxidants or oxygen scavengers, depending upon the type of food. However, it should be noted that α-tocopherol is not an effective antioxidant, as is propylgallate or BHA, in these systems.[6]

Influence of water activity

It has been shown by Kanner and Karel that the greatest loss in enzyme activity as a consequence of the interaction between oxidized linoleate and lyzozyme was at an A_w of 0.75, approaching 80% in 26 days (Figure 11). This loss parallels the changes seen (76%) in protein insolubilization. At A_w of 0, 0.11, and 0.43, the corresponding losses were between 16% and 33%. Nielson et al. investigated the effect on amino acid availabilities of two A_w in a whey-protein-oxidizing methyl linolenate model; they reported that considerably greater damage to lysine quality occurred at an A_w of 0.84 than of 0.33.[147] Major damage to ovalbumin by oxidized lipids[126] was seen at high water content rather than at low water content.

Nutritional effects

From the nutritional standpoint, the extensive destruction of amino acids due to this type of reaction is of concern. Data on amino acid yield and composition upon 6N HCl hydrolysis of oxidized lipid–protein complexes in several experiments[8,135,136] show that there is extensive damage in almost all amino acids (17–54%)[8] and in specific amino acids (22–40%); such damage is consistent with the mechanism of free-radical-induced protein polymerization. Matoba et al. reacted casein or egg albumin with methyl linoleate and observed amino acid losses in the range of 9–41% in methionine (MET), LYS, and histidine (HIS) based upon acid hydrolysis.[136] In the casein-oxidized linoleate model stored at 80% RH for 10 days at a high temperature of 50°C, they observed even greater losses, based upon enzymatic hydrolysis, in LYS, HIS, and MET (41–94%).

Early data by Johnson and Narayan (unpublished results, 1957) indicated that the growth of weanling rats was not reduced when the oxidized corn oil–egg albumin complexes (either 7% or 32% lipid) were given to them at a level corresponding to 30% of protein. However, when the protein complex level was reduced to 10% and the amount fed was restricted to 5 g/rat/day, there was a substantial loss in weight of rats fed the complex in contrast to the gain in weight in control rats.

Extensive investigations have been conducted by Nielson and colleagues on the effects of A_w, temperature, O_2 limitation, and storage time on amino acid losses in an oxidizing methyl linolenate–whey protein model system.[97,147,148] The destruction of LYS and TRY increased with increase in A_w storage time, and temperature, as indicated by chemical analysis, and approached losses of 71% and 31%, respectively, under extreme conditions. Methionine loss was very extensive (up to 100%) due to oxidation to methionine sulfoxide (see below). The losses in histidine approached 51%. Other amino acids were unaffected.

By the rat balance methods,[147] it was observed that at low A_w the bioavailabilities of LYS and TRY were essentially unaffected. At high temperature (55°C) and high A_w (0.84), there were extensive losses in LYS (88%) and TRY (71%). These losses are higher than indicated by chemical methods. When the chemical assay values were adjusted for lower digestibility (47–97%), the new values were similar to those obtained by rat assay. Methionine sulfoxide is as highly available as methionine in the rat and, therefore, there appears to be no problem with the bioavailability of methionine. The possible importance of cysteine loss, indicated by a 28% decrease in bioavailability in an oxidizing methyl linolenate–whey protein model limited to 1 mol O_2 uptake per mole lipid, and its role in oxidized lipid–protein interaction, remain to be explored. Karel et al. have also demonstrated the formation of sulfur radicals in cysteine, cystine, and in proteins containing free sulfhydryl groups that were exposed to oxidized methyl linoleate.[139] The conclusion drawn by Nielson et al. was that at low A_w and with limited access to O_2 (e.g., freeze dried foods), the losses in LYS, TRY are of minor significance in human foods.[147]

Protein interactions

Protein–protein interactions

The protein–protein interactions are important because they affect chemical, physical, and biochemical characteristics of foods. Such interactions are involved in gelling phenomenon[149]; they may adversely affect digestibility[150–152]; and they may influence spun characteristics[153] of food fibers. The desirable type of food fiber for use as meat analogs can be conveniently obtained by suitable selection of temperature, pH, and ionic strength, rather than by amino acid modification through either genetic or other means.

Protein modification by thermal treatments

Thermal treatments of proteins may cause considerable diminution of protein and amino acid digestibilities. Several reactions are known to occur[154]: oxidation of methionine to methionine sulfoxide[148]; formation of lactone ring through reaction of serine or threonine hydroxy groups with carboxyl groups[1]; isopeptide formation due to the reaction of the amino group of lysine with the carboxyl group of aspartic and glutamic acids[155]; and unfolding of proteins and attendant changes in conformational structure or, under extreme cases, gel formation due to ionic and hydrophobic interaction.[156] Additionally, it has been suggested that new three-dimensional crosslinks may be formed and could sterically hinder enzymatic activity. Accordingly, the availabilities[154] of several amino acids (aspartic acid, glutamic acid, threonine, serine, methionine, tyrosine, and lysine) were found to be reduced as a result of heating diafiltered low lactose (2.4%) whey protein to 121°C at $A_w = 0.97$ for 5000 s (Figure 12). The damage[157] was far greater at an A_w of 0.97 than at 0.7,

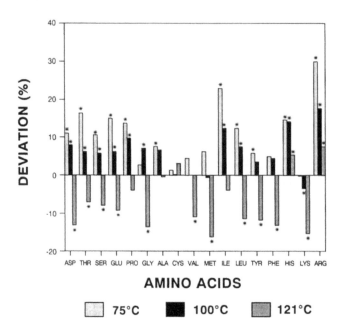

Figure 12 Effect of heat treatment for 5000 s on *in vitro* digestibility of totally delactosed whey protein (with 2.4% residual lactose). Results are expressed as percentage of deviation in digestibility from control (unheated) sample. For each amino acid (AA) shown with a three letter code, there are three bars shown in sequence and represent heating temperatures of 75°C, 100°C, and 121°C, respectively. Percentage of deviation in AA digestibility equals 100 × [(digestibility of heated sample − digestibility of control)/(digestibility of control)]. *Significantly different ($p < .05$) from the unheated control sample. Redrawn with permission from (154), Copyright, 1987, Institute of Food Technologists, Chicago, IL.

0.5, and 0.3 for most amino acids (Figure 13). Hakansson et al. have demonstrated that the biological value and available lysine of whole grain wheat were reduced (15–35% and 14–60%, respectively) during thermal processing, such as autoclaving or popping.[158] Substantial losses in thiamin (63–95%) and in folacin (16–71%) under these conditions were also reported. Severe popping decreased sulfur amino acids (30%), tryptophan (27%), and threonine (15%). Extrusion processing of white flour even under severe conditions had no effect upon either the biological value of the proteins or upon available lysine. Drumdried products unexpectedly exhibited an increase in biological value.

Protein modification by alkali treatment

An amino acid containing two primary amino groups, two carboxyl groups, and one secondary amino group (lysinoalanine, LAL) has been found in acid hydrolysates of proteins severely damaged by treatment with alkali.[159,160] Such damaged proteins are less digestible and have reduced net protein utilization values.

LAL and lanthionine (LAN) occur as a result of the β-elimination reaction in proteins containing specific seryl or cystinyl residues and the subsequent interaction of the intermediate dehydroalanine formed with the ε-NH_2 group of lysine residues. LAL has been shown to cause histological lesions in rat kidneys.[161] While the initial focus was upon alkali damage to proteins, later work demonstrated that alkali treatment was not a necessary prerequisite, but that heat treatment could also result in limited formation of LAL.[159] The ubiquitous presence of LAL in home-cooked and commercially processed foods is now recognized.[159]

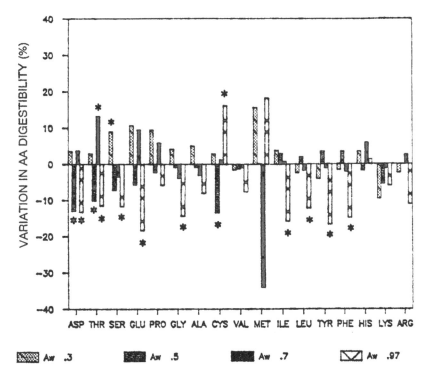

Figure 13 Effect of severe heat treatment at 121°C at four A_w for 5000 s on the *in vitro* amino acid (AA) digestibility of totally delactosed whey protein (2.4% residual lactose). Results are shown as variation in AA digestibility expressed as percentage of deviation from unheated control values. Variation in AA digestibility (%) = 100 × [(digestibility of heated sample − digestibility of control)/(digestibility of control)]. Shown are four bars for each amino acid, represented as a three-letter code for each of the four A_w 0.3, 0.5, 0.7, and 0.97, in sequence. (From Desrosiers, T., et al. 1987. J. Dairy Sci. 70: 2476-2485. With permission.)

Finley and Friedman have cautioned that the dehydroalanine could also react with other amino acid residues (containing histidine, arginine, etc.) to form histinoalanine (HAL) and arginoalanine (AAL).[162] However, their biological effects are not established. Finley et al. have shown that LAL formation can be suppressed by the inclusion of mercaptoamino acids.[163]

Racemization must also be considered. Due to the antimetabolic effects of some D-amino acids and due to unnatural peptides being formed that are resistant to hydrolysis (e.g. D-L, D-D, and L-D peptides), there could be nutritional consequences. Racemization is generally brought about by treatment with alkali, for example, by heating proteins at 80°C for 1 h in 0.2 *N* NaOH.[164] Even without alkali, partial racemization (15–40%) can occur upon heating for prolonged periods. In plasma albumin and in fat-extracted chicken muscle heated at 121°C for 27 h, aspartic acid, cysteine, and serine were partially racemized (e.g., 32%, 15%, and 40%, respectively, in the case of albumin).

Protein–polyphenol interactions

Phenolic compounds fall broadly into four chemical groups (benzoic acids, cinnamic acids, flavonoids, and terpenoids) and are widely distributed in plants foods.[165] The action of polyphenol oxidase in enzymatic browning and resulting loss in protein quality[166] are known. *o*-Diphenols, such as caffeic acid, can be oxidized at alkaline pH to *o*-quinone or *o*-semiquinone radicals. These quinones can react with the free amino groups and sulfhydryl groups under appropriate conditions. Hurrel et al. reacted casein–caffeic acid mixtures under several conditions (nonenzymatically at pH 10 or enzymatically at pH 7) and determined their nutritional properties.[167] They observed that this reaction reduced the protein quality of casein and has an effect similar to the Maillard reaction. Lysine and tryptophan were drastically reduced (50%). Histidine and tyrosine were also degraded to a lesser extent (14–22%). The determination of chemically available FDNB–lysine agreed fairly well with rat assay for lysine availability. The enzymatic reaction has a pH optimum of 7.0. For the nonenzymatic reaction, a pH of 10 or above is essential. The data obtained also established that the nonenzymatic reaction is more rapid and caused greater loss in lysine than the enzymatic reaction. Additionally, in the range of temperature of 20–60°C, maximum loss in available lysine (24–56%) was encountered at 60°C for the nonenzymatic reaction, while the loss in available lysine (13–21%) due to the enzymatic reaction was unaffected by temperature.

It is also evident that, in order to minimize losses in lysine quality and, more generally, in the protein quality of a plant food, alkaline treatment should be eliminated or, at the very least, the pH should be kept below 8.0. To suppress damage brought about by polyphenol oxidases in plant food being processed or stored, the pH should be kept below 6.0. Clearly, keeping the pH below 6.0 minmizes both types of degradation.

Protein–phytic-acid interaction

Phytic acid (meso-inositol hexaphosphoric acid) occurs extensively in cereals, nuts, and legumes. In bran products, the levels could be as high as 5%.[168] Based on its structure, it would be expected to interact with positively charged compounds, such as metal ions and proteins, in foods. At low pH, proteins have a maximum positive charge and readily bind the negatively charged phytic acid. As the pH is raised toward its isoelectric point, this binding becomes weaker because of two factors: (1) the net charge on the protein approaches zero[165]; and (2) proteins tend to become insoluble near this point. At neutral or basic pH, another kind of association occurs between phytic acid and proteins. Under these conditions, cations such as Ca^{2+}, Mg^{2+}, or Zn^{2+} can bind both the negatively charged proteins and phytate ions.[165] These complexes are important because of their effects on the bioavailability of proteins and of such physiologically critical elements. Stunted growth among the population in parts of Iran has been attributed to the zinc deficiency caused by excess ingestion of phytate in cereal diets. Salterlee and Abdul-Kadir have shown that the protein quality of high protein bran flour, but not soy isolate, is affected by their phytate content.[169]

Blocking undesirable side reactions

Protein–carbohydrate interaction

It is difficult to effectively block the first phase of the MR, although sulfites have long been well known to retard browning in foods.[170] Other suggestions that have been advanced to limit Maillard-induced LYS loss have included acetylation of free amino groups in proteins, use of aromatic aldehydes, manipulating A_w and pH, and adding specific amino acids or protein hydrolysates to *redirect* the reaction.[1, 170–171]

It has been shown that Schiff bases of aromatic aldehydes, such as benzaldehyde and salicylaldehyde, are as fully available as lysine in the rat.[172] This opens the possibility of utilizing these or other aromatic aldehydes, such as pyridoxal, to minimize reaction with aliphatic amines. Schiff bases of aromatic aldehydes are stable and will not undergo the Amadori rearrangement. However, even if this approach were to be successful, the stoichiometry of the reaction makes it impractical. It would also be objectionable from biological and sensory considerations to use an excess of aromatic aldehydes in the food.

Acetylation of the free amino groups in proteins is a way of preserving lysine[173] by blocking the amino–carbonyl interaction. However, this reaction may not be simple to accomplish in foods and is likely to adversely affect the functionality of the foods.

Freeze drying of foods to low A_w is yet another approach to minimize Maillard-induced lysine losses. However, it is recognized that serious losses in lysine can be encountered even in low A_w systems.[4] Shifts in pH are difficult to accomplish because normal foods have pHs in the range of 5–6, and a decrease in pH could lead to unacceptable tartness.

A unique suggestion to block the Maillard reaction through the use of antioxidants was made by Yen and Lai.[116] The rationale for this approach is based upon the proposed mechanism[117] in which free radicals are formed prior to the Amadori rearrangement, so it involves trapping these radicals with antioxidants (AO) before they start to propogate. Using a casein–glucose–antioxidant model at 37°C and a RH of 60% for 60 days, the observed losses in lysine were considerably decreased when comparing the model with and without AO (13% vs. 60%, respectively). However, it is clear from Figure 14 that there is a lag period before the lysine starts to fall precipitously. The lysine was estimated by the trinitrobenzene sulfonic acid method, which is generally recognized to underestimate lysine loss. An alternative explanation of these findings is that the small amounts of residual lipids in the casein autoxidized, formed lipid-free radicals during incubation, and contributed to the loss in lysine. This would account both for the lag period (which is uncommon for browning reactions) as well as for the effectiveness of the AO. Because of residual lipid, one must use hot-alcohol extracted, vitamin-free casein[174,175] to create an essential fatty acid (EFA) deficiency or a vitamin deficiency in animals. Further, the minor amount of residual linoleic acid in starch is sufficient to delay the onset of EFA deficiency in rats.[175] Whether or not the small amount of lipids in the casein is sufficient to sustain the propogation of lipid-free radicals needs to be determined. Despite these concerns about experimental procedures, the suggestion to use AO as a Maillard blocking agent is novel and should be further explored. A recent paper by Djilas and Milic[176] has described the inhibition of Maillard browning and the inhibition of the formation of free pyrazine cation radical by naturally occurring phenolic compounds. It would be worthwhile to use simple models such as the actyllysine–glucose models, under the relatively mild conditions used by Narayan and Cross,[77] and establish by HPLC procedures whether or not lysine losses can be eliminated through the inclusion of natural antioxidants, since it would have a profound impact upon the storage stability of foods.

The use of cysteine to reduce Maillard browning was demonstrated (Figure 15) in lysine–glucose–cysteine–Avicel model systems.[73] The mole ratio of cysteine to lysine was 1.0, which is excessive for food applications. Pham and Cheftel have observed reduced Maillard induced lysine loss and browning by adding cysteine, aspartic acid, glutamic acid, sodium chloride, ammonium chloride, or urea.[177] The blocking action of several of these compounds is impressive, but they were used in large amounts and several are unsatisfactory for human consumption. Cysteine and the dicarboxylic amino acids form stable Amadori compounds[178], which are less susceptible to browning reactions.[112]

Glutamic acid and aspartic acids were shown to inhibit the rate of browning in lysine-reducing sugar models at a pH of 8.0.[179] However, when these experiments were repeated using controlled pH conditions, it was found that the earlier observations of inhibition were due to a large decrease in pH caused by adding these dicarboxylic acids to phosphate buffers (Narayan, K.A., Splaine, M., Neidhart, A., and Jason, K., unpublished results).

The ability of sodium sulfite to prevent enzymatic and nonenzymatic browning in foods is well recognized. However, it has not been convincingly demonstrated whether

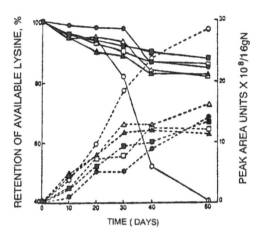

Figure 14 Effects of antioxidants on the retention of available lysine (top solid lines) and the formation of furosine in a casein–glucose (1.5:1, wt/wt) system at 37°C and 60% relative humidity. The antioxidants were added at 20–260 ppm to a paste of casein–glucose, homogenized, quick-frozen, and stored in a dessicator containing 40% sulfuric acid for 0–60 days, ○ control, ● ascorbic acid (260 ppm), □ - tocopherol (100 ppm), ■ propylgallate (20 ppm), △ BHT (200 ppm), ▲ BHA (200 ppm). (From Yen, G.-C. and Lai, Y.-H. 1987. J. Food Sci. 52: 1115-1116. With permission.)

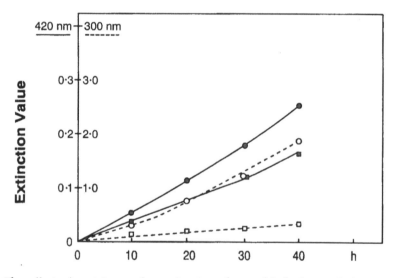

Figure 15 The effect of cysteine on browning in a freeze-dried glucose–lysine–cysteine model system (2:1:1 mol, 14 g Avicell per gram glucose) compared with a corresponding cysteine-free model at an A_w of 0.4. Data at 300 nm (UV-absorbing, browning intermediate) and at 420 nm (visible brown pigments) are shown. Circles, no cysteine added: ●, 420 nm and ○, 300 nm. Squares, cysteine added: ■, 420 nm and □, 300 nm. The predominant effect is on the intermediate compound. (From Eichner, K. 1975. In: Water Relations of Food (R.B. Duckworth, ed.), Academic Press, New York. With permission.)

sulfite also prevents the formation of the Amadori compound in which lysine is irreversibly bound or if it inhibits only the latter stages of the Maillard reaction. Early studies of Friedman and Kline suggested that sodium bisulfite drastically reduces Maillard browning in protein hydrolysate–glucose systems, but has only a minor effect upon the preservation of amino acids, which were determined by microbiological methods.[180]

Investigations of Friedman and Molnar-Perl have shown that far less than stoichiometric levels (0.02 to 0.2 M) of inhibitors were required for optimum inhibition of browning in the reaction between amino acids and glucose.[181-183] For high protein foods, either 0.1 M sodium bisulfite or 0.25 M N-acetylcysteine was sufficient to suppress browning. It was implied, but not stated, that these and other sulfur compounds (L-cysteine and reduced glutathione) may further help preserve amino acid and protein quality. Earlier work of Friedman and Kline,[180] based upon microbiological assay alone, as well as that of Farmar et al.[184] and Cerami[57] would appear to suggest that sodium bisulfite intervened after the first stage of the MR where lysine is already compromised. Hirsch and Feather have recently demonstrated the usefulness of aminoguanidine as an effective inhibitor of the Maillard reaction.[60] Unfortunately, aminoguanidine does not prevent the formation of Amadori compounds but interferes with the subsequent crosslinking reactions. Therefore, from the standpoint of nutrition and bioavailability, this compound is ineffective.

Because of the need for substances that are highly effective for inhibiting Maillard-induced lysine and protein quality loss in stored foods, it is important to further explore these approaches.

Protein-oxidized lipid interaction

Although it is well-recognized that lipid oxidation is favored at very low A_w it is interesting to note that the greatest damage to proteins by oxidized lipids has been shown to occur at $A_w = 0.75$.[139] This observation has been explained as due to the greater protein free radical mobility at $A_w = 0.75$, even though the concentrations of lipid and protein free radicals are higher at low A_w. On the basis of these and other data on optimum water content for this interaction,[126] it would seem that protein-rich canned foods and frozen entrees may be susceptible to protein insolubilization caused by lipid oxidation, which would adversely impact protein quality, food texture, and overall food acceptance. A low A_w wherever feasible would help to minimize these interactions, but favor lipid oxidation.

Since O_2 is essential for lipid oxidation, the use of antioxidants, vacuum packaging, and O_2 scavengers would be most helpful in retarding these reactions. As indicated above, tocopherol is not as effective as propyl gallate or BHA in inhibiting these reactions. Using chelators to bind metal ion catalysts in the food or in packaging systems would help to reduce the rate of lipid oxidation as well as its interaction with proteins. Their use must be balanced against the possibility of reducing the bioavailability of minerals.

Narayan and Kummerow demonstrated that thermally oxidized corn oil and autoxidized linoleic acid, but not thermally polymerized corn oil, could be tightly bound by alumina.[7] Roubal and Tappel[9] and Roubal[135] have concluded that oxidized lipids are unlikely to act as crosslinking agents, as initially suggested by Desai and Tappel,[8] and that oxidized lipids bound to insoluble protein polymers were in some way occluded or associated noncovalently in such complexes. This conclusion is consistent with the early observations of Narayan and colleagues concerning the nature of interaction between egg albumin and oxidized lipids.[7,126,127] The idea[9] that denatured polymeric proteins could provide a matrix to trap and stabilize lipid free radicals is important and may provide a means to block undesirable reactions compromising other proteins. The observation of Roubal,[185] which has been referred to by Karel,[10] that powder glass, quartz wool, and amino acids were ineffective for such trapping, whereas polysaccharides were partially effective as traps, is worthy of further study as a way to block protein polymerization as well as to inhibit other undesirable reactions.

Protein interactions

Racemization of amino acids, as well as the formation of abnormal amino acids, such as LAL, LAN, HAL, AAL, etc., can be greatly minimized by reducing the temperature and alkalinity of the food medium during processing. Finley et al. have shown that several mercaptoamino acids (such as cysteine and reduced glutathione) added at a 1% level can substantially reduce (13-fold) the formation of LAL during alkali treatment of soy proteins.[186] This area deserves further exploration and exploitation in all aspects (chemical, nutritional, and sensory). To minimize intramolecular reactions in food proteins, such as those between lysine and glutamic acid or between threonine and aspartic acid, the inclusion of small quantities of mixtures of carefully fractionated short peptides from partial protein hydrolysates should be considered.

Protein–polyphenol interactions

Protein–polyphenol interactions can be substantially reduced by keeping the pH of the foods being processed at 6 or less. This approach would counteract the damage caused by enzymatic (polyphenol oxidases) as well as by nonenzymatic reactions.

Intermediate moisture food interactions

For intermediate moisture foods (IMF), the suggestion has been made to increase A_w and decrease pH so that less of the humectant is required for the same degree of protection against microbiological degradation. As a means of boosting the morale of the troops in the front lines and as a way of enhancing the acceptance of carbohydrates in MRE and T-rations, Berkowitz and colleagues have developed an MRE pouch bread that is highly acceptable even after 3 years of storage at 27°C.[187,188] The successful outcome of this product was possible through the use of a combination of hurdle techniques and through the inclusion of an innovative antistaling compound (sucrose polyester). Pouch bread samples stored at 38°C and 49°C for 3 months and beyond were digested by α-amylase to a lesser extent than those stored at –18°C, which coincided with corresponding decreases in sensory acceptance scores (Narayan et al.[189]). From the discussion above, it is clear that reducing processing time, reducing temperature, and lowering pH of the food would significantly contribute to enhancing nutrient bioavailabilities.

Detection of protein quality losses and quantification

The principles upon which the detection and assessment of protein quality losses are based include (1) chemical estimation of reactive groups, especially the ε-amino group of lysine; (2) growth of microorganisms in the presence of test proteins or protein hydrolysates; (3) *in vitro* estimation of protein digestibility using protease mixtures; (4) *in vitro* determination of amino acids after enzymatic digestion; (5) chemical scores for essential amino acids corrected for digestibility; (6) true digestibility of protein and amino acids by rat balance methods; (7) biological availability of amino acids by rat growth; (8) protein efficiency ratio (PER) by determining protein contribution for rat growth and net protein ratio (NPR) by determining protein contribution for rat growth as well as for rat maintenance; (9) biological value (BV) by determining the proportion of the absorbed nitrogen retained for maintenance and/or growth; and (10) net protein utilization (NPU) by determining the proportion of the ingested nitrogen retained for maintenance and/or growth. Brief discussions of these methods are given below. A much-needed reference on the

Table 3 Comparison of Several Methods for Estimating Available Lysine in Eight Animal and Nine Vegetable Food Proteins

Food	Ratio A/E	Ratio B/E	Ratio C/E	Ratio D/E
A. Animal products:				
Casein	1.03	1.01	1.13	1.03
Tuna fish	1.03	0.98	1.03	1.26
Beef salami	0.93	0.84	0.97	1.80
Milk, nonfat	0.91	0.90	1.03	0.85
Beef stew	1.10	1.03	1.21	1.69
Chicken frankfurter	0.98	0.67	1.11	0.96
Milk, nonfat, heated	1.57	1.46	1.94	0.73
Breakfast sausage	1.12	1.07	1.32	1.08
B. Samples containing vegetable material:				
Pea protein concentrate	1.12	1.06	1.40	1.47
Pinto beans	1.52	1.45	1.62	1.44
Soy protein, isolated	1.11	1.05	0.99	1.05
Chick peas	1.09	1.05	1.37	1.20
Macaroni & cheese	1.09	1.09	1.31	1.49
Rolled oats	1.04	1.02	1.35	1.06
Peanut butter	1.00	0.94	0.96	—
Whole wheat cereal	1.04	1.01	1.25	0.90
Rice-wheat gluten cereal	1.50	1.31	2.30	0.15

Note: Values shown are ratios to rat assay values determined as part of a USDA collaborative study. A: total lysine by acid hydrolysis; B: fluorodinitrobenzene lysine; C: dye-binding lysine; D: *o*-phthalaldehyde lysine; E: rat assay values.

From Carpenter, K.J., et al. 1989. Plant Food Human Nutr. 39: 129–135. With permission.

bioavailability of nutrients for animals has recently appeared[40]; specific additional literature pertaining to amino acid and protein bioavailability include References 190–192.

Fluorodinitrobenzene method

This reagent was originally used by Sanger[193] and Sanger and Tuppy[194] in their classical experiment on the determination of free amino acid groups of insulin. Its application to foods to estimate the chemical availability of lysine was suggested by Carpenter[195] and is generally considered the method of choice. However, it is not without problems: (1) the fluorodintrobenzene (FDNB) interacts to some extent with the ACs, thereby overestimating available lysine; and (2) as much as 30% of dinitrophenyllysine is lost during acid hydrolysis of starchy foods.[3]

Carpenter et al. have shown (Table 3) that for 12 out of 16 foods the FDNB difference method provided estimates of lysine availability comparable to those obtained by rat assay.[196] In the case of pinto beans and heated nonfat milk, the chemically available lysine values were grossly overestimated by 45%. With chicken frankfurters, there was a 33% *underestimate* of the value obtained by rat assay. Thus, in foods in which the Maillard reaction occurs, the AOAC FDNB method[197] is likely to provide a more favorable view of lysine quality.

o-Phthalaldehye (OPA) is frequently used to estimate available lysine residues in proteins as well as to derivatize amino acids for fractionation by high-performance liquid chromatography. The data[196] shown in Table 3 illustrates that the OPA method *overestimates*

LYS availability in 7 out of 16 foods. In addition, LYS availability in the rice/wheat gluten cereal was 85% underestimated by this method.

Trinitrobenzene sulfonic acid method

Kakade and Liener have developed a simple method for estimating free amino groups in proteins, and they developed a means of distinguishing ε-amino groups in lysine from the terminal alpha-amino groups.[198] Unfortunately, the Amadori carbonyls react almost totally (85%) with this reagent.[1] Consequently, the data obtained might overestimate available lysine and should be validated by other methods.

Dye-binding method

Acid Orange 12 is an acidic dye used by the milk industry to estimate the protein content of milk, because it determines the basic groups (LYS, HIS, and ARG) in proteins. It has been made specific for lysine by measuring the reactivity before and after propionylation.[3] Of the three basic amino acids, only LYS is propionylated and upon acylation does not interact with the dye. Some of the data in the literature indicates that this method also overestimates available lysine by as much as 40%.[1,199]

Studies by Carpenter et al. showed that this dye-binding method overestimated lysine availability in heated nonfat milk by as much as 94% compared with rat assay (Table 3).[196] Additionally, this method[196] resulted in unexplained high values for beef stew (21%), sausage (32%), and several vegetable products (pea proteins, pinto beans, chick peas, macaroni and cheese, rolled oats, whole wheat cereal, and rice/wheat gluten cereal (25–130% increase, Table 3)). These results, therefore, call into question the general applicability of the dye method for lysine availability in foods.

Quanidation

The reactivity of the Amadori products led to the search of a lysine-specific reagent that would yield a product that would be stable during hydrolysis.[1] o-Methylisourea reacts with lysine to form homoarginine, which is detected by ion-exchange or gas chromatography. However, this reaction is not quantitative and requires further investigation.

Furosine

Maillard-reacted proteins, upon 6N HCl hydrolysis, release about 50% of the bound lysine together with two new compounds, furosine and pyridosine[2] to the extent of about 20% and 10%, respectively. Furosine formed correlated directly with the amount of Amadori compound formed[2] and with the loss in lysine (Figure 2) in model systems.[4] This method would clearly not be applicable to proteins altered by oxidized lipids, heat, alkali, or polyphenols.

Sodium borohydride reduction

A method developed by Hurrell and Carpenter stabilizes the Amadori product against subsequent acid hydrolysis.[3, 200] This stabilization is accomplished using sodium borohydride, which efficiently hydrogenates the keto group in the reducing sugar bound to the lysine in protein. Acid hydrolysis neither releases bound lysine nor produces furosine. Therefore, the free lysine liberated after hydrogenation and acid hydrolysis corresponds to the available lysine. One drawback of this procedure is that the Schiff bases are also hydrogenated, which could lead to a reduced estimate of lysine availability.

Lysine decarboxylase

A specific enzyme, lysine decarboxylase, has been explored for determining lysine availability.[201] Its use has been limited, because of the need to acid hydrolyze Maillard-damaged protein products, which will release a considerable amount of lysine from deoxyketosyllysine derivatives.

Microbiological methods

A number of microbiological methods are available for measuring either protein quality or amino acid bioavailabilities in foods.[202,203] Several organisms have been used with varying degrees of success: *Leuconostoc mesenteriodes*, *Streptomyces faecalis*, *Streptococcus zymogenes*, and *Tetrahymena pyriformis*. *S. zymogenes* is generally favored, because it needs only a short incubation time in comparison with *T. pyriformis*. However, this microorganism does not have an absolute requirement for lysine, threonine, and phenylalanine.

Methods to make use of protazoa *T. pyriformis* or *Tetrahymena WH$_{14}$* are attractive for several reasons: (1) their essential amino acid requirements are similar to those of humans, and (2) whole proteins can, in principle, be utilized by these microorganisms. In general, these methods have met with mixed success.[204,205] Anderson et al. have concluded that *Tetrahymena* is unsuitable for "early" Maillard products, because it can utilize Amadori compounds as a source of lysine.[205] Shepherd et al. have used a novel marker for estimating cell growth; they have extracted tetrahymenol from cell cultures and quantified it by gas liquid chromatography.[206] This procedure appears to have improved the reproducibility.

The use of *Escherichia coli* mutants to assess amino acid bioavailabilities was investigated by Hitchins et al.[207] While a reasonable correlation has been obtained between biological and chemical assays, it is important to recognize that deoxyketosyllysine formed in Maillard-reaction-altered foods may be a source of lysine for the mutants.

In vitro estimate of protein digestibility

There have been many attempts to correlate enzymatic digestibility of proteins with true digestibility as determined by animal assay procedures.[208–222] Satterlee et al.[208] and Hsu et al.[211] developed a simple *in vitro* method to estimate protein digestibility based upon a pH drop during enzymatic digestion. Later, Pederson and Eggum[212] and McDonough et al.[213] modified this procedure to function at constant pH. The enzymes used were a mixture of trypsin, chymotrysin, and intestinal peptidase. The volume, B, of 0.1 N NaOH required in the titration to maintain a pH of 7.98 for exactly 10 min was used to compute[212] the true digestibility (TD) using the equation

$$TD = 76.14 + 47.77B$$

This regression was derived by relating B to the TD obtained from rat balance studies. To correct for certain anamolies, McDonough et al.[213] have used a 30 min pretreatment with 0.1 N NaOH and a short enzymatic digestion (5 min).

While excellent correlation has been observed for 57 protein sources (vegetable proteins and vegetable–animal protein mixtures), it must be noted that there is a very large intercept in this equation, implying that the digestibility cannot fall below 76%. Data (Table 4) of Bodwell et al. indicate that heated nonfat dry milk is well digested[214]; values as high as 98% (range 84 –98%) digestibility have been reported using this *in vitro* method. However, this finding needs confirmation using other Maillard-reaction-altered proteins and food models, because extensive crosslinking is known to occur. Even *in vivo*

Table 4 Comparison of the Results of Protein Quality Determinations on Seven Foods

Reference Numbers	Digestibility of N, %					Relative PER		Relative NPR		Relative NPU	Ess. Am. Ac. Index		
	Rat Values		In vitro								Uncorrected	Corr. for N Dig. %	Corr. for Am. Acid Dig. %
	[1]	[2]	[1]	[6]	[10]	[3]	[5]	[3]	[5]	[1]	[3]	[3]	[3]
Casein	97	99	96	92	(100)[a]	(100)[a]	(100)[a]	(100)[a]	(100)[a]	(100)[a]	100	100	100
Tuna	93	97	92	100	99	96	104	98	106	101	100	100	100
NFDM	93	97	95	89	100	98	109	99	102	103	100	100	100
NFDM (heated)	90	90	92	84	98	59	58	67	79	95	89	80	78
Canned pinto beans	73	79	90	82	80	21	3	46	44	64	80	63	42
Chick peas	88	89	100	90	87	66	56	80	72	76	74	66	61
Pea protein concentrate	94	92	93	92	99	49	41	80	64	79	74	68	67

Note: Studies carried out in different laboratories as part of a USDA collaborative study. Reference numbers indicate the numbers for the laboratories. Values expressed as a percentage of the corresponding vlaue obtained for casein.

From Bodwell, C.E., et al. 1989. Plant Food Human Nutr. 39: 3-12. With permission.

digestibility values can be in error due to the metabolism of undigested protein residues by colon bacteria.

For most native proteins, the initial phase of the digestion proceeds rapidly. However, protein–sugar–protein crosslinked cores are likely to be poorly digested, so their presence might not be observed from titration data based on the first 10 min of reaction.

In vitro determination of enzymatically available amino acids

Several enzymatic assay procedures are available; they vary in the technique, the enzyme used, the selection of end point, and the type of data obtained. The early method of Mauron et al.[215] and Mottu and Mauron[216] utilized digestion, first with pepsin, and then with pancreatin, followed by ion-exchange chromatography to determine available lysine and other amino acids in foods. It has been claimed that the data obtained by this procedure was in good agreement with rat bioassay methods.

Savoie and Gauthier automated the digestion procedure of Mauron et al. using a digestion cell with two concentric cylindrical chambers the inner to digest the sample and the outer to continuously sweep the amino acids and peptides as they were transported through a tubular dialysis membrane (1000 molecular weight cutoff) into a flowing buffer.[217] Pepsin digestion for 30 min, followed by 1–24 h treatment with pancreatin was used. Savoie et al. have provided data on the percentage of amino acids (Table 5) released following digestion by pepsin, pancreatin, and 6N HCl (24 h at 110°C).[218] The data indicate that lysine, arginine, tyrosine, and phenylalanine were released rapidly, whereas cystine, proline, aspartic and glutamic acids, serine, and threonine were released slowly. During the 6-h digestion with pancreatin, only 31–51% of the protein–nitrogen was digested, leaving an average of 60% of the sample protein undigested.

Investigations of this type using an innovative continuous digestion cell are critical for mimicking gastrointestinal digestion and minimizing animal experimentation. Before accepting this technique, it is necessary to consider an inconsistency. Although rat balance studies (Table 6) have shown that most of these foods are well digested (79–100%), the data from the collaborative study[214] demonstrated considerable differences in rates of digestion of the various amino acids during the 6-h period. It is possible that the digestion is limited and that acid hydrolysis compromises the validity of the enzymatic process.

Rayner and Fox have developed a digestion procedure using a high level (5% of the substrate) of protease (pronase) from *Streptomyces griseus* and incubating at 40°C for 48 h.[219] In untreated muscle, 84% of the lysine was released upon pronase treatment. Experience with beef stew has shown that aspartic and glutamic acids, leucine, and isoleucine are poorly released by this procedure.

To overcome the problem of incomplete digestion of food proteins with pronase, a three-stage digestion procedure was used in which pronase treatment was followed by carboxypeptidase and aminopeptidase.[220] Lysine was found to be completely released by pronase alone. Carboxypeptidase vastly enhanced the release of aspartic and glutamic acids, while aminopeptidase increased the release of isoleucine and leucine (Figure 16). This procedure was successfully applied to beef stew thermally processed in flexible pouches and in No. $2\frac{1}{2}$ cylindrical cans and freeze dehydrated. The data (Figure 17) demonstrated that lysine and methionine were better preserved in beef stew processed in flexible pouches compared with those processed in cans. The preconcentration step that preceded the freeze dehydration was also shown not to cause significant damage to lysine and methionine. Storage at 100°F for 12 months preserved (Figure 18) the enzymatic availability of lysine and methionine in these beef stew products.

A very different and interesting approach is that of Porter et al.[221] and Culver and Swaisgord[221,222] who have used two separate reactors—one containing pepsin immobilized

Table 5 Percentage of Digestibility after Treatment with Pepsin and Pancreatin in a Digestion Cell followed by 6N HCl (24 h at 110°C)

Protein Sources	Asp	Thr	Ser	Glu	Pro	Gly	Ala	Val	Met	Ile	Leu	Tyr	Phe	His	Lys	Arg	Cys
ANRC casein	35[a]	36	35	35	30	52	50	36	50	35	59	72	68	45	66	57	50
NFD milk	34	35	32	29	29	57	47	41	52	44	60	80	64	50	70	88	42
NFD milk (heated)	32	36	30	30	33	55	48	39	45	39	59	82	45	41	61	73	28
Macaroni and cheese	32	34	31	32	24	46	52	43	51	45	57	76	56	47	72	83	20
Tuna	29	34	34	27	30	40	39	42	44	44	52	60	56	77	62	57	30
Chicken franfurters	33	41	36	33	28	36	46	54	55	55	60	80	70	56	67	70	30
Salami	39	49	47	33	31	40	49	51	57	48	60	98	66	67	70	73	23
Breakfast sausage	22	25	21	19	15	19	23	32	39	35	40	53	44	38	57	46	16
Beef stew	26	31	26	29	25	34	39	42	42	38	43	66	51	46	55	65	20
Pinto beans	26	27	28	28	30	33	35	34	47	27	37	63	47	34	51	72	31
Kidney beans	29	33	29	27	47	40	45	45	61	44	48	70	50	367	51	61	36
Lentils	28	34	31	26	28	38	42	38	70	38	45	92	68	31	50	49	17
Pea concentrate	34	37	38	31	29	39	50	54	42	52	54	70	61	45	51	62	16
Chick peas	26	30	31	28	27	38	42	41	37	36	44	72	56	41	52	60	20
Soy isolate	27	34	37	23	21	40	51	55	59	50	64	65	60	46	76	65	17
Peanut butter	34	44	39	23	35	39	52	60	53	57	61	77	69	52	70	73	32
Instant whole wheat	29	34	28	25	19	30	38	40	40	46	45	61	41	39	57	55	22
Rolled oats	21	27	24	25	21	29	33	31	43	33	39	60	41	32	47	63	10
Rice-wheat gluten cereal	27	32	28	22	17	39	39	40	35	44	45	59	39	40	61	65	13

[a] Mean of triplicate analyses.

From Savoie, L., et al. 1989. Plant Food Human Nutr. 39: 93-107. With permission.

Table 6   In Vivo Digestibility of Protein and Specific Amino Acids in
Various Food Products by the Rat Balance Method

Mixture	Protein	Lys	Met	Cys	Thr	Trp
Casein	99	100	99	100	100	100
Skim milk	95	96	92	94	95	98
Beef (roast)	100	100	100	100	100	100
Beef salami	99	99	99	100	100	100
Sausage	94	94	91	95	92	93
Egg white solids	98	97	98	97	96	97
Tuna fish	97	97	95	96	98	97
Chicken Franks	96	97	97	100	95	96
Pea flour	88	92	77	84	87	82
Pea, Century (autoclaved)	83	85	62	85	78	72
Pinto bean (canned)	79	78	45	56	72	70
Lentil (autoclaved)	85	86	59	75	76	63
Fababean (autoclaved)	86	85	59	75	76	63
Soybean	90	87	82	82	84	89
Soybean protein isolate	98	98	94	94	96	98
Rapeseed protein concentrate	95	91	92	93	91	93
Peanut	96	90	85	89	89	94
Peanut meal	91	88	89	89	87	—
Peanut butter	98	96	94	100	97	99
Sunflower meal	90	87	92	91	90	—
Wheat	93	83	94	97	91	96
Rolled Oats	94	90	92	98	90	97
Rice-wheat-gluten	93	85	81	95	88	92
Wheat flour-casein	95	91	91	89	90	90
Macaroni-cheese	95	95	93	98	92	98
Potatoes-beef	86	89	83	89	83	86
Rice-soybean	90	89	77	82	84	87
Corn-pea	83	85	84	86	82	80
Corn-soybean	93	93	87	94	93	98

From FAO/WHO. Report of the Joint FAO/WHO Expert Consultation on Protein
Quality Evaluations. 1990. Food and Agriculture Organization of the United Nations, Rome and World Health Organization, Geneva, pp 1–66. With permission.

on glass beads and the other containing trypsin, chymotrypsin, and intestinal peptidase
covalently attached to glass beads. This method has the potential of being very efficient,
except for insoluble food proteins or foods where the interaction, if any, is bound to be slow.

Corrected chemical scores for essential amino acids

In minimally processed foods where the change due to Maillard reaction or due to oxidized
lipids is relatively small or assumed to be small, the amount of amino acids in terms of
milligrams per gram protein released by chemical hydrolysis (6N HCl, performic acid
plus 6N HCl, and/or barium hydroxide) is used to estimate individual essential amino
acid scores.[223] Acid hydrolysis with 6N HCl under N_2 at 110°C for 20 h is the standard
procedure for determining most amino acids in proteins, except for sulfur amino acids
and tryptophan.[224] For branched-chain amino acids (valine and isoleucine), a 96 h hydrolysis in 6N HCl is recommended. The sulfur amino acids (methionine and cysteine) in the
intact protein are oxidized to methionine sulfone and cysteic acid residues, respectively,
with cold performic acid (16 h at 0°C), hydrolyzed with 6N HCl (18 h at 110°C), and

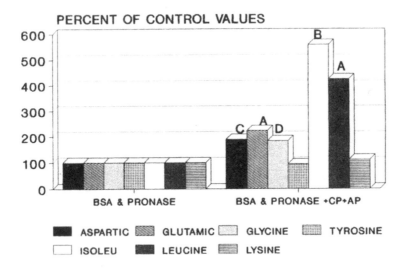

Figure 16 The influence of carboxypeptidase (CP) and aminopeptidase (AP) upon the digestion of proteins such as bovine serum albumin (BSA) after initial treatment with pronase. Bars with letters are statistically significant from corresponding controls. A: ($p < .001$); B: ($p < .005$); C: ($p < .025$); D: ($p < .05$).

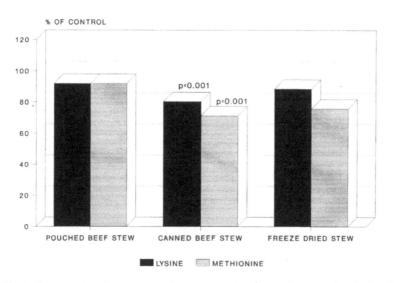

Figure 17 The influence of various processing treatments (thermal processing in flexible pouches; thermal processing in cylindrical cans (No. $2\frac{1}{2}$); preconcentration-freeze dehydration) of beef stew on the enzymatic availability of lysine and methionine. Bars marked $p < 0.001$ are statistically significant from corresponding controls. All other values are not statistically different from controls (i.e., preprocessed samples).

determined by HPLC.[224] Trytophan is determined after alkaline hydrolysis with lithium hydroxide, sodium hydroxide plus starch, or with barium hydroxide.[224]

The amino acid scores are derived from the ratios of the amino acid values for the test protein to a FAO/WHO reference pattern for amino acids.[223,225] The rules for assigning amino acid scores are as follows: (1) all ratios above 100 are considered as 100%; and (2) the lowest ratio represents the first limiting amino acid and is defined as the amino acid score for the test protein. Because all the protein may not be digested by mammalian

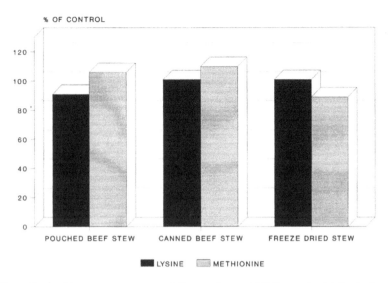

Figure 18 The effect of long-term storage (12 months at 100°F) on the biological availability of lysine and methionine in processed beef stew (see legend to previous figure). Statistical tests showed that none of those values were significantly ($p < .05$) altered from corresponding zero time controls.

proteolytic enzymes, a correction needs to be applied for true digestibility, obtained either by the rat balance method or by the *in vitro* method discussed above. The corrected amino acid scores give a reasonably good indication of the protein quality of foods and are in fair agreement with the NPR values (Table 4).

Carpenter et al. estimated available LYS by three chemical procedures in 17 test foods.[196] They observed that total LYS (determined by 6N HCl hydrolysis) was a reasonably good indicator of LYS availability, except for heated nonfat milk (Table 3) where by comparison with rat assay the LYS was overestimated (46–94%) by three methods: acid hydrolysis, FDNB, and dye binding. It is, therefore, necessary to interpret cautiously LYS data obtained using these for assessing damage to food proteins .

It is recognized that plant proteins are not complete proteins because they are low in one or more essential amino acids. For example, legumes and corn are deficient in tryptophan; cereals and millets are low in lysine and threonine; while legumes are deficient in sulfur amino acids (swaminathan[226]). Most nuts and oilseeds are low in methionine and lysine. Hence, the poor amino acid scores (42, 61, and 67) seen for pinto beans, chick peas, and pea protein is understandable (Table 4). On the other hand, legumes and soybean proteins are rich in lysine and threonine, while sesame and sunflower seeds have a high content of tryptophan and sulfur amino acids.

By mixing a low-quality vegetable protein source with a high-quality food protein (whether animal or vegetable) or with specific amino acids, the protein quality of the mixture can be increased substantially,[223,227] as seen in Table 7. The corrected amino acid score of wheat flour of 0.41 was increased to 0.67–0.91 by the addition of an equal part of soy protein, pea flour, beef, or casein. The corrected score of wheat gluten of 0.25 was greatly enhanced to 0.77 by adding beef. Similarly, beef increased the amino acid scores of peanut meal and sunflower protein isolates by 46–127% (Sarwar and McDonough[228]).

True digestibility of proteins by rat balance methods

This method is based upon feeding rats a diet containing 8% protein for a period of 28 days, measuring protein intake, fecal protein, and metabolic fecal protein, and then computing

Table 7 Amino Acid Scores for Some Protein Mixtures Illustrating the
Enhancement in Their Quality

Mixture[b] (50:50 Protein Basis)	True Protein Digestibility %	Amino Acid Score	Protein Digestibility-Corrected Score
Wheat flour (WW)	90	0.46	0.41
WW + beef	93	0.91	0.85
WW + egg white	95	0.83	0.79
WW + casein	95	0.96	0.91
WW + rapeseed concentrate	93	0.72	0.67
WW + pea flour	92	0.89	0.82
WW + soy protein	92	0.78	0.72
Beef + rapeseed concentrate	95	1.12	1.00
Beef + rapeseed isolate	96	1.12	1.00
Beef + soybean concentrate	96	1.17	1.00
Beef + soybean isolate	98	1.07	1.00
Beef + peanut meal	95	0.80	0.76
Beef + pea concentrate	95	0.84	0.80
Beef + sunflower isolate	95	0.88	0.84
Beef + wheat gluten	95	0.81	0.77

From FAO/WHO. Report of the Joint FAO/WHO Expert Consultation on Protein
Quality Evaluations. 1990. Food and Agriculture Organization of the United Nations,
Rome and World Health Organization, Geneva, pp 1–66. With permission.

the true percentage of the protein that is retained by the rat. Metabolic fecal protein is
determined in rats fed a nitrogen-free diet for 14 days. The amino acid digestibility values
are similarly computed except that amino acid values, determined after chemical hydrolysis, are substituted for the protein values. Sarwar et al. demonstrated that for many foods
the differences between protein and amino acid digestibilities were less than 5%.[229] However, for several vegetable products (beans and lentils), the true digestibility for methionine
and cysteine were considerably lower (12–43%) than the digestibilities for crude protein.
It is encouraging to note that the protein digestibility values were high for most products
except for peas, pinto beans, and lentils. In general, the lysine digestibility values were
similar to the protein digestibility values.

While the true lysine digestibilities of both unheated and heated nonfat milk were
close to 92%,[230] the lysine availability in the heated nonfat milk had a significantly lower
value of 66% compared to 108% for the unheated nonfat milk.[231] This difference between
digestibility and availability may suggest that some microbial modifications of unab-
sorbed protein residues occur in the colon and lead to an overestimate by the rat balance
method.[228,232]

Bioavailabilities by rat growth assay

Rat bioassay methods are available based upon dose-response curve for various amino
acids, such as lysine, methionine, and tryptophan.[231–234] Growth curve responses are
obtained using a basal diet that contains graded levels of the test amino acids, which at
the same time provides 125% of the requirements of all other essential amino acids. Growth
curves for rats fed experimental foods at two levels are then compared to the standard
dose-response curves in order to estimate the bioavailability of the selected amino acid.
As Table 8 shows, the lysine values were poor for pinto beans (73%) and for heated skim
milk (66%). The latter value clearly demonstrated the loss in lysine due to the Maillard

Table 8   Relative Bioavailability of Lysine in Test Foods as Determined by Rat Growth

Source	Lysine (400 mg/100 g diet)[a]			Lysine (600 mg/100 g diet)[a]		
	Gain (g)	Intake (g)	Availability (%)	Gain (g)	Intake (g)	Availability (%)
Nonfat milk	38.9 ± 5.0[b]	0.545 ± 0.07	115 ± 19	65.3 ± 4.5	0.869 ± 0.06	108 ± 10
Nonfat milk-heated	32.5 ± 6.7	0.536 ± 0.08	61 ± 29	50.0 ± 4.5	0.852 ± 0.06	66 ± 7
Tuna	36.0 ± 3.8	0.539 ± 0.03	91 ± 20	63.8 ± 8.3	0.897 ± 0.07	96 ± 13
Chicken frankfurters	35.0 ± 7.9	0.480 ± 0.09	128 ± 30	53.1 ± 5.8	0.732 ± 0.08	104 ± 8
Beef salami	35.9 ± 7.4	0.523 ± 0.08	101 ± 17	65.4 ± 4.3	0.879 ± 0.05	105 ± 8
Breakfast sausage	30.9 ± 6.1	0.487 ± 0.09	83 ± 20	59.4 ± 9.3	0.878 ± 0.11	88 ± 8
Beef stew	41.6 ± 6.7	0.584 ± 0.08	111 ± 31	64.3 ± 7.7	0.927 ± 0.07	91 ± 9
Macaroni/cheese	36.5 ± 6.2	0.550 ± 0.06	88 ± 25	60.5 ± 8.7	0.873 ± 0.10	92 ± 6
Chick peas	40.9 ± 5.0	0.617 ± 0.06	84 ± 33	63.4 ± 5.9	0.939 ± 0.08	88 ± 8
Pinto beans	33.0 ± 4.0	0.567 ± 0.06	51 ± 28	54.6 ± 4.6	0.824 ± 0.06[c]	73 ± 11
Pea protein	37.5 ± 5.0	0.561 ± 0.06	90 ± 21	56.6 ± 7.2	0.835 ± 0.10	89 ± 4
Soy isolate	34.9 ± 7.1	0.548 ± 0.08	73 ± 27	57.7 ± 4.1	0.823 ± 0.06	90 ± 8
Peanut butter	28.4 ± 5.8	0.430 ± 0.08	102 ± 52	41.9 ± 6.6	0.585 ± 0.06	108 ± 8
Whole wheat cereal	34.1 ± 8.0	0.548 ± 0.08	67 ± 30	49.5 ± 5.5	0.705 ± 0.08	99 ± 10
Rice-wheat gluten cereal	31.9 ± 3.7	0.515 ± 0.03	72 ± 25	35.2 ± 5.1	0.571 ± 0.08[d]	70 ± 8
Rolled oats	41.5 ± 6.6	0.606 ± 0.07	95 ± 21	53.9 ± 4.8	0.750 ± 0.05[d]	101 ± 10

[a] Basal casein provided 300 mg lysine for all diets; thus, 25% and 50% of lysine were from test foods.

[b] All results are mean (N = 8) ± standard deviation.

[c] Low lysine content required feeding at 550 mg/100 g diet.

[d] Low lysine content required feeding at 500 mg/100 g diet.

From McDonough, F.E., et al. 1989. Plant Food Human Nutr. 39: 67-75. With permission.

reaction. Lysine availability was high (90–101%) in several plant proteins such as rolled oats, whole wheat, and soy isolate; it was low (70–73%) in rice/wheat gluten cereal and pinto beans (Table 8).

Protein efficiency ratio

A common means of assessing protein quality is to determine the protein efficiency ratio (PER), which is defined as follows:

$$PER = \frac{W_G}{P},$$

where W_G = weight gain of test group; P = protein consumed by test group. The rats are generally fed diets with 10% protein. The PER values are usually expressed as a ratio of the PER of test protein to the PER of casein, and which is normalized to a value of 2.5. Unfortunately, the PER value takes into account the contribution of the protein for growth, but not for maintenance. It does not provide a linear response for proteins of varying nutritional value. Furthermore, a test protein may have a PER of 0 (i.e., no growth in rats) but still be adequate for maintenance.[223,228]

Net protein ratio

A major modification of the PER method is the net protein ratio (NPR) procedure, which is defined as

$$NPR = \frac{W_G + W_L}{P},$$

where W_G = weight gain by test group; W_L = average weight lost by rats fed basal (nonprotein) diet; and P = protein consumed by test group.

This method is a significant improvement over the PER method, because allowance is made for both growth and maintenance, by introducing an additional group—a nonprotein group. The weight lost by this group is added positively to the weight gained by the test group. In order to scale the protein quality values from 1 to 100, the NPR values are conveniently expressed as a percent of the NPR of a reference protein (such as 10% casein or 8% casein + 0.1% methionine). These are referred to as relative net protein ration (RNPR) values. Similar scaling is not possible for PER because PER values are not proportional to protein qualities as mentioned.

Biological value and net protein utilization

Among the many methods used, the biological value (BV) using either animals or humans[223-236] is a highly reliable and exact indicator of protein quality. However, this method is time-consuming, laborious, and expensive. On the basis of a six-laboratory collaborative study to estimate protein quality variability in seven protein sources, Sarwar et al. have concluded that NPR and RNPR methods are the most suitable for rat growth assay and are both economical and reproducible.[237]

BV is defined as the proportion of absorbed nitrogen that is retained for maintenance and growth.

$$BV = \frac{B}{A} = \frac{I \pm (F \pm F_K) \pm (U \pm U_K)}{I \pm (F \pm F_K)},$$

where B = body nitrogen; A = absorbed nitrogen; I = nitrogen ingested; F = fecal nitrogen; F_K = endogenous fecal nitrogen; U = urinary nitrogen; and U_K = endogenous urinary nitrogen.

Net protein utilization (NPU) is defined as the proportion of ingested nitrogen that is retained for maintenance and growth. NPU is, therefore, the product of biological value and true digestibility. True digestibility, TD, is then defined as the proportion of food nitrogen that is absorbed:

$$TD = \frac{I \pm (F \pm F_K)}{I}.$$

From equations for BV and TD, we see that

$$NPU = (BV)(TD) = \frac{I \pm (F \pm F_K) \pm (U \pm U_K)}{I}.$$

NPU can also be experimentally obtained from carcass nitrogen.

Table 4 also compares the relative NPU and relative NPR values of seven foods. In several instances, the NPU values are higher than those for NPR. As explained by Bodwell et al., this difference was due partly to the restriction in diet to 10 g/rat/day for the rats in the NPU assay and partly to the efficient recycling for maintenance of lysine, the limiting amino acid.[214] In any event, these kinds of data emphasize the difficulties in reaching conclusions from one protein quality detection method alone, be it the chemical or enzymatic, *in vitro* or *in vivo*, animal or human.

Conclusions

The vulnerability of micronutrients such as thiamin, ascorbic acid, pyridoxal, folacin, tocopherol, carotene, and retinol to stresses such as temperature, light, pH, and A_w is well recognized. In addition, the retention and biological availability of macro- and micronutrients may be severely compromised as a result of diverse nutrient–nutrient interactions. These interactions include those between proteins and reducing sugars; potentially deleterious interactions among autoxidizing lipids and proteins; between thiamin and carbohydrates, thiamin and tannins in brewed tea, among riboflavin or cobalamin and ascorbic acid; and between thiamin, nicotinamide, and cobalamin, to mention a few of the many reactions known.

The first stage of the Maillard reaction between reducing sugars and proteins has been known to render lysine unavailable for digestion and absorption in humans and animals due to the formation of Amadori compounds. Subsequent degradation of these in the second and third stages gives rise to a variety of highly desirable and undesirable flavor compounds and brown melanoidin pigments, as wells as to undesirable toughening of texture due to crosslinking reactions.

Studies with lysine–glucose model suggest that the Maillard reaction in dry foods could lead to substantial loss in lysine in a few months or years when it is not possible to block this reaction. High storage temperatures and "intermediate" moisture activities (0.5–0.7) favor the reaction, but even foods stored at normal or at low temperatures or low A_w are not immune from Maillard-induced losses. Amadori compounds in the presence of reducing sugars such as glucose, but not lactose, decompose to brown pigments.

Considerable attention has been focused upon lactose-hydrolyzed milk (LHM) and milk products for people with lactose intolerance. However, appropriate consideration should be given to the fact that LHM and LH dairy products are far more susceptible to Maillard-induced changes damage than is regular milk and regular dairy product due to the greater reactivity of monosaccharides compared with disaccharides. In this connection, the reactivity also differs, depending upon whether they are amorphous or crystalline; the former possibly leading to losses in hygroscopic foods higher than expected based upon A_w considerations. The use of humectants in intermediate moisture foods tends to shift downward the A_w for maximum rate of Maillard reaction, possibly due to increased mobility of the reactants in the presence of the humectant. To overcome the objections of high levels of humectants, such as poor flavor, tough texture, decreased acceptability, as well as deleterious nutritional consequences, a strategy involving a combination of pH, O_2 tension, antimicrobial agents, or mild heat treatments has been suggested and deserves further exploration.

It has been increasingly recognized that polyunsaturated lipids can readily autoxidize during processing and storage in the presence of air and then interact with food proteins to form tightly bound complexes under normal or physiological conditions (pH 7, 30°C, and high A_w), which could have a deleterious effect upon the biological availability of these proteins. In this respect, those interactions are similar to the Maillard reaction, but the nature of the associated reactions is far less understood. It is generally accepted that lipid free radicals leads to protein free radicals, resulting in protein polymerization and consequent loss in biological availability of several amino acids. Although the binding between the lipids and the proteins appears to be noncovalent, it is strong and could be the result of secondary forces, occlusion, and/or adsorption. Copolymerization of lipid peroxy radicals and proteins is not as widely accepted as the nonradical mechanism involving condensation of malonaldehyde formed by lipid peroxidation and scission. For foods devoid of linoleic acid or higher polyunsaturated fatty acids, this nonradical mechanism is unlikely.[7,132]

From a practical standpoint, the consequences of the reaction between oxidizing lipids and proteins in foods is important, because it is considered to be responsible for a wide variety of problems: the yellowing of bacon; the rusting of fish upon storage; the formation of ceroid pigments in human aortas; and possibly for the depression of growth of rats fed used corn oil from a 1957 commercial potato chip fryer. In addition, large losses in the biological availability of important amino acids such as lysine, tyrptophan, and cysteine have been reported in proteins interacted with oxidized lipids. Extensive losses in almost all amino acids (17–54%) upon 6N HCl hydrolysis is consistent with the mechanism of free radical induced protein polymerization.

Besides racemization caused by alkali treatment of food proteins, which results in incompletely utilizable D-amino acids and in unnatural peptides, toxic compounds such as lysinoalanine are formed. o-Diphenols, such as caffeic acid, are capable of interacting with proteins either enzymatically through the action of polyphenol oxidase or nonenzymatically at pH 10, which reduces their nutritional quality and gives rise to brown pigments.

Numerous efforts have been made to block the Maillard reaction in food, but these have met with only limited success. Among the promising avenues are freeze drying to very low A_w, acetylation of the amino groups of food proteins, and where feasible the use of sulfite and sulfur compounds (cysteine or glutathione) and aminoguanidine. Most of the reports have provided data on the suppression of browning rather than on the retention of lysine quality; additional work in this area is highly desirable.

On the basis of a new mechanism involving the formation of free radicals prior to the Amadori rearrangement, the addition of antioxidants to a casein–glucose system has been investigated. The antioxidants appear to have been effective in minimizing lysine losses. However, this requires confirmation from other sources to eliminate other possibilities from clouding the picture such as oxidized lipid–protein interactions.

To inhibit oxidized lipid–protein interactions, the use of antioxidants, vacuum packaging, metal chelating agents, and O_2 scavengers have been recommended. Protein–polyphenol interactions can be greatly reduced by keeping the pH of foods at 6.0 or lower. This approach would minimize damage due to polyphenol oxidase as well as to nonenzymatic reactions.

Several methods are available to detect and assess protein quality changes in foods due to processing or storage. These include the following: (1) chemical estimate of reactive groups, especially the ε-amino groups of lysine; (2) microbiological estimation of the availability of proteins or protein hydrolysates; (3) *in vitro* or *in vivo* estimate of protein digestibility; (4) enzymatic determination of available amino acids; (5) determination of protein efficiency ratio or net protein ratio; (6) determination of biological availability of specific amino acids; (7) determination of biological value and net protein utilization. Most of these methodologies were used in an interlaboratory study initiated by USDA, Beltsville, MD, to assess protein quality and amino acid availability of several foods. These data have further emphasized the difficulties in reaching a definititve conclusion about protein quality based on a single method of assessment.

References

1. Mauron, J. 1981. The Maillard reaction in food; A critical review from the nutritional stand-point. In: Progress in Food and Nutrition Science. Maillard Reactions in Food (C. Eriksson, ed.), Pergammon Press, Elmsford, NY, p. 5-35.
2. Finot, P.A., Deutsch, R., and Bujard, E. 1981. The extent of the Maillard reaction during the processing of milk. In: Progress in Food and Nutrition Science. Maillard Reactions in Food (C. Eriksson, ed.), Pergammon Press, Elmsford, NY, p. 345-355.
3. Hurrel, R.F. and Carpenter, K.J. 1981. The estimation of available lysine in foodstuffs after Maillard reactions. In: Progress in Food and Nutrition Science. Maillard Reactions in Food (C. Eriksson, ed.), Pergammon Press, Inc., Elmsford, NY, p. 159-176.
4. Narayan, K.A. and Andreotti, R.E. 1989. Kinetics of lysine loss in compressed model systems due to Maillard reaction. J. Food Biochem. 13: 105-125.
5. Labuza, T.P., Reineccius, G.A., Monnier, V., O'Brien, J., and Baynes, J. (eds.). 1994. Maillard Reactions in Chemistry, Food and Health. Royal Society of Chemistry, Special Publication No. 151, pp. 1-440.
6. Tappel, A.L. 1955. Studies of the mechanism of Vitamin E action. III. *In vitro* copolymerization of oxidized fats with protein. Arch. Biochem. Biophys. 54: 266-280.
7. Narayan, K.A. and Kummerow, F.A. 1958. Oxidized fatty acid-protein complexes. J. Am. Oil Chem. Soc. 35: 52-56.
8. Desai, I.D. and Tappel, A.L. 1963. Damage to protein by peroxidized lipids. J. Lipid Res. 4: 204-207.
9. Roubal, W.T. and Tappel, A.L. 1966. Polymerization of proteins induced by free-radical lipid peroxidation. Arch. Biochem. Biophys. 113: 150-155.
10. Karel, M. 1975. Free radicals in low moisture systems. In: Water Relations of Foods (R.B. Duckworth, ed.), Academic Press, New York, p. 435-453.
11. Pokorny, J. 1981. Browning from lipid-protein interactions. In: Progress in Food and Nutrition Science. Maillard Reactions in Food (C. Eriksson, ed.), Pergammon Press, Elmsford, NY, p. 421-428.
12. Duell, P.B. 1996. Prevention of atherosclerosis with dietary antioxidants: Fact or fiction. J. Nutr. 126, 1067S-1071S.
13. Jailal, I. and Devaraj, S. 1996. The role of oxidized low density lipoprotein in atherogenesis. J. Nutr. 126, 1053S-1057S.
14. Thomas, M.J. and Rudel, L. L. 1996. Dietary fatty acids, low density lippoprotein composition and oxidation and primate atherosclerosis. J. Nutr. 126, 1058S-1062S.
15. Lynch, S.M. and Frei, B. 1996. Mechanism of metal ion-dependent oxidation of human low density lipoprotein. J. Nutr. 126, 1063S-1066S.
16. Strapran, I., Rapp. J.H., Pan, X.-M., Hardman, D.A., and Feingold, K.R. 1996. Oxidized lipid in the diet accelerate the development of fatty streaks in cholesterol-fed rabbits. Arteriosclerosis, Thrombosis vascular biol. 16: 533-538.
17. Hayashi, T., Hoshii, Y., and Namiki, M. 1983. On the yellow product and browning of the reaction of dehydroascorbic acid with amino acids. Agric. Biol. Chem. 47: 1003-1009.
18. Mercier, C., Charbonniere, R., Grebaut, J., and de la Guerrivierre, J.F. 1980. Formation of amy-lose-lipid complexes by twin-screw extrusion cooking of manioc starch. Cereal Chem. 57: 4-9.
19. Bhatnagar, S. 1994. Amylose-lipid complex formation during single-screw extrusion of var-ious corn starches. Cereal Chem. 71: 582-587.
20. Bhatnagar, S. and Hanna, M.A. 1994. Extrusion processing conditions for amylose-lipid complexing. Cereal Chem. 71: 587-593.
21. Holm, J., Bjorck., I., Ostrowska, S., Eliasson, A.-C., Asp, N.-G., Larsson, K., and Lundquist, L. 1983. Digestibility of amylose-lipid complexes *in vitro* and *in vivo*. Starch 9: 294-297.
22. Krog, N. 1971. Amylose complexing effect of food grade emulsifiers. Die Starke 23: 206-210.
23. Ink, S.L. 1988. Fiber-mineral and fiber-vitamin interactions. In: Nutrient Interactions (C.E. Bodwell and J.W. Erdman, Jr., eds.), Marcel Dekker, New York, p. 253-264.
24. Bowering, J., Sanchez, A.M., and Irwin, M.I. 1980. I. A conspectus of research on iron re-quirements in man. In: Nutritional Requirements of Man. A Conspectus of Research (M.I. Irwin, ed.), The Nutrition Foundation, New York, p. 307-396.

25. Kelsay, J.L. 1982. Effects of fiber on mineral and vitamin bioavailability. In: Dietary Fiber in Health and Disease (G.V. Vahouny and D. Kritchevsky, eds.), Plenum Press, New York, p. 91-103.

26. Frost, D.V., Lapidus, M., Plaut, K.A., Scherfling, E., and Fricke, H.H. 1952. Differential stability of various analogs of cobalamin to vitamin C. Science 116: 119-121.

27. Hand, D.B., Guthrie, E.S., and Sharp, P.F. 1938. Effect of oxygen, light, and lactoflavin on the oxidation of vitamin C in milk. Science 87: 439-441.

28. Darby, W.J., McNutt, K.W., and Todhunter, E.N. 1976. Niacin. In: Present Knowledge in Nutrition (D.M. Hegsted, C.O. Chicester, W.J. Darby, K.W. McNutt, R.M. Stalvey, and E.H. Stotz, eds.), 4th Edition, The Nutrition Foundation, Washington, D.C., p. 162-174.

29. Doyon, L. and Smyrl, T.G. 1983. Interaction of thiamine with reducing sugars. Food Chem. 12: 127-133.

30. Yagi, N. and Itokawa, Y. 1979. Clevage of thiamine by chlorine in tap water. J. Nutr. Sci. Vitaminol. 25: 281-287.

31. Vimokesant, S.L., Nakornchai, S., Dhanamitta, S., and Hilker, D.M. 1974. Effect of tea consumption on thiamine status in man. Nutr. Rep. Int. 9: 371-376.

32. Blitz, M., Eigen, E., and Gunsberg, E. 1956. Studies relating to the stability of Vitamin B_{12} in B-complex injectable solutions. J. Am. Pharm. Assoc. Sci. Ed. 45: 803-806.

33. Gambier, A.S. and Rahn, E.P.G. 1957. The combination of B-complex vitamins in ascorbic acid in aqueous solutions. J. Am. Pharm. Assoc. Sci. Ed. 46: 134-140.

34. Baker, D.H. 1992. Biological availability of vitamins. Fed. Am. Soc. Exp. Biol., Anaheim, CA, April 5-8.

35. Bodwell, C.E. and Erdman, Jr., J.W. (eds.). 1988. Nutrient Interactions. Marcel Dekker, New York, p. 1-389.

36. Frolich, W. 1986. Bioavailability of minerals from cereals. In: Handbook of Dietary Fiber in Human Nutrition (G.A. Spiller, ed.), CRC Press, Boca Raton, FL, p. 173-191.

37. Munez, J.M. 1986. Overview of the effects of dietary fiber on the utilization of minerals and trace elements. In: Handbook of Dietary Fiber in Human Nutrition (G.A. Spiller, ed.), CRC Press, Boca Raton, FL, p. 193-200.

38. Irwin, M.I. (ed.). 1980. Nutritional Requirements of Man: A Conspectus of Research. The Nutrition Foundation, New York, p. 1-592.

39. Levander, O.A. and Cheng, L. 1980. Micronutrient Interactions: Vitamins, Minerals, and Hazardous Elements. Ann. N.Y. Acad. Sci. 355: 1-372.

40. Ammerman, C.B., Baker, D.H. and Lewis, A.J. (eds.). 1995. Bioavailability of Nutrients in Animals. Academic Press, San Diego, CA., pp. 1-441.

41. Aoyagi, S. and Baker, D.H. 1993. Nutritional evaluation of copper-lysine and zinc-lysine complexes for chicks. Poultry Sci. 72: 165-171.

42. Kattelmann, K.K. 1994. Methods to enhance iron availability/utilization by dietary manipulation of copper and studies in mechanism of enhancement by meat protein. Dissertation Abstr. Internat. B 54(11): 5608.

43. Burton, G.W. and Traber, M.G. 1990. Vitamin E: Antioxidant activity, biokinetics and bioavailability. Ann. Rev. Nutr. 10: 357-382.

44. Papas, A. M. 1993. Vitamin E and exercise: Aspects of biokinetics and bioavailability. World Rev. Nutr. Diet 72: 165-176.

45. Ferroli, C.E. and Trumbo, P.R. 1994. Bioavaiability of vitamin B-6 in young and old men. Am J. Clin. Nutr. 60: 68-71.

46. Pietrzik, K., Hages, M., and Remer, T. 1990. Methodological aspects of vitamin bioavailability testing. J. Micronutr. Anal. 7: 207-222.

47. Schelling, G.T., Roeder, R.A., Garber, M.J., and Pumfrey, W.M. 1995. Biovailability and interaction of vitamin A and vitamin E in ruminants. J. Nutr. 125: 1799S-1803S.

48. Kawamura, S. 1983. Seventy years of the Maillard reaction. In: The Maillard Reaction in Foods and Nutrition (G.R. Waller and M.S. Feather, eds.), ACS Symposium Series No. 215, American Chemical Society, Washington, D.C., p. 3-18.

49. Ikan, R. (ed.). 1996. The Maillard reaction; Consequences for the chemical and life sciences. John Wiley & Sons, New York, p. 1–214.

50. Rizzi, G.P. 1994. The Maillard reaction in foods. Roy. Soc. Chem. Special Publication No. 151: 11-19.
51. Hodge, J.E. 1953. Dehydrated Foods. Chemistry of browning reactions in model systems. J. Agric. Food Chem. 1: 928-943.
52. Labuza, T.P. 1994. Interpreting the complexity of the kinetics of the Maillard reaction. Roy. Soc. Chem. Special Publication No. 151: 176-181.
53. Pischetrsreider, M. and Severin, T. 1994. The Maillard reaction of disaccharides. Roy. Soc. Chem. Special Publication No. 151: 37-42.
54. Bailey, M.E. and Ki, W.U. 1992. Maillard reaction products and lipid oxidation. ACS Symposium Series. 500: 122-139.
55. Lin, L.J. 1994. Regulatory status of Maillard reaction flavors. ACS Symposium Series 543: 7-15.
56. Ames, J.M. 1991. Control of Maillard reaction in food systems. Trends Food Sci. Tech. 1: 150-154.
57. Cerami, A. 1994. The role of Maillard reaction *in vivo*. Roy. Soc. Chem. Special Publication No. 151: 1-10.
58. Van Boekel, M.A.J.S. and Berg, H.E. 1994. Kinetics of the early Maillard reaction during heating of milk. Roy. Soc. Chem. Special Publication No. 151: 170-175.
59. Sensidoni, A., Polini, C.M, Peressini, D. and Sari, P. 1994. The Maillard reaction in pasta drying: Study in model systems. Roy. Soc. Chem. Special Publication No. 151: 418-418.
60. Hirsch, J. and Feather, M.S. 1994. Aminoguanidine as an inhibitor of the Maillard reaction. Roy. Soc. Chem. Special Publication No. 151: 325-328.
61. Smith, J.S. and Alfawaz, M. 1995. Antioxidative activity of Maillard reaction products in cooked beef, sensory and TBA values. J. Food Sci. 60: 234-236.
62. Kim, S.W., Rogers, Q.R. and Morris, J.G. 1996. Maillard reaction products in purified diets induce taurine depletion in cats which is reversed by antibiotics. J. Nutr. 126: 195-201.
63. Khosrof, S.C. and Nagaraj, R. 1995. Advanced Maillard reaction and crosslinking of corneal collagen in diabetics. Biocehem. Biophys. Res. Comm. 214: 793-797.
64. Erbersdobler, H.F., Lohmann, M. and Buhl, K. 1991. Utilization of early Maillard reaction products by humans. Adv. Exp. Med. Biol. 289: 363-370.
65. Waller, G.R. and Feather, M.S. (eds.). 1983. The Maillard Reaction in Foods and Nutrition. ACS Symposium Series No. 215, American Chemical Society, Washington, D.C., p. 1-585.
66. O'Brien, J. and Morrissey, P.A. 1989. Nutritional and toxicological aspects of the Maillard browning reaction in foods. Crit. Rev. Food Sci. Nutr. 28: 211-248.
67. Eriksson, C. (ed.). 1981. Progress in Food and Nutrition Science. Maillard Reactions in Foods, Pergammon Press, Elmsford, NY, p. 1-501.
68. Cannon, P.R., Jones, D.B., Lewis, H.B., Rose, W.C. and Stare, F.J. 1950. The Problem of Heat Injury to Dietary Protein. National Research Council, Reprint and Circular Series, Washington, D.C., Vol. 131, 1-19.
69. McCollum, E.V. and Davis, M. 1915. The cause of the loss nutritive efficiency of heated milk. J. Biol. Chem. 23: 247-254.
70. Morgan, A.F. and King, F.B. 1925–1926. Changes in biological value of cereal proteins due to heat treatment. Proc. Soc. Exp. Biol. Med. 23: 353–355.
71. Tsen, C.C., Reddy, P.R.K. and Gehrke, C.W. 1977. Effects of conventional baking, microwave baking, and steaming on the nutritive value of regular and fortified foods. J. Food Sci. 42: 402-406.
72. Lee, T.-C., Kamigar, M., Pintauro, S.J. and Chichester, C.O. 1981. Physiological and safety aspects of Maillard browning of foods. In: Progress in Food Science and Nutrition Science. Maillard Reactions in Foods (C. Erikkson, ed.), Pergammon Press, Elmsford, NY, p. 243-256.
73. Eichner, K. 1975. The influence of water content on nonenzymatic browning reactions in dehydrated foods and model systems and the inhibition of fat oxidation by browning intermediates. In: Water Relations of Food (R.B. Duckworth, ed.), Academic Press, New York, p. 417-433.
74. Fors, S. 1983. Sensory properties of volatile Maillard reaction products and related compounds. In: The Maillard Reaction in Foods and Nutrition (G.R. Walker and M.S. Feather, eds.), ACS Symposium Series No. 215, American Chemical Society, Washington, D.C., p. 185-268.

75. Labuza, T.P. and Massaro, S.A. 1990. Browning and amino acid loss in model total parenteral nutritional solutions. J. Food Sci. 55: 821-826.

76. Hannan, R.S. and Lea, C.H. 1952. Studies of the reactions between proteins and reducing sugars in the "dry" state. The reactivity of the terminal amino groups of lysine in model systems. Biochem. Biophys. Acta 9: 293-305.

77. Narayan, K.A. and Cross, M.E. 1992. Temperature influence on acetyllysine interaction with glucose in model systems due to Maillard reaction. J. Food Sci. 57: 206-212.

78. Stamp, J.A. and Labuza, T.P. 1983. Kinetics of the Maillard reaction between aspartame and glucose in solution at high temperatures. J. Food Sci. 48: 543-547.

79. Kato, Y., Matsuda, T., Kato, N. and Nakamura, R. 1988. Browning and protein polymerization induced by amino-carbonyl reaction of ovalbumin with glucose and lactose. J. Agric. Food Chem. 36: 806-809.

80. Kato, Y., Matsuda, T., Kato, N., Watanabe, K. and Nakamura, R. 1986. Browning and insolubilization of ovalbumin by the Maillard reaction with some aldohexoses. J. Agric. Food Chem. 34: 351-355.

81. Burvall, A., Asp, N.-G., Bosson, A., San Jose, C. and Dahlquist, A. 1978. Storage of lactose-hydrolyzed dried milk: Effect of water activity on the protein nutritional value. J. Dairy Res. 45: 381-389.

82. Okomoto, M. and Hayashi, R. 1985. Chemical and nutritional changes of milk powder proteins under various water activities. Agric. Biol. Chem. 49: 1683-1687.

83. Holsinger, V.H., Posati, L., Devilbiss, E.D. and Pallansch, M.J. 1973. Variation of total and available lysine in dehydrated products from cheese wheys by different processes. J. Dairy Sci. 56: 1498-1504.

84. Desrosiers, T. Savoie, L., Bergeron, G. and Parent, G. 1989. Estimation of lysine damage in heated whey proteins by furosine determinations in conjunction with the digestion cell technique. J. Agric. Food Chem. 37: 1385-1391.

85. Hsu, K.-H. and Fennema, O. 1989. Changes in the functionality of dry whey protein concentrate during storage. J. Dairy Sci. 72: 829-837.

86. Huss, W. 1974. Damage to amino acids of wheys and whey powders during processing and storage. Tierphysiol., Tierernahrg. Futtermittelkde 34: 60-67.

87. Huss, W. 1970. Crystallization of lactose and availability of lysine after storage of dried skim-milk powder at different values of atmospheric humidity. Landwirtsch. Forsh. 23: 275-289.

88. Saltmarch, M., Vagnini-Ferrari, M. and Labuza, T.P. 1981. Theoretical basis and application of kinetics to browning in spray-dried whey food system. In: Progress in Food and Nutrition. Science. Maillard Reactions in Food (C. Eriksson, ed.), Pergammon Press, Elmsford, NY, p. 331-344.

89. Kim, M.N., Saltmarch, M. and Labuza, T.P. 1981. Nonenzymatic browning of hygroscopic whey powders in open versus sealed pouches. J. Food. Proc. Pres. 5: 49-57.

90. Karel, M. and Buera, M.P. 1994. Glass transition and its potential effects on kinetics of condendsation reactions and in particular on nonenzymatic browning. Roy. Soc. Chem. Special Publication No. 151: 164-169.

91. Karmas, R. and Karel, M., 1994. The effect of glass transition on Maillard browning in food models. Roy. Soc. Chem. Special Publication No. 151: 182-187.

92. Warmbier, H.C., Schnickels, R.A. and Labuza, T.P. 1976. Effect of glycerol on nonenzymatic browning in a solid intermediate moisture model food system. J. Food Sci. 41: 528-531.

93. Kramer, A. 1974. Storage retention of nutrients. Food Technol. 28(4): 50-60.

94. Hicks, E.W. 1944. Note on the estimation of the effect of diurnal temperature fluctuation on reaction rates in stored foodstuff and other materials. J. Counc. Sci. Ind. Res. (Australia) 17: 111-114.

95. Chen, J.Y., Bohnsack, K. and Labuza, T.P. 1983. Kinetics of protein quality loss in enriched pasta stored in a sine wave temperature condition. J. Food Sci. 48: 460-464.

96. Labuza, T.P., Bohnsack, K. and Kim, M.N. 1982. Kinetics of protein quality change in egg noodles stored under constant and fluctuating temperatures. Cereal Chem. 59: 142-148.

97. Nielson, H.K., DE Weck, D., Finot, P.A., Liardon, R. and Hurrell, R.F. 1985. Stability of tryptophan during processing and storage. 1. Comparative losses of trypophan, lysine, and methionine in different model systems. Brit. J. Nutr. 53: 281-292.

98. Labuza, T.P. and Saltmarch, M. 1981. The non-enzymatic browning reaction as affected by water in foods. In: Water Activity/Influences on Food Quality (L.B. Rockland and G.F. Stewart, eds.), Academic Press, New York, p. 605-650.

99. Poulsen, K. and Lindelou, P. 1981. Acceleration of chemical reactions due to freezing. In: Water Activity: Influence of Food Quality (L.B. Rockland and G.F. Stewart, eds.), Academic Press, New York, p. 651-678.

100. Ben-Gara, I. and Zimmerman, G. 1972. Changes in the nitrogenous constituents of staple foods and feeds during storage. Decrease in the chemical availability of lysine. J. Food Sci. Technol. 9: 113-118.

101. Loncin, M., Bimbennet, J.J. and Lenges, J. 1968. Influence of the activity of water on the spoilage of food stuff. J. Food Technol. 3: 131-142.

102. Narayan, K.A., McMullen, J.J., Butler, D.P., Wakefield, T. and Calhoun, W.K. 1975. Dietary glycerol-induced fat accumulation in rat livers. Nutr. Rep. Int. 12: 211-219.

103. Narayan, K.A. and McMullen, J.J. 1979. The interactive effect of dietary glycerol and corn oil on rat liver lipids, serum lipids, and serum lipoproteins. J. Nutr. 109: 1836-1846.

104. Narayan, K.A. and Ross, E.W. 1987. The interactive effect of glycerol and type and level of fat on rat tissue lipids. Nutr. Rep. Int. 36: 335-343.

105. Petriella, C., Resnik, S.L., Lozano, R.D. and Chirife, J. 1985. Kinetics of deteriorative reactions in model food systems of high water activity: color changes due to non-enzymatic browning. J. Food Sci. 50: 622-626.

106. Cerrutti, P., Resnik, S.L., Seldes, A. and Fontan, C.F. 1985. Kinetics of deteriorative reactions in model food systems of high water activity: Glucose loss, 5-hydroxymethylfurfural accumulation and fluorescence development due to non-enzymatic browning. J. Food Sci. 50: 627-630, 656.

107. Fox, M., Loncin, M. and Weiss, M. 1983. Investigations into the influence of water activity pH, and heat treatment on the velocity of the Maillard reaction in foods. J. Food Qual. 6: 103-118.

108. Leistner, L. and Gorris, L.G.M. 1995. Food preservation by hurdle technology. Trends Food Sci Tech. 6: 41-46.

109. Ohlsson, T. 1994. Minimal processing—Preservation methods of the future: An overview. Trends Food Sci. Tech. 5: 341-344.

110. Grijspaardt-Vink, C. 1994. Food preservation by hurdle technology. Food Tech. 48(12): 28-28.

111. Wolfrom, M.L., Kolb, D.K. and Langer, A.W., Jr. 1953. Chemical interactions of amino compounds and sugars. VII. pH dependency. J. Am. Chem. Soc. 75: 3471-3473.

112. Ashoor, S.H. and Zent, Z.B. 1984. Maillard browning of common amino acids and sugars. J. Food Sci. 49: 1206-1207.

113. Karel, M. and Labuza, T.P. 1968. Non-enzymatic browning in model systems containing sucrose. J. Agric. Food Chem. 16: 717-719.

114. Hashiba, H. 1982. The browning reaction of Amadori compounds derived from various sugars. Agric. Biol. Chem. 46: 547-548.

115. Warmbier, H.C., Schnickels, R.A. and Labuza, T.P. 1976. Non-enzymatic browning kinetics in an intermediate moisture model system. Effect of glucose to lysine ratio. J. Food Sci. 41: 981-983.

116. Yen, G.-C. and Lai, Y.-H. 1987. Influence of antioxidants on Maillard browning reactions in a casein-glucose model system. J. Food Sci. 52: 1115-1116.

117. Namiki, M. and Hayashi, T. 1983. A new mechanism of the Maillard reaction involving sugar fragmentation and free radical formation. In: The Maillard Reaction in Foods and Nutrition (G.R. Waller and M.S. Feather, eds.), American Chemical Society, Washington, D.C., p. 21-46.

118. Lingnert, H. and Eriksson, C.E. 1981. Antioxidative effect of Maillard reaction products. In: Progress in Food and Nutrition Science. Maillard Reactions in Food (C.E. Eriksson, ed.), Pergammon Press, Elmsford, NY, p. 453-466.

119. Lingnert, H. and Eriksson, C.E. 1980. Antioxidative Maillard reaction products. I. Products from sugars and amino acids. J. Food Proc. Preser. 4: 161-172.

120. Lingnert, H. and Eriksson, C.E. 1980. Antioxidative Maillard reaction products. II. Products from sugars and peptides or protein hydrolysates. J. Food Proc. Preser. 4: 173-181.

121. Patton, S. 1955. Browning and associated changes in milk and its products: A review. J. Dairy Sci. 38: 457-478.

122. Bohart, G.S. and Carson, J.F. 1955. Effects of trace metals, oxygen, and light on the glucose-glycine browning reaction. Nature 175: 470-471.

123. Rendleman, J.A., Jr. and Inglett, G.E. 1990. The influence of copper in Maillard reaction. Carbohydrate Res. 201: 311-326.

124. Rendleman, J.A., Jr. 1987. Complexation of calcium by melanoidin and its role in determining bioavailability. J. Food Sci. 52: 1699-1705.

125. Kato, Y., Watanabe, K. and Sato, Y. 1981. Effects of some metals on the Maillard reaction of ovalbumin. J. Agric. Food Chem. 29: 540-543.

126. Narayan, K.A. and Kummerow, F.A. 1963. Factors influencing the formation of complexes between oxidized lipids and proteins. J. Am. Oil Chem. Soc. 40: 339-342.

127. Narayan, K.A., Sugai, M. and Kummerow, F.A. 1964. Complex formation between oxidized lipids and egg albumin. J. Am. Oil Chem. Soc. 41: 254-259.

128. Filler, L.J., Rumery, R.E. and Mason, K.E. 1946. Specific unsaturated fatty acids in the production of acid-fast pigment in the vitamin E-deficient rat and the protective action of tocopherols. Transactions of the First Conference on Biological Antioxidants, Josiah Macy, Jr. Foundation, New York, p. 67-77.

129. Lillie, R.D., Ashburn, C.C., Sebrell, W.H., Daft, F.S. and Lowry, J.V. 1942. Histogenesis and repair of the hepatic cirrhosis in rats produced on low protein diets and preventable with choline. Public Health Rep. 57: 502-508.

130. Hartroft, W.S. 1953. Pathogenesis and significance of hemoceroid and hyaloceroid, two types of ceroid like pigments found in human atheromatous lesions. Gerentology 8: 158-166.

131. Hartroft, W.S. 1951. In vitro and in vivo production of a ceroid like substance from erythrocytes and certain lipids. Science 113: 673-674.

132. Chio, K.S. and Tappel, A.L. 1969. Inactivation of ribonuclease and other enzymes by peroxidizing lipids and by malonaldehyde. Biochemistry 8: 2827-2832.

133. Andrews, F., Bjorksten, J., Trenk, F.B., Henick, A.S. and Koch, R.B. 1965. The reaction of an autoxidized lipid with proteins. J. Am. Oil Chem. Soc. 42: 779-781.

134. Roubal, W.T. 1970. Trapped radicals in dry lipid-protein systems undergoing oxidation. J. Am. Oil Chem. Soc. 47: 141-144.

135. Roubal, W.T. 1971. Free radicals, malonaldehyde and protein damage in lipid protein systems. Lipids 6: 62-64.

136. Matoba, T., Yoshida, H. and Yonezawa, D. 1982. Changes in casein and egg albumin due to reactions with oxidizing methyl linoleate in dehydrated systems. Agric. Biol. Chem. 46: 979-986.

137. Gamage, P.T., Mori, T. and Matsushita, S. 1973. Mechanism of polymerization of proteins by autoxidized products of linoleic acid. J. Nutr. Sci. Vitaminol. 19: 173-182.

138. Kanner, J. and Karel, M. 1976. Changes in lysozyme due to reactions with peroxidizing methyl linoleate in a dehydrated model system. J. Agric. Food Chem. 24: 468-472.

139. Karel, M. Schaich, K. and Roy, R.B. 1975. Interaction of peroxidizing methyl linoleate with some proteins and amino acids. J. Agric. Food Chem. 23: 159-165.

140. Witting, L.A., Nishida, T., Johnson, O.C. and Kummerow, F.A. 1957. The relationship of pyridoxine and riboflavin to the nutritional value to polymerized fats. J. Am. Oil Chem. Soc. 34: 421-424.

141. Perkins, E.G. 1960. Nutritional and chemical changes occurring in heated fats: A review. Food Technol. 14: 508-513.

142. Sugai, M., Witting, L.A., Tsuchiyama, H. and Kummerow, F.A. 1962. The effects of heated fat on the carcinogenic activity of 2-acetylaminofluorene. Cancer Res. 22, 510-519.

143. Fouad, F.M., Shahidi, F. and Mamer, O.A. 1995. Comparison of thermally oxidized lipids and acetaminophen with concurrent consumption of ethanol as inducers of liver cirrhosis. J. Toxicol. Environ. Health 46: 217-232.

144. Suarna, C., Dean, R.T., May, J. and Stocker, R. 1995. Human atherosclerotic plaque contains both oxidized lipids and relatively large amounts of α-tocopherol and ascorbate. Arterioscler. Throm. Vasc. Biol. 15: 1616-1624.

145. Staprans, I., Rapp, J.H., Pan, X.-M. and Feingold, KR. 1996. Oxidized lipids in the diet are incorporated by the liver into very low lipoprotein in rats. J. Lipid Res. 37: 420-430.

146. Staprans, I., Rapp, J., Pan, X.-M., Kim, K.Y., and Feingold, K.R. 1994. Oxidized lipids in the diet are a source of oxidized lipid in chylomicrons of human serum. Arterioscler. Throm. 14: 1900-1905.

147. Nielson, H.K., Finot, P.A. and Hurrell, R.F. 1985. Reactions of proteins with oxidizing lipids. 2. Influence on protein quality and on the bioavailability of lysine, methionine, cyst(e)ine and tryptophan as measured in rat assay. Brit. J. Nutr. 53: 75-86.

148. Nielson, H.K., Loliger, J. and Hurrell, R.F. 1985. Reactionms of proteins with oxidizing lipids. 1. Analytical measurements of lipid oxidation and of amino acid losses in a whey protein-methyl linolenate model system. Brit. J. Nutr. 53: 61-73.

149. Peng, I.C. and Nielsen, S.S. 1986. Protein-protein interactions between soybean beta-congly-cinin (B_1-B_6) and myosin. J. Food Sci. 51: 588-590, 607.

150. Mauron, J. 1977. General principles involved in measuring specific damage of food components during thermal processing. In: Physical, Chemical, and Biological Changes in Foods Caused by Thermal Processing. (T. Hoyen and O. Kuale, eds.), Proceeding of International Union of Food Science and Technology, the Scandinavian Association of Agricultural Scientists and the Norwegian Agricultural Food Research Society, Applied Science Publishers, London, p. 328-359.

151. Desphande, S.S. and Nielsen, S.S. 1987. In vitro digestibility of dry bean (*Phaseolus vulgaris* L.) proteins: The rate of heat-stable protease inhibitors. J. Food Sci. 52: 1330-1334.

152. Nielsen, S.S. 1991. Digestibility of legume proteins. Food Technol. 45: 112-114, 118.

153. Gallant, D.J., Bouchet, B. and Cucoli, J. 1984. Ultrastructural aspects of spun pea and fababean proteins. Food Microstructure 3: 175-183.

154. Desrosiers, T., Bergeron, G. and Savoie, L. 1987. Effect of heat treatments on *in vitro* digestibility of delactosed whey protein as determined by the digestion cell technique. J. Food Sci. 52: 1525-1528.

155. Otterburn, M., Healy, M. and Sinclair, W. 1977. The formation, isolation, and importance of isopeptides in heated proteins. In: Protein Crosslinking: Nutritional and Medical Consequences (M. Friedman, ed.), Plenum Press, New York, Vol. 86B, p. 239-262.

156. de Rham, O. and Chanton, S. 1984. Role of ionic environment in insolubilization of whey protein during heat treatment of whey products. J. Dairy Sci. 67: 939-949.

157. Desrosiers, T., Bergeron, G. and Savoie, L. 1987. In vitro digestibility of thermally processed diafiltered whey as influenced by water activity. J. Dairy Sci. 70: 2476-2485.

158. Hakansson, B., Jagerstad, M., Oste, R., Akesson, B. and Jonsson, L. 1987. The effects of various thermal processes on protein quality, vitamins, selenium content in whole-grain wheat and white flour. J. Cereal Sci. 6: 269-282.

159. Sternberg, M. and Kim, C.Y. 1977. Lysinoalanine formation in protein food ingredients. In: Protein Crosslinking. Nutritional and Medical Consequences (M. Friedman, ed.), Plenum Press, New York, Vol. 86B, p. 73-84.

160. De Groot, A.P. and Slump, P. 1969. Effects of severe alkali treatment of proteins on amino acid composition and nutritive value. J. Nutr. 98: 45-56.

161. Gould, D.H. and MacGregor, J.T. 1977. Biological effects of alkali treated protein and lysinoalanine: An overview. In: Protein Crosslinking. Nutritional and Medical Consequences (M. Friedman, ed.), Plenum Press, New York, Vol. 86 B, p. 29-48.

162. Finley, J.W. and Friedman, M. 1977. New amino acid derivatives formed by alkaline treatment of proteins. In: Protein Crosslinking. Nutritional and Medical Consequences (M. Friedman, ed.), Plenum Press, New York, Vol. 86 B, p. 123-130.

163. Finley, J.W., Snow, J.T., Johnston, P.H. and Friedman, M. 1977. Inhibitory effect of mercaptoamino acids on lysino-alanine formation during alkali treatment of proteins. In: Protein Crosslinking, Nutritional, and Medical Consequences (M. Friedman, ed.), Plenum Press, New York, Vol. 86 B, p. 85-92.

164. Liardon, R. and Hurrel, R.F. 1983. Amino acid racemization in heated and alkali-treated proteins. J. Agric. Food Chem. 31: 432-437.

165. Anderson, P.A. 1985. Interactions between proteins and constituents that affect protein quality. In: Digestibility and Amino Acid Availability in Cereals and Oilseeds (J.W. Finley and D.T. Hopkins, ed.), American Association of Cereal Chemists, St. Paul, MN, p. 31-45.

166. Horigome, T. and Kandatsu, M. 1968. Biological value of proteins allowed to react with phenolic compounds in the presence of *o*-diphenol oxidase. Agric. Biol. Chem. 32: 1093-1102.

167. Hurrell, R.F., Finot, P.A. and Cuq, J.L. 1982. Protein-polyphenol reactions. 1. Nutritional and metabolic consequences of the reaction between oxidized caffeic acid and lysine residue in casein. Br. J. Nutr. 47: 191-211.

168. Maga, J.A. 1982. Phytate: Its chemistry, occurrence, food interactions, nutritional significance, and methods of analysis. J. Agric. Food Chem. 30: 1-9.

169. Satterlee, L.D. and Addul-Kadir, R. 1983. Effect of phytate content on protein nutritional quality of soy and wheat bran proteins. Lebensm. Wiss. Technol. 16: 8-14.

170. Legault, R.R., Hendel, C.E., Taburt, W.F. and Pool, M.F. 1951. Browning of dehydrated sulfited vegetables during storage. Food Technol. 5: 417-423.

171. Kim, H.-J., Loveridge, V.A., and Taub, I.A. 1984. Myosin cross-linking in freeze-dried meats. J. Food Sci. 49: 699-703, 708.

172. Finot, P.A., Bujard, E., Mottu, F. and Mauron, J. 1977. Availability of the true Schiff's base between lysine and lactose in milk. In: Protein Crosslinking. Nutritional and Medical Consequences (M. Friedman, ed.), Plenum Press, New York, Vol. 86 B, p. 343-365.

173. Mohammad, A., Frankel-Conrat, H. and Olcott, H.S. 1949. The "browning" reaction of proteins with glucose. Arch. Biochem. Biophys. 24: 157-178.

174. Burr, G.O. and Burr, M.M. 1929. A new deficiency disease produced by the rigid exclusion of fat from the diet. J. Biol. Chem. 82: 345-367.

175. Narayan, K.A. and McMullen, J.J. 1980. Accelerated induction of fatty livers in rats fed fat-free diets containing sucrose or glycerol. Nutr. Rep. Int. 21: 689-697.

176. Djilas, S.M. and Milic, B. Lj., 1994. Naturally occurring phenolic compounds as inhibitors of free radical formation in the Maillard reaction. Roy. Soc. Chem. Special Publication No. 151: 75-81.

177. Pham, C.B. and Cheftel, J.C. 1990. Influence of salts, amino acids, and urea on the non-enzymatic browning on the protein sugar system. Food Chem. 37: 251-260.

178. Eichner, K. and Ciner-Doruk, M. 1981. Early indication of the Maillard reaction by analysis of reaction intermediates and volabile decomposition products. In: Progress in Food and Nutrition Science. Maillard Reactions in Food (C. Eriksson, ed.), Pergammon Press, Elmsford, NY, p. 115-135.

179. Nafisi, K. and Markakis, P. 1983. Inhibition of sugar-amino browning by aspartic and glutamic acids. J. Agric. Food Chem. 31: 1115-1116.

180. Friedman, L. and Kline, O.L. 1950. The amino acid-sugar reaction. J. Biol. Chem. 184: 599-606.

181. Friedman, M. and Molnar-Perl, I. 1990. Inhibition of browning by sulfur amino acids. 1. Heated amino acid-glucose systems. J. Agric. Food Chem. 38: 1642-1647.

182. Molnar-Perl, I. and Friedman, M. 1990. Inhibition of browning by sulfur amino acids. 2. Fruit juices and protein-containing foods. J. Agric. Food Chem. 38: 1648-1651.

183. Molnar-Perl, I. and Friedman, M. 1990. Inhibition of browning by sulfur amino acids 3. Apples and potatoes. J. Agric. Food Chem. 38: 1652-1656.

184. Farmar, J.G., Ulrich, P.C. and Cerami, A., 1988. Novel pyrroles from sulfite-inhibited Maillard reactions: Insight into the mechanism of inhibition. J. Org. Chem. 53, 2346-2349.

185. Roubal, W.T. 1971. Nature of free radicals in freeze-dried fishery products and other lipid-protein systems. Fishery Bull. 69, p. 371-377.

186. Finley, J.W., Snow, J.T., Johnston, P.H. and Friedman, M. 1977. Inhibitory effect of mercaptoamino acids on lysino-alanine formation during alkali treatment of proteins. In: Protein Crosslinking. Nutritional and Medical Consequences (M. Friedman, ed.), Plenum Press, New York, Vol. 86B, p. 85-121.

187. Berkowitz, D. and Oleksyk, L. 1991. Leavened breads with extended shelf-life. U.S. Patent No. 5,059,432.

188. Berkowitz, D., Shults, G.W. and Powers, E.M. 1990. Advances in shelf-stable operational rations. Proc. Third Natick Science Symposium. Natick/TR-90/039.

189. Narayan, K.A., Porcella, C. and Shaw, C. 1995. The effect of high temperature storage on the digestibility of carbohydrates in pouch bread. FASEB J. 9: A454.

190. Moon, H.K., Kim, J.W., Heo, K.N., Kim, S.W., Kwon, C.H., Shin, I.S. and Han, I.K. 1994. Growth performance and amino acid digestibilities affected by various plant sources in growing-finishing pigs. Asian-Austral. J. Anim. Sci. 7: 537-546.

191. Van Barneveld, R.J., Batterham., E.S. and Norton, B.W. 1994. The effect of heat on amino acids for growing pigs. 3. The availability of lysine from heat-treated field peas (*Pisum sativum* cultivar Dundale) determined using the slope-ration assay. Brit. J. Nutr. 72, 257-275.

192. Sarwar, G., Peace, R.W. and Botting, H.G. 1993. Effect of amino acid supplementation on protein quality of soy-based infant formulas fed to rats. Plant Foods Human Nutr. 43: 259-266.

193. Sanger, F. 1945. The free amino groups of insulin. Biochem. J. 39: 507-515.

194. Sanger, F. and Tuppy, H. 1951. The amino-acid sequence in the phenylalanyl chain of insulin. 1. The identification of the lower peptides from partial hydrolysates. Biochem. J. 49: 463-490.

195. Carpenter, K.J. 1960. The estimation of available lysine in animal protein foods. Biochem. J. 77: 604-610.

196. Carpenter, K.J., Steinke, F.H., Catignani, G.L., Swaisgood, H.E., Alfred, M.E., McDonald, J.L. and Schelstraete, M. 1989. The estimation of available lysine in human foods by three chemical procedures. Plant Food Human Nutr. 39: 129-135.

197. Couch, J.R. 1975. Collaborative study of the determination of available lysine in proteins and feeds. J. Assoc. Off. Anal. Chem. 58: 599-601.

198. Kakade, M.L. and Liener, I.E. 1969. Determination of available lysine in proteins. Anal. Biochem. 27: 273-280.

199. Friedman, M. 1982. Chemically reactive and unreactive lysine as an index of browning. Diabetes 31(Suppl. 3): 5-14.

200. Hurrell, R.F. and Carpenter, K.J. 1974. Mechanisms of heat damage in protein. 4. The reactive-lysine content of heat damaged material as measured in different ways. Br. J. Nutr. 32: 589-604.

201. Roy, R.B. 1979. An improved semiautomated enzymatic assay of lysine in food stuffs. J. Food Sci. 44: 480-487.

202. Sheffner, A.L. 1967. *In vitro* protein evaluation. In: Newer Methods of Nutritional Biochemistry, A.A. Albanese, ed., Academic Press, New York, Vol. 3, p. 125-195.

203. Wells, P., McDonough, F., Bodwell, C.E. and Hitchens, A. 1989. The use of *Streptococcus zymogenes* for estimating trytophan and methionine bioavailability in 17 foods. Plant Food Human Nutr. 39: 121-127.

204. Baker, H., Frank, O., Rusoff, I.I., Morck, R.A. and Hutner, S.H. 1978. Protein quality of foodstuffs determined with *Tetrahymena thermophilia* and rat. Nut. Rep. Int. 17: 525-536.

205. Anderson, T.R., Robbin, D.J. and Erbersdober, H.F. 1984. The stimulation and inhibition of *Tetrahymena pyroformis* W cultures by Maillard products during lysine availability determinations. Nutr. Rep. Int. 30: 493-500.

206. Shepherd, N.D., Taylor, T.G. and Wilton, D.C. 1977. An improved method for the microbiological assay of available amino acids in proteins using *Tetrahymena pyriformis*. Br. J. Nutr. 38: 245-253.

207. Hitchins, A.D., Wells, P.A. and McDonough, F.E. 1987. Use of *Escherichia coli* mutants to assay amino acid biological availability. Fed. Proc. 46: 890 (3344, Abstract).

208. Satterlee, L.D., Marshall, H.F. and Tennyson, J.M. 1979. Measuring protein quality. J. Am. Oil Chem. Soc. 56: 103-109.

209. Lowry, K.R., Fly, A.D., Izquierdo, O.A. and Baker, D.H. 1990. Effect of heat processing and storage on protein quality and lysine bioavailability of a commercial enteral product. J. Parenter. Enter. Nutr. 14: 68-73.

210. Coelho, R.G. and Sgarvbieri, V.C. 1995. Methionine liberation by pepsin-pancreatin hydrolysis of bean protein fractions: Estimation of methionine bioavailability. J. Food Biochem. 18: 311-324.

211. Hsu, H.W., Vavak, D.C., Satterlee, L.D. and Miller, G.A. 1977. A multienzymatic technique for estimating protein digestibility. J. Food Sci. 42: 1269-1273.

212. Pederson, B. and Eggum, B.O. 1981. Prediction of protein digestibility by in vitro procedures based on two multi-enzyme systems. Z. Tierphysiol. Tierernahrg. U. Futtermittelkde 45: 190-200.

213. McDonough, F.E., Sarwar, G., Steinke, F.H., Slump, P., Garcia, S. and Boisen, S. 1990. In vitro assay for protein digestibility: Interlaboratory study. J. Assoc. Off. Anal. Chem. 73: 622-625.

214. Bodwell, C.E., Carpenter, K.J. and McDonough, F.E. 1989. A collaborative study of methods of protein evaluation: Introductory paper. Plant Food Human Nutr. 39: 3-12.

215. Mauron, J., Mottu, F., Bujard, E. and Egli, R.H. 1955. The availability of lysine, methionine, and tryptophan in condensed milk and milk powder. In vitro digestion studies. Arch. Biochem. Biophys. 50: 433-451.
216. Mottu, F. and Mauron, J. 1967. The differential determination of lysine in heated milk. II. Comparison of the in vitro method with the biological evaluation. J. Sci. Fd. Agric. 18: 57-62.
217. Savoie, L. and Gauthier, S.F. 1986. Dialysis cell for the in vitro measurement of protein digestibility. J. Food Sci. 51: 494-498.
218. Savoie, L., Charbonneau, R. and Parent, G. 1989. In vitro amino acid digestibility of food proteins as measured by the digestion cell technique. Plant Food Human Nutr. 39: 93-107.
219. Rayner, C.J. and Fox, M. 1978. Measurement of available lysine in processed beef muscle by various laboratory procedures. J. Agric. Food Chem. 26: 494-497.
220. Narayan, K.A., Senecal, A., Atwood, B., Branagan, M., Salupu, P., McClure, P., Neidhardt, A., Robertson, M., Kim, H.-J. and Jarboe, J. 1990. Nutritional quality of beef stew processed in flexible pouches. Inst. Food Technol. Annual Meeting. Program and Abstracts, p. 131, abstract 150.
221. Porter, D.H., Swaisgood, H.E. and Catignani, G.L. 1984. Characterization of an immobilized enzyme system for the determination of protein digestibility. J. Agric. Food Chem. 334-339.
222. Culver, C.A. and Swaisgood, H.E. 1989. Change in digestibility of dried casein and glucose mixtures occurring during storage at different temperatures and water activities. J. Dairy Sci. 72: 2916-2920.
223. FAO/WHO. Report of the Joint FAO/WHO Expert Consultation on Protein Quality Evaluation. 1990. Food and Agriculture Organization of the United Nations, Rome and World Health Organization, Geneva, p. 1-66.
224. Finley, J.W. 1985. Reducing variability in amino acid analysis. In: Digestibility and Amino Acid Availability in Cereals and Oilseeds (J.W. Finley and D.T. Hopkins, eds.), American Association of Cereal Chemists, St. Paul, MN, p. 15-30.
225. Henley, E.C. and Kuster, J.M. 1994. Protein quality evaluation by protein digestibility-corrected amino acid scoring. Food Technol. 48(4): 74-77.
226. Swaminathan, M. 1967. Availability of plant proteins. In: Newer Methods of Nutritional Availability. (A.A. Albanese, ed.), Academic Press, New York, Vol. 3, p. 197-241.
227. Sarwar, G. 1987. Digestibility of proteins and bioavailability of amino acids in foods. World Rev. Nutr. Diet 54: 26-70.
228. Sarwar, G. and McDonough, F.E. 1990. Evaluation of protein digestibility — corrected amino acid method for assessing protein quality of foods. J. Assoc. Off. Anal. Chem. 73: 347-356.
229. Sarwar, G., Peace, R.W., Botting, H.G. and Brulé, D. 1989. Digestibility of protein and amino acids in selected foods as determined by rat balance methods. Plant Food Human Nutr. 39: 23-32.
230. Eggum, B.O., Hansen, I. and Larsen, T. 1989. Protein quality and digestibile energy of selected foods determined in balance trials with rats. Plant Food Human Nutr. 39: 13-21.
231. McDonough, F.E., Bodwell, C.E., Hitchins, A.D. and Staples, R.S. 1989. Bioavailability of lysine in selected foods by rat growth assay. Plant Food Human Nutr. 39: 67-75.
232. Elwell, D. and Soares, J.H., Jr. 1975. Amino acid bioavailability: A comparative evaluation of several assay techniques. Poultry Sci. 54: 78-85.
233. McDonough, F.E., Bodwell, C.E., Staples, R.S. and Wells, P.A. 1989. Rat bioassays for methionine availability in 16 food sources. Plant Food Human Nutr. 39: 77-84.
234. McDonough, F.E., Bodwell, C.E., Wells, P.A. and Kamula, J.A. 1989. Bioavailability of tryptophan in selected foods by rat growth assay. Plant Food Human Nutr. 39: 85-91.
235. Mitchell, H.H. 1923-1924. A method of determining the biological value of protein. J. Biol. Chem. 58: 873-903.
236. MacLaughlan, J.M. 1972. Effects of protein quality and quantity on protein utilization. In: Newer Methods of Nutritional Biochemistry (A.A. Albanese, ed.), Academic Press, New York, Vol. 5, p. 33-64.
237. Sarwar, G., Blair, R., Friedman, M., Gumbman, M.R., Hackler, L.R., Pellet, P.L. and Smith, T.K. 1984. Inter- and intra-laboratory variability in rat growth assays for estimating protein quality of foods. J. Assoc. Off. Anal. Chem. 67: 976-981.

chapter seven

Color: origin, stability, measurement, and quality

F. M. Clydesdale

Contents

Introduction

Color is the major appearance attribute of most foods and as such is an important characteristic of food quality. There are many reasons for its importance, chief among them being standardization of the product (the consumer is suspicious of the same brand having widely variable colors); utilization as a measure of quality and economic worth; and utilization as an indicator of biological and/or physicochemical breakdown, and as a predictor of other quality characteristics.[1] Color is also critically important in the many dimensions of food choice[2] and influences the perception of other sensory characteristics by the consumer. In fact, Clydesdale has suggested that it may be unwise to evaluate a quality characteristic such as color as only a function of appearance.[3] Such an evaluation, if it ignored the influence of color on other sensory attributes, would tend to limit its total contribution and limit one's ability to maximize acceptability.

Color in food is a result of a variety of factors both endogenous and exogenous to the food that may be affected by genetics and pre- and postharvest treatments. The exogenous factors consist of such things as packaging films, display lights, and processing, while the endogenous factors involve pigments within the food, added colors, and physical characteristics that affect glossiness and haze.[4]

In this chapter, we discuss endogenous factors such as pigments and the physical characteristics of food that affect color perception and we describe as well systems for color evaluation and measurement.

Pigments

The major causative factor of color in most foods is the presence of a broad array of natural pigments. There are some notable exceptions, such as the carmelization and browning reactions that occur in many foods. The causative factors for these exceptions are beyond the scope of this discussion and are discussed in other chapters.

The major food-related pigments in plants may be broadly classified into four groups: flavonoids (anthocyanins and flavonoids), chlorophylls, carotenoids, and betalaines, while those in meats are based upon hemoglobin.

Anthocyanins and flavonoids

The anthocyanins are a broad and important group of blue to red, water-soluble pigments, which are part of the flavonoid family and as such are characterized by the flavylium nucleus. There are over 4000 flavonoids with some 260 of these being anthocyanins.[5]

The basic flavylium cation structure is shown in Figure 1. When this flavylium nucleus is substituted at the 3', 4', or 5' positions with hydrogen, hydroxy, or methoxy groups, an aglycone, or anthocyanidin, is formed. In the case of the flavonoids, the 7, 5, 4', and 3' positions are often substituted and, unlike the anthocyanins, the 7 position is most frequently substituted.

There are some 20 anthocyanidins, but only the 6 shown in Table 1 are important in food; cyanidin, delphinidin, malvidin, pelargonidin, peonidin, and petunidin. The esterification of these anthocyanidins with one or more sugars forms the anthocyanins. Depending on the number of sugars involved (limited to glucose, rhamnose, galactose, xylose, and arabinose), the anthocyanin is classified as a monoside (3 position), bioside (both in the 3, in the 3 and 5 or, rarely, in the 3 and 7 positions), or trioside (two on the 3, one on the 5; three in a branched or linear structure on 3; or, rarely, two on the 3 and one on the 7 position). No anthocyanins with four sugars have been reported,[5] but there has been one with five sugars reported.[6] Para-coumaric, ferulic, caffeic, malonic, vanillic, and acetic acids may also be esterified to the sugar molecules to form the so-called acylated anthocyanins.

It is obvious with these structures that both the anthocyanins and the yellow flavonoids will be greatly affected by the reactivity of the resonating structure of the flavylium nucleus as well as by the type and degree of substitution. Detailed discussions on anthocyanin chemistry are available elsewhere,[5,7,8] but it is noted here that stability is directly related to the degree of methoxylation, while it is inversely related to hydroxylation. Further, the diglucosides are generally more stable than the monoglucosides, but the former may contribute more to browning due to the extra sugar available for Maillard

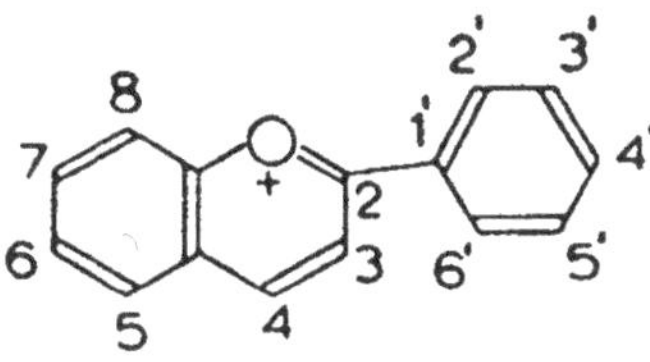

Figure 1 The structure of the basic flavylium cation. (From Schwartz, S.J., et al. *J. Agric. Food Chem.* 1981, 29, 533-537. With permission.)

Table 1 The Major Anthocyanidins in Food

Anthocyanidin	Substitution on the Flavylium Nucleus[a]		
	3	4	5
Cyanidin	OH	OH	H
Delphinidin	OH	OH	OH
Malvidin	OMe	OH	OMe
Pelargonidin	H	OH	H
Peonidin	OMe	OH	H
Petunidin	OMe	OH	OH

[a] Carbons 3, 5, 7 have OH groups and hydrogen on the other carbon atoms.

reactions. Packaging may also accelerate nonenzymatic browning and decrease anthocyanin concentration. Lin et al. have found that packaging that induces high concentrations of CO_2 decreases the concentration of anthocyanins.[9] They propose that a high CO_2 atmosphere damages tissue and releases free amino acids, which then react with the anthocyanins.

The hydroxyl group at position 4 is particularly important, since it may shift the color from orange to red as well as destabilize the molecule. As a result, several investigators have suggested deoxyanthocyanins or 4-substituted anthocyanins as stable food colorants. Monomeric anthocyanins may also polymerize to form more stable forms,[13] but the polymerization as yet is an uncontrolled reaction, which occurs at times in some food products such as strawberry jam. Anthocyanins in foods are most susceptible to changes in pH, heat, light, metal ions, ascorbic acid, and oxygen. All are of great concern in processed and stored foods.

The striking changes in the color of anthocyanins from blue to red with pH has been well known for many years. The red flavylium cation converts to a blue quinoidal base as pH is raised. The actual mechanisms of the transformations of anthocyanins with pH are much more complicated and includes equilibria involving the flavylium cation, the quinoidal base, a colorless carbinol pseudo base, and a colorless chalcone.[14] However, for our purposes, it is especially necessary to note the color shifts to blue, which preclude the use of most anthocyanins as red pigments above pH 4.

The instability of the anthocyanins is of great concern in food processing and has necessitated the development of some unique color measuring systems in order to assess and control the color quality. Interestingly, certain acylated anthocyanins have been found to have unusual stability in neutral or weakly acidic solutions. Brouillard has attributed this stability to two acyl groups above and below the pyrrilium ring.[15] He postulates that this arrangement protects the oxonium ion from hydration, thereby preventing the formation of the pseudo base and the chalcone.

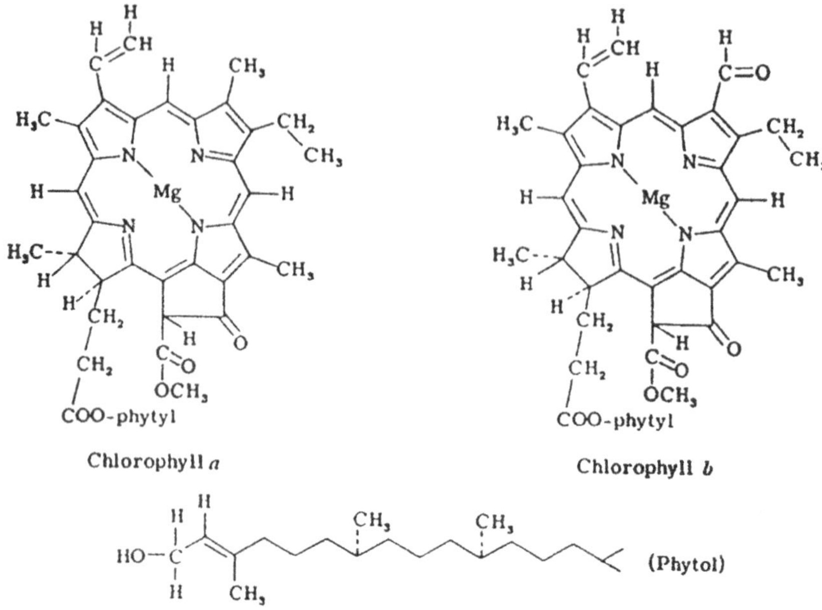

Figure 2   The structures of chlorophyll a and b. (From Aronold, S. In: The Chlorophylls; Vernon, L.P. and Seely, G.R., Eds.; Academic Press: New York, 1966. With permission.)

Chlorophylls

A variety of chlorophylls have been described, but the major ones of interest in food are chlorophylls a and b, which exist in plants in an average ratio of about 3 to 1. These pigments are very important in food color, since they are responsible for the green color in fruits and vegetables.

The structures of chlorophyll a and b are shown in Figure 2.[16] Chlorophyll a is a magnesium-chelated tetrapyrrole with methyl substitutions at the 1, 3, 5, and 8 positions; a vinyl substitution at the 2 position; an ethyl group at the 4 position; propionate, esterified with phytyl alcohol, at the 7 position; a keto group at the 9 position; and a carbomethoxy substitution at position 10. Chlorophyll b simply has the methyl group on the carbon 3 position replaced with a formyl group.

The chlorophylls convert to a variety of products, and Figure 3 shows those that may be of most interest in food.

The absorption spectra of the chlorophylls is due to the poryphyrin structure and shows four major absorption bands between 500 and 700 nm and a large "Soret" band in the UV region at 400 nm. A simple test for the integrity of the porphyrin structure is the presence of the Soret band.

The biochemistry of chlorophyll degradation during ripening, although extremely important to food quality, is largely unknown,[17] but the picture is clearer for food processing. The bright green chlorophylls generally degrade to a dull olive-brown during processing. This degradation during heat processing is thought to be due to the formation of organic acids,[18] which replace the Mg in chlorophyll to form pheophytin. Further degradation of the pheopytins to pyropheophorbides, involving decarboxymethylation from carbon 10, has been shown by Schwartz and colleagues[19] who suggest that the mechanism for chlorophyll decomposition during canning is a two-step process with chlorophyll first being converted to pheophytins and then pheophytins converted to pyropheophorbides.[17]

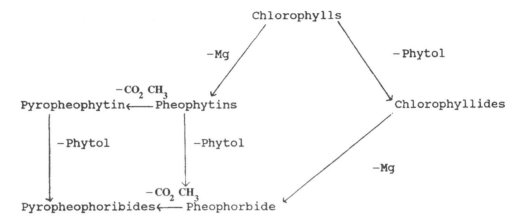

Figure 3 The chlorophylls and some of their conversion products.

Several reviews have summarized the many methods proposed to prevent the degradation of chlorophylls and to maintain the bright-green color in processed foods.[17,20,21] Since the major pathway of degradation involves the replacement of magnesium with hydrogen (pheophytinization), a popular approach to prevent degradation is to block this replacement. One of the methods commonly used is to raise the pH of the food with approved alkalinizing agents such as $MgCO_3$. Several patents have been issued on this basis, with the "Blair process" being one of the best known. Other patents have utilized the naturally occurring enzyme chlorophylase, which converts chlorophylls to chlorophyllides. High-temperature, short-time heat treatments have been used as well, but advantages gained in such processing appear to be lost after about 3 months storage. Clydesdale and Francis[22] and Clydesdale et al.[21] combined these treatments in an attempt to maximize their effects, but met with limited success, as did the individual treatments.

A patent was issued based on the fact that certain metal ions such as copper and zinc are able to displace magnesium from the tetrapyrrole ring and form stable bright-green complexes.[23] This process, Veri-Green, incorporates a metal ion blanch along with the use of metal ions in the can coating.

Myoglobin and hemoglobin

Meat color is due to the heme pigments and their derivatives. In a live animal, hemoglobin is the predominant pigment, but in a slaughtered and bled animal, myoglobin accounts for some 95% of the remaining heme pigments. Both hemoglobin and myoglobin are complex proteins consisting of a protein moiety, globin, and a nonpeptide component, heme. Heme is composed of an iron atom and tetrapyrrole, or porphyrin, a large planar ring similar to that found in the chlorophylls but with iron at the center rather than magnesium. In myoglobin, the heme is attached to globin, while hemoglobin may be viewed simply as four myoglobin molecules linked together.

The reactions of the heme pigments involve complexes of heme, globin, and ligands in which iron is either in the oxidized state (Fe^{3+}) or in the reduced state (Fe^{2+}). In fresh meats, there is a dynamic, reversible cycling among these major pigments: oxymyoglobin, myoglobin, and metmyoglobin. Myoglobin (Mb) is purple in color and in the presence of oxygen can become oxygenated (i.e., producing a covalent complex between myoglobin with Fe^{2+} and molecular oxygen) to form the bright-red oxymyoglobin (O_2Mb) corresponding to the familiar bloom of fresh meats or can be oxidized to metmyoglobin (MMb), containing Fe^{3+}, with a resulting undesirable brown color. There are many other complexes formed both in fresh and cured meat,[20,24,25] but a detailed discussion of meat pigment chemistry is beyond the scope of this chapter.

Red O_2Mb is stabilized by the formation of a highly resonant structure and, as long as the heme remains oxygenated, no further color changes will take place. However, the oxygen is continually associating and dissociating from the heme complex, a process which is influenced by a number of conditions, including low oxygen pressures. When this happens, the reduced form is oxidized to brown MMb.

Packaging is obviously very important in preventing this condition, but some studies have indicated that light may have a greater effect on discoloration than oxygen penetration during storage.[26] Another study found that the elimination of oxygen by using pure carbon dioxide gas in the head space of packaged restructured steaks produced the best color.[27]

Mitsumoto et al. found that metmyoglobin formation could be delayed by treating the meat with a 10% solution of vitamin C, a reducing agent.[28]

Carotenoids

Carotenoids are a group of yellow to red, lipid-soluble pigments, which are varied and widespread in nature. They include carotenes and their oxygenated derivatives xanthophylls. Structurally, they consist of eight isoprenoid units arranged symmetrically around a pair of carbon atoms and may be cyclized at the ends.[20] Such a structure allows an enormous number of derivatives, with some 500 carotenoids already known.[29] Carotenoids have been well studied over the years, more for their pro-vitamin-A activity than for their color. However, interest in color has been increasing, and beta carotene, beta-apo-8¹-carotenal, and canthaxanthin have been synthesized.[30,37]

As might be suspected from the structure, the main cause of degradation of the carotenoids is oxidation, the severity of which depends in large part on whether the pigment is *in vivo* or *in vitro*. For instance, lycopene is quite stable in tomatoes, but is unstable when extracted and purified. The enzymes in plant tissues, however, often degrade carotenoids rapidly, with lipoxygenase interacting in a number of foods to cause its degradation.

In processed foods the oxidative mechanism is more complex and depends on many factors such as light, heat, and the presence of prooxidants and antioxidants, since the reactions are believed to be caused by free radical formation.[20]

Betalaines

Mabry and Dreiding introduced "betalaine" as a term to describe the class of nitrogenous red and yellow, water-soluble plant pigments.[32] Due to their similarity to anthocyanins, there was some confusion in their nomenclature and Wyler and Dreiding proposed that the red pigments be called betacyanines and the yellow pigments, betaxanthines.[33]

The chromophore of the betalaine pigments can be described as a protonated 1,2,4,7,7-penta substituted 1,7-diazaheptamethin system. If the conjugated system is not extended, it is yellow, and if it is extended by a substituted aromatic system, it is red.[34]

Unlike anthocyanins, the hue of the betalaines is unaffected by the pH of most food systems, from pH 3.5–7.0. Maximum absorbance occurs at 538 nm in this pH range and shifts to 535 nm at pH 2.0, and to 544 nm at pH 9.0. The hue stability of the betalaines imparts to them some major advantages over the anthocyanins as colorants in many foods. However, as with other pigments, the stability of the betalaines is affected by many factors, and their sensitivity to heat, oxygen, and water activity has limited their food usage.

Physical factors

In general, when we assess color either visually or objectively by instrumental techniques we are influenced by factors other than those which result from chemical chromophores. This is obvious if it is remembered that a food or beverage may be transparent, translucent, or opaque and may have surface fibers or striations as in meat or a cob of corn. Food may also be homogenous or nonhomogeneous and may be uniformly or nonuniformly colored.

All of these factors will affect the manner in which light reflects from the food and, consequently, the color that we perceive and measure. The physical phenomena responsible for these effects may be classified optically as follows:

- Reflectance from the surface
- Refraction into the object
- Transmission through the object
- Diffusion

Instrumentally, some of these effects are compensated for by designing instruments with a standard angle such that reflecting materials are viewed or measured 45° from the angle of illumination. This procedure provides a measure of diffuse reflection, which is mainly a measure of color, and since the angle is constant, the effects of surface aberrations are minimized.

Reflectance

When light strikes an object, reflection at the surface occurs, the type of reflection being dependent on the surface that the light strikes. Where the boundary consists of a series of small interfaces oriented at all possible angles to the normal, such as occurs when light strikes a rough surface, the distribution of the reflected energy follows the Lambert Cosine Law.[35] Such an incident is shown in Figure 4. This might be thought of as a rough metallic surface where any radiation that enters the medium is quickly absorbed by free electrons, so essentially the only reflectance is *regular* reflectance. As shown in the figure, some of the rays strike the surface more than once before being reflected, thus losing some energy. As a result, a rough surface would be expected to have a lower total reflectance than a smooth surface.[36]

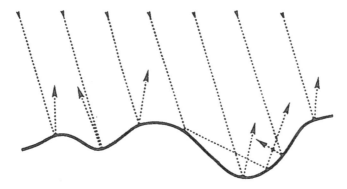

Figure 4   A rough surface reflects light in random directions and thus has a lower total reflectance than a smooth surface.

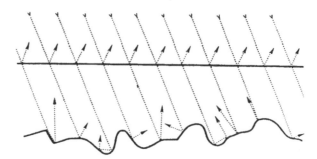

Figure 5 Light striking an initially smooth surface with a rough layer underneath undergoes regular reflection at the smooth surface and diffuse reflection at the rough layer. The sample appears glossy.

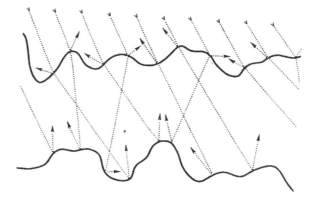

Figure 6 Light striking an initially rough surface with another rough layer underneath undergoes diffuse reflection at both layers and appears dull.

Figure 5 represents a very common situation in which two optical paths can be traced[35]: (1) The incident light encounters an initial smooth surface, and the regular reflection produces gloss or glare, commonly referred to as the *specular* component of reflectance. This type of reflection is not normally considered as a function of color and is measured at an angle of 90° from the illuminating source. That is, the angle of incidence equals the angle of reflection and specialized instruments, called goniophotometers, based on this principle, are available. (2) The light that is transmitted through the first interface undergoes absorption according to the Beer–Lambert Law, which is an exponential function of the distance, and then is reflected at randomly oriented internal interfaces; a fraction of this reflected radiation is transmitted back through the initial interface. The radiation that follows the second path constitutes *diffuse* reflection, which is normally considered a function of color. However, it should be stressed that the absorption of the material affects both these types of reflectance differently.[36]

Figure 6 represents another situation where the smooth surface shown in Figure 5 is replaced by a rough surface. Here, both the regular reflectance and diffuse reflectance are Lambertian; therefore, the two forms of reflectance are not so easily separated.

These phenomena have major implications in assessing quality in biological systems. For instance, the apple as harvested has a thin wax coat that is not naturally smooth, so the regular reflectance is Lambertian; however, since the eye sees the sum of both regular and diffuse reflectances, it is the diffuse reflectance that conveys the characteristic color to the eye. Accordingly, a low color value is perceived in an unpolished apple. Polishing the apple makes the wax coating smooth so that the regular reflectance, or gloss, can be distinguished from the diffuse reflection and imparts to the polished apple the bright glossy appearance. However, a polished fruit seems to have a higher color value, because the observer unconsciously orients it so that the specular component of reflectance is not observed and one sees only that radiation which has undergone absorption by the pigment in the fruit skin. Thus, one must decide whether or not to include gloss in a color measurement when quality is being assessed.

Refraction

Refraction measures the degree to which light traveling in the material is slowed down relative to its travel in air. At every boundary between two materials of different refractive index, light changes its speed and a small fraction of it is reflected according to Snell's Law.

This process usually accounts for about 4% reflection for most materials and, when repeated continuously in a medium containing particles of a different refractive index, can result in thorough diffusion and up to 80–90% reflectance. Thus, when dealing with such materials, it is important to recognize the effect that scattering will have on the composite energy reflected.

Transmission

Transmitted light is that light which is transmitted through an object and is a continuation of the incident beam.

Diffusion

Diffusion is an optical phenomenon that encompasses both reflection and refraction and is a function of surface roughness as well as particle size. It is the term often used to identify what happens to a ray of light that is affected by contact either with a smooth surface or with particles within the body of a granular or fibrous material. The first phenomenon is called *surface diffusion* and the second *internal diffusion*.

Surface diffusion is a function of both the smoothness (or roughness) of a surface and the angle at which the surface is viewed. This dependence is due to the fact that both the surface and the angle will affect the degree of interference of the light rays as they are emitted from the surface.

Internal diffusion represents the phenomenon that occurs when light enters the surface of the object and is subjected to reflection and refraction due to particulate matter. Obviously, one of the requirements for internal diffusion to occur is the presence of randomly oriented particles with a refractive index different from that of the surrounding material.

The size of the particles is also important to diffusion. As particle diameter decreases, the scattering or diffusion of the material increases inversely with the particle diameter, reaching a maximum, at approximately one quarter the wavelength of light. At this point, the reverse effect occurs and scattering or diffusion decreases in proportion to the diameter cubed. Thus, it is important to control particle size when measuring a sample.[36]

All of these effects are extremely important in assessing quality. Consider the case where the particle size varies from sample to sample: the amount of reflectance and the amount of "color" we measure will vary extensively from sample to sample, so that reproducible measurements cannot obtained from any given set of such samples. Therefore, if a particulate material such as cereals or potato chips is being measured, one must use a sieve or some other means whereby uniform particle size can be presented to the instrument every time a measurement is made. Another suggested method is to compress granular materials into a compact "cake" such that the effect of particle size will be minimized.[36]

Most food materials are generally turbid or translucent and as such may be considered multiple scattering medium to which the comprehensive theory of Kubelka–Munk may be applied.[36] In its simplest form, this theory assumes that the optical properties of turbid/translucent materials can be described at any wavelength of light by a scattering coefficients and an absorption coefficient K. This simply quantifies the manner in which the relationship between the physical characteristics (scatter) and the chemical characteristics (absorption) affects reflectance and, therefore, color. This effect was clearly shown in a comparison of the color of canned and fresh carrots,[39] which did not correlate well. The difference in color was probably due to the thermal tratment associated with canning, which would likely isomerize carotenoids, change the ratio of pigment to protein in complexes, affect the degree of esterification of xanthophylls, and modify the physical state of the starch (which would dramatically affect scatter (K) and thus color).

Measurement of color

Principles of color measurement

All too often the measurement of color is thought to be the measurement of the concentration of pigments present. With this in mind the chemist would tend to classify color measurement under either absorption spectroscopy or reflectance spectroscopy at a wavelength of maximum absorbance. It must be remembered, however, that color is the composite energy reflected from an object over the entire visual spectrum, not from just a single wavelength. Thus, specification may only be achieved by comparison to a similar color in a visual color solid *either* by using a visual colorimeter or by converting the energy from the entire visible spectrum via appropriate filters or mathematical models to data that are meaningful in terms of visual perception.

The specification of color by visual comparison or visual colorimeters is common in some parts of the world, but is rapidly being replaced by reflectance spectrophotometers or tristimulus colorimeters. This changeover is not difficult to understand when one considers the problems of color memory, biased judgments, human frailty, and physiological differences among individuals as well as within individuals on any given day or with age. Nonetheless, with appropriate care, visual color systems can be quite effective in certain applications.

Comparisons are generally made with established visual color solids, which may be classified into four groups: random collections, colorant mixture systems, color-mixture systems, and color-appearance systems. The first three systems deal with colors in a nonuniform manner, while the last is based on a color solid where each color chip is visually equidistant to its neighbors. The most well-known solid in the color-appearance system and probably in all the systems is the Munsell color solid. In the spherical Munsell solid, each color is located in three-dimensional space by a specification of "Hue,Value/Chroma," written as shown. If the solid is viewed as an orange, the vertical core would represent Munsell value going from light to dark. Each segment of the orange would represent hue, i.e., green, yellow, red, etc., as the circumference of the orange was traversed, and the distance from the core to the exterior would represent Munsell chroma, or intensity, of color.

Visual colorimeters are distinct from these systems and from color comparators that utilize the Beer–Lambert Law. A visual colorimeter produces a color based on the addition or subtraction of colored light that is visually matched with the color to be specified.

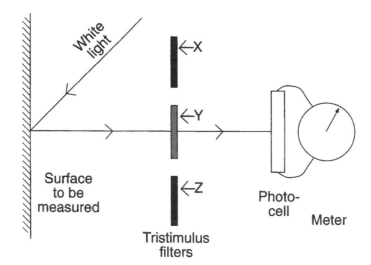

Figure 7 The essential features of a tristimulus colorimeter.

In additive colorimetry, a color is produced by the addition of three primary lights: red, green, and blue. In subtractive colorimetry, the final color is achieved by the selective substraction of light from white light by three filters, red, yellow, and blue. If all three filters are used the resultant color would obviously be dark gray or black.

Instrumental color measurement employs filters or mathematical models to simulate the response of the eye and it also collects and measures energy from the sample reflected across the entire visible spectrum. Both the filters and the mathematical models rely on "standard observer curves" that define the amount of red, green, and blue primary lights required to match a series of colors across the visible spectrum. Filters can then be made to simulate these curves or they may be used to mathematically modify the reflectance spectrum such that it corresponds to how the eye sees color.

The essential features of a tristimulus colorimeter are shown in Figure 7.[37] Colorimeters in use today are much more sophisticated, but the fundamental operation is the same. In Figure 7 it may be seen that reflectance at 45° from the source of illumination (diffuse reflectance) is modified by each of three filters prior to being measured. This modification produces three readings per sample that locate and specify the color in three dimensional space, much as the Munsell system does. One of the most common mathematical color solids was developed by Hunter and is in use in several colorimeters. It provides a readout in terms of L, a, and b. L measures whiteness or darkness, similar to Munsell value, and the chromatic portion of the solid is defined by: a (red); $-a$ (green); b (yellow); and $-b$ (blue). These data may be reduced to produce functions that define hue and chroma as shown in Figure 8[38] in a chromatic plane from L, a, and b color solid drawn on Cartesian coordinants. A triangle has been drawn from point P_s, representing a color. Within this triangle it can be seen that the distance of P_s from the origin is $(a^2 + b^2)^{1/2}$ and represents chroma, while the angle θ_s defines the hue or distance around the circumference. Thus colors may be specified in terms of tristimulus values L, a, and b, or hue, θ_s, chroma $(a^2 + b^2)^{1/2}$, or some combination of these.

When reflectance spectrophotometry is used, the reflectance at wavelengths across the visible spectrum is modified mathematically from values obtained from each of the standard observer curves. The area under each of the curves is then calculated via programs supplied with the instrument. The most common solid in use for reflectance spectrophotometry is that originally defined by the CIE (Commission Internationale de l'Eclairage)

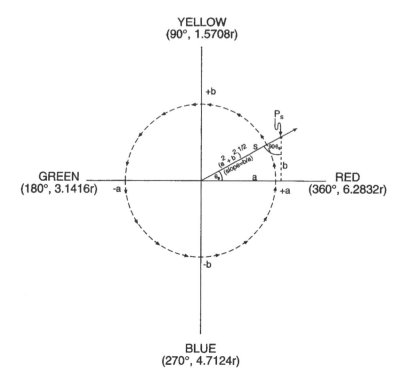

Figure 8 A unit plane from the L, a, b Hunter color solid. Point P_s is located at (a,b) in rectangular coordinates and (θ_s, s) in polar coordinates. θ_s is the hue angle and may be expressed as a tangential or cotangential function. Chroma may be expressed as $(a^2 + b^2)^{1/2}$. The broken curve, with arrows running counterclockwise, indiates the direction of change in hue from red to yellow as the slope increases for 0 to ∞, and from yellow to green as the slope decreases from ∞ to 0, etc. (From Little, A.C. *J. Food Sci.* 1975, 40, 410-411. With permission.)

in 1931. This system recommends a standard system for measurement in terms of the tristimulus values X, Y, and Z. Y represents lightness and darkness while the chromaticity coordinates x, y, and z calculated from the tristimulus values, as fractions of their total, represent a triangle in the chromatic plane. Therefore, in this solid a color may be specified by Y, x, and y, while hue and chroma maybe specified by the dominant wavelength and purity functions, respectively.[37]

There are many other color solids but the Hunter and CIE are the ones most commonly used and are widely accepted in food color measurement. There are other reflectance instruments that simply measure energy, unmodulated by filters or mathematical models. These reflectometers do not measure color, but the energy readouts provided may at times correlate with color changes. As a result, they are quite useful in some applications.

Further information on the principles of color measurement may be found in a variety of sources.[4,36,37,39]

Color evaluation

The measurement of color utilizing the principles discussed in the previous section is dependent upon the kind and number of pigments present and the amount of scatter provided by the food medium. Consideration of both of these factors is essential in color measurement, whether a food or a beverage is being measured.

The simplest measurement involves a completely transparent or opaque food. In these cases, either reflectance or transmittance colorimetry may be used and data obtained in terms of tristimulus values, chromaticity coordinants, or hue, value, and chroma. If such foods or beverages have solid colors and no directional effects then these data will provide a correlation with visual color high enough to use as a measure of quality.

Unfortunately, many foods do not fit this category. For example, orange juice, which has 40 out of 100 points allocated to color in grading, is a translucent food with both a scatter and absorption component. As such, simple measurements will not suffice, so Hunter developed a citrus colorimeter that recognized these components.[40] Data obtained from citrus products showed that the function A/Y correlated well with visual color assessment. A is an amber filter measuring absorption (K) and Y represents CIE Y, which measures lightness or darkness and therefore is a function of scatter.[5] Consequently, the ratio of A/Y represents a measure of K/S, which according to Kubelka–Munk theory provides a measure of color. The final readout of the colorimeter was CR (citrus red) = $200 (A/Y - 1)$, which expanded the A/Y scale.

Even when scattering is not present, problems might exist in color measurement such as occur in very dark colored juices such as grape juice. The problem with dark samples was that an "area of confusion"[41,42] developed below a certain level of lightness. The confusion arose because the a and b scales became nonlinear between L values of approximately 19 and 54. In order to compensate for this nonlinearity, Eagerman et al. mathematically distorted the L, a, and b color scale in order to expand it at low L values in a manner similar to that shown diagrammatically in Figure 9.[41,42]

These manipulations resulted in the development of a color scale where a and b remained linear until an L value of 2.5 was reached. This linearized scale provided a high level of correlation of measured color with visual assessment and allowed samples to be measured directly as consumed, thus providing the same color as the consumer sees. Interestingly and rather unexpectedly, this scale also correlated highly with pigment content. Color is not often a good indication of pigment content, because there are a variety of pigments in most foods, these pigments are not uniformly distributed but localized, there are contributuons from the browning products formed, and the scattering component of the food modifies the absorption characteristics.

There are, however, some useful exceptions as noted above. Girard et al. proposed a spectrophotometric method that relates with cooked meat color.[43] It involves using thin slices of pork and making transmission spectrophotometric measurements at 414, 520, and 550 nm. The measurement of meat color has been a particular problem due to the oxygen-induced, dynamic equilibrium among pigments and the directional structure of the meat surface, so any techniques that allow color to be evaluated are welcome. Palombo and Wijngaards have designed a predictive model for assessing porcine lightness during heating based on a regression analysis of L vs. time to obtain model parameters.[44] A description of the temperature dependence of these parameters was then obtained using linear and Arrhenius models.

Colorimetric estimation of other quality parameters

It has been suggested[1] that color might be used to define quality parameters other than appearance. In some cases it has been used as a predictor of pigment content as indicator of the degree of processing, as an indicator of maturation and ripeness, for sorting, for assessing moisture content, and for the identification of debris and unwanted material.

Expanded scales that correlated with pigment content in dark-colored juices have already been discussed. Other studies in our laboratory have shown that color provides

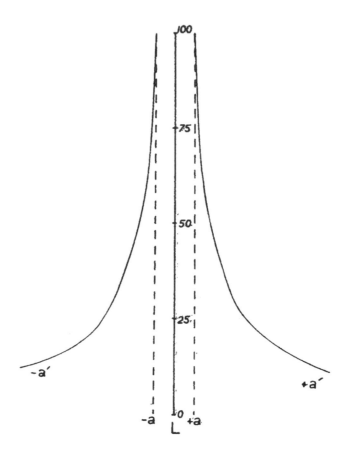

Figure 9 A representation of the *L*, *a*, and *b* color scale expansion through *L* = 100 when *a* is held constant at +1.0 and −1.0.

an accurate assessment of moisture content and degree of doneness. Color has many on-line applications in the food industry for sorting and measuring degree of doneness, such as in baking lines.

An interesting application was proposed by Little,[45] who showed the value of color in both grading and estimating overprocessing of tuna. The color of tuna was measured before and after treatment with sodium dithionite, which reduced the oxymyoglobin. This treatment caused a shift in color values that was indicative of the amount of reducible pigment present. Accordingly, a large amount of pigment would show a greater change in color on reduction then a small amount, as represented by the difference in chromaticity coordinants, expressed as Δx and Δy. Furthermore, a sample with a high concentration of pigment (large Δx, Δy) would be dark and have a low *Y* value (*Y* is a measure of lightness and darkness).

Thus, a white or light tuna would have a low Δx, Δy and a high *Y* value, and a dark tuna would correspond to the reverse situation. This approach was found to be suitable for grading and an excellent indicator of overprocessing or stackhurn. If a sample was found to have a low *Y* value (dark) and a small color shift (Δx, Δy) it was indicative of a sample with a low pigment concentration, which must have been a consequence of the sample being mistreated in some way.

Interesting and valuable applications of color measurement to quality assessment are often possible when the systems under study are well understood, which allows the data to be viewed in the proper perspective.

Conclusion

As we progress into the 21st century, the necessity for high-quality, safe, and nutritious foods will increase. We will be modifying some of the familiar sensory attributes of food as the composition of our diet changes and we strive to reduce both calories and fat. This will require an increasing knowledge of the role of such factors as color play in food acceptance, so that we may fully utilize them in order to compensate for other changes in the sensory profile and maintain maximum acceptability.

We will also see a growing number of new processes for manufacturing natural food colors,[46] and it is hoped that the present worldwide trend of increasing patents pertaining to food colors will continue and not follow the trend in the U.S.[47]

Color is an integral part of food quality and the enjoyment of eating. Unfortunately, an understanding of the optical effects that underlie color and its measurement is often overlooked. It is hoped that this overview will help to remedy that situation by providing a stimulus to further study in this exciting and challenging area.

References

1. Clydesdale, F.M. *J. Food Qual.* 1991, 14, 61-74.
2. Clydesdale, F.M. In: Developments in Food Colours-2; Walford, J., Ed.; Elsevier Applied Sciences: London, 1984; Chapter 3.
3. Clydesdale, F.M. *J. Food Qual.* 1991.
4. Hunter, R.S. and Harold, R.W. The Measurement of Appearance, 2nd Edition; John Wiley and Sons: New York, 1987.
5. Francis, F.J. *Crit. Rev. Food Sci. Nutr.* 1989, 28, 273-314.
6. Yoshimata, K. *Phytochem.* 1977, 16, 1857-1861.
7. Timberlake, C.F. *Food Chem.* 1980, 5, 57-71.
8. Timberlake, C.F. and Bridle, P. In: Developments in Food Colours-1; Walford, J., Ed.; Applied Sciences: London, 1980; Chapter 5.
9. Lin, T.Y., Koehler, P.E. and Shewfeld, R.L. *J. Food Sci.* 1989, 54, 405-407.
10. Jurd, L. *Food Technol.* 1964, 18, 559-563.
11. Mazza, G. and Bouillard, R. *J. Agric. Food Chem.* 1987, 35, 422-427.
12. Timberlake, C.G. and Bridle, P. *J. Sci. Food Agric.* 1967, 18, 473-478.
13. Taylor, A. In: Developments in Food Colours-2; Walford, J., Ed.; Elsevier Applied Sciences: London, 1984; Chapter 5.
14. Brouillard, R. and Dubois, J.E. *J. Am. Chem. Soc.* 1977, 99, 1359-1368.
15. Brouillard, R. *Phytochem.* 1981, 20, 143-148.
16. Aronold, S. In: The Chlorophylls; Vernon, L.P. and Seely, G.R., Eds.; Academic Press: New York, 1966; Chapter 1.
17. Schwartz, S.J. and Lorenzo, T.V. *Crit. Rev. Food Sci. Nutr.* 1990, 29, 1-17.
18. Lin, Y.D., Clydesdale, F.M. and Francis, F. J. *J. Food Sci.* 1970, 35, 641-644.
19. Schwartz, S.J., Woo, S.L. and von Elbe, J.H. *J. Agric. Food Chem.* 1981, 29, 533-537.
20. Clydesdale, F.M. and Francis, F.J. In: Principles of Food Science. Part I. Food Chemistry; Fennema, O.R., Ed.; Marcel Dekker: New York, 1976; p. 385.
21. Clydesdale, F.M., Fleischnan, D.L. and Francis, F.J. *J. Food Prod. Dev.* 1970, 4, 127-131.
22. Clydesdale, F.M. and Francis, F.J. *Food Technol.* 1968, 22 (6), 135-138.
23. Segner, W.R., Ragusa, T.J., Nauk, W.K. and Hoyle, W.C. U.S. Patent No. 4,473,591, 1984, Sept. 25.
24. Price, J.F. and Schweigert, B.S. The Science of Meat and Meat Products, 2nd Edition; W. H. Freeman: San Francisco, 1971.
25. Fox, J.B., Jr. *J. Agric. Food Chem.* 1966, 14, 207-210.
26. Jen, J.R., Brown, R.B., Dick, R.L. and Acton, J.C. *J. Food Sci.* 1988, S3, 1043-1046.
27. Chu, Y.H., Huffnan, D.L., Egbert, W.R. and Trout, G.R. *J. Food Sci.* 1988, 53, 705-710.
28. Mitsumoto, M., Cassens, R.G., Schaeffer, D.M. and Scheller, K.K. *J. Food Sci.* 1991, 56, 857-858.

29. Clydesdale, F.M., Ho. C.T., Lee, C.Y., Mondy, N.I. and Shewfelt, R.L. *Crit. Rev. Food Sci. Nutr.* 1991

30. Counsell, J.N. In: Developments in Food Colours-1; Walford, J., Ed.; Elsevier Applied Sciences: London, 1980; Chapter 6.

31. Gordon, H.T. In: Current Aspects of Food Colorants; Furia T.E., Ed.; CRC Press: Boca Raton, FL, 1977; Chapter 4.

32. Mabry, T.J. and Dreiding, A.S. In: Recent Advances in Phytochemistry; Mabry, T.J., Alston, R.E. and Runeckles, V.C., Eds.; Appleton Century Crofts: New York, 1968.

33. Wyler, H. and Dreiding, A.S. *Experientia* 1961, 17, 23-26.

34. von Elbe, J.H. In: Current Aspects of Food Colorants; Furia,T.E., Ed.; CRC Press: Boca Raton, FL, 1977; Chapter 3.

35. Birth, G.S. and Zachariah, G.L. Paper 71-328, ASAE Meeting, St. Josephs, MI, 1971.

36. Clydesdale, F.M. In: Food Analysis Principles and Techniques, Vol. I, Physical Characterization; Gruenwedel, D.W. and Whitaker, J.R., Eds.; Marcel Dekker: New York, 1984; Chapter 3.

37. Francis, F.J. and Clydesdale, F.M. Food Colorimetry: Theory and Applications; AVI: CT, 1975.

38. Little, A.C. *J. Food Sci.* 1975, 40, 410-411.

39. Clydesdale, F.M. In: Current Aspects of Food Colorants; Furia, T.E., Ed.; CRC Press: Boca Raton, FL, 1977; Chapter 1.

40. Hunter, R.S. *Food Technol.* 1967, 21, 906-911.

41. Eagerman, B.A., Clydesdale, F.M. and Francis, F.J. *J. Food Sci.* 1973, 38, 1051-1055.

42. Eagerman, B.A., Clydesdale, F.M. and Francis, F.J. *J. Food Sci.* 1973, 38, 1056-1060.

43. Girard, B., Vanderstoep, J. and Richards, T.F. *J. Food Sci.* 1990, 55, 1249-1254.

44. Palombo, R. and Wijngaards, *J. Food Sci.* 1990, 55, 601-603.

45. Little, A.C. *Food Technol.* 1969, 23, 1466-1471.

46. Stafford, A. *Trends Food Sci. Tech.* 1991, 2, 116-122.

47. Francis, F.J. Handbook of Food Colorant Patents. CRC Press: Boca Raton, FL, 1986.

chapter eight

Effect of storage on texture

Alina S. Szczesniak

Contents

Introduction

Texture can be defined as "that group of physical characteristics that arise from the structural elements of the food, are sensed primarily by the feeling of touch, are related to the deformation, disintegration and flow of the food under a force, and are measured objectively by functions of mass, time, and length".[1] This broad definition, which comprises the concepts of "consistency" and "body," will be used here.

The above definition points to three important dimensions of texture: (1) that it is a sensory property; (2) that it has its roots in the structure of the food—macroscopic, microscopic, and molecular; and (3) that it is a multidimensional property comprising a number of characteristics (e.g., tenderness, chewiness, crispness, softness, juiciness, etc.). These have been grouped into mechanical, geometrical, and parameters related to fat and moisture release/absorption.[2]

With few exceptions, the effect of storage on texture has not been studied extensively because texture deterioration is related to quality decrease, rather than to potentially health-hazardous spoilage, and because the importance of texture in consumer acceptance of foods has only recently gained general recognition.[3,4] Usually, unpleasant odor and appearance, signifying microbial spoilage, provide signals that the food is unsafe to eat long before textural changes become evident.

Several food systems, however, have been studied in depth from the standpoint of chemical, physicochemical, and physiological alterations in the molecular and microscopic structure responsible for textural changes on storage (e.g., fish, bread, muscle meat). This work has helped to move the field of texture from the realm of food technology to food science. Unfortunately, much of it was done without any attention to reliable quantitative texture measurements. Engineering studies have also been done on the susceptibility of fruit (e.g., oranges, apples) to textural damage through bruising during handling.

Textural changes on storage are generally caused by enzymatic action, changes in moisture, or reactions in food polymers leading to crosslinking and toughening. Enzymatic action can be beneficial, such as occurs in ripening of fruits, meat, and cheese, or detrimental, such as occurs in senescence of fruits. The effect of changes in moisture depends on the original moisture level in the food and, thus, on its textural character. Moisture can be lost, absorbed, or rearranged. Loss of moisture from high moisture foods (e.g., apples) leads to loss of juiciness and crispness (lack of turgor). Loss from medium moisture systems (e.g., cakes) leads to the development of dryness; in the mouth, moisture will be absorbed by the food rather than released from it. Moisture rearrangement is believed to be involved in the staling of bread. Moisture gain by dry systems valued for their crispness (e.g., snacks, cold cereal) results in loss of crispness, development of toughness and, ultimately, sogginess.

Baked goods

Baked goods are farinaceous products with an expanded structure produced through incorporation of gas (either through yeast fermentation, decomposition of baking powders, whipping in of air, or steam production) and solidification of the matrix by high temperature coagulation and moisture removal. The means of gas production and the chemical composition of the matrix (especially with respect to sugar and fat content) define the many existing types of baked products. Only bread and some typical cakes will be considered here in some detail.

Bread

Many types of bread are available in the marketplace. The white, wheat pan bread and its firming on storage have been the subject of most studies. The main constituents of bread are flour, water, yeast, and salt, to which are added other ingredients such as fat, emulsifiers, sugar, and flour improvers. Briefly, it is prepared by hydrating the ingredients, producing CO_2 through yeast fermentation, developing gluten in the flour by chemical and mechanical means, and baking.

The baked bread consists of the crust, the often crisp outer shell, and the crumb, the soft inside portion. The crumb may be likened to a biochemically inactive colloidal system with physicochemical instability.[5] It may also be likened to a firm foam system in which the air cells are surrounded by a viscoelastic system composed of flour and other ingredients and heat-set during the baking process. A detailed description of bread structure has been published by Taranto.[6] The major component of the crumb is wheat starch which contains 25% amylose (linear glucose polymer connected by 1,4 glucosidic linkages) and 75% amylopectin (branched glucose polymer connected by 1,4 and 1,6 glycosidic bonds). The backbone of bread is gluten, the wheat protein, and its ability to form an expanded, elastic structure following hydration and mechanical manipulation. Typically, bread contains about 35% water.

The staling of bread is probably the most researched storage-induced textural deterioration because of the high economic impact. Although it does have a flavor component, the effect of staling on consumer acceptance of bread is texture-driven. The mechanism of staling is still not completely understood after more than a century of research. It has been reviewed by Willhoft,[7] Kulp and Ponte,[5] and others. The accepted definition of staling is that it is "a term which indicates (decreased) consumer acceptance of bakery products caused by changes in the crumb other than those resulting from the action of spoilage organisms".[8]

The staling process starts the moment the bread is removed from the oven and becomes sensorially detectable in 1–2 days. From the sensory standpoint, staling reduces the springback and softness of the crumb and increases its crumbliness and dryness in the mouth. Firming and reduction in springiness are detectable instrumentally (Figures 1 and 2). Staling also reduces drastically the crispness of the crust and increases its toughness,[9] a phenomenon recently confirmed to be related to moisture migration from the

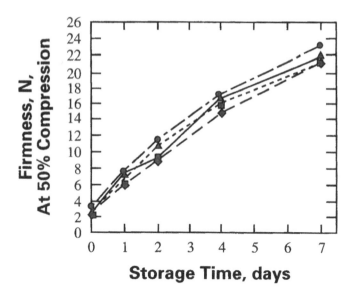

Figure 1 Effect of storage on instrumental assessment of bread crumb firmness. (From Stollman, U.; Lundgren, B. Cereal Chem. 1987, 64, 230-236. With permission.)

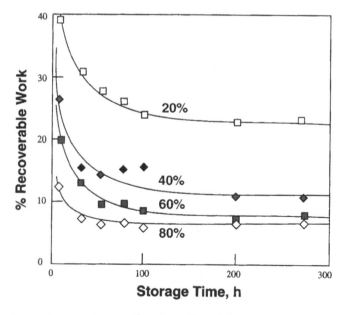

Figure 2 Change in crumb springiness with aging of bread. Figures represent percent deformation. (From Nussinovitch, A., et al. J. Texture Stud. 1992, 23, 13-24. With permission.)

crumb to the crust.[11] From the physical standpoint, staling reflects the following changes in the crumb: increases in firmness, crumbliness, and starch crystallinity; and decreases in absorption capacity, susceptibility of starch to enzymatic attack, and soluble starch content.[5] Fundamental studies on the mechanism of staling, for many years involving chemical and technological approaches, have recently benefitted from the physicochemical approach and the application to food polymers of the knowledge developed in the field of synthetic polymers,[12] especially when dealing with the plasticizing role of water and the transition between glassy (immobile) and rubbery (mobile) states of molecules.

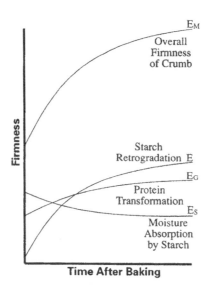

Figure 3 Factors contributing to crumb firming on staling. (From Kulp, K.; Ponte, J.G, Jr. CRC Crit. Rev. Food Sci. Nutr. 1981, 15, 1-48. With permission.)

According to Kulp and Ponte[5] and Willhoft[7] the mechanism of bread staling appears to involve the starch component, the gluten component, and the transfer of moisture between starch and gluten. Originally ascribed to retrogradation, changes in starch are now considered to be due to the crystallization of amylopectin preceded by formation of networks of amylose in which amylopectin crystals are entangled. Crystals of amylose are also formed. The process is accompanied by moisture redistribution and involves transformation to glassy state through a continuum of rubbery states. Nucleation and crystal melting explain some of the temperature effects on staling.[13] The concepts of glass and water dynamics advanced in this physicochemical approach stress the nonequilibrium nature of the staling process.

According to the physical chemistry of high polymers approach advanced by Slade and Levine,[13] the retrogradation of starch is a recrystallization process involving three sequential steps: nucleation, propagation, and maturation. Nucleation is inhibited by storage at –11°C and accelerated by storage at 0°C. The theoretical maximum overall crystallization rate (i.e., nucleation and propagation rates) for wheat starch occurs at about room temperature. A detailed treatment of the role of starch in bread staling, considered from the standpoint of the classical polymer science approach to structure/property relations and kinetics of phase transition, can be found in a recent review.[13]

The evidence regarding the involvement of gluten in bread staling is less abundant. A model has been proposed by Hoseney and coworkers which relates bread firming to formation of H-bond crosslinks between the continuous three-dimensional protein structure and the discontinuous remnants of starch granules.[14,15] It is possible that the formation of these crosslinks is related to the release by gluten of structurally bound moisture that occurs first during baking and then during storage at room temperature. A recent study by Cocero and Kokini demonstrated that glutenin is very sensitive to the plasticizing effect of water in the 4–14% moisture range and that its glass transition temperature shifts to lower regions with increasing moisture content.[16] According to Willhoft, moisture migrates from gluten to starch, firming the former and softening the latter.[17] Willhoft's theory of bread staling is summarized in Figure 3. There is some evidence that moisture also migrates from starch to gluten.[5] Pentosans, natural wheat flour hydrocolloids, have been implicated in moisture interchange among bread constituents.[18]

The severity of staling is decreased by the presence of fat, emulsifiers and bacterial amylases. Enzymes act by breaking up the continuity of the three-dimensional network, while fat acts as a hydrophobic plasticizer. The mechanism by which emulsifiers affect bread firming on aging appears to be related to their role in maintaining the crumb's viscous properties.[19] Staling is accentuated by low temperature storage and temporarily reversed by heating.[20] The temporary reversal may be explained by the melting and subsequent reforming of amylose and amylopectin crystals.

Cakes

Compared to bread, cakes contain more fat and sugar. They usually contain eggs and a "weaker" flour (i.e., lower in protein content), because the development of gluten is less important for structure formation. For the most part, their shelf life is limited by the microbial spoilage of the fillings, but staling also occurs. Its mechanism is believed to be similar to the staling of bread, except that dehydration and changes in the proteins appear to be more important in the sensory aspects of cake staleness.[7] Because of the importance of moisture in the consumer acceptance of cakes, humectants such as corn syrup, honey, and simple sugars are often used as ingredients.

The role of other ingredients appears to be as follows: gluten and egg white provide the mechanical strength during gas expansion in the early stages of baking; starch increases batter viscosity and its gelatinization and, together with the coagulation of egg proteins, contributes to the setting of the structure; lower gluten level is responsible for the less cohesive, crumblier crumb in cakes. The two key textural attributes of cakes valued by consumers are moistness and tenderness. Sugar and fat play key roles in tenderizing the structure.

Guy reported that cakes stale in two stages: (1) an intrinsic firming process and (2) a macroscopic migration of moisture.[20] The mechanism of the intrinsic crumb firming is similar to that believed to occur in bread. Moisture migration, a process that continues for several weeks at ambient temperature, is said to be responsible for the loss of the moist-eating quality. Similar to bread, the firming of cake crumb is not related to moisture loss and can occur in high humidity storage.[21] Sugars are considered to have an antistaling effect[13] because of their inhibitory effect on starch crystallization (glucose homologs) and reduction of water mobility (other sugars). Sucrose has been reported to be more effective than glucose or fructose.

Reference is made to the review by Willhoft[7] for details on the limited work done on cake staling and to the review by Taranto[6] for information on cake ultrastructure.

Others

Two other types of baked goods should be mentioned in reference to textural changes on storage: crackers and pies.

Crackers are highly valued for their crispness detected as mechanical (stiffness and ease of fracture) and auditory (sound production) sensations. Vickers and Bourne recognized wet/crisp and dry/crisp foods and related their behavior to the rheological properties of the structure.[22] It is interesting that dry/crisp foods lose their crispness on gaining moisture, while wet/crisp foods lose their crispness on losing moisture. When stored under high humidity conditions without proper package protection, crackers and similar products will absorb moisture and become deformable and less prone to fracture.[23,24] Sensorially, this will be felt as softness, sogginess, or toughness and will lead to product rejection. The mechanism involves plasticization by water of the amorphous starch structure.[13,25]

Pies are multitextured products in which moisture gradients exist between the different components. The crust is a dry/crisp component that loses its crispness on gaining moisture from the filling. Because of its high fat content, the structure and crispness of pie crust are different from those of crackers. The moisture transfer between the components is governed not so much by their actual moisture contents, but by the difference in their water activity (a_w) values (i.e., the relative amounts of free water). It may be stopped either by formulating the components to identical (or closely similar) a_w levels[26] or by separating them with a moisture-impermeable barrier.

Recently, frozen yeasted bread dough became of interest as an article of commerce. Its serious drawback is a gradual loss of baked loaf volume on frozen storage apparently related to rheological changes in the dough and the decreased yeast viability. Dough strength was found to decrease on freezing and frozen storage, the damage being dependent on the starting strength of the original dough.[27,28] An increase in the onset temperature of starch gelatinization was also reported.[29]

Dairy products

Dairy products represent a very diversified class of foods from the textural and structural standpoints and are grouped here for convenience. Only the most important categories will be considered.

Butter and margarine

Structurally, both butter and margarine consist of a continuous fat phase, in which are dispersed droplets of water and of liquid fat and small fat crystals. The fat in butter is derived from milk, while that in margarine is usually of plant origin. Thus, margarine is not a true dairy product. Because of their textural and usage similarity, however, the two products can be considered together.

Firmness, spreadability, smoothness, and proper melt-down are the key textural characteristics of these products. Firmness and spreadability are influenced by the solid-to-liquid-fat ratio, which in turn is influenced by the product temperature. This ratio can be varied over a fairly wide range in the production of margarine; it can be varied only little in the production of butter. Thus, margarines are generally more spreadable than butter. Some countries allow manufacturers to lower the solid-to-liquid-fat ratio in butter by incorporation of vegetable oil. The plastic character of butter and margarine is the result of a three-dimensional network of fat crystals. Mechanical treatment can be used to soften the product through the destruction of this structure.[30]

The effect of storage on butter and margarine involves growth of fat crystals and strengthening of the network bonds. Fat crystals growth in margarine leads to increasing firmness[31] and development of sensory graininess,[32,33] as illustrated in Figure 4. Low-temperature storage accentuates the development of graininess in margarine.[34] Vaisey-Genser et al. reported that for smoothness fat crystals should be in the β' form (<1 μ long) rather than in the β form (>25 μ long). The threshold for sensory detection of fat crystals is 22 μ. Good correlations were obtained between graininess and crystal size ($r = 0.89$) and crystal size and storage time ($r = 0.84$). Butter was also shown, via electron microscopy studies, to grow fat crystals in storage.[35] They were described as uncurved platelets and believed to represent peripheric fragments of disrupted fat globules.

Shama and Sherman used rheological techniques to study the structural changes on work softening and subsequent storage of butter and margarine.[36] They found that, compared to butter, margarine undergoes a greater structure recovery. An analysis of the results led to the conclusion that the textural properties of butter, particularly the spread-

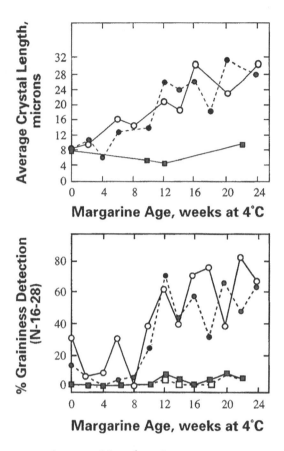

Figure 4 Effect of storage on fat crystal length and sensory graininess of canola oil margarine produced by two manufacturers. O, ● — manufacturer A; □, ■ — manufacturer B. (From Vaisey-Genser, M., et al. J. Texture Stud. 1989, 20, 347-361. With permission.)

ability, are governed by the ratio of primary to secondary bonds.[37] A later rheological study by Kawanari et al. using different techniques showed that butters stored at –30°C fractured more easily than those stored at 5°C.[38] The flow properties obeyed the Casson equation. The relationship involves a yield value (i.e., stress needed to initiate flow) that reflects the butter structure. Storage had a slight lowering effect on the yield value and made the flow of butter slightly less shear-dependent. Both effects suggest some weakening of the structure; however, they could not be related to sensory texture.

Cheese

Cheese is made from milk coagulated with rennin and acid, followed by cutting the curd and pressing out some of the water. Production of soft, unripened cheese (e.g., cottage cheese) stops essentially at this point. Most cheeses, however, are ripened by the presence of bacteria (cheddar) or mold (Roquefort) to produce different textures and flavors. The growth of these microbes is controlled by storage temperature and humidity, salt concentration, moisture, and pH.

Cheeses represent a wide gamut of textures, from very soft and runny, to soft and spreadable, to firm and sliceable, to hard and gratable. Although no publications have appeared on the subject, the consumer must have a set of mental standards for optimum texture in each cheese type, and the processors must be aware of what these are. It would seem that tolerance for textural variations should be the least for soft cheeses and the greatest for hard cheeses.

Unripe soft cheese has a very limited storage life and must be refrigerated. Any textural changes would be immaterial, since microbial spoilage soon makes the product unacceptable. The situation is different with ripened cheeses. Here, the early part of storage (up to several months) involves the ripening process and, thus, brings about desirable changes in texture as well as in flavor.

Texture development in cheese on ripening has been reviewed by Lawrence et al.[39] Adda et al.,[40] and Green and Manning.[41] Structurally, cheese is a protein network in which are entangled fat globules and water. The exact nature of the structure is determined by milk composition and concentration,[42] pH, salt level, and ripening conditions, as well as the structure of the curd. The continuous protein phase is produced by proteolytic enzymes, the nature and activity of which are determined by the starter culture. The casein micelles are subjected to specific proteolysis that removes their hydrophillic portion. This allows the micelles to aggregate by a random, diffusion-controlled mechanism. Chains of micelles link together to form a three-dimensional network.[41] The coarseness of this network causes crumbliness, granularity, and firmness in the cheese. It is affected by the structure of the coagulum and the presence of fat. Higher fat levels lead to a softer, more elastic, and smoother cheese. A softer texture is also produced by a higher degree of fat unsaturation. The size of the fat globules and the lipolysis do not appear to have significant effects on cheese texture.[40] The most noticeable texture changes on ripening involve firmness and fracturability. These parameters decrease with increasing proteolysis, higher fat, and moisture content.

At the end of the ripening period, cheese texture is at its best. Depending upon the cheese type, subsequent storage may cause rapid texture deterioration (as with soft cheeses, such as Brie), a slow change (as with semisoft cheeses, such as Muenster), or essentially no change (as with hard cheeses, such as Parmesan). Soft cheeses may undergo additional softening, or may become firmer, chewier, and less spreadable due to further proteolytic changes and/or moisture loss. The texture of semisoft cheeses usually deteriorates because of desiccation in refrigerated storage. When poorly protected from moisture loss, such cheeses may become very firm, crumbly, and chewy.

Fredrick and Dulley studied the effect of above-refrigeration temperature (15°–20°C) on cheddar cheese texture using instrumental texture profiling analysis.[43] Storage at higher temperature was found to make the cheese more fracturable and less springy. Cheese hardness was negatively and significantly correlated with proteolysis.

Dieffes et al. reviewed the published studies on the effect of frozen storage on the texture of mozzarella cheese used to stop ripening and prolong shelf life during distribution.[44] Decreases in hardness, gumminess and springiness, and changes in cohesiveness and meltability were reported. The extent and type of texture deterioration varied significantly with differences in freezing conditions and timing of post-thawing evaluation. Controlled studies by these authors using dynamic rheologic testing showed that part-skim milk mozzarella cheese became harder and more elastic when stored at –20°C, while that stored at 5°C became softer and more elastoviscous.[44] Under both storage conditions, longer storage resulted in cheeses that were less elastic and less viscous.

Ice cream

Ice cream is basically frozen foam, the cell walls of which are composed of milk proteins, soluble solids, fat globules, and ice crystals. It is made by simultaneously whipping and freezing a fat-in-water emulsion stabilized by milk proteins, nonionic emulsifiers, and stabilizers (usually vegetable hydrocolloids). It is then kept for at least 12 h at low temperature (about −25°C) to harden. The product contains about 50% air and about 10–20% fat depending on the type.[45] Ice milk, which is similar to ice cream in structure, texture, and usage (but less creamy and "rich"), contains 1–6% fat.[45,46]

The structure of ice cream is created during the freezing and aeration process, which causes partial destabilization (flocculation) of fat globules.[47] A small initial fat globule size is desirable.[48] The membrane around the fat globule is formed by milk proteins, which influence fat coalescence.[49,50] The texture of ice cream depends largely on the resistance of air cells to deformation, a property governed by the strength of the lamella, which in turn is affected by ice crystal size. A mechanical model for frozen ice cream has been developed by Sherman.[51] It equates the structural factors with specific rheological properties.

The ice cream producers have recognized early in the history of the industry that the mechanical and mouthfeel characteristics play a paramount role in the consumer acceptance of the product. Ice cream is particularly valued for its pleasing smoothness and meltdown. A considerable amount of information was generated in this area and applied to quality grading as well as to product/process improvement. Traditionally, ice cream makers have considered separately *body* (defined as "that quality which gives weight and substance to the product and enables it to stand up well") and *texture* (defined as "referring to the grain or the finer structure of the product").[45]

Fluctuating and high storage temperatures, such as those encountered in household refrigerator freezer compartments, cause textural degradation in ice cream. The defects are accentuated in poorer ice cream grades and minimized in better quality mixes and stabilizers. Traditionally, the better quality ice creams contain higher fat levels. Responding to consumer demands for low-fat products, the food industry has produced excellent quality low-fat ice creams by judicious selection of the stabilizing hydrocolloids. The following defects are recognized as storage-prompted:

coarseness—presence of large/nonuniform ice crystals
butteriness—clumping of destabilized fat globules
sandiness—presence of large insoluble lactose crystals
crumbliness—poor protein hydration due to moisture loss and other factors

Stabilizers and emulsifiers are most effective in reducing or eliminating the textural degradation of ice cream on storage. Stabilizers promote the formation of small ice crystals by immobilizing and binding the water. Emulsifiers strengthen and stabilize the colloidal system in the film surrounding the air cells. "Sandiness" in ice cream has been reported to be related to the glass transition temperature (T_g) of the sucrose–lactose glass,[52] and the plasticizing effect of water on T_g is also implicated in ice recrystallization on storage (see Chapter 15). The T_g of standard ice cream is about −34°C, while the normal freezer temperature is about −20°C. The larger the difference between the storage and the glass transition temperatures, the greater is the likelihood of ice cream developing sandiness. The reader is referred to a review by Hartel for a detailed discussion of the role of T_g and sugar crystallization in the textural deterioration on storage of ice cream and candies.[53]

A number of other textural defects recognized in ice cream (e.g., "doughy," "soggy," "fluffy," "snowy") are known to be caused more by improper formulation/processing than by storage.[45]

Others

Milk, fermented milks, and cream

Milk is an essentially Newtonian (i.e., shear-independent) product of low viscosity, only slightly higher than that of water. Unless subjected to microbial fermentation, which can be held in check by pasteurization and low temperature storage, the consistency of milk does not change on storage. It does change drastically on fermentation, which destabilizes the casein network and converts the thin fluid to a soft semisolid. When uncontrolled, fermentation of milk causes spoilage and consumer rejection. When controlled, it leads to the production of highly accepted products, such as buttermilk and yogurt in the U.S. and "sour milk" and kefir in Eastern Europe. Similar to cheese, during the production period these products develop an optimum texture that then deteriorates on subsequent storage (with a mechanism related to cheese production), unless the nature of the product allows it to be pasteurized.

Gelation and emulsion breakdown can occur on storage of some heat-processed fluid milk products. It has been shown that destabilization of the emulsion and formation of a gel-like consistency occur on 4°C storage in unhomogenized cream of 30% fat pasteurized at 105–135°C.[54] A similar phenomenon was observed in high (up to 10.5%) fat milks processed with injected steam at 138–149°C.[55] Thickening and gelation were related to process residence time (3.4–20.3 s), storage temperature (4–40°C), time (up to 60 weeks), and fat level, but not to process temperature.[55]

Chocolate milk

Three types of textural instability were reported to occur on storage: sedimentation of cocoa particles, formation of large flocs, and segregation (i.e., formation of light and dark colored layers).[56] Rheological measurements characterized chocolate milk as a fluid with a weak network formed by an interaction of milk proteins with the stabilizer, such as carrageenan. The cocoa particles are incorporated in the network structure. Segregation was reported to be due to gravitational forces[56] and could be prevented when the product of the elastic modulus multiplied by the deformation gradient of the network exceeded their magnitude.

Milk powder

Dried milk usually contains 3–4% moisture and 38% (whole milk) or 50% (skim milk) lactose, which exists in glassy form. When its equilibrium relative humidity (ERH) is exceeded on storage by the relative humidity (RH) of the surrounding atmosphere, milk powder will absorb moisture and become sticky. Adherence of the particles to one another will lead to caking. Solid lumps will form when the moisture content reaches about 9%, causing crystallization of lactose.[52]

Eggs

Immediately after laying and during storage, eggs lose CO_2 by diffusion through the shell, making them more alkaline and affecting the proteins. This causes the yolks and whites of such eggs to spread over a wide area when the shell is cracked and lowers their quality.[46]

Freezing of liquid whole eggs is a common method of preservation. Pasteurization prior to freezing was reported to cause the product to undergo a change in flow behavior on storage, from Newtonian to Bingham plastic.[57] This means that molecular changes leading to yield formation occur in storage of pasteurized eggs. Because this yield stress has to be overcome before the product can flow, this textural change will make more difficult the removal of eggs from their container. Apparent viscosity was reported also to increase with pasteurization and frozen storage.

Whole eggs and egg whites do not require any additives prior to freezing. This is not the case with egg yolks, which will undergo gelation on freezing, unless salt or sugar is added. Gelation produces a product that is thick and gummy, and difficult to handle.

Fruits and berries

The botanical definition of fruits and berries is very complex and not always clear.[58] The common usage of the term "fruit" reflects the definition in Webster's dictionary as "a succulent plant part used chiefly in a dessert or sweet course," and of "berry" as "an edible fruit (as a strawberry, raspberry, or checkerberry) of small size irrespective of its structure." These products are highly valued for their appealing sensory properties (aroma, color, taste, and texture) and for their vitamins (especially C and A) and fiber.

From the food scientist's viewpoint, the structural feature of fruits and berries is the parenchyma, the tissue in which the plant deposited the nutrients. Parenchyma has a very acceptable texture and is easily broken down to release its aqueous contents. The morphology and chemistry of this tissue play an essential role in determining its texture. They will be described here only briefly and in a simplified manner to facilitate the understanding of textural changes on storage. For a deeper treatment, reference is made to the reviews by Reeve,[59] Northcote,[60] Van Buren,[61] and Ilker and Szczesniak.[62]

Parenchyma cells are confined by a thin primary cell wall consisting of an organized network of similar quantities of pectic substances, hemicelluloses and cellulose, and a small amount of proteins. Some plants and some woody tissues develop lignified secondary cell walls; these are virtually absent from mature fruits. In a very simplified sense, the cellulose in the primary cell wall provides rigidity and resistance to tearing, while the pectic substances are responsible for plasticity and the ability of the cell wall to stretch.

The middle lamella, composed of pectic and other polysaccharides, cements together the walls of adjacent cells. It is sometimes considered to be an extension of the primary cell wall lacking the cellulose fibrils. The living parenchyma tissue is rigid because of the mechanical strength of the cell wall and because of turgor pressure (i.e., the osmotic pressure exerted against the wall by the aqueous cell contents) that keeps it extended. The turgidity of the cell plays an important role in the maintenance of crispness.

Pectic substances form the basic structure of parenchyma cells and have been studied extensively.[63] They consist of chains of galacturonic acid residues linked by α–1,4-glycosidic bonds; normally, 50–90% of the carboxyl groups are methylated. Attached to the main chain is a wide array of side chains composed of neutral sugars (galactose, xylose, arabinose), which may merge with the hemicellulose portion of the cell wall.[61] Enzymes capable of catalyzing the demethoxylation and depolymerization of pectic materials are widespread in fruits and vegetables. The depolymerizing enzymes are considered to be mainly hydrolases causing rupture of the glycosidic bond between the uronide residues. There is some evidence for the presence of enzymes that catalyze the splitting of the chain by β-elimination.

Loss of intermolecular bonding between the cell wall polymers and changes in the amount and character of the middle lamella leading to increased solubility of pectic substances are regarded to be the causes of tissue softening and loss of cohesiveness occurring on fruit ripening and senescence. A weak lamella causes separation between the cells, while a strong lamella causes tissue failure across the cell wall. The lamella is believed to exist as a polyelectrolyte gel with divalent ions (Ca^{2+} and Mg^{2+}) that provide bridges and facilitate the organization of the pectic structure into fibrils. Chelating agents (phytic acid, citric acid) that remove the divalent ion or monovalent salts (Na^+ and K^+) that substitute for the divalent ions lead to textural softening. Pectin demethylation, which provides additional sites for calcium bridges, can lead to textural firming.

Water is the main component (80–90%) of the parenchyma tissue. Much of it is present in the vacuoles (i.e., fluid reservoirs containing various solutes), but some is located in the cell wall. It has been suggested that its role is as a wetting agent, preventing direct H-bonding between polymers, as a stabilizer for the conformation of polymers, and as a medium for enzymatic activity.[60] Water also acts as a structural component in the matrix gel.

Table 1   Storage Life of Fruits

Fruit	Approximate Storage Life
Fruits That Soften Greatly	
Apricots	1–2 weeks
Blackberries	2–3 days
Blueberries	2 weeks
Raspberries	2–3 days
Strawberries	5–7 days
Cherries, sweet	2–3 weeks
Figs	7–10 days
Nectarines	2–4 weeks
Peaches	2–4 weeks
Plums	2–4 weeks
Fruits That Soften Moderately	
Apples	3–8 months
Cranberries	2–4 months
Quinces	2–3 months

From Bourne, M.C. J. Texture Stud. 1979, 10, 25-44. With permission.

Fresh

Physiological and chemical changes following harvest of perishable fruits and vegetables have been the subject of numerous studies and are treated in detail in Chapter 2.

Storage has very interesting effects on the texture of fresh fruit because it can be beneficial or detrimental, depending on fruit type and the point in the life of the product. Fruit that on ripening has a delicate texture and is easily damaged on handling (e.g., peaches) is harvested mature but unripe. Ripening (i.e., softening of texture and development of optimum flavor and color) is allowed to occur on storage, which then has a beneficial effect. This ripe fruit easily goes into the overripe stage, and continuing storage will have a detrimental effect on texture. Such fruit has a soft, melting texture at the peak of its quality, which changes rapidly into an overly soft, mushy texture. Fruit that on ripening is still firm and fairly resistant to mechanical stress (e.g., apples, oranges), for the most part, will not rely on storage for significant beneficial textural effect, although some desirable reduction in hardness may occur. Fruit that softens greatly on ripening has a much shorter life unless, as is the case with pears, it is kept in controlled atmosphere storage that delays the onset of ripening.[64]

Bourne divided temperate fruit into two texture groups: (1) those that soften greatly as they ripen and (2) those that soften moderately as they ripen.[58] Table 1 lists the typical fruits within each of these categories and their approximate storage life. Table 2 summarizes the changes in texture on storage of ripe fruit resulting in overripeness and, ultimately, senescence with the accompanying deterioration in texture.

The chemistry of textural changes in fruit during storage has been reviewed by Bartley and Knee.[65] The most important textural change, softening of the tissue, is brought about by the structural alterations in both the middle lamella and the primary cell wall caused by enzymatic degradation and solubilization of pectic materials. These lead to cell separation and decreased resistance to applied forces. The observed physical changes[66] primarily involve moisture loss and redistribution leading to a shrivelled appearance and texture deterioration. Moisture loss is slowed down by the waxy skin coating on some fruits (e.g., apples).

Table 2 Some Textural Characteristics of Stored Fruit

Ripe	Overripe/Senescent
Fruits That Soften Greatly on Ripening	
Soft and yielding	Excessively soft, may not retain its shape
Juicy and/or pulpy	Watery (some may be dry)
Forms a pulpy bolus on mastication	Little or no resistance to mastication
Fruits That Soften Moderately on Ripening	
Firm, crisp, fracturable	Not crisp, flabby
Juicy, not pulpy	Mealy, dry
Cellular	Cells less noticeable

The rate and extent of textural changes is controlled by keeping the fruit in modified or controlled atmosphere storage (Chapter 16). Because of its high economic importance, the apple has been the subject of most studies. Low oxygen levels were demonstrated to reduce its softening on storage.[67] Several investigators have demonstrated that a postharvest dip in sucrose polyester solution combined with refrigerated storage may offer the same benefit as controlled-atmosphere storage.[68] The effect, however, appears to be somewhat inconsistent and cultivar-dependent.

Cold storage will slow down the metabolic processes in overripening, but could result in tissue injury with cold-sensitive fruit (e.g., banana). Common symptoms of injury include collapse of cells below the skin surface leading to pitting, enzymatic browning, and waterlogging. The mechanism is believed to involve the solidification of membrane lipids, the critical temperatures for which depend on whether the fruit is of tropical or temperate origin. Chilling injury occurs at higher temperatures (10–15°C) for tropical fruit whose lipids may be expected to have a higher solidification temperature. Immobilization of lipids will affect membrane properties. It has also been suggested that toxic compounds, such as ethanol and acetaldehyde, which can accumulate at low temperature and lead to cellular disruption, play a role in chilling injury.[69,46] In peaches, the internal breakdown on cold storage results in a dry and mealy or wooly texture.[70] The molecular basis for this defect, expecially the involvement of pectic substances, has been studied,[71] but a clear understanding has not yet been developed. A detailed treatment of the biochemistry of chilling injury is presented in Chapter 2, including the potential of increasing the tolerance to low temperature through genetic engineering.

Mechanical damage

Serious textural damage can occur to fruit tissue on slow compression (as in bulk storage) or impact compression (as in drop on handling). Apples and citrus fruit (oranges and grapefruit) have been the main subjects of such studies because of large volumes shipped, long storage, and frequent transport over long distances made possible by their relatively resistant texture. Biochemical and engineering approaches have been used to study this textural damage with the ultimate objective of designing better packages and handling procedures. A suitable package will cushion the fruit and absorb much of the mechanical energy that causes fruit injury.

Tissue failure leading to textural damage can occur as cleavage (i.e., cracking) caused by normal stresses and as slip or bruising caused by shear stress. Tissue resistance to normal and shear stresses may vary with ripening conditions and fruit type.[72] It has been reported that, in contrast with most other fruit, the strawberry is more easily damaged by slow, than by impact, compression.[73] Apples appear to be more susceptible to impact injury after cold storage.[74] In 18 weeks of cold storage, tensile strength decreased from 200–300 to 10 kPa, and fracture resistance from 200–300 to 10 Jm^{-2}.[75] Cleavage depends on the toughness of the tissue structure (i.e., the energy required to break across a given cross section). Slip causes cell separation along defined surfaces with relatively no damage to the tissue on either side of the fracture. Bruising is initiated by the cells bursting when the shear stress exceeds their mechanical strength (i.e., when it is greater than the yield stress). A theory that predicts the tissue failure mode has been proposed by Holt and Schoorl.[72] A discussion of the mechanical properties of agricultural products having a bearing on this subject can be found in the texts by Mohsenin.[76,77]

Mechanical load causes visible bruising in apples and internal injury culminating in fruit rupture in oranges. It has been shown by Peleg and Calzada that fruits and vegetables undergo substantial irreversible deformation when subjected to mechanical forces of certain magnitude.[78] Impact damage is usually studied by dropping the fruit on different surfaces and evaluating the bruise volume, which is a quantitative measure of the amount of tissue softening, and the discoloration caused by cell bursting. Impacting the fruit with a pendulum is also practiced. Cellular softening is caused by mechanical damage followed by enzymatic degrading of the cell wall. Browning is an oxidative reaction involving phenolic compounds.[79]

Strong correlations were obtained between bruise volume and energy absorbed on impact. These correlations indicate that bruise formation dissipates the absorbed kinetic energy and could be minimized by using packaging material that could absorb that energy instead. The energy dissipative nature of bruising accounts for the distortion and bruising of cells. The severity of bruising in apples can be predicted by a multiple regression model involving the parameters of fruit firmness, maximum acceleration, total velocity change, drop height, and apple diameter,[80] and by a model based on the Fourier transformed acceleration data.[81] Mathematical modeling of the mechanical properties of apple flesh under compressive load has also been done.[82]

Peaches are susceptible to mechanical damage by impact, vibration, and compression, each causing somewhat different bruise types. Impact bruises appear as tear lines and voids suggesting a brittle fracture. They have a caret shape and occur under an undamaged area beneath the skin. Vibration bruises relate to the viscoelastic nature of peaches and originate at the skin. Compression bruises originate at the pit with browning and sometimes fiber tearing. They are located at the opposite sides of the pit and follow the direction of the applied force.[83] Bruise susceptibility of peaches increases with maturity and storage time.[84] Preharvest cultural practices, especially irrigation, have been reported to affect textural quality and bruise resistance, but the data are inconsistent and may depend on fruit cultivar and interrelationships between various factors.[85]

In citrus fruit, impact loading leads to oil sac rupture and subsequent skin discoloration, juice sac crushing, and voids in the flesh. Ultimately, rupture or bursting of the fruit may occur. This damage results in softer texture as the cells lose their turgidity and spew out their fluid contents. It also results in microbial decay and serious quality deterioration when the product is intended for the fresh market. Firmness decreases and nonrecoverable (upon load removal) deformation increases with drop height and number of drops.[86]

An example of the application of the mechanical analysis to the evaluation of the effect of the shipping container on fruit damage is provided by the work of Chuma et al. on oranges.[87] They used a modified Burgess rheological model with a slider to explain the time-deformation curve under static load and defined the maximum safe stacking height. Oranges were found to suffer permanent deformation under a 2 kg per fruit load and to rupture when the load exceeds 4.2 kg per fruit.

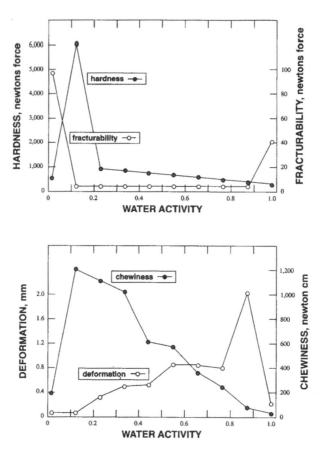

Figure 5 Effect of a_w on the textural parameters of apples. (From Bourne, M.C. J. Texture Stud. 1986, 17, 331-340. With permission.)

Processed

Fruit is preserved by dehydration, canning, and freezing. Two types of dehydrated fruit are available commercially. The so-called dried fruits (raisins, figs, dates, and sliced apples, pears, apricots) have a moisture content of about 23–26% and an a_w of about 0.75. Thus, they are considered intermediate moisture (IM) foods. They are often sun-dried and are valued as high-energy snacks. The slow removal of moisture causes collapse of the cellular structure. Texture-wise, dried fruit is pliable, cohesive, chewy, and somewhat adhesive because of concentrated sugars. It lacks crispness, or fracturability, and juiciness of its fresh counterparts. The so-called dehydrated fruits (sliced apples, pears) are of a lower moisture content (2.5–3.5%, $a_w < 0.1$) and are often prepared by freeze-dehydration. This method, which involves quick removal of moisture by sublimation of ice, avoids structure collapse and produces an open, porous, firm, and crisp texture. In contrast to the fresh fruit that is wet/crisp, freeze-dried fruit is dry/crisp (somewhat similar to a cracker). Instead of releasing moisture and being juicy in the mouth, it absorbs moisture from the mouth and is drying. Neither dehydrated nor dried fruit can replace fresh fruit in its usage and appeal. They are used primarily as snacks or recipe ingredients, with or without prior rehydration.

Figure 5 illustrates what is happening to hardness, fracturability, chewiness, and deformability along the a_w or moisture content spectrum. This information has a bearing on the effect on texture of moisture loss from fresh fruit (the right side of the spectrum) and of moisture pick-up by dehydrated fruit (the left side of the spectrum). Both can occur on storage. Even a small loss of moisture from fresh fruits reduces crispness (fracturability) and increases the deformability of the fruit, due to the loss of turgor. Moisture pick-up by the dehydrated fruit also reduces crispness and increases hardness and chewiness, due to the plasticizing effect of moisture.[88]

Commercial or home canning is a popular way of preserving fruit. Because it involves heat processing, the resulting texture resembles that of cooked fruit in terms of cellular disintegration and softening. The more delicate the cellular structure and the thinner and weaker the cell wall, the more textural damage will occur on canning. Strawberries will suffer much damage and pineapple slices only little. Textural damage on canning manifests itself in the development of softness, a process that sometimes continues on storage. The problem appears to be pronounced with apricots and has been the subject of several studies. All fruits benefit from a pretreatment with calcium, which makes the tissue firmer by strengthening the cell wall. The extent of the effect depends on fruit type and its physiology. A firming calcium dip is a standard practice in commercial production of apple slices for use in apple pie. In contrast to apples, apricots have a very low pectinesterase activity and, thus, little low methoxyl pectin with which calcium could react to produce a firming gel.[89] However, addition of calcium to the canning liquid does have a beneficial effect on apricot texture. The effect may be counteracted on storage by the presence of chelators, such as citrate ions, that remove the structural calcium from the cell wall once the membranes have been lysed by heating.[90] The action of pectolytic enzymes of mycotic origin, not destroyed in the heat treatment normally practiced in canning, has been reported as a possible cause of softening in canned apricots in Australia.[91]

Although freezing preserves more of the original sensory attributes of fruit than drying or canning, its effect on textural degradation has been well recognized.[92] It is due to the loss of turgor caused by intra- and extracellular ice formation and membrane injury by solute concentration (see section on fish for a more detailed treatment of this topic). Little texture deterioration on freezing is suffered by blueberries and cranberries, because of their high solids content and lack of organized internal structure. The extent of textural degradation is diminished by rapid freezing rates, low freezing temperatures, and addition of sugar syrup. It is increased by fluctuating storage temperatures (see Chapter 15 for more details) such as are normally encountered in transport and home freezers.

Textural degradation of fruits and berries upon freezing and frozen storage is caused by the process disrupting the cells and all their life-supporting, protective mechanisms. As with canning, the more delicate the structure, the more damage the cell suffers. Loss in the mechanical strength of the cell wall causes loss of crispness and firmness and leads to excessive softness. Loss of cell membrane integrity causes seepage of cell fluids, which leads to excessive drip. Strawberries, which contain 90% water, suffer extensive damage to their structure on freezing and thawing (Table 3) as determined by histochemical, trained profile panel, and instrumental assessment.[93] Much of the damage is said to occur on thawing.[59]

The effect of freezing/thawing on the texture of fruit is so pronounced that it commanded the primary attention of research done on the effect of frozen storage. Also, in contrast to the bulk storage of fresh fruit, the storage conditions of frozen fruit are not controllable, since most of the storage is outside the jurisdiction of the processor. Thus, any strategies for quality improvement would be difficult to implement.

Meat

The animal flesh used for food is called "meat" by consumers and "muscle tissue" by food scientists. Texture is its main quality characteristic; it affects its price as well as

Table 3 Textural Changes in Strawberry Fruit on Ripening and Processing as Determined by Sensory, Instrumental, and Microscopic Methods

Changes On	Method of Evaluation		
	Sensory	Instrumental	Microscopic
Ripening and overripening	decreased firmness decreased cohesiveness decreased fibrousness decreased crispness increased rate of juice release ripe berries break down into a liquid	decreased firmness decreased crispness increased juiciness larger percentage of samples showing adhesiveness	thinning of cell wall some folding of cell wall cell elongation and expansion cytoplasm disarrangement liquefaction of cell contents degradation of pectin thickening and deposition of lignin in xylem vessels
Freezing and thawing	onset of softness limpness wet, easy juice release toughening of fibers increased slipperiness loss of cellularity, onset of mushiness	decreased firmness decreased crispness greater juiciness faster rate of liquid release	some disorientation of cellular organization cell plasmolysis folding of cell walls, some rupture disorganization of cytoplasm
Freeze-drying and rehydration	greater softness limpness no apparent juice release loss of fibrousness greater slipperiness greater mushiness sponginess	decreased firmness loss of crispness, increased adhesiveness lower juiciness lower rate of liquid release	complete disorganization cell plasmolysis much rupture of cell walls, degradation of cellulose further disorganization of cytoplasm leaching of lignin from xylem elements

From Szczesniak, A.S.; Smith, B.J. J. Texture Stud. 1969, 1, 65-89. With permission.

consumer acceptance. The textural parameters of prime importance are tenderness and juiciness. *Tenderness* refers to the ease with which the meat yields on chewing and how much energy is required to masticate it to the state ready for swallowing. *Juiciness* refers to the amount and manner of liquid (water and liquefied fat) released on mastication. Of all the food products, meat has received the most attention from the textural standpoint.

Although much knowledge has been gained, the detailed mechanism of tenderization has not yet been uncovered. It is generally agreed that tenderness is influenced by factors in the raw material (animal type and breed, age, nutritional regime, degree of marbling, etc.), muscle contraction (cold shortening, rigor shortening), amount of connective tissue, ripening, and cooking treatments. An extensive review of factors affecting meat tenderness and of methods of measurements was published by Szczesniak and Torgeson.[94] Recently, a new dimension has been added to meat texture research by the technological require-ments of meat patties and restructured steaks.

Even more so than with other products, the texture of meat cannot be discussed without first establishing its structural basis. Excellent reviews on the subject have been published by Stanley,[95–97] to which reference is made for details.

For the most part, the term "meat" refers to skeletal muscles. In contrast to plant materials, where the cells are round or oblong, muscle cells are extremely long, 10–100 μm thick, polynucleated, and exhibit striations (i.e., alternate light and dark crossbonds). Because of their geometry, they are usually referred to as "fibers".[97] Muscle fibers are composed of many smaller myofibrils built up from sarcomeres (i.e., repeating transverse subunits, about 2–3 μm long in live muscle) joined at the Z-line and lying parallel to the major fiber axis. Two contractile, rod-shaped proteins, actin and myosin, occur in the myofibrils. The sheath surrounding each fiber is called the sarcolemma. In addition to the myofibrils, the second major structural feature affecting meat texture is the connective tissue. Composed mostly of collagen in the form of highly branched fibers, the connective tissue forms a network giving support to the muscle structure. It covers the sarcolemma of muscle fibers (endomysium), bundles of fibers (perimysium), and the entire muscle (epimysium).[96]

Fresh

Muscle is converted to meat through the post-mortem biochemical changes. Beef has been the subject of most research. However, with the current trends toward reducing the intake of red meats, increased research activity is noted on storage-induced changes in pork and poultry meat. In general, these are similar to those occurring in beef and will not be discussed in detail.

The effect of storage on the texture of muscle may be divided into three stages: (1) prerigor, where toughening due to cold storage shortening occurs, (2) rigor mortis, where substantial toughening occurs due to extensive biochemical changes, and (3) aging, where the meat relaxes and develops the desired tenderness. Perhaps a fourth stage, that of spoilage, should be added. However, as mentioned in the Introduction to this chapter, meat signals its spoiled state primarily through the olfactory sense, and consideration of textural changes would be irrelevant.

After slaughter, the meat is usually chilled to control microbial spoilage. If the tem-perature is reduced below 15°C while the meat is still in the prerigor state, toughness develops because of muscle contraction and shortening of the sarcolemma. The lower the storage temperature, the greater the shortening. Restraining the muscles on the carcass and avoiding rapid chilling will prevent the contraction of muscle fibers. Cold shortening can also be prevented by the electrical stimulation of the carcass immediately after death. This accelerates the rate of glycolysis, causes the pH to fall below 6.4, and shortens the time to the onset of rigor mortis. There is no agreement on the mechanism of cold shortening, although it is generally accepted that its cause is the stimulation of adenosine-5-triphosphate (ATP) breakdown. Release of calcium ions, which enhances the contractile actomyosin ATPase, has been implicated.[98] Evidence is available that white muscles (i.e., those with less myoglobin) are less prone to cold shortening than red muscles. In the bovine species, younger animals, which contain fewer red muscles than older animals, exhibit less cold shortening.

Rigor mortis is the stage where the muscles become stiff and inextensible. The length of this stage depends on the species. During rigor, glycogen is converted to lactic acid until the pH of 5.4–5.5 is reached when glycolytic enzymes become activated. This pH is also the isoelectric pH of myofibrillar proteins at which their shape is less extended and their water-holding capacity is diminished. The mechanism of rigor involves disappearance of ATP and creatine phosphate, which act as plasticizers between myosin and actin filaments. Their absence prevents these filaments from sliding past one another and causes muscle rigidity[98] (see section on fish and Chapter 2 for additional treatment of the subject).

Following rigor, tenderness of meat is achieved on aging through the enzymatic action during the process of autolysis. Beef is usually aged for 10–14 days at 5°C. Pork, lamb, and poultry generally are not aged because toughness is not a significant factor in their case.[95] Figure 6 shows the effect of beef aging on tenderness as determined sensorially and instrumentally. Figure 7 illustrates the structural changes believed to be responsible for the tenderizing effect. It will be seen that at the start of the aging process the myofibrils are intact and crossed by continuous and prominent transverse elements at the sarcome ends (Z-line). These elements are less pronounced by the sixth day. A general deterioration of the muscle microstructure is seen by the 12th day.[97] A decrease in binding between actin and myosin is another well-documented change on aging. Differences among muscles in their behavior on aging have been reported.[99]

The following enzymes appear to be involved in meat tenderization on aging[95]:

- endogenous catheptic enzymes leading to structural breakdown
- a specific Ca-activated enzyme weakening the Z-lines
- a proteolytic enzyme fragmenting collagen
- nonspecific proteases of bacterial origin.

Lanari et al. studied aging in vacuum-packed raw and cooked Argentinian beef and found the process to follow first order kinetics.[100] The activation energy for raw beef was 45–47 kJ/mol and for cooked beef 62–66 kJ/mol, depending on whether a shear or a tensile test was used to quantify tenderness (Figure 8).

Frozen

There are two kinds of textural changes on storage of frozen meat: those induced by a faulty freezing process and those induced by the ice crystals forming in the product. In the first category is the freezer burn, the phenomenon of surface desiccation developed on storage due to a condensed layer of muscular tissue near the surface produced by blast freezing of unprotected meat. It manifests itself as whitish or amber-color spots on the surface of frozen meat,[98] especially evident on poultry. They are caused by the sublimation of ice crystals into the atmosphere of the cold room.

The main textural defect of properly frozen meat is the "drip," i.e., the reduced water-holding capacity of the thawed product due to the damage caused by the formation of large ice crystals. These are produced during slow freezing or upon subsequent storage at high or fluctuating temperatures from recrystallization seeded by small ice crystals formed

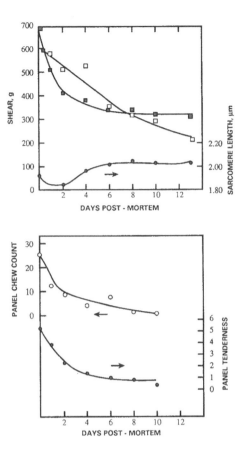

Figure 6 Effect of postmortem aging on tenderness of beef *psoas major* muscle. (From Stanley, D.W.; Tung, M.A. in Rheology and Texture in Food Quality. ■, ● — uncooked; □, ○ — cooked. deMan, J.M., et al., Eds.; AVI Publishing, Westport, CT, 1976, Chapter 3. With permission.)

during fast freezing. In meat, formation of large ice crystals is particularly damaging, because of the barrier effect of the fibers constraining the free displacement of water.[101] The effect of ice crystal size on textural deterioration of frozen beef has been researched by Martino and Zaritzky,[101] who developed a kinetic model for ice recrystallization and proposed a theoretical equation to predict the limit ice diameter applicable to any product.

The value for beef was calculated to be 61.63 ± 0.83 µm. It depends only on the tissue structure. The presence of ice crystals leads to a redistribution of water and to protein–protein interactions that result in the elimination of water from the tissues as exudate. The time at which the maximum amount of exudate is produced has been calculated to be very close to the time at which the ice limit diameter is reached.[101] Drip is reduced by the high ultimate pH in meat and a fast freezing rate.[98]

Toughening often occurs on thawing of meat that has been rapidly frozen without prior chilling. Called "thaw rigor," it is caused by marked shortening and moisture exudation resulting from a high ATPase activity.[98] This phenomenon appears to be related primarily to the biochemistry of prefreezing meat and is unrelated to storage.

It is currently accepted that myofibrillar proteins are denatured during frozen storage of meat. This process has been researched in greater detail with respect to frozen fish muscle. More recently, protein denaturation in bovine muscle has begun to be studied. Wagner and Añon reported that the denaturation of myofibrillar proteins on storage of frozen beef proceeds as two consecutive first-order reactions.[102] The first, faster reaction

(A)

(B)

(C)

Figure 7 Scanning electron microscope micrographs of beef *psoas major* muscle aged for (A) 0 days, (B) 6 days, and (C) 12 days. (From Stanley, D.W.; Tung, M.A. in Rheology and Texture in Food Quality. deMan, J.M., et al., Eds.; AVI Publishing, Westport, CT, 1976, Chapter 3. With permission.)

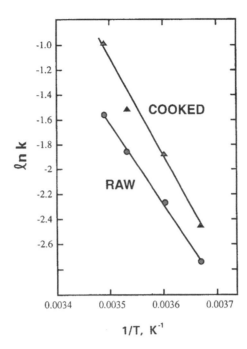

Figure 8 Effect of temperature (0–13°C) on the rate constants of the tenderizing process on aging of vacuum-packed cooked and raw *semitendinosus* bovine muscle. (From Lanari, M.C., et al. J. Food Process. Preserv. 1987, 11, 95-109. With permission.)

involves denaturation mostly of the myosin head, although some denaturation of the myosin tail also appears to occur. This stage, which begins during freezing, manifests itself in the diminished ATPase activity in the presence of Ca^{2+} and Mg^{2+}. The second, slower reaction involves aggregation of the denatured protein. A high correlation ($r = 0.93$) was obtained between myofibrillar ATPase activity and the enthalpy of thermal denaturation of frozen beef muscle stored at different temperatures. In general, the rate of the denaturation reaction increases with higher storage temperatures.[102]

Intermediate moisture

Lowering the water activity (a_w) to below that supporting microbial growth by infusion of water binders (e.g., glycerol, NaCl, sucrose) or by simple water removal has been researched for several decades as a promising approach to preserving meat for storage at ambient temperatures. Storage at tropical temperatures is of particular interest, especially for the military.

It has been reported that storing precooked intermediate moisture (IM) beef of $a_w = 0.83$ at 38°C resulted in textural degradation due to simultaneous protein breakdown and crosslinking, the latter reaction being overriding.[103–104] Similar results were obtained with other meat types (pork, chicken, goat, ewe), with the reaction rates being muscle type and source dependent.[105] The samples were prepared by desorption of water through soaking in glycerol. An extension of these studies confirmed textural degradation using instrumental methods of texture characterization.[106] Hardness in compression, shear force, and adhesion between fibers decreased on storage, with the decrease being greater for samples stored under aerobic conditions which promote protein denaturation. The storage life of IM meats stored at high temperature, however, appears to be limited by browning rather than by textural changes. It was reported to vary from 3 to 12 weeks depending on the sample.[105]

Different results were obtained when precooked IM beef of lower a_w (0–0.75), and prepared by humidifying the freeze-dried material, was stored at about 38°C and 4°C for up to 6 months.[107] Hardness and chewiness were reported to increase at the higher storage temperature to a maximum at 5 months and to decrease after that. These changes were minimized at the low storage temperature and for lower a_w for shorter storage times. The plate temperature at which the freeze drying was accomplished affected the texture of the stored product. It was also found that chewiness, being the product of hardness, cohesiveness, and springiness,[2] changed more than hardness on storage. The method of sample preparation (desorption vs. adsorption) might possibly account for the difference in results obtained between this study and those discussed in the preceding paragraph. Changes observed on storage of IM beef prepared by the adsorption technique are similar to those reported to occur in stored freeze-dried beef.

The texture of precooked IM beef over the entire range of a_w was studied by Kapsalis et al. using mechanical measurements and thermodynamic theory to gain an insight into structure alterations due to moisture removal.[108] Compression testing revealed important textural changes in the range a_w = 0.15–0.30, while cutting/extrusion testing showed a maximum force reading at a_w = 0.85. The effect of storage was not considered in this work. Another report from the same senior author revealed that samples of a_w = 0–0.66 stored for 5.5 months at 20°C exhibited hardness and cohesiveness that increased with a_w. Again, the textural characteristics were not related to storage.[109]

Currently, there is practically no interest in IM products, due to flavors imparted by water binders and chemical deterioration on storage.

Freeze-dried

Textural changes can occur on storage of freeze-dried meat, especially at high temperatures. The rehydrated product exhibits increased toughness and decreased juiciness.[110] It has been demonstrated that a Maillard-type reaction is responsible for this deterioration.[111,112] It causes myosin to crosslink and aggregate. A negative correlation has been shown between the rehydratability of freeze-dried meat and the nonoxidative browning.[112] It has been suggested that toughening of freeze-dried meat due to the formation of myosin aggregates may be minimized by treatment with hydrolyzed vegetable protein, N-acetyl-L-lysine, which inhibits the crosslinking by competing with the ϵ amino group of lysine residues in myosin,[111] or by bromelain which degrades myosin leaving actin essentially intact.[113] Heat applied during freeze-drying to accelerate moisture removal may induce crosslinking and protein insolubility in the dried portion thus causing texture degradation.[110]

Meat products

Of the different commercial meat products, frankfurters are economically the most important in the U.S. They are prepared according to the following key steps:

1. *Chopping of meat*, which destroys the cellular structure, puts the connective tissue proteins in suspension, sarcoplasmic proteins in solution, and actomyosin in a sol state.
2. *Chopping with adipose tissue*, which suspends fat particles less than 200 μm in the water/protein matrix.
3. *Stuffing*, which immobilizes the meat batter in the casing.
4. *Cooking*, which forms an ordered three-dimensional protein network.
5. *Smoking*, which introduces a desirable flavor and aroma.

It is generally recognized that myosin gelation is the key factor in producing the desired textural properties in frankfurters in specific, and in meat products in general. Thus, any changes on storage of fresh or frozen meat leading to myosin denaturation/aggregation will be detrimental to the texture of frankfurters. In a recent review on factors that control muscle protein functionality, Whiting stated that "prerigor muscles have superior water-holding capacities and batter-forming abilities because their myosin is readily extractable at the higher pH values and ATP contractions."[114] Batter stability requires some water and salt addition for myosin extraction and binding characteristics. Process specifics are based on experience and preference of individual processors.

Cocktail frankfurters are typically packed in a vinegar pickle, which causes texture deterioration on storage. The frankfurter emulsion can be stabilized against acid deterioration by incorporation of anionic gums, especially xanthan.[115]

Seafood

Fish

Fish meat consists of muscle sheets extending from head to tail on both sides of the body. These long muscles are divided into segments by transverse sheets of uniformly distributed connective tissue, forming a structure very different from that encountered in meat of terrestrial animals. The amount of connective tissue in fish is much less than in warm-blooded animals (about 3%). This lack of connective tissue, coupled with the high thermal instability of its collagen, accounts for the delicate, tender texture of cooked fish. Fish collagen liquefies readily on heating, thus causing the connective tissue to lose its binding power. The structure of muscle fibers (i.e., the basic elements of the musculature), however, is quite similar between fish and warm-blooded animals.[116]

A quantitative study of the relationship between the collagen content of raw and cooked dorsal muscles of five fish species and instrumentally determined firmness showed that connective tissue contributes to raw fish texture, while the muscle fiber characteristics define the cooked fish texture.[117] It has also been shown for squid mantle that collagen contributes to tensile strength in the longitudinal direction, while muscle cells contribute to it in the transverse direction.[118]

Fresh

As soon as possible after the catch, fish is placed on ice and stored at that temperature for the fresh market to minimize enzymatic and microbial quality deterioration associated with spoilage. Textural changes on cold storage occur in three consecutive stages: prerigor, rigor, and postrigor. They are influenced by the size, maturity, nutritional status, gross chemical composition, morphological structure, as well as the activity of the fish at the time of the catch. They are also influenced by the condition of death and post-mortem handling. Some of the morphological factors may govern the ratio of connective tissue to myofibrillar proteins.

The early post-mortem textural changes in fish are due to changes in the physico-chemical state of the myofibrillar proteins. The muscles of freshly killed fish are soft, plastic, and extensible.[116] If the fish did not struggle much during the catch and still has some glycogen reserves, the pH may fall in the first several hours post-mortem due to the anaerobic conversion of glycogen to lactic acid. As the pH drops and approaches the isoelectric point of the myofibrillar proteins, the altered charges on the protein chain lead to the tightening of the protein structure. Fish cooked at that stage will suffer increased cooking losses and will be tough and dry.[116] When much struggling occurs during the catch, fish will use essentially all of the glycogen in the muscles, which then will be unavailable for the outlined texture-altering mechanism. Such fish after few days of iced storage was shown to be softer after cooking than anaesthetized fish.[119] However, fish cooked immediately after catch will be very tough, because heating accelerates the development of rigor.[116]

The onset of rigor mortis causes the most severe textural changes in post-mortem fish. It usually reaches its peak in 1–2 days after catch and is accompanied by a contraction of the muscle leading to toughness. This is caused by the depletion of ATP, which results in formation of bonds between rods of actin and myosin. This reaction prevents myosin and actin filaments from sliding passively past each other and makes the muscle tough, hard, and inextensible.[116] However, the elevated temperature of cooking brings about a partial resolution of rigor and a decrease in toughness. In contrast to beef, fish muscles do not appear to exhibit cold shortening.

Biochemical changes in the postrigor period lead to softening and, when excessive, to mushiness and tissue disintegration. It has been reported that the softening is affected more by changes in the structure of muscle tissue than by changes in the proteins.[120] These are said to be caused by a rise in pH due to formation of basic compounds, such as trimethylamine and ammonia, and to bacterial action.[121] Bacteria living on the surface of the fish are accustomed to low temperatures and will grow in refrigerated storage.[116] Proteolytic breakdown of structural proteins due to endogenous enzymes may be expected and has been postulated to occur. The effect of high pH appears to be reversible upon soaking of the fish in low pH buffer.[122] However, no conclusive proof is available[121] on proteolysis occurring during storage of fish that is still sensorially acceptable.

Kuo et al. reported that, after more than 4 days in refrigerated (4°C) storage, the tensile strength of squid mantle decreased significantly in the longitudinal but not in the transverse direction.[118] This also resulted in a greater strength loss on cooking than observed for nonstored fish. It was concluded that refrigerated storage produces changes in collagen that make it more susceptible to high temperature. However, some collagen structure is still believed to remain. Montero and Mackie showed that in cod stored on ice degradation of collagen occurs with the development of rigor mortis and continues as the meat ages.[123]

A mechanical model has been proposed to characterize the texture of fish in rheological terms. It accounts for the effects of time and temperature and expresses toughening on rigor mortis as well as the postrigor softening.[124]

Frozen

Fresh fish is highly subject to microbial spoilage and, even when kept on ice, has very limited storage life. Potter cites 14 days for cod and halibut kept at 0°C, assuming that the fish were immediately iced and never allowed to warm up.[46] When stored at 16°C, the shelflife of this fish decreases to 1 day. Drying is an old, effective preservation method that, however, results in drastic and irreversible textural changes, making it currently unpopular. Dried fish does not resorb water and yields a firm, tough, and dry product when cooked. Freeze-drying, as opposed to air-drying, gives a better product but one that does not hold water well and has a poor texture. Smoking and canning yield very acceptable products that, however, have limited uses.

Freezing is the most important commercial process for preserving fish, especially in the fillet form, in technologically advanced countries. For retail trade, the fillets are wrapped or packaged individually. For wholesale and further processing, fish is frozen in blocks. The shelf life of frozen fish may be as long as 2 years,[46] depending on fish type (low-fat fish keeps better than high-fat fish), pretreatment, rate of freezing, and storage conditions. Texture and, to a smaller extent, color and flavor are the main quality attributes undergoing undesirable changes on storage of frozen fish. Because of its economic significance, the problem has been studied in some depth by physical chemists and food scientists and the reaction mechanism proposed. Cod has been used most widely in this research,[125-128] although other fish such as whitefish,[129,130] turbot,[130] ocean perch,[131] mackerel,[132,133] yellowtail rockfish,[134] cuttlefish,[135] ice fish,[136] and squid[137] were also studied. Several studies involved instrumental texture measurements (e.g., References 134, 127, 128), which provided useful quantification of textural changes. Minced samples were often employed to eliminate variations within and among fish.

Provided the product is protected from desiccation, the main textural change in frozen fish is toughening leading to a rubbery mouthfeel,[135,125] although some species develop an undesirably soft, mushy texture.[134] Dryness is often observed due either to dehydration or to loss of moisture binding ability. Working with frozen cod, Kelly observed that texture deterioration limits the acceptability of low pH fish, while off-flavor development limits the acceptability of high pH fish.[125] This limit of acceptability was supported by Kramer and Peters, who found a very good negative correlation between pH and instrumentally measured toughness of yellowtail rockfish.[134]

Although there is no firm evidence that the freezing temperature affects the rate of textural changes in fish kept in frozen storage,[126] it is generally accepted following the work of Love[138] that the rate of freezing has a significant effect on texture deterioration on storage. As will be discussed under the mechanisms of deterioration, a faster freezing rate is less damaging because it results in finer ice crystal size and less disruption to the structure. There is much evidence that texture deterioration is very dependent on the storage temperature. Because of its greater instability, the optimum frozen storage temperature for fish is lower (about $-30°C$ to $-23°C$) than for meat (about $-18°C$).[121] Texture deterioration and water loss occur faster at higher temperatures.[127,131,121] Fluctuations in storage temperature are very detrimental. As may be expected, the deterioration is a function of both temperature and time. Using the Instron to quantify instrumentally the textural changes, Kim and Heldman found that they correspond to first order reactions.[127] The changes in frozen cod were described as increases in toughness and decreases in cohesiveness (see Figure 9). Toughening was more temperature-dependent than cohesiveness, as evidenced by larger activation energy (25.4 ± 0.16 kJ/mole and 10.0 ± 5.25 kJ/mole, respectively).

Frozen storage was shown to affect the microscopic structure of fish meat. Thought to contribute to textural changes is the degradation of the sarcoplasmic reticulum that appears to hold together the individual myofibrils.[121] Reduction in interfilament spacing and disorders within and between myofibrils may account for the impaired ability to hold water.[139] Determination of centrifuge drip was suggested as a useful indicator of textural quality of frozen hake fillets. It correlated well with sensory assessment ($r = 0.86$) and the relationship was independent of frozen storage temperature.[140]

Textural changes on frozen storage are also influenced by the condition and freshness of the fish at the time of freezing and by prefreezing treatments. Beneficial effects have been reported for holding the fish in refrigerated seawater[134] and for dipping in sodium pyruvate[141] or in tripolyphosphate[130] prior to freezing. Addition of antioxidants, cryoprotective agents, or enzyme inactivation treatments appear to be beneficial as well as elimination of oxygen from the package,[133] vacuum packaging,[142] and prevention of moisture loss. The effect of cryoprotectants (polyphosphates or sugars) on preserving the quality of frozen minced fish was found to be enhanced by the presence of lecithin or vegetable oil.[143] The length of time that fish is held on ice prior to freezing has been the subject of several studies. Lowered sensory texture scores on subsequent frozen storage were shown to correlate significantly with decreased protein solubility that occurred during storage on ice.[144] A detailed treatment of the prevention of protein deterioration on frozen storage of fish can be found in the review paper by Sikorski et al.[145]

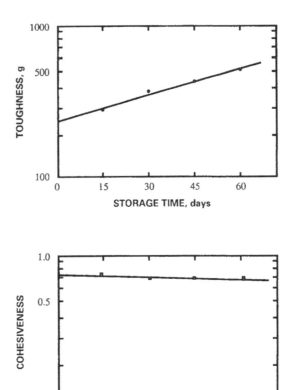

Figure 9 First-order plots for toughness (upper frame) and cohesiveness (lower frame) of cod muscle stored at –7°C. Rate constants: K(toughness) = 1.31 × 10^{-2} day^{-1}; k(cohesiveness) = 0.173 × 10^{-2} day^{-1} (From Kim, Y.J.; Heldman, D.R. J. Food Process. Eng. 1985, 7, 265–272. With permission.)

Mechanisms of deterioration

It is generally accepted that texture deterioration of fish flesh on frozen storage is caused by protein denaturation, particularly that of the myofibrillar proteins.[122,138,142] This is usually measured by the decrease in salt-extractable proteins. The subject has been studied extensively and reviewed expertly by Dyer,[146] Shenouda,[147] and Sikorski et al.[145] Shenouda grouped into three categories factors causing protein denaturation: (1) changes in fish moisture, (2) changes in fish lipids, and (3) activity of the trimethylamine oxidase. These categories are illustrated in Figure 10, which also outlines the possibility of secondary influences that, however, have not been studied.

It is well-recognized that structural damage on freezing is caused by ice crystals formation. Fast freezing leads to the intracellular formation of small ice crystals. In contrast, slow freezing leads to the intercellular formation of large ice crystals. The growth and size of ice crystals is increased by fluctuations in storage temperature. The state of fish at the time of freezing affects the location of ice formed. Prerigor fish will develop ice intracellularly regardless of the rate of freezing. Rigor fish, on the other hand, will develop ice inter- and intracellularly depending on the rate of freezing.

In addition to the physical damage due to the formation and growth of ice crystals, changes in moisture on freezing and frozen storage cause damage because of dehydration and increase in salt concentration. Ice formation robs the protein network of water, result-

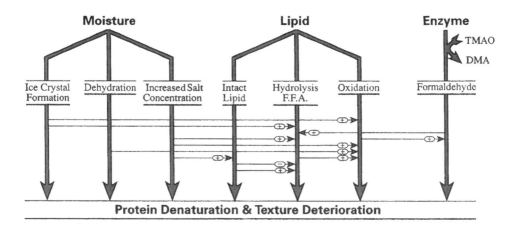

Figure 10   Factors that affect directly (vertical pathways) or indirectly (horizontal arrows) fish protein denaturation during frozen storage. (From Shenouda, S.Y. Adv. Food Res. 1980, 26, 275-311. With permission.)

ing in its disruption by weakening the system of hydrogen bonds. This weakening leads to the disintegration of the three-dimensional protein structure and to aggregation of the proteins. Consequently, the protein network cannot bind the water, which is held merely through capillary forces and is easily expressed by pressure from thawed fish.[138] It is believed that at common freezing temperatures ($-10°C$ to $-20°C$) there is a tenfold increase in the concentration of soluble solids.[147]

Myofibrillar proteins appear to be most susceptible to denaturation by solute concentration. Of these, myosin and actin were studied the most. Myosin is the most abundant and the most sensitive to freeze denaturation. Frozen cod was reported to suffer an 80% loss in myosin extractability, while under the same conditions actin solubility was changed little.[147] Actin undergoes polymerization in high ionic strength buffer solutions that leads to insoluble gel formation and to an increased ability to bind lipids forming insoluble lipoprotein complexes. Complexes with neutral lipids are stronger than those with polar lipids. This effect may be related to the greater textural stability of fatty fish and may explain the enhancing effect of lipids on the action of cryoprotectants.[143]

Free fatty acids (FA), derived from enzymatic and nonenzymatic hydrolysis of lipids located in the cell membrane, have a detrimental effect on the textural quality of frozen stored fish. This well-documented phenomenon is due to the insolubilization of myofibrillar proteins, primarily actomyosin.[147] Polyunsaturated and shorter FA are more effective in forming insoluble protein complexes. The free-FA–protein reaction appears to begin before freezing and is related to lipid hydrolysis in fish stored on ice. The reaction mechanism has not been totally explained but is believed to involve secondary forces. As postulated by Sikorski et al., the final result of the FA attachment to proteins is the creation of a more hydrophobic microenvironment around the protein surface, leading to a decrease in aqueous solubility or to an increase in further intermolecular linkage formation.[145] In fatty fish, the reaction of oxidized lipids with proteins (presumably through a free radical mechanism) contributes to texture deterioration. Oxidized lipids appear to attack specific susceptible functional groups on proteins entangling individual filaments into aggregates.[147]

The activity of trimethylamine oxidase (TMAOase) contributes significantly to textural deterioration of fish in the gadoid family (cod, pollack, haddock, whiting, hake, and cusk), which contain high amounts of this enzyme. Flatfish (plaice, flounder, sole, etc.) contain low amounts. Ocean perch and rockfish and some members of the flatfish group (halibut, some flounders) were reported to have none. The highest activity of the enzyme appears to be located in the species with the largest amount of dark lateral muscle in the fillets. Also, significant amounts of TMAOase are found in the eviscera, which present an important source of contamination: TMAOase catalyzes the reaction as follows:

$$\text{trimethylamine oxide} \rightarrow \text{dimethylamine} + \text{formaldehyde}$$

It is universally accepted that formaldehyde produced by this reaction leads to texture degradation in frozen fish even at low concentrations. Childs showed through both sensory and instrumental texture assessment that fish tissue containing formaldehyde became tougher and eventually rubbery with a sponge-like structure.[148] It had a decreased ability to hold water and felt dry in the mouth. Formaldehyde decreased the extractability of myofibrillar proteins; it affected myosin more than actin. The mechanism of formaldehyde action appears to be covalent bonding to various functional groups in the protein followed by crosslinking via methylene bridges.[147] A dissenting opinion was recently voiced by Yoon et al., who reported that removal of TMAO by washing did not prevent hardening of minced red hake fillets on frozen storage.[149] They believe that freeze-induced contraction and crosslinking of myofibrils is the main cause of hardening and that water-soluble sarcoplasmic proteins play a role in retarding this reaction.

Protein crystallization has been demonstrated to occur on processing of fish by freezing, cooking, or dehydration. Further increases in crystallinity were shown to occur during storage of frozen or dehydrated muscles.[150]

Shellfish

Information on textural changes in shellfish is less abundant and less focused. Similar to fish, but to a greater extent, the flesh of shellfish is highly perishable due to microbial spoilage. This spoilage can be prevented by keeping the crustaceans alive up to the point of cooking or freezing.

It is generally recognized that refrigerated postharvest raw shellfish will become soft and, with time, mushy due to proteolysis. This has been demonstrated for blue summer crab using both sensory and instrumental measurements.[151,152] Trawl-caught crabs deteriorated faster than pot-caught specimens, and postmolt crabs were more susceptible to mushiness than those between molt. The development of mushiness was related to the presence of heat-labile proteolytic enzymes in the hepatopancreas and could be prevented by adequate and timely cooking. However, deterioration in other sensory qualities made the shelf life of cooked, iced crab quite limited. A similar study on crawfish[153] also related the development of mushiness to hepatopancreatic proteolytic enzymes and to the length of blanch time.

The shelf life of refrigerated seafood (both fish and precooked crustaceans) may be extended with modified atmosphere. The bacteriostatic effect increases with increasing CO_2 level, and 80% CO_2 and 20% air appears to be optimum for a number of products.[154]

Freezing is the prevalent preservation method. As with other products, freezing of crustacean muscle forms ice crystals. Their growth on subsequent storage causes structural damage that leads to textural degradation. This is illustrated in Figure 11, where structural damage in shrimp on frozen storage is evidenced by drastically enlarged spacings in fiber bundles. Also in analogy to other foods, slow freezing promotes extracellular growth of large ice crystals causing tissue dehydration, shrinkage, and compactness.[156] Toughening is the most severe texture change. However, it cannot be attributed to the effect of formaldehyde, since TMAOase was shown to be absent in scallops, lobster, and shrimp.[147]

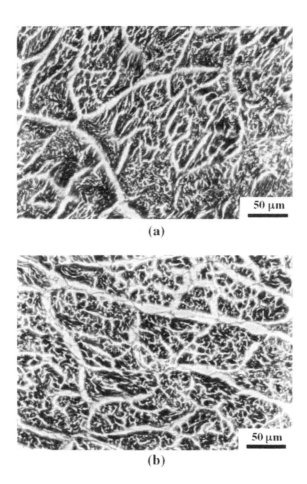

(a)

(b)

Figure 11 Micrographs of shrimp muscle frozen in liquid nitrogen at –100°C (a) immediately after freezing, (b) after 4 weeks of storage at –20°C (× 100). (From Pan, B.S.; Yeh, W.-T. J. Food Sci. 1983, 48, 370-374. With permission.)

Textural attributes of whole crayfish frozen in different ways and then stored were studied by Godber et al.[157] Toughness increased in the following order: individually quick-frozen < packaged and conventionally still-frozen < seasoned, packaged, and conventionally still-frozen. The decrease in extractable protein followed a similar order. The toughest product showed most lipid oxidation and least moisture retention.

Storage temperature was shown to be a crucial parameter in determining the rate of toughening of precooked, frozen lobster. Significant textural changes occurred after 3–4 months at –12°C but only after 9–10 months at –27°C.[158] The pH increased and the water-holding capacity decreased on frozen storage,[159] a phenomenon that has also been observed with other products, such as raw shrimp.

The effect of storage on the texture of cocktail shrimp was studied by Ahmed et al.[160,161] The product consists of precooked, deveined shrimp packed in tomato-based sauce of low pH. It is normally stored at refrigeration temperature in glass containers. Storage for up to 3 weeks resulted in the toughening of the shrimp texture, a phenomenon that was also dependent on the length of cooking time and the low pH. Pretreatment with sodium tripolyphosphate or NaCl reduced the toughening effect and increased the acceptability of the product.

Shellfish analogs

Shellfish analogs made from fin fish based on surimi technology have become a significant article of commerce. Crab legs are the most popular, but lobster tail, scallop and shrimp analogs are also being produced.

Surimi technology was developed in Japan and is a take-off on the old Japanese process for making kamaboko. Various fin fish can be used, with Alaskan pollack being the main raw material in the West. Surimi is produced through the solubilization of the fish myofibrillar proteins and subsequent formation of a protein gel network. The product is stored frozen before further processing. Freezing and frozen storage are detrimental to the gel-forming ability, and large-scale production of surimi became possible only with the discovery and use of cryoprotectants (sucrose, sorbitol, and phosphates). Cryostabilization and textural improvement are additionally effected by leaching of fish mince to remove Ca^{2+} and Mg^{2+}, which accelerate freeze-denaturation of myofibrillar proteins, to remove formaldehyde formed on storage of fish, and to concentrate the gel-forming myofibrillar proteins by removing sarcoplasmic proteins.[162]

The quality of surimi is also highly affected by the storage conditions of fish. There is a general decrease in hardness and cohesiveness of gels formed from fish kept on ice for increasing length of time due to denaturation and proteolysis of myofibrillar proteins. The denaturation rate increases with decreasing pH and is thus faster in low pH, active fish in which significant losses of the gel-forming ability may occur in as little as 1 to 2 days of chilled storage. In Alaskan pollack the postmortem pH is usually close to 7.0 and, thus, protein denaturation on storage occurs much slower.[163] Hardness of surimi gels made from pollack was reported to decrease significantly on storage at 10°C,[162] but to hold quite well in frozen storage even after 15 freeze–thaw cycles.[163]

The manufacture of crab meat analog involves, in brief, the flaking of frozen surimi, blending in of salt (to solubilize proteins), starch (to hold water and, as a dispersed phase, modify the elasticity of the surimi gel), and comminuted crab meat. The formed paste is extruded as a sheet and heat-set to form a gel, which is then slit to achieve parallel fibrous appearance and texture. The gelled sheet must have proper cohesiveness and elasticity to withstand the stretching, cutting, and pulling action of the slitting process. Strands of surimi are bundled diagonally into a rope, wrapped in a red-colored sheet to more closely simulate the appearance of real crab or lobster meat, heat-set, cut to desired lengths and shapes, retort-pasteurized (HTST), and packaged.[164]

Microbial contamination and spoilage are the primary causes of deterioration on storage of shellfish analogs. For this reason, controlled atmosphere packages, aseptic packaging and low-temperature storage are used.[165]

Vegetables

Various parts of the plant may be consumed as vegetables: tubers (potato, yam), petioles (celery, asparagus), flowers (broccoli, cauliflower), bulbs (onions, garlic), roots (carrots, beets), leaves (spinach, cabbage), and seeds (peas, beans, corn). Even some plant parts considered botanically as fruits (tomatoes, olives, cucumbers, eggplant) are used as vegetables. Webster's dictionary defines a vegetable as an edible part of a plant usually eaten with the main course of the meal. In contrast to fruits, vegetables are usually not sweet.

From the textural standpoint, vegetables represent a very diversified group of products. In the form in which they are consumed, they can be hard and crisp (raw carrots); firm, crisp, and juicy (cucumber); soft and mealy (cooked peas); or soft, pulpy, and fibrous (cooked asparagus). Texture is determined by the structural features of the tissue. In contrast to fruits, which as a general rule are picked mature but unripe, many vegetables are picked and consumed in the immature stage when they are tender and succulent. Exceptions are legume seeds, tubers, bulbs, and roots, which are harvested fully mature.

Fresh

In contrast to fruits, which soften on maturation and storage, the young vegetables become tougher. This toughening is due to the lignification of the primary cell wall and formation of the secondary cell wall. Lignin is a heterogeneous, high-molecular-weight aromatic polymer composed of randomly linked *p*-hydrocinnamyl alcohols. Its formation is mediated by oxidizing enzymes, although nonenzymatic lignification of vegetable tissue has been reported.[166]

Many of the morphological and physiological considerations discussed in connection with fruit texture also apply to vegetable texture. The main difference is that, while on maturation the fruit tissue stops growing and suffers enzymatic degradation, the vegetable tissue proceeds to differentiate and cell growth and enlargement continue. The amount of fibrous tissue increases. Toughening may reach the point where the vegetable is no longer suitable for human consumption.[61] Toughening is due to lignification, and fiber formation is a particularly important problem with asparagus. Smith et al. found significant increases in the width of the mechanical tissue layer, in the thickness of its cells, and in the thickness of the xylem vessel wall of the vascular bundles of stored asparagus that paralleled instrumental toughness measurements.[166] Lignification was decreased, but not stopped, by blanching and proceeded irrespective of storage temperature. It has been reported that treatment with glyphosate reduced postharvest lignification in asparagus spears.[167] A kinetic model has been developed for predicting the influence of storage conditions and spear characteristics on toughness in stored asparagus.[168] Asparagus is one of the most perishable commodities from the textural standpoint; it has a respiration rate 15 times that of apples.[169]

Most vegetables are cooked before consumption, a process that softens their texture (unless there is much lignification) and gelatinizes the starch deposited in the vacuoles as food for the plant. Starch and protein deposits are particularly abundant in beans, peas, and corn picked at maturity.

The texture-affecting sugar↔starch reaction occurs often on storage of vegetables. Low-temperature storage will favor the equilibrium going to the left (i.e., to sugar), while higher temperature storage will favor the equilibrium going to the right (i.e., to starch). Sweet corn and sweet peas when picked immature have much sugar and a succulent texture. Within hours the texture of sweet corn can become tougher and drier due to sugar conversion to starch. White potatoes, which contain much starch at harvest, will undergo a starch to sugar conversion when stored at below 10°C. This conversion leads to textural deterioration in both raw and cooked potatoes, a sweet taste, and a proneness to nonenzymatic browning on dehydration particularly troublesome in the manufacture of potato chips.

In general, lowering the temperature at which fresh vegetables are stored will decrease their rate of textural changes by slowing down the metabolic processes. In addition to the aforementioned potatoes, an important exception to this rule are cucumbers. When stored below 7°C, the cucumbers will suffer a total textural breakdown (excessive softening and exudation of liquid) due to tissue structural collapse followed by microbial spoilage. The texture of tomatoes is also susceptible to low-temperature injury. Reversible ultrastructural changes in organelles (mitochondria and plastids) in unripe tomato fruit were demonstrated.[170] (See section on Fruits and Berries for the mechanism of chill injury.)

The benefits of controlled atmosphere storage have not been proven with most vegetables.[171] Lowering the oxygen content in storage of tomatoes delayed ripening and softening as with other fruit,[172] but degraded the texture of green bell peppers.[173] Recent work by Gariepy et al. found that modified atmosphere storage of Canadian green asparagus (CO_2 levels of 2–10%) improved the keeping quality over refrigeration alone, apparently by decreasing the respiration rate by about 50%.[169] Asparagus kept in modified atmosphere storage had a firmer texture, better color, and fresher appearance. Adegoroye et al. published a quantitative study on the effect of storage and ripening on the texture of tomatoes, with special reference to sunscald injury.[174] Instrumental measurements (puncture probe fitted into an Instron) yielded six parameters as texture descriptors. Ripening on storage decreased firmness, locular resistance, and epicarp strength, and increased deformation. Sunscalding increased the ease of deformation, the effect becoming more pronounced with ripening and storage. Epicarp strength increased on storage of fruit with advanced injury. Thome and Alvarez developed an equation to predict the storage life of tomato fruit under irregular temperature regimes (12–27°C) based on the relationship of firmness and color to storage temperature and time.[175]

Loss of water due to respiration and physical drying is another contributor to textural changes on storage of fresh vegetables. It results in wilting of leafy vegetables and loss of plumpness in fleshy vegetables. This is prevented in commerce by waxing some vegetables, such as cucumbers, eggplant, and rutabaga.

Processed

From the textural standpoint, vegetables may be divided into those that are eaten raw and valued primarily for their crispness (lettuce, radishes), those that are eaten cooked and valued for their softness/mealiness (potatoes, eggplant), and those that can be consumed in either form (cauliflower, carrots). Products in the first group lose their firm/crisp texture on heating because of loss of turgor and cannot be fully preserved by the common processes of canning (involving a heat sterilization step) or freezing (involving blanching, i.e., an enzyme inactivation heating pretreatment). Some, such as cucumbers, are preserved by brining/fermentation and turned into a crisp new pickled product. The vegetables in the third group have relatively thick cell walls that provide rigidity and crispness in the absence of turgor.[92]

The consumer generally takes for granted that the texture of canned vegetables resembles that of overcooked vegetables. Just like with fruit, the softening effect is due to pectin degradation and calcium displacement. It can be reduced by addition of calcium to the canning liquid. Huang and Bourne reported that the rate of vegetable softening on canning proceeds according to two simultaneous first-order kinetic mechanisms.[176] Once degraded, the texture of canned vegetables is fairly stable on storage. Vegetables stored for a significant period of time prior to processing are more prone to softening on canning.[177]

Frozen vegetables have a higher-quality image and their texture is expected to resemble that of fresh, firm-cooked vegetables. The effect of blanching, freezing, and frozen storage on the textural properties of vegetables has been a popular research subject. Green peas have often been the product of choice because of their economic importance and uniform cellular structure.[155]

The texture of vegetables is less susceptible to damage on freezing than that of fruits.[92] The early work by Woodruff in the 1930s (quoted in Reference 92) showed that the main cause of texture alteration on freezing was the irreversible separation of water from the protoplasm. Cell breakage was also demonstrated and related to the rate of freezing. The blanched tissue is more susceptible to freezing damage. As with other high moisture products, slow freezing causes formation of large ice crystals leading to cell membrane rupture, flocculation of the plasm, tearing, detachment of cell layers, and leaching out of cell contents. The damage is less severe in young tissues because of the elasticity of the cell wall and the presence of fewer vacuoles in the cytoplasm. Small ice crystals do not rupture the cell membrane, thus the texture of fast-frozen products is less degraded.[178–180]

The damage done to the texture of vegetables on blanching and freezing is much more severe than that occurring on subsequent storage. Textural changes on frozen storage are primarily due to ice crystals growth and desiccation. Both processes are accentuated by fluctuating temperatures and lead to chewier products. Much damage can be done by thawing and slow refreezing of the vegetables, resulting in much ice formation and the development of a spongy texture that does not hold water.

Pickles

Softening on storage of cucumber pickles preserved by brine fermentation is a serious problem that has been commanding much research attention. It is believed to be caused by degradative enzymes of endogenous (e.g., from cucumber blossoms) or microbial origin that bring about solubilization of cell wall pectic substances. This degradation can be attenuated by removal of blossoms, refrigerated storage (of unbrined pickles), or pasteurization,[181] but subsequent microbial recontamination can occur. Softening may be retarded by protection from excessive demethylation, addition of calcium ions, and addition of calcium ions with alum.[182] Calcium added at the beginning of the fermentation was most effective in preventing pectin solubilization.[183] Of the different organic acids investigated, acetic acid was most effective in firmness retention.[184]

Ultrastructural studies indicated that in the softened tissue the cell walls were swollen and striated, and the cell-to-cell junctions at the middle lamellae were weakened and poorly defined. These changes are consistent with cellulolytic and/or pectinolytic degradation of cell wall components.[185] The mechanism by which calcium ions inhibit the softening of the pickles is not yet clear. It has been postulated, based on thermodynamic considerations, that it may involve binding to a protein.[186]

Softening on storage is also a recognized texture defect in brine-fermented olives. Prior to fermentation, olives are given a dilute lye treatment to eliminate bitterness. Endogenous celluloses are believed to be responsible for the softening of olives.[187] Brenes Balbuena et al. compared the effect of different storage conditions on the texture and color of ripe black olives.[188] They concluded that greater stability is achieved with aerobic storage and addition of low levels of acetic acid. They also reported that black olives processed after a certain postharvest storage period are less prone to texture degradation.

Mechanical damage

Seeds of low moisture content (e.g., soybeans, corn) handled in bulk are often subjected to conditions in storage and transport that result in kernel breakage, which is of substantial concern to agricultural engineers. Factors affecting the susceptibility to breakage as well as equipment to quantify it continue to be researched.

Breakage is due to internal stresses produced by drying and temperature/moisture gradients that lead to crack formation and propagation on mechanical impact.[189] High moisture removal rates are more detrimental than slow drying. The amount of seed fragmentation is also related to the severity of forces on handling and storage.[190]

Some vegetables of high moisture content (e.g., potatoes) are also subject to mechanical damage on bulk storage because of high static loading. Brusewitz et al. found that preloaded potato tissue was less stiff, less recoverable after deformation, tougher, and lost water.[191] These changes are consistent with loss of turgor, loss of cell wall elasticity, and cell reorientation.

Hard-to-cook defect in legumes

The hard-to-cook defect, i.e., the inability to soften during a reasonable cooking time, can develop in beans when they are stored at high temperature and high relative humidity. This defect results in poor-quality products and increased fuel costs, thus lowering the commercial value of the beans. The importance of the problem is connected with the fact that beans are a major source of nutrients for much of the world population, especially in the poorer countries of Latin America. Two excellent reviews have been published by Stanley and Aguilera on the influence of structure, composition, storage, and processing on textural defects in beans.[192,193] The interested reader is referred to them for details, as only a brief summary of the problem and its possible mechanisms will be presented here. Although beans have been the subject of most research, the problem also occurs with other legumes (e.g., yellow peas and cowpeas).

Bean seeds are harvested as mature seeds at 20% moisture, field dried to 10% moisture, and stored under ambient conditions. They are composed of two major structural parts: the seed coat (testa) and the cotyledons (embryonic leaves). The seed coat constitutes only less than 1% of the seed volume, but plays an important role in water absorption.[194] It is composed of cellulose, hemicellulose, and lignin. Its cell walls contract on ripening. The cotyledons are composed of parenchyma cells, which differ from those in other plants in that storage proteins and starch accumulate and cell organelles are reduced as cell desiccation occurs during maturation.

Figure 12 illustrates the effect of 25°C storage on hardness of beans based on composite data from different investigators.[193] It indicates that storage hardening follows first-order kinetics. Hardening on storage increases almost exponentially at bean moisture levels above 10% (Figure 13) and at higher temperatures. It may be controlled by artificially drying the beans to a safe moisture level or by storing them under refrigeration. The possibility has been demonstrated of reversing the hard-to-cook defect in beans and cowpeas by storing them at low temperature (6.5°C) and high humidity (71%).[195] These solutions, however, are impractical in the tropics, where high humidity and high temperature are the prevailing conditions. Some evidence is available that hardening may be partially inhibited by very high storage temperatures. This inhibition may suggest enzymatic pathways of deterioration. Figure 14 shows that the activation energy for the hardening process decreases with increasing bean moisture. The value of the activation energy is typical of diffusion-controlled and enzymatic reactions.[196]

Recently, Del Valle et al. developed two first-order kinetic models that adequately predict the storage-induced hardening of black and white beans.[197] The models are based on a very extensive and controlled set of data, and use the activation energy and the equilibrium constant of the hardening reaction as parameters. They demonstrate that the reaction occurs in three phases: the initial lag period, fast hardening, and a phase of declining rate of hardening leading to an equilibrium value. Hardening rates increased with storage temperature and water activity of the beans. Equilibrium values were determined by temperature for white beans and legume water activity for black beans.

Beans are prepared for consumption by first soaking them in water to soften the seed coat and then cooking at atmospheric or elevated pressures to inactivate antinutritional factors and soften the texture. It has been postulated[193] that softening of the beans on cooking occurs in two stages: breakdown of the middle lamella, leading to cell separation, and gelatinization of starch granules. A study of starch granules isolated from hard-to-cook beans indicated that they were more resistent to amyloglucosidase attack and gelatinized at higher temperature.[198]

Much research has been and continues to be done on the mechanism of bean hardening, resulting in the accumulation of much pertinent data. However, there is no general

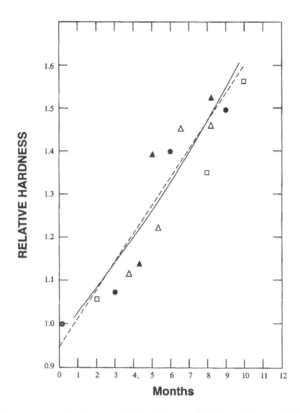

Figure 12 Effect of storage on hardness of black beans (12% moisture, 25°C) following cooking. Relative hardness is given by H_1/H_0. Solid line is for reaction order N = 1; dotted line fro N = 0. Data: □ Mora 1982; ● Molina et al. 1976; △ Ballivian 1984; ▲ Aguilera et al. 1984b. (From Aguilera , J.M.; Stanley, D.W. J. Food Process. Preserv. 1985, 9, 145-169. With permission.)

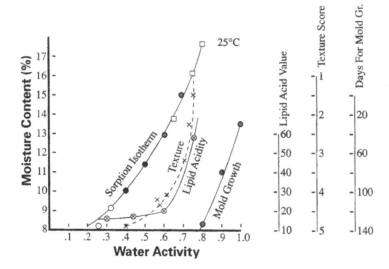

Figure 13 Sorption isotherm and stability map for beans after prolonged storage. ● Morris et al. 1980; □ McCurdy et al. 1984b; × Morris and Wood 1956; ○ Weston and Morris 1954. (From Aguilera, J.M.; Stanley, D.W. J. Food Process. Preserv. 1985, 9, 145-169. With permission.)

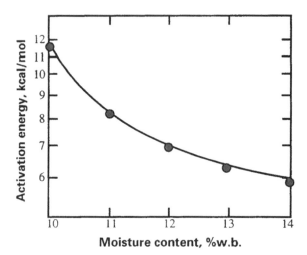

Figure 14 Effect of moisture content on the activation energy of the bean hardening process. Line is defined by $\ln E_A = 9.825 - 1.150W + 0.041W^2$. (From Aguilera, J.H.; Ballivian, A. J. Food Sci. 1987, 52, 691-695, 718. With permission.)

agreement on the exact mechanism. The following major factors[192,194] are recognized or proposed[198]:

1. *Failure of phytate to chelate and remove calcium and magnesium from the pectate structure in the middle lamella.* This prevents the pectic substances from solubilizing and allowing the cells to separate during cooking. It is believed to be caused by the degradation of phytate by phytase, whose action is accelerated by elevated storage temperature and higher humidities. However, there is still a debate on the validity of the hypothesis involving the role of phytate, with recent work presenting evidence for[200,201] and against it.[202,203]
2. *Lignin formation in the cell walls of the cotyledons and the testa through oxidation and polymerization of phenols; this lignification may occur through enzymatic and nonenzymatic pathways.* These factors relate the impaired texture-softening on cooking to the fact that beans with harder seed coats will absorb less water during soaking, thus making less water available for starch gelatinization and protein swelling. Secondly, lignified cell walls in the cotyledons will also be less water-permeable and mechanically stronger.
3. *Crosslinking of products from protein hydrolysis and/or polyphenolics.* Appearance of additional bonds with lesser electrophoretic mobility suggests the possibility of some protein denaturation and/or association.[204]
4. *Loss of microsomal membrane functionality* caused by lipoxygenase-catalyzed oxidation of polyunsaturated fatty acids liberated from membrane lipids by phospholipases,[199] which may allow cations to migrate into the cell and bind intracellular components.[205]

Dispersions, emulsions, and colloidal systems

This section considers dispersions, emulsions, and colloidal systems not covered under specific product categories. A dispersion is a two-phase system consisting of a continuous and a dispersed phase. Colloids are dispersions in which the dispersed component is too small to be detected under the light microscope, yet much larger than ordinary molecules.

Emulsions are colloidal dispersions of liquids in liquids[206] or liquid crystals in a liquid.[207] The physicochemical laws and factors governing the stability of these systems are well-known, and the interested reader should consult several excellent sources of in-depth information on the subject (e.g., References 208 and 209).

Creaming, flocculation, and drop coalescence are key textural defects occurring on storage of emulsions. Creaming refers to the separation into a concentrated layer of dispersed drops and a small volume of continuous phase. It can be upward or downward, depending on the density difference between the two liquid phases. Creaming can be prevented by eliminating these density differences. Flocculation is aggregation of droplets, which entraps the continuous phase liquid. The aggregate breaks into individual drops under high shear conditions, except in the case of very concentrated systems. Drop coalescence follows flocculation when, during storage, the continuous phase film between the drops drains away and the protective emulsifier layer around the drops ruptures. It is associated with fat separation and leads to product rejection. Flocculation increases emulsion viscosity, while drop coalescence decreases it.[210] Total breakdown leads to two distinct layers.

Mayonnaise

Mayonnaise is an oil-in-water emulsion consisting of 77–82% winterized salad oil as the dispersed phase and a continuous aqueous phase consisting of vinegar, egg yolk, salt, sugar, and spices. Egg yolk and mustard act as emulsifying agents. Normally, in an emulsion, the phase in greater quantity is the continuous phase; this is reversed in mayonnaise to give the product its characteristic texture. Such an "unnatural" emulsion tends to be unstable and difficult to prepare.[46]

Electron microscope studies showed that the periphery of the oil droplets in mayonnaise was covered by a speckled layer presumably made up of coalesced low-density lipoproteins of egg yolk and of microparticles from the yolk granules.[211] In addition, a fibrous membrane was detected on the surface of the water-washed droplets, which led to the postulate that the stability of mayonnaise is due to the high degree of plasticity contributed by the particulate matter in the speckled layer and to the properties of the fibrous membrane surrounding the emulsified oil droplets.

Mayonnaise tends to be unstable in cold and in frozen storage. If the oil used is not properly winterized, fat crystals formed at refrigeration temperature will break the emulsion. The emulsion will also break on freezing, due to the puncturing by the ice crystals of membranes covering the oil droplets.

Salad dressings

Salad dressings may be divided into several categories: spoonable, i.e., mayonnaise-like, and pourable (bottled) in either an emulsified or separating form. The first two types are oil-in-water emulsions. The third type is an unstable mixture of oil and water that must be shaken prior to use. Only the emulsion-type salad dressing will be considered here. The spoonable salad dressing typically contains about 50% less oil than mayonnaise. The desired consistency is achieved through the presence of gelatinized starch or other water-binding carbohydrates. The pourable salad dressings contain 40–60% oil, depending on the type. A number of reduced, low, and zero fat salad dressings are currently available in the marketplace, responding to the consumer demands for less fat in the diet.

Hydrocolloid gums (tragacanth, locust bean, xanthan, propylene glycol, alginate, etc.) are used as emulsifiers and also as thickening agents and substitutes for the rheological properties of fat in pourable salad dressings. Their use spurred research on the textural/rheological properties of these systems (e.g., References 212 and 213) and of model emulsion systems (e.g., References 214 and 215).

In general, the texture of properly prepared salad dressings is quite stable on storage. Any instabilities will involve structural changes in the emulsion and the gum system. Paredes et al. found that major textural changes take place in bottled salad dressing during the initial week of storage.[216] The product became more pseudoplastic and showed increases in the yield stress and the consistency index. These changes were attributed to structure buildup due to hydration of xanthan gum and coalescence of oil droplets. This finding relates occasional problems experienced by consumers with "pourable" dressings not being "pourable" due to improper formulation/processing and to quality control that takes place too soon after processing.

Mustard

Prepared mustard, the form considered here, is a smooth, spreadable paste consisting of ground mustard seed (or other form of mustard), vinegar, spices, and other flavoring materials. It contains about 79% water and about 5% oil derived from mustard seed. Rheologically, prepared mustard is a heterodisperse Bingham plastic, i.e., it exhibits a yield stress that must be overcome before flow can be initiated.[217]

The stability of the mustard system on storage, as affected by the size and size-distribution of mustard seed particles, was studied recently by Aquilar et al. via rheological methods that reflected its functionality.[218] Samples were prepared containing fine, medium, and slightly coarse particles. After 3 months storage, syneresis was observed in fine particle samples kept at 25°C and in all samples kept at 45°C. It was postulated that syneresis was caused by the aggregation of the colloidal particles, which changed the packing array and reduced the voids in the network where the liquid phase was trapped. Thus, it should be possible to minimize syneresis by controlling the particle size distribution during processing. Other changes detected on storage included a decrease in yield stress and consistency as well as a change to a more Newtonian (i.e., less shear rate dependent) flow. Because the yield stress is a measure of the strength of forces stabilizing a network of interacting particles, its decrease indicates that storage weakens the internal structure of mustard. Functionally, this translates into greater flowability and spreadability.

Chocolate

Chocolate is a dispersion of sugar and cocoa particles in a continuous phase of cocoa butter. In milk chocolate, nonfat milk components are also present in the dispersed phase and butter fat is present in the continuous fat phase. The solid particles must be less than 20 μm in diameter for the chocolate to have a smooth texture. This smoothness is accomplished in the conching step in which pressure rollers grind and aerate the melted chocolate mass. Another requirement for proper texture is controlled crystallization of cocoa fat, leading to the production of stable polymorphs of β crystals of glycerides. This crystalization contributes to the hardness and sharp melting of chocolate, attributes much appreciated by the consumer. When the liquid product is not properly tempered and cooled, higher melting point glycerides solidify within an oily mass. The resulting product has poor texture and a tendency to develop fat bloom on cool storage.[46]

The texture of chocolate is generally quite stable on storage unless the product is subjected to extreme temperature fluctuations, which cause melting and solidification. This leads to textural deterioration as ascribed above to uncontrolled solidification of cocoa butter during processing. This can, to some extent, be prevented by addition of lecithin. Sometimes chocolate develops a sugar bloom,[219] which has recently been related to the collapse phenomenon occurring as a consequence of a structural relaxation process[220] in materials stored above their glass transition temperature.

Methods of texture measurement

Texture can be measured by direct (sensory and instrumental) and indirect methods. The latter include physical (e.g., specific gravity, drained weight) and chemical (e.g., collagen content, alcohol-insoluble solids) properties that have been shown to relate to texture.[221] Because texture is a sensory property, the best method of characterizing and quantifying it is a well-executed sensory evaluation. However, instrumental methods are often more precise and more reproducible. An excellent, extensive review of the principles and applications of texture measuring methods was published by Brennan.[222]

Sensory methods

Traditional sensory methods of texture evaluation involved assessment and grading by "expert" tasters who worked either for the producers, trade associations, or the government. Specific terms and their definitions were developed for specific products, with special attention being paid to textural defects. Many years of training and experience were usually needed to become an expert grader of cheese, ice cream, or other products. This type of texture evaluation was later adopted by the universities for student training and also found its way into research. Simple judging for specific parameters, e.g., tenderness, chewiness, softness, was also used. Hedonic scales (like/dislike) were often employed incorrectly instead of intensity scales.

These approaches were critically examined in the 1960s when texture began to be studied as a subject itself, rather than as a part of commodity research. Today, probably the best sensory method of texture description is the sensory texture profiling technique, a form of descriptive analysis. It recognizes the multiparameter nature of texture and the dynamic character of its evaluation in the mouth. Developed in the 1960s by Brandt et al.,[223] based on Szczesniak's classification of textural characteristics[2] (Table 4) and the development of standard scales,[224] the method can be applied to any product. It is not limited to in-mouth evaluation, although this is most common.

Depicted schematically in Figure 15, the sensory texture profile technique evaluates the order of appearance of textural characteristics according to successive stages, from visual inspection and the first bite, through mastication, to swallowing and residual feel in the mouth. The intensity of each characteristic is quantified based on the standard scales.[224,225] These encompass the entire intensity range of a given characteristic encountered in foods; they can be truncated as needed and expanded in the segment pertinent to the specific food of interest. Shorter versions of the technique may be used in situations (e.g., shelf life studies) that require detailed evaluation of only few characteristics.[226] The scaling methodology often varies depending on the preference of the panel leader. Category, line, or magnitude estimation scales have been used. The results may be expressed numerically in a table or as two-dimensional area profiles.

The proper use of the technique requires the availability of a well-trained and well-maintained panel[227,228] composed of about four to eight members. It also requires the understanding and agreement by the panelists on the definitions of the textural characteristics present in the tested food(s) and total familiarity with the standard scales. For the details of the sensory texture profile technique, the reader is referred to the text by Meilgaard et al.[226] and the cited literature references.

A modified sensory texture profile technique may also be used with untrained consumer panels to provide "fingerprints" of specified products and of their "ideal" coun-

Table 4   Relationship between Textural Parameters and Popular Nomenclature

	Mechanical Characteristics	
Primary Parameters	Secondary Parameters	Popular Terms
Hardness		Soft→Firm→Hard
Cohesiveness	Fracturability	Crumbly→Crunchy→Brittle
	Chewiness	Tender→Chewy→Tough
	Gumminess	Short→Mealy→Pasty→Gummy
Viscosity		Thin→Viscous
Elasticity		Plastic→Elastic
Adhesiveness		Sticky→Tacky→Gooey

	Geometrical Characteristics
Class	Examples
Particle size and shape	Gritty, Grainy, Coarse, etc.
Particle shape and orientation	Fibrous, Cellular, Crystalline, etc.

	Other Characteristics	
Primary Parameters	Secondary Parameters	Popular Terms
Moisture content		Dry→Moist→Wet→Watery
Fat content	Oiliness	Oily
	Greasiness	Greasy

From Szczesniak, A.S. J. Food Sci. 1963, 28, 386-389. With permission.

terpart. The latter advantage makes this a powerful technique for product improvement work by providing the description of a meaningful target and specific guidance for meeting it.[229] Figure 16 illustrates the type of results obtained.

Instrumental methods

A very large number of instruments have been developed for testing food texture since the first puncture tester for quantifying firmness of jellies was constructed in 1861.[230] Over a hundred became commercially available.[1,231] They differ in size from hand-held instruments to floor units. They also differ in principle, complexity, and ruggedness of design depending on the intended application and test products. They can be divided after Scott Blair into imitative, empirical, and fundamental classes.[232] Bourne divided them into force, distance, time, energy, ratio, and multiple variable measuring units, and provided a detailed description of instruments and tests within each category.[230]

Most of the instrumental texture measurements involve mechanical tests quantifying the resistance of the food to applied forces greater than gravity. The realization that the various instruments used over the years consisted of some common basic elements (a probe contacting the sample, a driving mechanism for imparting motion to it, a sensing element for detecting the food's resistance, and a read-out system),[231] coupled with the birth of a generalized, intercommodity approach to texture, led to the use and development of universal testing machines. These units supply the common three basic elements, provide interchangeable probes for specific tests (shear, compression, puncture, tensile, etc.), and give the operator flexibility in selecting the force range and the rate and amount of deformation. Quite often, they can also provide static load and measure the rate of recovery. Practically all are recording instruments, providing force vs. deformation tracings, and can be computerized with respect to testing regime and data collection/analysis. They are used mostly for research purposes but can be simplified for use in quality control.

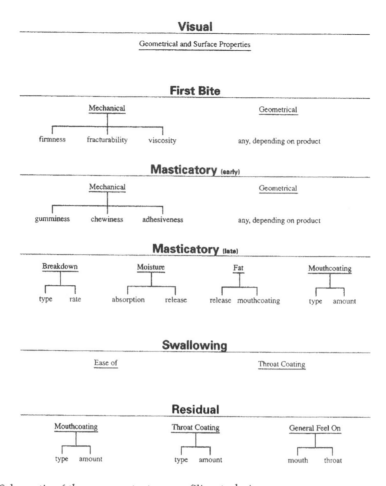

Visual

Geometrical and Surface Properties

First Bite

| Mechanical | | | Geometrical |
| firmness | fracturability | viscosity | any, depending on product |

Masticatory (early)

| Mechanical | | | Geometrical |
| gumminess | chewiness | adhesiveness | any, depending on product |

Masticatory (late)

| Breakdown | | Moisture | | Fat | | Mouthcoating | |
| type | rate | absorption | release | release | mouthcoating | type | amount |

Swallowing

| Ease of | Throat Coating |

Residual

| Mouthcoating | | Throat Coating | | General Feel On | |
| type | amount | type | amount | mouth | throat |

Figure 15 Schematic of the sensory texture profiling technique.

The Instron was the first instrument of this type to be used for food testing. It was introduced to the food industry by M.C. Bourne in 1966[234] and has since become a very popular unit available in different models. It has been joined in the field by several other universal testing instruments (e.g., Food Technology's Texture Test System, Ottawa Texture Measuring System, Texture Technologies' TA.XT2 Universal Texture Analyzer). Similar to the situation in the mouth, this instrumental assessment usually involves destructive testing. However, the destruction is not as extensive, the temperature is usually ambient (i.e., lower than body temperature), and there is no saliva present.

The principle of texture profiling has been applied to instrumental texture measurements with universal testing machines[235–237] using the same classification and definitions of textural characteristics as the sensory profiling method. The main difference is that the instrumental assessment does not involve as many evaluation stages and is usually confined to two "bites." With time, both sensory and instrumental texture profiling began to build on and deviate from the original system[2,224,235,223] by including detailed descriptive terms outside the original nomenclature (Table 4). These terms need the development of further definitions and of correlations between the two methods of texture assessment using psychophysical approaches. With temperature-sensitive foods, such as chocolate, ice cream, or gelatin gels, instrumental measurements should be performed at different temperatures.[238] A review of instrumental texture profiling was published by Breene.[239] Basically, the test is performed by compressing the food twice

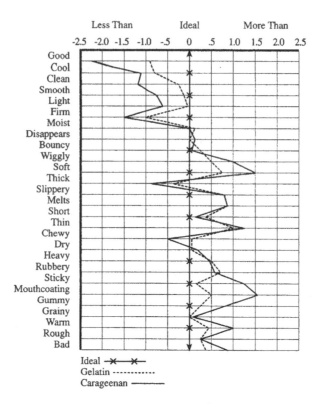

Figure 16 Consumer texture profiles for dessert gels, with results represented as deviations from the ideal product. (From Szczesniak, A.S., et al. J. Food Sci. 1975, 40, 1243-1246. With permission.)

between two parallel flat plates and quantifying the mechanical parameters of texture, as shown in Figures 17 and 18.

Texture profiling may be classified as a multipoint measurement. Another technique in this category involves quantifying from one force-distance curve several parameters reflecting structural or other features of the product. Figure 19 depicts the resistance to penetration of a ripe tomato fruit.[240] The various portions of the curve can be equated with the resistance (in unit force) of the various structural entities in the tomato. An extensive treatment of the interpretation of force curves from instrumental texture tests has been published by Bourne.[241]

Liquid and semisolid foods are tested with viscometers and consistometers, which again can be one-point or multipoint measurement types. Available are capillary, orifice, falling ball, vibrating reed, and rotational (including coaxial cylinder, cone-and-plate, and other types) viscometers.[230] Rotational viscometers are the most popular and are available in the recording, computerized forms that allow one either to vary the shear rate (i.e., speed of rotation) or to keep it constant and to record how stress changes with the increasing/decreasing shear rate or with time.

The study of the stress/strain or force/deformation relationship, which underlies the mechanical parameters of texture, constitutes the realm of rheology. Foods are rheologically very complex and nonideal in that their behavior is often both time and strain rate-dependent. They usually exhibit some elastic (i.e., recoverable) and some viscous (i.e., unrecoverable) deformation and are called viscoelastic bodies. Consequently, the conditions of instrumental tests, in terms of the rate of deformation (i.e., speed of crosshead movement), will often affect the measurement and must be properly selected.

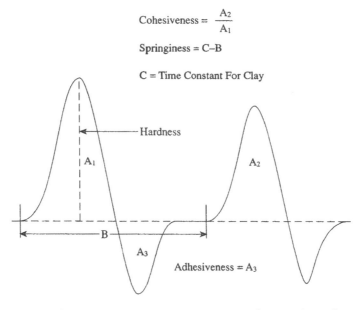

$$\text{Cohesiveness} = \frac{A_2}{A_1}$$

$$\text{Springiness} = C{-}B$$

$$C = \text{Time Constant For Clay}$$

Figure 17 Schematic of the instrumental texture profiling technique. (Not shown is the fracturability, or brittleness, which is detected as the force at the break in the left-hand side of the first peak; see Figure 18.)

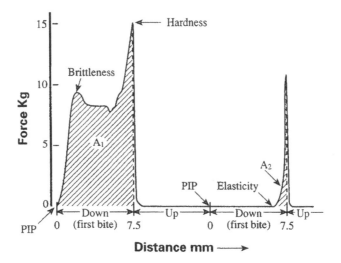

Figure 18 Instrumental texture profiling of cylindrical pear sample using the Instron. (From Bourne, M.C. J. Food Sci. 1968, 33, 223-226. With permission.)

The viscoelastic nature of foods can be separated into the component parts and characterized in depth by means of dynamic viscoelastic measurements. These nondestructive tests, introduced to the field of foods from the field of high polymers, are gaining popularity in basic research on the molecular structure of food polymers and homogeneous food products, especially as related to network formation. Their disadvantages are high equipment cost, requirement of high operator skills, and lack of demonstrated correlations with sensory texture assessment. They are based on the principle that, when an oscillary motion is imparted to a material placed between two surfaces, its elastic component will oscillate in phase and its viscous component 90° out of phase with the stimulus.

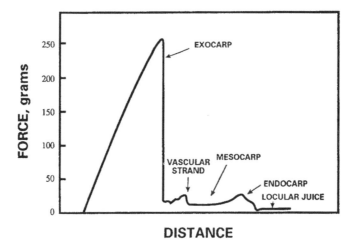

Figure 19 Force–distance curve for a ripe tomato obtained with a punch probe fitted into an Instron. (From Holt, C.B. J. Texture Stud. 1970, 1, 491-501. With permission.)

Present-day mechanical spectrometers are equipped with a computer program that separates the overall oscillation of the test substance (0–90° out of phase) into its elastic (G') and viscous (G'') components moduli. The moduli may be plotted against varying oscillation frequency, percent strain, or temperature.

Figure 20 illustrates this type of measurement using 15, 20, and 25% soy isolate/water mixtures and plotting the values of G' and G'' against percent deformation.[242] It will be seen that the concentration of soy isolate increases the strain level at which $G'' > G'$ (i.e., the material starts behaving more like a liquid). The measurements also indicate that the 15% sample is different from the other two in that it behaves more like a dispersion than a dough.

Summary

The effect of storage on the textural properties of foodstuffs is very diverse and varies with the product type and storage conditions. For some products, such as meat, controlled storage has a beneficial (tenderizing) effect. For others, such as fruit that is picked mature but unripe, storage develops proper (softer) texture followed by its deterioration (senescence). For still others, storage is always detrimental, but the rate of deterioration varies from very rapid (e.g., bread) to very slow (e.g., butter). With ripened cheeses, storage is part of processing that develops proper flavor and texture.

Frozen storage is always detrimental to texture due to ice crystal formation and growth, a process that is usually controlled with formulation in products especially designed for frozen storage, such as ice cream. The severity of textural damage depends upon the structure of the product. It can vary from total product failure (e.g., mayonnaise) to a reduction in quality, such as the development of toughness and drip in meat and fish.

With frozen and dehydrated products, texture deterioration on storage is a continuation of quality-degrading events initiated in the processing step. Drying of corn and soybeans causes formation of fissures that lead to mechanical damage when forces are applied during handling and bulk storage.

Fresh produce, such as potatoes, apples, and oranges, may also suffer structural damage on impact or static loading, resulting in cell rupture followed by enzymatic degradation.

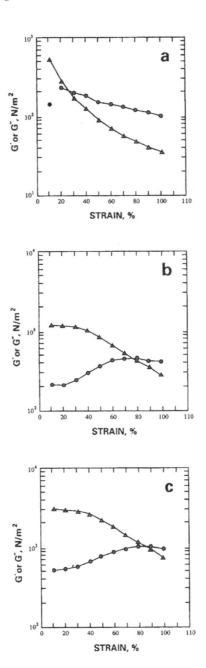

Figure 20 Dynamic viscoelastic measurements on 15% (a), 20% (b), and 25% (c) soy isolate/water mixtures. Symbols ▲ — G'; ● — G". (From Baird, D.G. J. Texture Stud. 1981, 12, 1-16. With permission.)

Staling in bread, toughness and poor water-holding capacity in frozen fish, and hard-to-cook phenomenon in beans are probably the most researched storage-induced textural changes. Much has been learned about the potential causes and mechanisms of these reactions.

Behavior of foods on storage has been studied mostly from the biochemical and structural viewpoints. Some studies included sensory texture assessments; few were based on quantitative instrumental measurements. Advanced sensory, mechanical, and rheological test methods available today are beginning to be applied to developing a better understanding of textural changes occurring on storage and transport of food and to constructing appropriate kinetic models.

Acknowledgment

The author wishes to thank her colleague, Prof. Malcolm C. Bourne of Cornell University, Ithaca, NY, for reviewing the manuscript and offering constructive comments, specifically in sections dealing with agricultural products. She also wishes to thank the late Prof. Donald D. Hamann of North Carolina State University for sharing the information on surimi. Their help and friendship were deeply appreciated.

References

1. Bourne, M.C.; Szczesniak, A.S. in Encyclopedia of Food Science, Food Technology and Nutrition; Meinae, R.; Robinson, R.K.; Sentler, M.J., Eds.; Academic Press, London, 1993, pp. 4059-4065.
2. Szczesniak, A.S. J. Food Sci. 1963, 28, 386-389.
3. Szczesniak, A.S. J. Food Quality 1991, 14, 75-85.
4. Szczesniak, A.S. Food Technol. 1990, 44 (9), 86-88, 90, 92, 95.
5. Kulp, K.; Ponte, J.G, Jr. CRC Crit. Rev. Food Sci. Nutr. 1981, 15, 1-48.
6. Taranto, M.V. in Physical Properties of Foods; Peleg, M.; Bagley, E.B., Eds.; AVI Publishing, Westport, CT, 1983, Chapter 8.
7. Willhoft, E.M.A. J. Texture Stud. 1973, 4, 292-322.
8. Bechtel, W.G.; Meisner, D.F.; Bradley, W.B. Cereal Chem. 1953, 30, 160–168.
9. Stollman, U.; Lundgren, B. Cereal Chem. 1987, 64, 230-236.
10. Nussinovitch, A.; Steffens, M.; Chinachoti, P.; Peleg, M. J. Texture Stud. 1992, 23, 13-24.
11. Piazza, L.; Masi, P. Cereal Chem. 1995, 72, 320-325.
12. Slade, L.; Levine, H. CRC Crit. Rev. Food Sci. Technol. 1991, 30 (2&3), 115-360.
13. Slade, L.; Levine, H. in Industrial Polysaccharides, The Impact of Biotechnology and Advanced Methodologies; Stivala, S.S.; Crescenzi, V.; Deo, I.C.M., Eds.; Gordon and Breach Science Publishers, New York, 1987, pp. 387-430.
14. Martin, M.L.; Zeleznak, K.J.; Hoseney, R.C. Cereal Chem. 1991, 68, 498-503.
15. Martin, M.L.; Hoseney, R.C. Cereal Chem. 1991, 68, 503-507.
16. Cocero, A.M.; Kokini, J.L. J. Rheol. 1991, 35, 257-270.
17. Willhoft, E.M.A. J. Sci. Food Agric. 1971, 22, 176-183.
18. Kim, S.K.; D'Appolonia, B.L. Cereal Chem. 1977, 54, 207-229.
19. Persaud, J.W.; Faubian, J.M.; Ponte, J.G., Jr. Cereal Chem. 1990, 67, 182-187.
20. Guy, R.C.E. J. Sci. Food Agric. 1983, 34, 477-491.
21. Sych, J.; Castaigne, F; Lacroix, C. J. Food Sci. 1987, 52, 1604-1610.
22. Vickers, Z.; Bourne, M.C. J. Food Sci. 1976, 41, 1158-1164.
23. Kim, H.S.; Lee, S.R. Korean J. Food Sci. Technol. 1988, 20, 553-557.
24. Schaller, A.; Mohr, E.L. und E. 1976, 29, 120-121.
25. Roos, Y.; Karel, M.J. Food Sci. 1991, 56, 38-43.
26. Shanbag, S.P.; Szczesniak, A.S. U.S. Patent 4 256 772, 1981.
27. Inoue, Y., Bushuk, W. Cereal Chem. 1991, 68, 627-631.
28. Inoue, Y., Bushuk, W. Cereal Chem. 1992, 69, 423-428.
29. Autio, K., Sinda, E. Cereal Chem. 1992, 69, 409-413.
30. deMan, J.M. J. Texture Stud. 1969, 1, 109-113.
31. deMan, J.M. Can. Inst. Food Sci. Technol. J. 1978, 11 (4), 194-202.
32. Juriaanse, A.C.; Heertje, I. Food Microstruc. 1988, 7, 181-188.
33. Vaisey-Genser, M.; Vabe, B.K.; Johnson, S. J. Texture Stud. 1989, 20, 347-361.
34. McBride, R.L.; Richardson, K.C. Lebensm.- Wiss. u. Technol. 1983, 16, 198-199.
35. Precht, D.; Buchheim, W. Milchwissenschaft 1980, 35, 399-402.

36. Shama, F.; Sherman, P. J. Texture Stud. 1970, 1, 196-205.
37. Sherman, P. in Rheology and Texture in Food Quality; de Man, J.M.; Voisey, P.W.; Rasper, V.F.; Stanley, D.W., Eds.; AVI Publishing, Westport, CT, 1976, Chapter 15.
38. Kowana, M.; Hamann, D.D.; Swartzel, K.R.; Hansen, A.P. J. Texture Stud. 1981, 12, 483-505.
39. Lawrence, R.C.; Creamer, L.K.; Gilles, J. J. Dairy Sci. 1987, 70, 1748-1760.
40. Adda, J.; Gripon, J.C.; Vassal, L. Food Chem. 1982, 9, 115-129.
41. Green, M.L.; Manning, D.J. J. Dairy Res. 1983, 49, 737-748.
42. Green, M.L.; Turrey, A.; Hobbs, D.G. J. Dairy Res. 1981, 48, 343-359.
43. Fredrick, I.A.; Dulley, J.R. J. Dairy Sci. Technol. 1984, 14 (2), 141-150.
44. Dieffes, H.A.; Rizvi, S.S.; Bartsch, J.A. J. Food Sci. 1993, 58, 764-769.
45. Arbuckle, W.S. Ice Cream; 3rd edition; AVI Publishing, Westport, CT, 1977.
46. Potter, N.N. Food Science; 3rd edition; AVI Publishing, Westport, CT, 1978.
47. Musselwhite, P.R.; Walker, D.A. J. Texture Stud, 1971, 2, 110-116.
48. Sherman, P. J. Food Sci., 1965, 30, 201-211.
49. Berger, K.G.; Bullimore, B.K.; White G.W.; Wright, W.B. Dairy Industries 1972, 37, 419-425, 493-497.
50. Berger, K.G. in Food Emulsions; Larsson, K.; Friberg, S., Eds.; 2nd edition; Marcel Dekker, New York, 1990, pp. 367-444.
51. Sherman, P. J. Food Sci. 1966, 31, 707.
52. White, G.W.; Cakebread, G.W. J. Food Technol. 1966, 1, 73-82.
53. Hertel, R.W. Food Technol. 1993, 47, 99-107.
54. Fink, A.; Kessler, H.G. Milchwiss. 1985, 40, 326-329.
55. Swartzel, K.R.; Hamann, D.D.; Hansen, A.P. J. Food Process. Eng. 1980, 3, 143-159.
56. Vanden Boomgard, Th.; Van Vliet, T.; Van Hooydonk, A.C.M. J. Food Sci. Technol. 1987, 22, 279-291.
57. Herald, T.J.; Osorio, F.A.; Smith, D.M. J. Food Sci. 1989, 54, 35-38, 44.
58. Bourne, M.C. J. Texture Stud. 1979, 10, 25-44.
59. Reeve, R.M. J. Texture Stud, 1970, 1, 247-284.
60. Northcote, D.H. Ann. Rev. Plant Physiol. 1972, 23, 113-132.
61. Van Buren, J.T. J. Texture Stud. 1979, 10, 1-29.
62. Ilker, R.; Szczesniak, A.S. J. Texture Stud. 1990, 21, 1-36.
63. Dull, G.G., Leeper, E.F. in Postharvest Biology and Handling of Fruits and Vegetables, Haard, N.F.; Salunkhe, D.K., Eds.; AVI Publishing, Westport, CT, 1975, pp. 55-61.
64. Bourne, M.C. J. Texture Stud. 1979, 10, 83-94.
65. Bartley, L.M.; Knee, M. Food Chem. 1982, 9, 47-58.
66. Perring, M.A.; Pearson, K. J. Sci. Food Agric. 1988, 44, 193-200.
67. Kramer, G.F.; Wand, C.Y.; Conway, W.S. J. Amer. Soc. Hortr. Sci. 1989, 114, 942-946.
68. Chai, Y-L.; Ott, D.B.; Cash, J.N. J. Food Proc. Preserv. 1991, 15, 197-214.
69. Willis, R.H.H.; Lee, T.H.; Graham, D.; McGlasson, W.B.; Hall, E.G. Postharvest—An Introduction to the Physiology and Handling of Fruit and Vegetables; AVI Publishing, Westport, CT, 1981.
70. Lurie, S. J. Food Quality 1993, 16, 57-65.
71. Cantor, S.; Meredith, F.I.; Wicker, L. J. Food Biochem. 1992, 16, 15-29.
72. Holt, J.E.; Schoorl, D. J. Texture Stud. 1982, 13, 83-97.
73. Holt, J.E.; Schoorl, D. J. Texture Stud. 1982, 13, 349-357.
74. Schoorl, D.; Holt, J.E. J. Texture Stud. 1977, 8, 409-416.
75. Holt, J.E.; Schoorl, D. J. Texture Stud. 1984, 15, 377-394.
76. Mohsenin, N.N. Physical Properties of Plant and Animal Materials, vol. 1, Structure, Physical Characteristics and Mechanical Properties; Gordon and Breach Science Publishers, New York, 1970.
77. Mohsenin, N.N. J. Texture Stud. 1977, 8, 169-193.
78. Peleg, M.; Calzada, J.F. J. Food Sci. 1976, 41, 1325-1329.
79. Rodriguez, L.; Ruiz, M.; De Felipe, M.R. J. Texture Stud. 1990, 21, 155-164.
80. Siyami, S.; Brown, G.K.; Burgess, G.J.; Gerrish, J.B.; Tennes, B.R.; Burton, C.L.; Zapp, R.W. Trans. ASAE 1988, 31, 1038-1046.

81. Chen, P.; Yazdeni, R. Trans. ASAE 1991, 34, 956-961.
82. Tscheuschner, H.-D.; Du, D. J. Food Eng. 1988, 8, 173-186.
83. Vergano, P.J.; Testin, R.F.; Newell, W.C., Jr. J. Food Quality 1991, 14, 285-298.
84. Hung, Y.-C.; Prussia, S.E. Trans. ASAE 1989, 32, 1377-1382.
85. Zhang, X.; Brusewitz, G.S.; Huslig, S.M.; Smith, M.W. J. Food Quality 1993, 16, 57-65.
86. Fluck, R.C.; Ahmed, E.M. J. Texture Stud. 1974, 4, 494-500.
87. Chuma, Y.; Shiga, T.; Iwamoto, M. J. Texture Stud. 1978, 9, 461-479.
88. Bourne, M.C. J. Texture Stud. 1986, 17, 331-340.
89. Souty, M.; Bevils, L.; Andre, P. Sciences des Aliments 1981, 1, 265-282.
90. French, D.A.; Kader, A.A.; Labavitch, J.M. J. Food Sci. 1989, 54, 86-89.
91. Harper, K.A.; Beattie, B.B.; Best, D.J. J. Sci. Food Agric. 1973, 24, 527-531.
92. Brown, M.S. J. Texture Stud. 1977, 7, 391-404.
93. Szczesniak, A.S.; Smith, B.J. J. Texture Stud. 1969, 1, 65-89.
94. Szczesniak, A.S.; Torgeson, K.W. Adv. Food Res. 1965, 14, 33-165.
95. Stanley, D.W. in Rheology and Texture in Food Quality; deMan, J.M.; Voisey, P.W.; Rasper, V.F.; Stanley, D.W., Eds.; AVI Publishing, Wesport, CT, 1976, Chapter 11.
96. Stanley, D.W. in Physical Properties of Foods; Peleg, M.; Bailey, E.B., Eds.; AVI Publishing, Westport, CT, 1983, Chapter 6.
97. Stanley, D.W.; Tung, M.A. in Rheology and Texture in Food Quality. deMan, J.M.; Voisey, P.W.; Rasper, V.F.; Stanley, D.W., Eds.; AVI Publishing, Westport, CT, 1976, Chapter 3.
98. Lawrie, R.A. Meat Science, 3rd edition; Pergamon Press, New York, 1979.
99. Olson, D.G.; Parrish, F.C., Jr.; Stromer, M.H. J. Food Sci. 1976, 41, 1036-1041.
100. Lanari, M.C.; Bevilacqua, A.E.; Zaritzky, N.E. J. Food Process. Preserv. 1987, 11, 95-109.
101. Martino, M.N.; Zaritzky, N.E. J. Food Sci. 1988, 53, 1631-1637, 1649.
102. Wagner, J.R.; Añon, M.C. J. Food Technol. 1986, 21, 9-18.
103. Obanu, Z.A.; Ledward, D.A.; Lawrie, R.A. J. Food Technol. 1975, 10, 657-666.
104. Obanu, Z.A.; Ledward, D.A.; Lawrie, R.A. J. Food Technol. 1975, 10, 667-674.
105. Obanu, Z.A.; Ledward, D.A.; Lawrie, R.A. J. Food Technol, 1976, 11, 187-196.
106. Ledward, D.A.; Lymn, S.K.; Mitchell, J.R. J. Texture Stud. 1981, 12, 173-184.
107. Heldman, D.R.; Reidy, G.A.; Palnitka, M.P. J. Food Sci. 1973, 38, 282-285.
108. Kapsalis, J.G.; Walker, J.E., Jr.; Wolf, M. J. Texture Stud. 1970, 1, 464-483.
109. Kapsalis, J.G.; Drake, B.; Johansson, B. J. Texture Stud. 1970, 1, 285-308.
110. MacKenzie, A.P.; Luyet, B.J. Nature 1967, 215, 83-84, July 1.
111. Kim, H.-J.; Loveridge, V.A.; Taub, I.A. J. Food Sci. 1984, 49, 699-703, 708.
112. Regier, L.W.; Tappel, A.L. Food Res. 1956, 21, 630-639.
113. Kim, H.-J.; Taub, I.A. Food Chem. 1991, 40, 337-343.
114. Whiting, R.C. Food Technol. 1988, 42 (4) 104, 110-114, 210.
115. Fox, J.B., Jr.; Ackerman, S.A.; Jenkins, R.K. J. Food Sci. 1983, 48, 1031-1035.
116. Dunajski, E. J. Texture Stud. 1979, 10, 301-318.
117. Hatae, K.; Tobimatsu, A.; Takeyama, M.; Matsumoto, J. Bull. Jap. Soc. Scientif. Fisheries 1986, 52, 2001-2007.
118. Kuo, J.-D.; Hultin, H.O.; Peleg, M.; Atallah, M.T. J. Food Process. Preserv. 1991, 15, 125-133.
119. Izquierdo-Pulido, M.L.; Hetze, K.; Haard, N.F. J. Biochem. 1992, 16, 173-192.
120. Hatae, K.; Tamari, S.; Miyanaga, K.; Matsumoto, J. Bull. Jap. Soc. Sci. Fisheries 1985, 51, 1155-1161.
121. Howgate, P. in Sensory Properties of Foods; Birch, G.G.; Brennan, J.G.; Parker, K.J., Eds.; Applied Science Publishers, London 1977, Chapter 15.
122. Love, R.M.; Muslemuddin, M. J. Sci. Food Agric. 1972, 23, 1229-1238.
123. Montero, P.; Mackie, I.M. J. Sci. Food Agric. 1992, 59, 89-96.
124. Johnson, E.A.; Peleg, M.; Sears, R.A.; Kapsalis, J.G. J. Texture Stud. 1981, 12, 413-425.
125. Kelly, T.R. J. Food Technol. 1969, 4, 95-103.
126. Kelly, T.R.; Dunnett, J.S. J. Food Technol. 1969, 4, 105-115.
127. Kim, Y.J.; Heldman, D.R. J. Food Process. Eng. 1985, 7, 265-272.
128. Samson, A.D.; Regenstein, J.M. J. Food Biochem. 1986, 10, 259-273.
129. Krivchenia, M.; Fennema, O. J. Food Sci. 1988, 53, 999-1003.

130. Krivchenia, M.; Fennema, O. J. Food Sci. 1988, 53, 1004-1008.
131. Hsieh, Y.L.; Regenstein, J.M. J. Food Sci. 1989, 54, 824-826, 834.
132. Yareltzis, K.; Zetou, F.; Tsiaras, I. Lebensmittel Wissenschaft und - Technologie 1988, 21 (4), 206-211.
133. Santos, E.E.M.; Regenstein, J.M. J. Food Sci. 1990, 55, 64-70.
134. Kramer, D.E.; Peters, M.D. J. Food Technol. 1981, 16, 493-504.
135. Jose, J.; Perigreen, P.A. Fishery Technol. 1988, 25 (1), 32-35.
136. Rehbein, H.; Orlick, H. Int. J. Refrig. 1990, 13 (5), 336-341.
137. Nitisewojo, P. ASEAN Food J. 1987, 3 (2), 72-73.
138. Love, R.M. Nature 1956, 178, 988-989.
139. Jarenback, L.; Liljemark, A. J. Food Technol. 1975, 10, 229-239.
140. Giannini, D.H.; Ciaro, A.S.; Boeri, R.L.; Almando's, M.E. Lebensmittel Wissenscaft und-Technologie. 1993, 26, 111-115.
141. Tran, V.D. J. Food Sci. 1975, 40, 888-889.
142. Sirois, M.E.; Slabyj, B.M.; True, R.H.; Martin, R.E. J. Muscle Foods 1991, 2, 197-208.
143. Akiba, M.; Motohito, T.; Tankiawa, E. J. Food Technol. (Japan) 1967, 2, 69-73.
144. Reddy, G.V.S.; Srikar, L.N. J. Food Sci. 1991, 56, 965-968.
145. Sikorski, Z.E. CRC Crit. Rev. Food Sci. Nutr. 1976, 8 (1), 97-129.
146. Dyer, W.J. Food Res. 1951, 16, 522-527.
147. Shenouda, S.Y. Adv. Food Res. 1980, 26, 275-311.
148. Childs, E.A. J. Food Sci. 1973, 38, 1009-1011.
149. Yoon, K.S.; Lee, C.M.; Hufnagel, L.A. J. Food Sci. 1991, 56, 294-298.
150. Mao, W.-W.; Sterling, C. J. Texture Stud. 1970, 1, 338-341.
151. Dionysius, D.; Slattery, S.; Smith, R.; Deeth, H. Aust. Fisheries 1988, 47(10), 26-29.
152. Slattery, S.L.; Diorysius, D.A.; Smith R.A.D.; Deeth, H.C. Food Aust. 1989, 41 (4), 698-703, 709.
153. Marshall, G.A.; Moody, M.W.; Hackney, C.R.; Godber, J.S. J. Food Sci. 1987, 52, 1504-1505.
154. Parkin, K.L.; Brown, W.D. J. Food Sci. 1983, 48, 370-374.
155. Pan, B.S.; Yeh, W.-T. J. Food Biochem. 1993, 17, 147-160.
156. Giddings, G.G.; Hill, L.H. J. Food Proc. Preserv. 1978, 2, 249-264.
157. Godber, J.S.; Wand, J.; Cole, M.T.; Marshall, G.A. J. Food Sci. 1989, 54, 564-566.
158. Dagbjartsson, B.; Solberg, M. J. Food Sci. 1972, 37,185-188.
159. Dagbjartsson, B.; Solberg, M. J. Food Sci, 1973, 38, 242-245.
160. Ahmed, E.M.; Koburger, A.; Mendenhall, V.T. J. Texture Stud. 1972, 3, 186-193.
161. Ahmed, E.M.; Mendenhall, V.T.; Koburger, A. J. Food Sci. 1973, 38, 356-357.
162. Matsumoto, J.J.; Noguchi, S.F. in Surimi Technology; Lanier, T.C.; Lee, C.M., Eds.; Marcel Dekker, New York, 1992, Chapter 15.
163. Hamann, D.D., MacDonald, G.A. in Surimi Technology; Lanier, T.C.; Lee, C.M., Eds.; Marcel Dekker, New York, 1992, Chapter 17.
164. Wu, M.-C., in Surimi Technology; Lanier, T.C.; Lee, C.M., Eds.; Marcel Dekker, New York, 1992, Chapter 10.
165. Yokoyama, M. in Surimi Technology; Lanier, T.C.; Lee, C.M., Eds.; Marcel Dekker, New York, 1992, Chapter 13.
166. Smith, J.L.; Stanley, D.W.; Baker, K.W. J. Texture Stud. 1987, 18, 339-358.
167. Saltveit, H.E., Jr. J. Am. Soc. Hort. Sci. 1989, 113, 569-572.
168. Sharma, S.C.; Wolfe, R.R.; Wang, S.S. J. Food Sci. 1975, 40, 1147-1151.
169. Gariepy, Y.; Raghavan, G.S.V.; Castaigne, F.; Arul, J.; Willemot, C. J. Food Proc. Preserv. 1991, 15, 215-224.
170. Moline, H.E. Phytopathology 1976, 66, 617-619.
171. Bernang, M.E.; Brackett, R.E.; Beuchat, L.R. J. Food Protection 1990, 53, 391-395.
172. Kim, B.D.; Hall, C.B. Hort. Sci. 1976, 11, 466.
173. Cappellini, M.C.; Lachance, P.A.; Hudson, D.E. J. Food Quality 1984, 7, 17-25.
174. Adegoroye, A.S.; Jolliffe, P.A.; Tung, M.A. J. Sci. Food Agric. 1989, 49, 95-102.
175. Thome, S.; Alvarez, J.S.S. J. Sci. Food Agric. 1982, 33, 671-676.
176. Huang, Y.T.; Bourne, M.C. J. Texture Stud. 1983, 14, 1-9.
177. He, F.; Purcell, A.E.; Huber, C.S.; Hess, W.M. J. Food Sci. 1989, 54, 315-318.

178. Huang, Y.-C.; Thompson, D.R. J. Food Sci. 1989, 54, 96-101.
179. Brown, M.S. J. Sci. Food Agric. 1967, 18, 77.
180. Crivelli, G.; Buonocore, C. J. Texture Stud. 1971, 2, 89-95.
181. Howard, L.R.; Buescher, R.W. J. Food Biochem. 1990, 14, 31-43.
182. Buescher, R.W.; Burgin, C. J. Food Sci. 1988, 53, 296-297.
183. Hudson, J.M.; Buescher, R.W. J. Food Biochem. 1985, 53, 211-229.
184. Bell, T.A.; Etchells, J.L.; Turney, L.J. J. Food Sci, 1972, 37, 446-449.
185. Walter, W.M. Jr.; Fleming, H.P.; Trigiano, R.N. Food Microstructure 1985, 4, 165-172.
186. McFeeters, R.F.; Fleming, H.P. J. Food Sci. 1990, 55, 446-449.
187. Heredia, A.M.; Fernandez-Balanos, J.G. Grasas y Aceites (Seville) 1985, 36, 130-133.
188. Brenes Balbuena, M.; Garcia Garcia, P.; Garrido Fernandez, A. Grasas y Aceites 1986, 37, 301-306.
189. Tubil, L.G.; Noomhorm, A.; Verma, L.P. J. Food Preserv. Eng. 1991, 14, 69-82.
190. Evans, M.D.; Holmes, R.G. Trans. ASAE 1990, 33, 234-240.
191. Brusewitz, G.H.; Pitt, R.E.; Gao, Q. J. Texture Stud. 1989, 20, 267-284.
192. Stanley, D.W.; Aguilera, J.M. J. Food Biochem. 1985, 9, 277-323.
193. Aguilera, J.M.; Stanley, D.W. J. Food Process. Preserv. 1985, 9, 145-169.
194. Stanley, D.W.; Wu, X.; Plhak, L.C. J. Texture Stud. 1989, 20, 419-429.
195. Hentges, D.L.; Weaver, C.M.; Nielsen, S.S. J. Food Sci. 1990, 55, 1474-1476.
196. Aguilera, J.H.; Ballivian, A. J. Food Sci. 1987, 52, 691-695, 718.
197. Del Valle, J.M.; Aguilera, J.M.; Hohlberg, A.I.; Richardson, J.C.; Stanley, D.W. J. Food Process. Preserv. 1993, 17, 119-137.
198. Garcia, E.; Lajolo, F.M. J. Agr. Food Chem. 1994, 42, 612-615.
199. Richardson, J.C.; Stanley, D.W. J. Food Sci. 1991, 56, 590-591.
200. Mafuleka, M.M.; Ott, D.B.; Hosfield, G.L.; Mebersax, M.A. J. Food Process. Preserv. 1993, 17, 1-20.
201. Kilmer, O.L.; Seib, P. A.; Hoseney, R.C. Cereal Chem. 1994, 71, 476-482.
202. Berrnal-Lugo, I.; Prado, G.; Parra, C.; Moreno E.; Ramirez, J.; Velazco, O. J. Food Biochem. 1990, 14, 253-261.
203. Bernal-Lugo, I.; Castillo, A.; Diaz de Leon, F.; Moreno, E.; Ramirez, J. J. Food Biochem. 1991, 15, 367-374.
204. Hussain, A.; Watts, B.M.; Bushuk, W. J. Food Sci. 1989, 54, 1367-1368, 1380.
205. Liu, K.; Phillips, R.D.; McWatters, K.H. Cereal Chem. 1993, 70, 193-195.
206. Krieger, I.M. in Physical Properties of Foods; Peleg, M.; Bagley, E.B., Eds.; AVI Publishing, Westport, CT, 1983, Chapter 14.
207. Friberg, S.E.; El-Nokaly, M. in Physical Properties of Foods; Peleg, M.; Bagley, E.B., Eds.; AVI Publishing, Westport, CT, 1983, Chapter 5.
208. Sherman, P. in Emulsion Science; Sherman, P., Ed.; Academic Press, London, 1968, Chapter 3.
209. Tadros, T.F.; Vincent, B. in Encyclopedia of Emulsion Technology, Vol. 1; Becher, P., Ed.; Marcel Dekker, New York, 1983, pp. 129-286.
210. Sherman, P. in Rheology and Texture in Food Quality; deMan, J.M.; Voisey, P.W.; Rasper, V.F.; Stanley, D.W., Eds.; AVI Publishing, Westport, CT 1976, Chapter 15.
211. Chang, C.M.; Powrie, W.D.; Fennema, O. Can. Inst. Food Sci. Technol. J. 1972, 5 (3), 134-137.
212. Elliott, J.H.; Ganz, A.J. J. Texture Stud. 1977, 8, 359-371.
213. Paredes, M.D.C.; Rao, M.A.; Bourne, M.C. J. Texture Stud. 1989, 20, 235-250.
214. Vernon Carter, E.J.; Sherman, P. J. Texture Stud. 1980, 11, 351-365.
215. Coia, K.A.; Stauffer, K.R. J. Food Sci. 1987, 52, 166-172.
216. Paredes, M.D.C.; Rao, M.A.; Bourne, M.C. J. Texture Stud. 1988, 19, 247-258.
217. Aquilar, C.; Rizvi, S.S.H.; Ramirez, J.F.; Inda, A. J. Texture Stud. 1991, 22, 59-84.
218. Aquilar, C.; Rizvi, S.S.H.; Ramirez, J.F.; Inda, A. J. Texture Stud. 1991, 22, 85-103.
219. Chevalley, J.; Rostagno, W.; Egli, R.H. Revue Int'l du Chocolat 1970, 25, 3-6.
220. Levine, H.; Slade, L. in Food Structure—Its Creation and Evaluation; Blanshard, J.M.V.; Mitchell, J.R., Eds.; Butterworths, London, 1988, pp. 149-180.
221. Szczesniak, A.S. in Texture Measurement of Foods; Kramer, A.; Szczesniak, A.S., Eds.; D. Reidel Publishing, Dordrecht, Holland, 1973, Chapter 7.

222. Brennan, J.G. in Developments in Food Analysis Techniques—2; King, R.D., Ed.; Applied Science Publishers, Essex, England, 1980, Chapter 1.

223. Brandt, M.A.; Skinner, E.Z.; Coleman, J.A. J. Food Sci. 1963, 28, 404-409.

224. Szczesniak, A.S.; Brandt, M.A.; Friedman, H.H. J. Food Sci. 1963, 28, 397-403.

225. Munoz, A. J. Sensory Stud. 1986, 1, 55-83.

226. Meilgaard, M.; Civille, G.V.; Carr, B.T. Sensory Evaluation Techniques; CRC Press, Boca Raton, FL, 1987, Chapter 8.

227. Civille, G.V.; Szczesniak, A.S. J. Texture Stud. 1973, 4, 204-223.

228. Civille, G.V.; Liska, I.H. J. Texture Stud. 1975, 6, 19-31.

229. Szczesniak, A.S.; Loew, B.S.; Skinner, E.Z. J. Food Sci. 1975, 40, 1243-1246.

230. Bourne, M.C. Food Texture and Viscosity: Concept and Measurement; Academic Press, New York, 1982 (reprinted in 1994).

231. Szczesniak, A.S. in Texture Measurement of Foods; Kramer, A.; Szczesniak, A.S., Eds.; D. Reidel Publishing, Dordrecht, Holland, 1973, Chapter 6.

232. Scott Blair, G.W. Adv. Food Res. 1958, 8, 1-61.

233. Bourne, M.C. J. Food Sci. 1966, 31, 1011-1015.

234. Bourne, M.C.; Moyer, J.C.; Hand, D.B. Food Technol. 1966, 20, 522-526.

235. Friedman, H.H.; Whitney, J.; Szczesniak, A.S. J. Food Sci. 1963, 28, 390-416.

236. Bourne, M.C. J. Food Sci. 1968, 33, 223-226.

237. Bourne, M.C. Food Technol. 1978, 32(7), 62-67.

238. Szczesniak, A.S. J. Texture Stud. 1975, 6, 139-156.

239. Breene, W.M. J. Texture Stud. 1975, 6, 43-82.

240. Holt, C.B. J. Texture Stud. 1970, 1, 491-501.

241. Bourne, M.C. in Rheology and Texture in Food Quality; deMan, J.M.; Voisey, P.W.; Rasper, V.F.; Stanley, D.W., Eds.; AVI Publishing, Westport, CT, 1976, Chapter 6.

242. Baird, D.G. J. Texture Stud. 1981, 12, 1-16.

chapter nine

Water migration and food storage stability

Pavinee Chinachoti

Contents

Introduction

Instability of food is a major issue in food storage, and solutions to this problem include modification of formulation, processing, packaging, and storage conditions. Even though great efforts are made to minimize deterioration, food scientists know that the inherent nature of food dictates the reaction rates and the approach to an equilibrium state. Investigators of reaction kinetics of food deterioration, therefore, have been much

0-8493-2646-X/97/$0.00+$.50
© 1998 by CRC Press LLC

Table 1 Moisture Migration Tendencies in Mixed Confections

Components	a_w	Moisture Content (%)	Direction of Migration of H_2O
Jelly	0.77	20.80	From fondant to jelly
Fondant	0.83	12.00	
Marshmallow (42 De glucose syrup)	0.80	22.50	From marshmellow to jelly
Jelly	0.77	20.80	
Marshmallow (63 De glucose syrup)	0.76	22.50	May move either way
Jelly	0.77	20.80	
Marshmallow (63 De glucose syrup)	0.77	22.50	From marshmallow to caramel
Caramel	0.73	13.00	
Nougat	0.59	9.50	May move either way
Toffee	0.59	9.40	

Note: Adapted from Reference 2.

interested in seeking the solution to solve or deal with the food instability problem. One of the key considerations attracting much interest lately is the role of reactant and solvent diffusion. It is already known with respect to basic principles of chemical and biochemical reactions that their rates are highly dependent on the reactant and catalyst (including enzyme) concentration and associated mobility. Thus, an understanding of the theory and its practical implications in food processing and for shelf life stability could lead to a better and more effective way of improving food quality and safety. As water is the most common solvent, this chapter is devoted to water migration, in particular to its transnational mobility and to the influence of physicochemical properties such as viscosity and glassy–rubbery transitions on such mobility and in turn on reaction kinetics.

Water migration and diffusion is considered to be one of the most important subjects in drying[1] and packaging technology.[2] Most of the engineering aspects of water diffusion have been studied in terms of diffusivity and the mechanism of diffusion, mathematically modeling the water diffusion rate, and the practical problems in drying processes. Osmotic dehydration technology is based on the migration of water from a food or fruit piece to an osmotic solution. Other aspects of water migration include water migration between different parts of a food, such as from crumb to crust, from pizza sauce to its shell, from intermediate-moisture fruit pieces to cereal flakes, and from a pie filling to the crispy crust. Also important are the drying out of a food during storage to an extent that may compromise quality and yield, such as in a "freezer burn" situation, and the drying out of baked goods (e.g., donuts and breads) or even fruits that do not have adequate protective coatings. In confections, migration of water between two different ingredients is common (Table 1).

Adsorption and desorption of water involve the ability of water molecules to penetrate into or diffuse within the food. Water sorption isotherm data provide valuable information about water migration during adsorption or desorption processes. The rate of diffusion decreases as the driving force (the water chemical potential difference between the food and the environment) becomes extremely small. According to Duckworth, "... information on the rates of movement of water between different parts of a system may be of great practical importance; ... gradients develop on either water content or water vapor pressure leading to moisture transfer which may or may not be accompanied by change in state."[3]

These problems are related to macroscopic moisture migration due to water molecules diffusing across a boundary under the influence of a driving force (e.g., the macroscopic moisture gradient). Such moisture migration is actually microscopic and molecular short-range in nature in food and is difficult to determine and control. Molecular mobility of water over relatively short ranges could have many implications: (1) physicochemical, such as staling and retrogradation of starch, gelatinization, syneresis, sugar crystallization, ice crystal formation (and antifreezing phenomena), and encapsulation properties (including time release of compounds by hydrophilic components or emulsion or vesicle systems); (2) biological, such as microbial growth in a diffusion-limited environment (e.g., surface degration of biodegradable packaging materials by molds, microbial activity in intermediate moisture foods, and solid-state fermentation); and (3) chemical, such as browning reactions, oxidation of lipids and pigments, antioxidative activity of water-soluble antioxidants, and protein as well as enzyme stability. These are only some of the many moisture-related phenomena encountered in a wide range of foods. It is fair to say that most food problems are *related* to molecular diffusion and many of them are *limited* by water diffusion (migration). Some of these are diagrammatically illustrated in Figure 1.

Theory

Overview of the fundamental aspects of diffusion

Water diffusion most simply can be described empirically in terms of an as effective diffusivity (D_{eff}), which is generally expressed by an Arrhenius-type equation with an activation energy that is inversely proportional to moisture content. D_{eff} typically decreases with decreasing moisture content, but it normally falls in the order of 10^{-10} to 10^{-12} m^2/s.[1] However, the mechanisms of water diffusion can be complex, and simple diffusion models (such as a Fickian type for diffusion through nonporous solids) may not apply. Variations in D_{eff} for similar materials are common and often attributed to violating the assumptions underlying the application of Fick's law. As described by Okos et al., several mechanisms have been proposed to explain water diffusion through food and packaging materials; these are water binding to the matrix components, surface diffusion, hydrodynamic or bulk flow, and capillary flow. Loncin[4] described the difference between a diffusion and a flow process as follows: diffusion occurs when the driving force is the difference in concentrations or in partial pressure, while flow occurs when the driving force is the difference in total pressures.[4] Mass transfer by a flow process takes place faster than by molecular diffusion. Drying of a hydrophilic product may cause shrinkage and hydration of some polymers may result in swelling.[5] The physical *state* of water is said to be *bound* when retained within capillaries of a hydrophilic solid, dissolved (in a matrix component), or physically adsorbed on the surface of the solid. This binding of water results in lowered vapor pressure and may affect greatly the barrier property of a packaging material.[6] Mass transfer properties of small molecules, such as water, flavorants, and gases, through a food system are discussed elsewhere.[7]

The theory of transmission through edible films follows that of any packaging material. The rate of gas transmission is controlled by a number of factors related to the properties of the film, the gas or vapor, and the interaction between the film and the gas or vapor. A crystalline structure presents a significant barrier to diffusion, diffusion would occur around the crystals and thus the rate would depend on the tortuosity of the matrix. In amorphous materials, the existence of pores allows a more possibility for a vapor or gas to permeate. If the disordered structure is more or less rigid, permeation would become

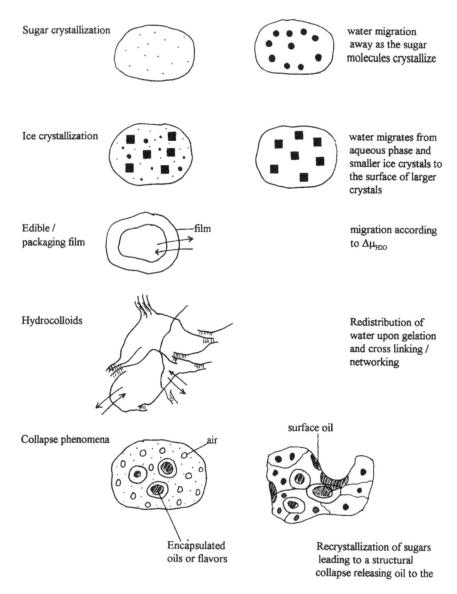

Figure 1 Diagram of some moisture migration and related physicochemical changes in foods.

selective, i.e., the pores would only allow molecules smaller than the pore size to pass through but would block larger molecules. In nonrigid amorphous structure, permeation through pores is highly dependent on the molecular motion due to thermal energy of the network. At or above a transition point, the pores will be constantly disappearing and reforming as a result of this thermal motion. Consequently, diffusion or movement of molecules from pore to pore would occur along the concentration gradient. Molecules can diffuse not only through intermolecular spaces, but also through cracks and microscopic capillaries. These different mechanisms are controlled by different physical and physicochemical factors. Thus, in understanding the nature of edible film, one must realize what would be the main contribution of permeation.

Under a steady state condition, Fick's Law of Diffusion can be used to describe diffusion of inert gas through films:

$$Q = \pm \left(\frac{AtD}{l} \right) \frac{dc}{dx}, \tag{1}$$

where Q = the quantity of gas diffusion through area A in time, D = the diffusion constant, dc/dx = the concentration gradient, and l = film thickness. After integration, the equation becomes

$$Q = \frac{AtD}{l} \left(c_1 \pm c_2 \right), \tag{2}$$

where c_1 and c_2 are the concentrations of the gas at the two surfaces of the film and are normally measured in terms of vapor pressure (p). According to Henry's Law, $c = S \cdot p$, where S is the solubility coefficient of the gas in the film. Thus,

$$Q = \frac{AtDs}{l} \left(p_1 \pm p_2 \right). \tag{3}$$

By definition, the permeability constant, P, is equal to $D \cdot S$. Assuming that D and S are independent of concentration, P can be calculated as follows:

$$P = DS = \frac{lQ}{At \left(p_1 \pm p_2 \right)}. \tag{4}$$

For elastic and nonelastic barriers, P, D, and S obey the Arrhenius relation:

$$P = P_o e^{\pm E/RT}$$

$$D = D_o e^{E_d/RT}$$

$$S = S_o e^{\pm \Delta H/RT}$$

where E and E_d are the activation energy for permeability constant and the diffusion process, respectively, and ΔH corresponds to the heat of solution. Consequently,

$$E = \Delta H + E_d$$

This relationship holds in gases, such as O_2, N_2, and CO_2. In such cases when there is no interaction between the diffusing species and the film, the permeability will generally be exponentially dependent on the temperature. The most resistant barriers will be mostly temperature dependent and the poorest barrier film will often be almost independent of temperature. All of the above theory applies only to gases or vapors that do not interact with films.

Physicochemical aspects of moisture migration

Unfortunately, most biopolymers are hydrophilic and interact extensively with water (as well as with organic volatiles). In such cases, this relationship does not apply. Normally, an increase in temperature would result in an increase in D (diffusion coefficient);

D_0 is a constant indicating the degree of "looseness" in the polymer structure. However, when the polymer is plasticized, the presence of the plasticizer results in a decrease in E_d and an increase in D_0. Thus, the diffusion of water is affected by the fact that the water molecules themselves are adsorbed by the film, decreasing E_d and increasing D_0 due to the increase in looseness, which allows a higher rate of permeation. Similarly, O_2 diffusion through a water-plasticized film will be also affected, since the pore size increases, allowing more gas to diffuse far more rapidly than when the film is in a dry state.[8] This concept has led to many suggestions that transport through films is a function of the glassy–rubbery transition temperature (T_g) of the polymer.[7,9]

Crosslinking and partial crystallinity lead to restricted diffusion in the film, whereby gas and vapor molecules travel along tortuous paths in the matrix.

The water vapor transmission rate (WVTR), defined as

$$\text{WVTR} = \frac{Q}{tA},\tag{5}$$

increases with moisture content of a hydrophilic material and has a strong correlation with the water sorption behavior. In measuring the WVTR, a film is placed between two compartments with different water vapor pressure (a_w). Once a steady state has been established, WVTR is measured. Thus, WVTR is a function of temperature and Δp (and also of the range of a_w in the two compartments, since the transmission rate depends on the amount of water sorbed onto the film). It has been suggested for the case of methyl cellulose films that Henry's law could not be applied, because the constant S (solubility) is calculated from the water sorption isotherm and varies with water concentration.[10] Permeability (P), in this case, cannot be defined simply as DS (i.e, diffusivity × solubility).

Because of such a strong relationship between water vapor transmission rate and water sorption, BET monolayer values have been used. The glassy–rubbery state of the polymer is also suggested to play a key role in water vapor transmission, thus T_g has been said to be a determining factor.

Studying plasticization of gluten films with various plasticizers, including water, glycerol, sorbitol, and sucrose, Cherian et al. found those gluten films show a marked degree of incompatibility in the system unless glycerol is used and at a high level.[11] DMA (dynamic mechanical analysis) and DSC (differential scanning calorimetry) showed that glycerol leads to a single transition. Water vapor transmission rate, tensile strength, and film elongation, however, were found independent of T_g, but rather highly dependent on the magnitude of the transition (i.e., the amount of glycerol present). It was hypothesized that, in these films, glycerol was present in excess and partially phase-separated, creating pores or channels for water transmission. Thus, the higher glycerol content, the greater the degree of phase separation and thus the number or size of the pores through the water could diffuse.

It remains questionable whether or not the system viscosity would have a direct influence on water vapor transmission, since the glassy–rubbery transition often leads to many orders of magnitude of change in viscosity. Upon transition from a glassy to a rubbery state, there is a spontaneous and significant increase in intermolecular space due to the increase in free volume. In a study on aroma diffusion through matrices probed by ESR, the correlation time (τ_c) was found inversely proportional to the transnational diffusivity value; the relationship between D_{tran} and τ_c conformed extremely well to the theoretical equation[12]:

$$D_{\text{trans}} = \frac{2a^2}{9\tau_c},\tag{6}$$

where a is a constant. It was concluded that aroma diffusivity depends on the molecular weights of nonvolatile solutes (glycerol, sorbitol, etc.), but not on viscosity. Diffusivity is rather dependent on the mobility of the solutes (τ_c).

Water vapor permeability and diffusivity through methyl cellulose edible films follows a non-Fickian model.[10] Theoretically speaking, in homogeneous solutions, the rotational diffusivity (D_{rot}) of a Brownian particle is directly related to the transnational diffusivity (D_{trans}) from the Stokes–Einstein relationships:

$$D_{rot} = \frac{kT}{8\pi\eta r^3}, \tag{7}$$

$$D_{trans} = \frac{kT}{6\pi\eta r}, \tag{8}$$

where k is the Boltzmann constant, T is the absolute temperature, η is the solution viscosity, and r is the radius of the particle. (Note that from the aroma diffusivity work discussed above, no correlation was found between diffusivity and matrix viscosity.[12])

In heterogeneous systems, such as in edible films with a polymer network and with small molecules in intermolecular spaces, D_{rot} obtained from ESR is representative of the local environment of the probe. The fact that D_{rot} is found to increase linearly with moisture content of the methyl cellulose film has been interpreted as due to the decrease in local viscosity caused by an enlargement of the free space in the polymer matrix, i.e., a plasticization of the film by water. At steady state,

$$D = \frac{\int_{c_2}^{c_1} D_c dc}{\left(c_1 \pm c_2\right)}, \tag{9}$$

then

$$D = \frac{WVTR \cdot l}{\left(c_1 \pm c_2\right)}, \tag{10}$$

where D_c is the real diffusion coefficient and c_1 and c_2 are the water concentrations at the film surfaces.

Diffusion coefficient of water was calculated to be $0.3–3.0 \times 10^{-12}$ m^{2}s^{-1} (over 2–20% moisture content, dry film basis). This value falls in the range normally found for water in many food products such as in gelatin and starch. At <3% moisture, which is associated with (monolayer, in a so-called glassy state), vapor phase diffusion of water and sorption of water occur simultaneously. It is assumed that in porous materials and at a very low moisture content, the diffusivity of water is equal to that of water vapor in air, 10^{-6} m^{2}s^{-1}. As moisture increases, the water molecules fill the pores and vapor diffusion is replaced by liquid diffusion through filled pores diffusion.

In biopolymers, as the water is sorbed, there is a considerable swelling in the polymer matrix as it starts to get into the glass–rubbery transition range. Since some regions might start to swell first and others later (depending on moisture and time–temperature relationship in this kinetically controlled process), diffusion of water through such a system becomes extremely complicated and the Fickian equation does not apply. As water is being absorbed, the T_g of that domain is depressed and that particular region becomes rubbery. Such a change should lead to a decrease in network density, an increase in large molecular chain motion, and rearrangement of chain segments. These chains, in turn, are controlled by relaxation phenomena and have been used to explain many diffusion-controlled processes.[13] However, the mechanism of diffusion is complex, and since there can be a significant heterogeneity in the system in terms of domains (T_g can vary on the order of 100°C in some cases, particularly in relatively dry systems), predicting diffusion by using T_g alone is questionable and is yet to be proven. Additionally, in systems with paracrystallinity or systems blended with lipids and waxes, diffusion is complicated by restricted diffusion due to tortuosity. Applying the appropriate model is very critical.

Analyses of non-Fickian transport have been described by Peppas and Brannon-Peppas[14] as follows. In amorphous polymers, diffusion in a glassy state is non-Fickian. Diffusion in rubbery polymers that sob the diffusing species is also non-Fickian due to swelling and other mechanisms discussed above. One can categorize the transport phenomena of water in biopolymers as Case I and Case II transport. In Case I (Fickian) diffusion, water mobility is much lower, i.e., on a longer time scale than the relaxation times of the molecular segments of the system (i.e., segments of the polymer in question), and the diffusion rate decreases with time as the water concentration gradient decreases with time. In Case II transport, water mobility is much higher, i.e., on a shorter time scale than the relaxation times of segmental motion of the polymer, so the boundary between swollen and nonplasticized polymer becomes a rate-limiting step. There is still another type of diffusion: is one that exhibits some behavior intermediate between Case I and Case II. Characterization of water diffusion in amorphous polymers by using the Deborah (De) number has been theoretically useful. Deborah number is the ratio between the characteristic relaxation time of the polymer, λ, and the characteristic diffusion time, θ:

$$De = \lambda/\theta.$$

The relationship between De and diffusion can be demonstrated in Figure 2. For De ≫ 1, the relaxation time is much longer that the diffusion time and thus there is no change in the polymer structure with time during the diffusion process. Transport of water follows the Fickian model. For De ≪ 1, the relaxation time is much shorter than the diffusion time and thus there is a conformational change in the polymer occurring spontaneously as diffusion takes place. It has been said that this type of diffusion is Fickian (although there are many exceptions in swelling rubbery polymers, etc.) When De = O(1), the relaxation and diffusion times are of the same order of magnitude; as the water penetrates into the polymer network, structural rearranging does not occur immediately.

Case II transport is described by a characteristic relaxation constant. This non-Fickian behavior requires a number of parameters obtained experimentally to describe the coupling of diffusion and relaxation phenomena. Assuming that a polymer film is subjected to water sorption equilibration at a given relative humidity, there is an advancing front of swollen region defined by $X \leq x \leq l/2$, and there is a uniform water concentration of C_o, the water concentration at equilibrium. In the glassy region defined by $0 \leq x \leq X$, there is essentially no water. For thin polymer films,

$$\frac{M_t}{M_\infty} = \frac{2k_o t}{C_o l},$$

where C_o is the concentration of water at equilibrium, K_o is defined as Case II relaxation constant. This equation shows that the Case II transport in a polymer is characterized

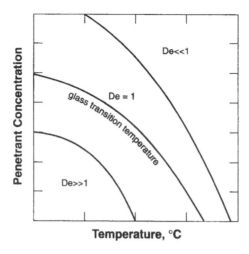

Figure 2 Regions of transport of penetrant in polymer/penetrant system. (From Peppas N. A., Brannon-Peppas L., in *Water in Foods*, Elsevier Applied Science, New York, 1994, 189. With permission.)

by a linear time-dependent water uptake until two diffusing fronts meet at the center of the slab.

It has been described that for diffusion in glassy state of polymers, diffusion of water follows the intermediate characteristics of Case I (Fickian) and Case II (non-Fickian).[14]

Inhibition of moisture migration in food has been a major problem. Recent research has focused on developing new materials and processes that would inhibit such migration. Possibilities include modifying the polymer network by changing the plasticizer, by crosslinking, and by laminating. The key is to strike a balance between maintaining desirable mechanical properties (tensile strength and elongation) and processing ability (casting and extrusion molding) and introducing resistance to the permeation of gases, vapors, and small solutes.

Upon moisture loss, there are also considerable changes at the microscopic level, such as internal cracks (due to tensile stresses developed on the shrinking surface), shrinkage, loss of cellular structure, and change in porosity.[15] These are factors that would influence the mass transport of water from a food.

Theoretical description of migration through a polymer network (homogeneous) has been given elsewhere.[2] If solid (discrete) particles are packed together in an ordered fashion, the resultant solid is crystalline. If they are randomly arranged, the result is an amorphous solid (disordered state). Amorphous or crystalline solids cannot form completely continuous arrangements, but they can be arranged in such a way that a network is formed. Water molecules migrate around the crystalline domains and penetrate through the amorphous matrix, depending on the size and dimension of cavities present, which in turn depend on the vibrational motion and flexibility of the molecules forming the network. According to polymer science theory, cavities in an entangled, amorphous polymer control the rate of movement of diffusing molecules. Above a transition point, these cavities are continually disappearing and reforming due to thermal motion, and the activation energy for diffusion of smaller molecules is related to the energy required to form a cavity against the forces holding the polymer chains together.[2] Therefore, to understand permeation through synthetic polymers one must understand the nature of the polymer itself. Properties such as chemical structure, degree of crystallinity, chain length, density, molecular weight and degree of polymerization, degree of branching, double bonds, additives, plasticizers, etc., are all relevant.[16] This concept of using our knowledge of glassy–rubbery transitions and the free volume theory to describe diffusion-controlled

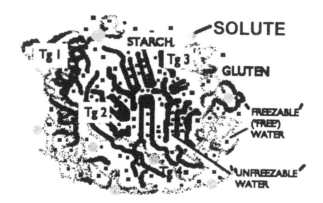

Figure 3 Diagram of a complex systems with separated regions of polymers, water, and solutes, where cooperativity among regions may be limited, leading to heterogenous domains.

reactions in foods has been applied by many food scientists in recent years.[13,17–21] Basic principles of glassy–rubbery transition can be found in many text books.[22,23]

Characterization of a glass transition

There is a great challenge to food scientists in exploiting this approach. It involves the difficulty in applying a concept developed for synthetic polymers to complex and heterogeneous food systems containing polymers, small solutes, and incompatible constituents that may not exhibit a glassy–rubbery transition. Additionally, with the exception of some carbohydrates such as pure sugars, starch, and starch by-products, literature reporting thermal analysis of T_g has been limited to one instrument (DSC), and the critical physical properties (e.g., mobility and local viscosity) are not clearly defined. The concept of a glassy–rubbery transition in some food carbohydrates is not new. Measurements have shown that even dynamic mechanical analysis (DMA), which is one of the most sensitive methods, may be difficult to interpret. In part, this difficulty arises because changes in storage modulus (E') involve at most 1–2 orders of magnitude, and the transition occurs not at a single, clearly recognizable temperature, but over a *range* of temperatures, in most cases over 50–100°C. Figure 3 represents an example of a heterogeneous food system with phase separations. At high enough moisture contents to include "freezable" water, where a clear phase separation due to ice interferes with and dominates other transitions, characterization is difficult. Additionally, food components are highly heterogeneous, and a molecule of a protein can have subregions that go through many extremely small transitions over a wide range of temperature, which are practically impossible to detect by DSC. Figure 4 compares thermograms of vital wheat gluten obtained using DSC to relative storage modulus (E') determined using DMA.[24,25]

Biopolymers

Biopolymers are produced to function in living systems and as such to have unique properties. Consequently, in contrast to synthetic polymers, biopolymers must be precise in chemical structure (from their primary to quaternary arrangements). Thus, different side chains on a protein could be characterized by different onsets of mobility, activation energies, and motional correlation times. Concerning the forces that stabilize the native conformations, association and dissociation of subunits by inter- and intramolecular

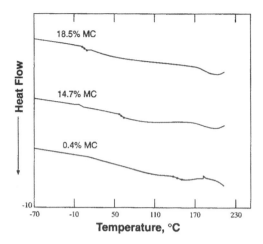

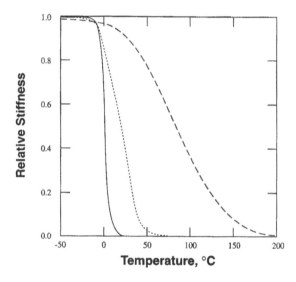

Figure 4 Thermograms from DSC (upper frame) and DMA (lower frame) for vital wheat gluten.[24] The relative stiffness was calculated from a model.[25] "MC" corresponds to moisture content.

interactions are very specific to the characteristic biological systems of having specific solvent composition and properties. Thus, thermal characterization of such polymers deals with thermal transitions of bond breaking and force breaking that may reversibly or nonreversibly transform the biopolymer in question. Additionally, mechanical characterization of a protein may be relevant to some functionally important rheological properties (e.g., DMA analysis of gluten and muscle proteins), but probably not so useful or appropriate in describing other functional properties (e.g., surface-active proteins and enzyme specificity). Lastly, one cannot lose sight of the fact that food systems are mostly heterogeneous and include phase separations, incompatible as well as compatible solvents, solutes and polymers, emulsions, micelles, cellular structures, and varying levels of arrangement and structuring. The *structural relaxation* phenomena associated with these constituents are expected to be composed of a combination of narrow and broad distribution of relaxation times.

Relaxation phenomena

Different types of analysis provide information reflecting different levels of distance or time dimensions. For instance, *structural relaxation* obtained by thermal analyses and rheological methods yields some information on long-range order (suprastructure, on a time scale of minutes, hours, or days), while *molecular relaxation* obtained by nuclear magnetic resonance (NMR), electron spin resonance (ESR), and dielectric relaxation yields information on short-range order (molecular, on a time scale of picoseconds to milliseconds). Perturbation of water molecules on surfaces is short-range, such as hydration of proteins. Characterization of water surrounding biological molecules has been useful but complicated. To what extent water molecules are affected by solute surfaces (and how many water molecules are affected by solutes in terms of local viscosity) is a question that cannot be answered easily. Nevertheless, a basic understanding of how short-range interactions among molecules are affected by the suprastructural relaxation (such as glass transition) or vice versa is most critical. This is particularly true in the area of diffusion-controlled phenomena where the mechanism of diffusion of water and solutes may or may not be controlled by T_g.

In a system containing macromolecules, no direct correlation is normally expected between the molecular mobility of the water and the rheological or other macroscopic properties. This limitation stems from the fact that relaxation processes for water (and small solutes) occur at a much higher frequency and shorter time frame (generally 10^{-9}–10^{-12} s) than macromolecules (generally 10^{-3}–10^{-6} s or even longer). Thus, methods used to discern these two contrasting relaxation processes differ by many orders of magnitude in terms of the time frames of events (and spatial dimensions probed). While local viscosity (on a molecular level) is likely to have a direct impact on the mobility of small molecules, macroscopic or bulk viscosity may not.

In a small solute system, local viscosity and bulk viscosity tend to change in similar ways. Thus, much of the local viscosity data (and correlation times) observed by techniques such as ESR agree quite well with a long-range relaxation process, such as bulk viscosity and glassy–rubbery transition.[26] This leads to correlations between T_g and reaction rates, such as lipid oxidation in amorphous sugars, recrystallization of sugars, and enzymatic browning in relatively small solutes (see below). It is necessary, however, to keep in mind the fundamental differences among various molecular and structural relaxation processes and what they truly mean in a molecular sense. It has been much argued in the polymer science field that what appears to be a compatible system from T_g analyses by DSC, TMA, and DMA may be incompatible (phase-separated) from the molecular (e.g., NMR) point of view. It is necessary to be cautious then in the food field (due to presence of a vast range of molecules with so many molecular weights, differences in nature, and differences in chemical and physical properties) in applying the glass transition concept without corroborating, independent molecular investigations. Since the use of the glass transition concept (which is based on long-range, structural relaxation) is becoming ever more common in the food science field, it becomes increasingly important to undertake more fundamental research on molecular relaxations.

A glassy-rubbery transition itself (and melting for that matter) is a kinetic phenomenon and thus its transition temperature would greatly depend on the moisture–temperature history of the samples before and during the analysis. Polymer relaxation upon a glassy–rubbery transition occurs over a narrow or a broad range in temperature depending on the characteristic distribution of relaxation times.[22] This distribution of polymer relaxation times is broadened markedly in a *nonequilibrium* glassy state as the temperature is lowered below T_g, because the activation energies are different at different domain sizes. Thus, experiments that vary in cooling rates should result in different T_g distributions and correspondingly different activation energies for diffusion and reaction rates. Different methods measure the unique properties (e.g., heat flow for DSC, mechanical change for DMA, dielectric properties for dielectric relaxation) that may or may not exhibit the same rate of change upon undergoing glassy-rubbery transition. These factors have made it difficult to compare data and even correlate reaction rates with such parameters, although such comparison have been successful in polymer science. Despite such success, there is much disagreement on how to define glassy–rubbery transitions in synthetic polymers.[27] Food polymer science is still in a very early stage of development, and efforts have to be made to better understand the measurements and interpretations associated with such a concept.

Applications

WLF kinetics

There is only a relatively small amount of data currently available to show *independently* that the WLF (Williams, Landel, Ferry)[29] kinetics adequately describes local diffusion of water and small solutes in food and is otherwise applicable to foods. Additionally, only extremely limited data from food systems are available on the local mobility of small molecules.

For the case of moisture migration, water and solute diffusion rates increase with increasing temperature, but increase by many orders of magnitude as the temperature passes through the glassy–rubbery transition temperature (T_g).[13] Thus, many believe that reaction rates are controlled by $(T-T_g)$.[19-21,28] If so, it might be possible to predict food quality from the knowledge of glassy–rubbery transition. According to some,[17,18,21] reaction rates that are diffusion-controlled are influenced by the amorphous components and might be expressed as a second-order type of kinetics (WLF),[29] when $T_g < T_{food} < T_g + 100$ K,[21]

$$\log \frac{D_s}{D} = \frac{\pm c_2 \left(T \pm T_s \right)}{c_2 + \left(T \pm T_s \right)} \tag{11}$$

where D_s and D are diffusion coefficients at T_s and T; T_s and T are reference and system temperatures, respectively; and c_1 and c_2 are constants.

If the reference temperature chosen is T_g then D_s would be D_g, the diffusion coefficient at the glassy–rubbery transition. There has been much evidence indicating (in synthetic polymers, mostly) that D, in general, is very small at temperatures below T_g, and that at temperatures above T_g its value increases with temperature in accordance with a different activation energy. For instance, in the case of poly(ethylene terepthalate) (PET) and poly(butylene terephthalate) (PBT),[30] D of water was found to fit two Arrhenius equations, one below T_g and one above T_g; the calculated activation energies were increased above T_g by more than 100% for PET and by >33% for PBT, as compared to the corresponding values at temperatures below T_g. Rossler[31] reviewed the data on diffusion in simple supercooled liquids showing a change in dynamics, related to viscosity and correlation times taking place at a temperature (T_c) well above T_g; these data suggest a discrimination in such diffusion corresponding to different Arrhenius parameters at high and low temperature ranges and perhaps reflect different mechanisms of diffusion.

Roos also described many reaction kinetics (such as nonenzymatic browning reaction, collapse and crystallization, and reactions in a frozen state) that may be restricted by diffusion and suggested that reaction rates may be influenced by the temperature or moisture where a glassy–rubbery transition occurs.[21] He went on to calculate the diffusion coefficient from the WLF equation above and showed, based on an Arrhenius plot (ln D vs. $1/T$), that D at a higher temperature range (common for food processing) would be much higher than could be reasonably expected. Unfortunately, there was no experimental values for D to support this hypothesis.

Physicochemical implications

There are many known or hypothetical physicochemical implications for food of moisture migration within the food. Some have been covered in a great detail elsewhere, so the discussion of such related topics will not be exhaustive, but is intended to highlight some aspects certain to be of interest to anyone involved with food processing and storage.

Drug release

In an amine/epoxy system, chemical and diffusion-controlled kinetics have been studied.[32] Accordingly, T_g was found to increase during curing, and the chemical reaction became more and more diffusion-controlled over time. Thus, if one waited for a sufficiently long time, T_g increased to a level temperature from the experimental temperature and the sample became vitrified. Before verification occurs, the reaction rate follows first-order Arrhenius kinetics, but after verification it becomes diffusion-controlled and follows second-order WLF kinetics. However, a diffusion phenomenon can also follow a zero-order type reaction, as in a diffusion-controlled drug release system.[33] One of the explanations given relates this to the mechanism of diffusion at the swelling front. Due to swelling, the system is not at equilibrium and the diffusion kinetics follows a non-Fickian type. Imbibition of water at the swelling front facilitates a drug release as the hydrophilic polymer relaxes upon swelling and undergoes glassy–rubbery transition.[34] The fraction of drug release (M_t/M_∞) can be empirically equated to kt^n (where k is the rate constant, t is time, and n is the order of the release). Due to a rapid glassy–rubbery transition during swelling of hydrophilic polymers, n may approach 1, resulting in a zero-order release.

Subzero temperatures

Kerr et al. reported kinetic data for the hydrolysis of disodium-p-nitrophenyl phosphate (DNPP, catalyzed by alkaline phosphatase) in partially frozen food polymer solutions containing sucrose, polydextrose (of varying molecular weights), and maltose.[35] The rate mostly fitted an Arrhenius equation fairly well at above T_m or the melting point of ice (R^2 in the range of 0.95), but showed different activation energies for the different solutes used. However, the rate data at temperatures between T and T_g' (where diffusion was supposed to be rate-limiting), failed to follow the WLF equation, although the data points clustered around straight lines when log [hydrolysis], in μmol DNPP h^{-1}, was plotted against $T-T_g'$. They concluded that the kinetics of the reaction changed as the temperature fell below T_m, but cautioned that partial ice melting over the (T_g-T_m) range should be taken into account. This issue was raised earlier by a number of researchers.[36] There are limited number of reports on diffusion and reaction rates at subzero temperatures, but it seems that reaction rates have been shown to be of an Arrhenius type.[21,35]

Phase separations

Water molecules are more likely to diffuse into amorphous regions. The presence of phase separated crystalline structures (e.g., ice crystals, starch or sugar crystals) can restrict the diffusion process. If there is a great degree of water–solid interaction that leads to swelling of the amorphous region, it further enhances the diffusion process. Swelling may also change the phase separated crystalline region.[37] In Nylon 6, as water enters the interlamellar regions of the amorphous structure, swelling increases and results in a decrease in T_g, a decrease in modulus, and an increase in the crystallinity of preexisting crystals. A report on maltodextrins in relation to gel formation[38] concluded that, in order to form a gel, two-phase systems must be formed: (1) disc-like crystalline domains (300 nm in diameter) and (2) amorphous polymer domains with water. In the absence of such crystalline structures, no gel formation was possible. In this case, the crystalline domains serve as the junction zones of the polysaccharide chains. These examples indicate that changes in the amorphous and crystalline domains may be interrelated and could have an impact on the three-dimensional architecture of the structure and on the corresponding mechanism of diffusion.

In edible gluten films plasticized by glycerin in comparison with other solutes,[11] it was found that the films plasticized by glycerin showed one dominating glassy–rubbery-like transition at ~ –58°C suggested to be due to glycerin-rich regions. This transition temperature agreed with that of glycerin. It is speculated that such films might have a microregion of phase-separated, glycerin-rich domains that contribute to the film functionality. Tensile strength, elongation, and water vapor transmission rates of the films, however, did not correlate with T_g but did correlate with the magnitude of the glycerin-rich transition (tan δ peak height, Figure 5). Because tan δ represents the ratio between E'' (loss modulus or the viscous component) and E' (storage modulus or the elastic component), the peak height increasing with glycerin content indicates the increasing ratio between the viscous and elastic properties, parameters that dictate the many film functionalities. Being hydrophilic, these glycerin-rich cavities act as "holes" in the film allowing or facilitating water diffusion. The glycerin ability to plasticize the most effective domains in the gluten also results in a lower tensile strength and a higher elongation values.

Microscopic and molecular migration

As mentioned earlier, moisture migration and diffusion can be accounted for at a macroscopic level (such as transport of water through edible films, loss of moisture from a pizza sauce to the shell part, or moisture loss through a donut glaze). These are events that are most obvious and are of practical importance to food processors and formulation specialists. Thus, control of moisture migration can be done by controlling water activity or relative vapor pressure, which determine the driving force and factors that contribute to the resistance to water transport. Knowledge about diffusion, glassy–rubbery transition, engineering properties, and other properties are helpful and can contribute to solving the problems. However, moisture migration also occurs at microscopic and molecular levels not easily characterized and detected by simple instrumentation. Combined methods, such as molecular mobility measurements (e.g., NMR imaging, NMR and ESR relaxation) are quite useful. The following discussion deals mostly with moisture migration and redistribution at the microscopic and molecular levels.

Water migration and bread staling

Kim-Shin et al., based on an ^{17}O NMR study of breads containing 5% sorbitol or glycerol, found a slight decrease in the average molecular mobility of the relatively free fraction of water and that about 20% of the mobile water became more "bound" (i.e., became undetected by NMR). This finding supports earlier work showing that water loses mobility in going from a "freezable" to "unfreezable" fraction,[13,40] which suggested that in stale bread a significant part of this water is associated with water of crystallization in the retrograded starch molecules. However, it was later found that of the water that becomes immobilized most

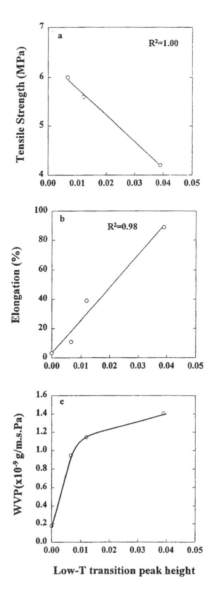

Figure 5 Edible gluten film properties as related to the glycerol-rich transition peak amplitude. (From Cherian G., Gennadios A., Weller C., Chinachoti P., Cereal Chem., 72, 1995. With permission.)

resides in the amorphous regions and only a small part can be accounted for as water of crystallization in the starch.[39] It was also found that bread with varying degrees of amylopectin crystallization did not show differences in the degree of change in water mobility, supporting the argument that the redistribution takes place in the amorphous regions.

Redistribution of water molecules among the amorphous components is expected, because after baking the bread structure is partially stabilized, but aging bread is never in a true equilibrium state. Slade and Levine hypothesized that the amorphous polymers in fresh bread might change in time in such a way that their three-dimensional network become more "mature."[13] In keeping with this hypothesis, the "effective glass transition temperature" increases from –5°C in fresh bread to a value close to 60°C in staled bread.

DMA and DSC analyses of aging military bread over a very long duration with no accompanying moisture loss indicated that there was no significant change in the thermal transitions (3-year storage at ambient temperature).[41] In contrast, the storage of an unstaled white bread,[42,43] resulted in an emergence of a broad transition (overlapping with ice melting) increasing in temperature over the first month period. This was found to be accompanied by decreasing "freezable" water. Once the "freezable" water was depleted (19 months storage, water content 17.8% wet basis), the transition remained broad and high in temperature range (at and above room temperature).[42] Thus, moisture loss contributed in part to the increase in transition temperature. Without moisture loss during aging of the military bread, no network formation was observed according to the DMA studies.

Collapse phenomena

Water absorption and diffusion into unstable supersaturated, low molecular weight carbohydrate solids (such as sucrose and lactose) normally leads to a decrease in local and over all viscosity, facilitators conversion to a more stable state. During this conversion period, the rate and the extent of water migration into the samples determine the rate of change in solute diffusion rates, which in turn affects the rates of rearrangement and recrystallization. The studies of collapse phenomena in freeze-dried systems containing sugars have been reviewed.[19,44] A concept has been developed that relates collapse phenomena of sugar or similar systems to glass transition.[13,19] It is based on the premise that water diffusion is controlled by the glassy–rubbery state of the system. Since sugars and similar systems show several orders of magnitude increase in rotational mobility as they are transformed from a glassy to rubbery state, translational motion or diffusion is expected to change accordingly.[45]

Lipid oxidation

Karel and his coworkers studied the rate of collapse and its relation to lipid oxidation.[44,46] They reported that the structural failure in the encapsulating matrix leads to exposure of the lipids to the air. Although the data are not complete enough to directly prove the role of T_g, they hypothesize that the glassy–rubbery transition plays a critical role in destabilizing the physical structure, leading to a rapid lipid oxidation. Note that this phenomenon has very little relationship to the rate of oxygen transfer through the matrix, but rather is directly related to the plasticizing effect of water on the structural integrity (i.e., collapse) of the matrix itself.

Diffusion of oxygen through encapsulating matrices has been studied by many. Imagi et al. reported that lipid oxidation depends on many parameters.[47] Methyl linoleate encapsulated with α-cyclodextrins, gelatin, arabic gum, pullulan, egg albumin, and sodium caseinate all showed different results. Estimation of oxygen diffusion coefficients through these films indicated a strong dependence on oxygen diffusivity in some but not all of the films. Sugars, for instance, has been suggested to have additional oxidative inhibitory effect.[48]

Similarly, recent reports have dealt with the role of glassy–rubbery transition in other chemical reactions, such as nonenzymatic browning reaction[49-51] and aspartame degradation.[52] No definite conclusions have yet been reached in this area of study.

Barrier films

Diffusion of a gas through a barrier material depends on (1) interactions between the gas and the barrier component and (2) inter- and intrachain dynamics of the barrier molecules.[53] For water, the diffusivity increases with moisture content as the materials are being plasticized[1] (Figure 6). Based on the "free volume" concept, gas diffusion is perceived to depend on voids or spaces in the barrier though which diffusing gas molecules can move.

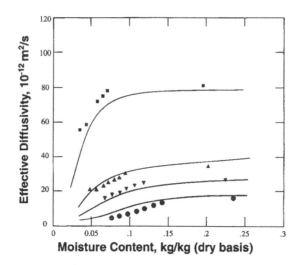

Figure 6   Effective diffusivity of water as a function of moisture content and temperature. ■ — 105°C, ▲ — 71°C, ▼ — 55°C, ● — 44°C. (From Okos M. R., Narsimhan G., Singh R. K., Weitnauer A. C., in *Handbook of Food Engineering*, Marcel Dekker, New York, 1992, 437. With permission.)

In a closer analysis, the mechanism of gas diffusion though a polymer may be viewed to be dependent on the transport properties of small solutes (which may fill the voids or change various intra- and intermolecular interactions) and on plasticizers (which may facilitate segmental motion on a long time scale). Such detailed analysis for food has yet to be done in order to determine the molecular events that controls oxygen diffusion. Data are needed that would show the influence of plasticizers (e.g., water) and small solutes (e.g., sugars) on the change in activation energy for oxygen diffusion. It is important to address the heterogeneity of such systems, since many water sorption phenomena are dependent on the presence of local environments or domains having unique properties that could influence swelling and other diffusion related properties.

In a study of water vapor transmission rate in an edible film,[10,54] rotational diffusivitiy of a nitroxyl radical probe and diffusion coefficient was found to be a function of moisture content (Figure 7). In a completely dried system, gaseous diffusion and sorption of water occurs simultaneously, and the mean diffusion coefficient (D) corresponds to water vapor in air, ~10^{-6} m^{2}s^{-1}; however, at a moisture content of 1.6%, D decreases to only 2.75×10^{-12} m^{2}s^{-1}. Additional moisture would progressively "clog up" the pores, so water diffusion through liquid-filled pores then replaces gaseous diffusion.[10] In this process, the diffusion coefficient of water is expected to change by four orders of magnitude, and in the case of starch[55] the very low diffusivity at very low moisture has been explained in terms of strongly adsorbed water. There is also a strong relationship between the diffusivity of water and the porosity and tortuosity of the systems. The presence of supersaturated sugar glass in freeze-dried and other porous foods tends to increase the degree of collapse upon hydration or warming, changing the diffusivity of water.

According to Schwartzberg, diffusion rates of various diffusing species are strongly affected by moisture content of the barrier film.[56] Figure 8 depicts diffusion in a biodegradable starch–plastic film with liquid-filled pores; amylase enzyme diffuses through the water from the mold filament to the starch and fungal germination is dependent on diffusion of digestion products to the mold filament.[57]

At a high moisture content, a hydrophilic film is not likely to be selective, since the diffusivities of low molecular weight solutes will correspond to their diffusivities in water

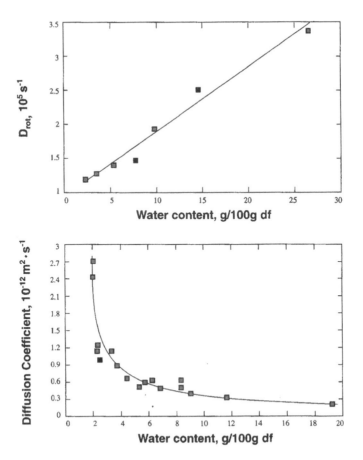

Figure 7 D_{rot} and diffusion coefficient as a function of moisture content. (From Debeaufort F., Voilley A., Meares P., *J. Membrane Sci.*, 91, 125, 1994. With permission.)

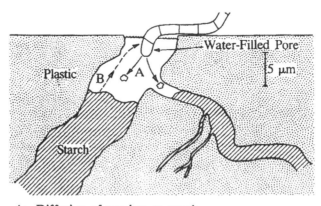

A: Diffusion of amylase to starch

B: Diffusion of digestion products to fungus

Figure 8 Diagram for diffusion in a biodegradable starch–plastic film with liquid-filled pores. (From Cole M. A., in *Agricultural and Synthetic Polymers: Biodegradability and Utilization*, American Chemical Society, Washington, DC, 1990, 76. With permission.)

and will depend on the volume fraction of water in the film. At a low moisture content, films are more selective, the diffusivities of solutes decreasing more rapidly than that of water. Thus, for reactions in systems involving small solutes and water, it would be helpful to relate rates to the diffusivities of the small molecules.

NMR

Typically, diffusion coefficients reported in literature are estimated gravimetrically, and many discrepancies are found that are due to variations in methods, differences in sample structure, and differences in composition. In contrast to the gravimetric method, pulsed-field gradient NMR (PFG-NMR) provides a way of measuring water diffusion on a local scale. It involves using the characteristic precessing frequency of relaxing nuclei in a magnetic field, and applying a gradient of magnetic field over the sample space. After a first and then a second pulse, nuclei that change in spatial position due to diffusion are detected as they fail to refocus, which result in an attenuation of the spin-echo signal.[58]

Belton and Colquhorn described the potential use of deuterium and ^{17}O NMR relaxation methods to determine diffusion rate of water in a food with various microenvironments.[59] This approach would further our understanding of diffusive exchange of water between regions with different relaxation rates.

Studying diffusion processes using NMR imaging has also become an important alternative approach, as can be seen from results from various systems.[60-65]

Summary

Inhibition of moisture migration in food has been a major challenge to food scientists. Recent research has focused on developing new materials and processes that would inhibit such migration. Several key factors such as changing plasticizer, crosslinking, glassy–rubbery state, temperature, and local mobility influence diffusion and migration. Moisture migration is also likely to influence reactant mobility and permeability of gases, vapors, and small solutes. Some reactions limited by diffusion can be named in a similar context.

References

1. Okos M. R., Narsimhan G., Singh R. K., Weitnauer A. C., Food Dehydration, in *Handbook of Food Engineering*, Heldman D. R., and Lund D. B., Eds., Marcel Dekker, New York, 1992, 437.
2. Paine F. A., Paine H. Y., Using barrier materials efficiently, in *A Handbook of Food Packaging*, Paine F. A., and Paine H. Y., Eds., Chapman & Hall, London, 1992, 390.
3. Duckworth R. B., Future needs in water sorption in foodstuffs, in *Physical Properties of Foods*, Jowitt R., Escher F., Hallstrom B., Meffert H. F. T., Spiess W. E. L., and Vos G., Eds., Applied Science Publishers, New York, 1983, 93
4. Loncin M., Mass transfer and permeability, in *Food Packaging and Preservation (theory and practice)*, Mathlouthi M., Ed., Elsevier Applied Science Publishers, New York, 1986, 1.
5. Menon A. S., Mujumdar A. S., Drying of solids: principles, classification, and selection of dryers, in *Handbook of Industrial Drying*, Mujumdar A. S., Ed., Marcel Dekker, New York, 1987, 3.
6. Mathlouthi M., Water interactions and food preservation, in *Food Packaging and Preservation (Theory and Practice)*, Mathlouthi M., Ed., Elsevier Applied Science Publishers, New York, 1986, 137.
7. Saravacos G. D., Mass tranfer properties of foods, in *Engineering Properties of Foods*, Rao M. A., and Rizvi S. S. H., Eds., Marcel Dekker, New York, 1986, 89.

8. McHugh T. H., Krockta J. M., Sorbitol- vs. glycerol-plasticized whey protein edible films: integrated oxygen permeability and tensile property evaluation, *J. Agric. Food Chem.*, 42, 841, 1994.

9. Koelsch C. M., Labuza T. P., Functional, physical and morphological properties of methyl cellulose and fatty acid-based edible barriers, *Lebensm.-Wiss. u.-Technol.*, 25, 404, 1992.

10. Debeaufort F., Tesson N., Voilley A., Aroma compounds and water vapour permeability of edible films and polymeric packagings, *International Symposium on Interaction: Food-Food Packaging Materials*, June, Lund, Sweden, 1994.

11. Cherian G., Gennadios A., Weller C., Chinachoti P., Thermomechanical behavior of wheat gluten films: effect of sucrose, glycerin, and sorbitol, *Cereal Chem.*, 72, 1995.

12. Voilley A., Le Meste M., Aroma diffusion: the influence of water activity and of molecular weight of the other solutes, in *Properties of Water in Foods*, Simatos D., and Multon J. L., Eds., Martinus Nijhoff Publishers, Boston, 1985, 357.

13. Slade L., Levine H., Beyond water activity: recent advances based on an alternative approach to the assessment of food quality and safety, *Crit. Rev. Food Sci. Nutr.*, 30, 115, 1991.

14. Peppas N. A., Brannon-Peppas L., Water diffusion and sorption in amorphous macromolecular systems and foods, in *Water in Foods*, Fito P., Mulet A., and McKenna B., Eds., Elsevier Applied Science, New York, 1994, 189.

15. Aguilera J. M., Stanley D. W., Simultaneous heat and mass transfer: dehydration, in *Microstructural Principles of Food Processing and Engineering*, Aguilera J. M., and Stanley D. W., Eds., Elsevier Science Publishers, New York, 1990, 291.

16. Pascat B., Mass transfer and permeability, in *Food Packaging and Preservation (Theory and Practice)*, Mathlouthi M., Ed., Elsevier Applied Science Publishers, New York, 1986, 7.

17. Levine H., Slade L., Water as a plasticizer: physico-chemical aspects of low-moisture polymeric systems, *Water Science Reviews*, Vol. 3, Cambridge University Press, Cambridge, 1988.

18. Levine H., Slade L., Principles of "cryostabilization" technology from structure/property relationships of carbohydrate/water systems—a review, *Cryo-Letters*, 9, 21, 1988.

19. Roos Y., Karel M., Applying state diagrams to food processing and development, *Food Technology*, 66 December, 68, 1991.

20. Roos Y., Karel M., Water and molecular weight effects on glass transitions in amorphous carbohydrates and carbohydrate solutions, *J. Food Sci.*, 56, 1676, 1991.

21. Roos Y., *Phase Transitions in Foods*, Academic Press, New York, 1995.

22. Matsuoka S., *Relaxation Phenomena in Polymers*, Hanser Publishers, New York, 1992.

23. Turi E. A., *Thermal Characterization of Polymeric Materials*, Academic Press, New York, 1981.

24. Cherian G., *Effect of plasticizers on glass transition behavior and functional properties of vital wheat gluten and gluten film*, Ph.D. Dissertation, University of Massachusetts, Amherst MA, 1994.

25. Peleg M., Mathematical characterization and graphical presentation of the stiffness-temeprature-moisture relationship of gliadin, *Biotechnol. Prog.*, 10, 652, 1994.

26. Hemminga M. A., Roozen M. J. G. W., Walstra P., Molecular motions and the glassy state, in *The Glassy State in Foods*, Blanshard J. M. V., and Lillford P. J., Eds., Nottingham University Press, UK, 1993, 157.

27. Seyler R. J., *Assignment of the Glass Transition*, American Society for Testing and Materials, Philadelphia, PA, 1994.

28. Roos Y., Karel M., Amorphous State and Delayed Ice Formation in Sucrose Solution, *Int. J. Food. Sci. Technol.*, 26, 553, 1991.

29. Williams M. L., Landel R. F., Ferry J. D., The temperature dependence of relaxation mechanisms in amorphous polymers and other glass-forming liquids, *J. Am. Chem. Soc.*, 77, 3701, 1955.

30. Bastioli C., Guanelia I., Romano G., Effects of water sorption on the physical properties of PET, PBT, and their long fibers composites, *Polym. Composites*, 11, 1, 1990.

31. Rossler E., Corresponding states concept for simple supercooled liquids identifying a change of diffusion mechanism above the glass transition temperature, *Ber. Bunsenges. Phys. Chem.*, 94, 392, 1990.

32. Wisanrakkit G., Gillham J. K., The glass tranition temperature (Tg) as an index of chemical conversion for a high-Tg amine/epoxy system: chemical and diffusion-controlled reaction kinetics, *J. Appl. Polym. Sci.*, 41, 2885, 1990.

33. Hsieh D. S., Rhine W. D., Langer R., Zero-order controlled-release polymer matrices for micro- and macromolecules, *J. Pharm. Sci.*, 72, 17, 1983.

34. Lindhart R. J., Biodegradable polymers for controlled release of drugs, in *Controlled Release of Drugs: Polymers and Aggregate Systems*, Rosoff M., Ed., VCH Publishers, New York, 1989, 315.

35. Kerr W. L., Lim M. H., Reid D. S., Chen H., Chemical reaction kinetics in relation to glass transition temperatures in frozen food polymer solutions, *J. Sci. Food Agric.*, 61, 51, 1993.

36. Simatos D., Blend G., Le Meste M., Relation between glass transition and stability of a frozen product, *Cryo-Letters*, 10, 77, 1989.

37. Murthy N. S., Stamm M., Sibilia J. P., Krimm S., Structural changes accompanying hydration in Nylon 6, *Macromolecules*, 22, 1261, 1989.

38. Reuther F., Damaschun G., Gernat C., Schierbaum F., Kettlitz B., Radosta S., Nothnagel A., Molecular gelation mechanism of maltodextrins investigated by wide-angle X-ray scattering, *Colloid Polym. Sci.*, 262, 643, 1984.

39. Kim-Shin M.-S., Mari F., Rao P. A., Stengle T. R., Chinahoti P., Oxygen-17 nuclear magnetic resoance studies of water mobility during bread staling, *J. Agric. Food Chem.*, 39, 1915, 1991.

40. Leung H. K., Magnuson J. A., Bruinsma B. L., Water binding of wheat flour doughs and breads as studied by deuteron relaxation, *J. Food Sci.*, 48, 95, 1983.

41. Hallberg L. H., Chinachoti P., Dynamic mechanical analysis for glass transitions in long shelf-life bread, *J. Food Sci.*, 57, 1201, 1992.

42. Hallberg L. M., *The staling of long shelf-life bread as determined by thermal analysis*, Ph.D. Dissertation, University of Massachusetts, Amherst MA, 1996.

43. Chinachoti, P., NMR dynamic properties of water in relation to thermal characteristics in bread in *Proceedings of the ISOPOW VI Conference*, Santa Rosa, CA, 1997, Chapman Hall/Blackie, New York,

44. Labrousse S., Roos Y., Karel M., Collapse and crystallization in amorphous matrices with encapsulated compounds, *Sci. Aliments*, 12(4), 757, 1992.

45. Karel M., Role of water activity, in *Food Properties and Computer-Aided Engineering of Food Processing Systems*, Kluwer Academic Publisher, New York, 1989, 135.

46. Shimada Y., Roos Y., Karel M., Oxidation of methyl linoleate encapsulated in amorphous lactose-based food model, *J. Agric. Food Chem.*, 39, 637, 1991.

47. Imagi J., Muraya K., Yamashita D., Adachi S., Matsuno R., Retarded oxidation of liquid lipids entrapped in matrixes of saccharides or proteins, *Biosci. Biotech. Biochem.*, 56, 477, 1992.

48. Sagone A. L., Greenwald J., Kraut E. H., Bianchine J., Singh D., Glucose: a role as a free radical scavenger in biological systems, *J. Lab. Clin. Med.*, 101, 97, 1983.

49. Buera M. P., Karel M., Effect of physical changes on the rates of non-enzymatic browning and related reactions, *Food Chem.*, 52, 167, 1995.

50. Karmas R., Buera M. P., Karel M., Effect of glass transion on rates of non-enzymatic brwoning in food systems, *J. Agric. Food Chem.*, 40, 873, 1992.

51. Roos Y. H., Himberg M. J., Nonenzymatic browning behavior, as related to glass transition, of a food model at chilling temperatures, *J. Agric. Food Chem.*, 42, 893, 1994.

52. Bell L. N., Hageman M. J., Differentiating between the effects of water activity and glass transition dependent mobility on a solid state chemical reaction: aspartame degradation, *J. Agric. Food Chem.*, 42, 2398, 1994.

53. Gao Y., Baca A. M., Wang B., Ogilby P. R., Activation barriers for oxygen diffusion in polystyrene and polycarbonate glasses: effects of low molecular weight aditives, *Macromolecules*, 27, 7041, 1994.

54. Debeaufort F., Voilley A., Meares P., Water vapor permeability and diffusivity through methylcellulose edible films, *J. Membrane Sci.*, 91, 125, 1994.

55. Marousis S. N., Karathanos V. T., Saravacos G. D., Effect of physical structure on starch materials on water diffusivity, *J. Food Process. Preserv.*, 15, 183, 1991.

56. Schwartzberg H. G., Modelling of gas and vapour transport through hydrophilic films, in *Food Packaging and Preservation: Theory and Practice*, Mathlouthi M., Ed., Elsevier Applied Science Publishers, New York, 1986, 115.

57. Cole M. A., Constraints on decay of polysaccharide-plastic blends, in *Agricultural and Synthetic Polymers: Biodegradability and Utilization*, Glass J. E., and Swift G., Eds., American Chemical Society, Washington, DC, 1990, 76.

58. Watanabe H., Fukuoka M., Measurement of moisture diffusion in foods using pulsed field gradient NMR, *Trends Food Sci. Technol.*, 3, 211, 1992.
59. Belton P. S., Colquhoun I. J., Nuclear magnetic resonance spectroscopy in food research, *Spectroscopy*, 4(9), 22, 1989.
60. Komoroski R. A., Nonmedical applications of NMR imaging, *Anal. Chem.*, 65, 1068, 1993.
61. McCarthy M. J., Lasseux D., Maneval J. E., NMR imaging in the study of diffusion of water in foods, *J. Food Eng.*, 22, 211, 1994.
62. McCarthy M. J., McCarthy K. L., Quantifying transport phenomena in food processing with nuclear magnetic resonance imaging, *J. Sci. Food Agric.*, 65, 257, 1994.
63. Ruan R., Lithcfield J. B., Eckhoff S. R., Simultaneous and nondestructive measurement of transient moisture profiles and structural changes in corn kernels during steeping using microscopic nuclear magnetic resonance imaging, *Cereal Chem.*, 69, 600, 1992.
64. Ruan R., Schmidt S. J., Schmidt A. R., Litchfield J. B., Nondestructive measurement of transient moisture profiles and the moisture diffusion coefficient in potato during drying and absorption by NMR imaging, *J. Food Process. Engr.*, 14, 297, 1991.
65. Schmidt S., Lai H.-M., Use of NMR and MRI to study water relations in foods, in *Water Relationships in Food*, Levine H., and Slade L., Eds., Plenum Press, New York, 1991, 405.

chapter ten

Factors affecting permeation, sorption, and migration processes in package–product systems

R. J. Hernandez and J. R. Giacin

Contents

0-8493-2646-X/97/$0.00+$.50
© 1998 by CRC Press LLC

Introduction

Function of package

Packaging has become an integral part of the processing, preservation, distribution, marketing, and even the cooking of foods. Initially, packages served simply to contain products and protect them from outside contamination during distribution and storage. The concept of protection includes the presence of a physical barrier to isolate the packaged product from external environmental influences, as well as to protect the external environment from the product.

According to their functional characterization, packages may be classified into two general categories. Primary or contacting packages are those in direct contact with the product and function to contain and protect the product. Secondary or distribution packages are those used to contain one or more primary packages and are used in general for distribution and storage; they also function as shipping containers. Many different materials, including plastic, glass, metal, and cellulosic materials are employed in the manufacture of the primary package, which may be further subdivided into categories based on structural rigidity to include flexible, rigid, or semirigid structures for enhanced mechanical support.

In general, rigid containers are made of glass, ceramics, or metal, while flexible and semirigid containers are made of polymeric materials alone or in combination with paper, foil, or metallized structures. In this regard, there has been tremendous growth in the development and design of new packages to fit specific needs, and while some of that growth has involved adapting traditional packaging materials, e.g., paper, glass, and metal, much of it has been due to the development of plastics packaging materials.[1] The variety, convenience, and economy of plastic materials have also contributed to this transformation.

In addition to the traditional functions of protection and containment, packages must also provide a means of communication to the user. The function of communication is critical in the design of today's package, as the package must convey information regarding the manufacturer, product preparation and use, as well as product ingredients and nutritional quality, in accordance with present federal regulations. In addition, when designing a package system the importance of the utility and convenience features of the package must not be overlooked. Such is the case of aseptic packages with easy-open features such as tear-strip pull tabs, or the convenience use of microwaveable packages. When needed, the package must also accomodate physical, chemical, and human effects.

Product quality and environmental influences

Product quality or shelf life is determined by the following parameters: (1) the product's physical, chemical, and biological characteristics; (2) processing conditions; (3) package characteristics and effectiveness; and (4) the environment to which the product is exposed during distribution and storage. The specific properties of the package required to adequately protect the product and, therefore, to maintain or extend product shelf life is a function of the product's susceptibility to environmental conditions. In order to properly select packaging materials, it is necessary to know the shelf life requirements for a given product/package/environment situation. Shelf life is defined as the length of time that a container or a product in a container will remain in an acceptable condition for its use or application, under specific conditions of storage.[2]

The influence of the product's physical, chemical, and biological characteristics on quality and shelf life, as well as the effect of processing conditions, are discussed in other chapters. This chapter will focus primarily on package characteristics and effectiveness in terms of interaction with the product and/or the external environment. Since there are a number of variables that can affect the overall quality of a product over time, the product/package/environment should be viewed as a dynamic system that is continually changing, from the time the product is packaged until the time the product is consumed.

Packaged products are subjected to a number of environmental influences, including moisture, oxygen, light, dust, and temperature. Biological elements include microorganisms, rodents, and insects. In addition, dropping products, incorrect storage and mishandling during distribution, as well as human tampering and other misuse result in damage to products amounting to many billions of dollars annually.

While the package serves as a barrier between the product and the environment to which the product/package system is exposed, the degree of protection varies. This variation is particularly important in connection with the transport of gases, vapors, water vapor, or other low molecular weight compounds between the external environment and the internal package environment, which is controlled by the packaging material. Unlike glass, metals, and ceramics, plastics packaging materials are relatively permeable to small molecules such as permanent gases (i.e., carbon dioxide, oxygen, or other gases that are considered "ideal" in behavior), water vapor, organic vapors, and liquids. In terms of their barrier properties, plastics packaging materials provide a broad range of mass transfer characteristics, ranging from excellent to low barrier values. The specific barrier requirements of the package system will be dependent upon the product's characteristics and the intended end-use application.

For the purpose of package design and optimization, it is important to consider how the mass transfer characteristics of the packaging material determine the transport of low molecular weight compounds both into and through the package. If the package has a polymeric structure, then water vapor, oxygen, flavor and aroma compounds, additives, and low molecular weight residual moieties may transfer from either the internal or external environment through the polymeric package wall, resulting in a continuous change in product quality. For example, in the case of a packaged product whose physical or chemical deterioration is related to its equilibrium moisture content, the barrier properties of the package relating to water vapor will be of major importance in maintaining or extending shelf life. Similarly, the change in oxygen concentration in a permeable package will directly affect the rate of oxidation of oxygen-sensitive nutrients such as vitamins, essential fatty acids, and proteins.[3] Analogously, the transport of low molecular weight organic volatiles such as aroma and flavor compounds from the product through the package, or even the migration of constituents from the package to the product, can impact on both the safety and quality of the packaged product.

Packaging thus affects the quality of food products, mainly by controlling moisture, oxygen, and light transfer. The loss of specific aroma or flavor constituents or the gain of off-odors due to permeation can also lead to a reduction in product quality, resulting in a reduction in shelf life for products, where quality is associated with the retention of flavor volatiles.

While there is a myriad of environmental factors that can impact the quality of a packaged product, the focus here is on product–package interactions associated with polymeric packaging materials. It is a broad based topic that includes transport of gases, vapors, water vapor, or other low molecular weight components (1) from the food product through the package; (2) from the environment through the package to the food; (3) from the food product into, but not through the package; and (4) from the package itself into the food product. Although not addressed, consideration should also be given to any interaction between the food product and package that induces a chemical change in the food, in the package, or in both.

Packaging materials

Plastics

Currently, plastic packaging materials are widely used in the food packaging industry due to their low cost and outstanding functional service properties. Plastic containers are much lighter in weight than comparable metal and glass containers, and require less energy to fabricate, convert, and transport.

Although packaging is used by virtually all industries, the markets for most end-use applications have become saturated. As a result, intermaterial shifts among packaging materials and containers have been and will continue to be emphasized.

Competition among packaging materials is growing, mainly due to the ever-increasing use of plastics. For example, expanded polystyrene foam and shrink film overwraps are beginning to impede the growth of corrugated containers, as these and other plastics are being used more frequently because of their cost-effectiveness. Further, plastics containers for such dairy products as milk, cottage cheese, and yogurt have penetrated the traditional paperboard market. Paperboard milk containers have probably been most affected by the successful entry of plastic containers into this market. For economic reasons, larger sizes are more in demand, and the plastic container is lighter, easier to carry, and leak-proof.

In addition to paper packaging, metal packaging is also facing very strong competition from both plastics and glass. Use of glass packaging has begun to level off due to inroads made by plastic bottles and flexible packaging into the market sector normally occupied by glass, as well as by a reduced demand for food packaging in general as a consequence of consumers increasingly taking meals away from home.

This high and steady growth rate for plastic packaging is attributed in part to consumer appeal. Plastic packaging systems are lightweight, break-resistant, and convenient to use. Bulky, hard-to-handle beverage and food containers made of paperboard, glass, and metal are now being replaced with plastic containers and/or plastic-based composite containers, aseptic packages, and bag-in-box aseptic package systems. Plastics are also easily molded into various shapes and configurations and are used for consumer point-of-sale packaging.

In package manufacturing, plastics are extremely cost competitive and thus become attractive to producers and users of packaging materials and package systems. As a result, the coextrusion of packaging materials with specific properties, the fabrication of plastics packaging systems, and the utilization of plastics in food packaging have become highly innovative over the past few years.

A detailed account of the use and application of plastics packaging materials for food packaging has been given by Jenkins and Harrington[4] and Brown,[5] and references cited therein. A listing of selected properties of polymer films is given in Table 1.

Glass

In spite of the high growth rate of plastic packaging, glass for food and beverage packaging continues to find utility, since glass has good inherent qualities such as relative inertness,

Table 1 Properties of Plastic Film

	LDPE	HDPE	LLDPE	>12% VA EVA	IONOMER	ORIENTED PET	OPP	PVC[a]	PS	PVDC	NYLON	EVOH[b]	BON
Density, g/cc	0.91–0.925	0.945–0.967	0.918–0.923	0.94	0.94–0.96	1.4	0.905	1.21–1.37	1.05	1.6–1.7	1.14	1.12–1.19	1.14
Yield, in²/lb-mil × 10⁻³	30	29.0	30.0	29.5	28.6–29.5	20–22	30.6	20–22.5	26	16.2–16.8	23.5–24.5	23–24	23.5–24.5
Tensile Strength, kpsi	1.2–2.5	3.0–7.5	3.5–8.0	3–5	3.5–5.5	25	25–30	2–16	5.0–8.0	8–20	7–18	1.2–1.7	25–30
Elongation at break, %	225–600	10–500	400–800	300–500	300–600	70–100	60–100	5–500	2–3	40–100	250–500	220–280	70
Impact strength, kg-cm	7–11	1–3	8–13	11–15	6–11	25–30	5–15	12–20	N/A	10–15	4–6	N/A	N/A
Elmendorf tear strength, g/ml	100–400	15–300	80–800	50–100	15–150	13–80	4–6	N/A	N/A	10–20	20–50	N/A	N/A
WVTR, g-mil/100 in²-day @ 100°F and 90% R.H.	1.2	0.3–0.65	1.2	3.9	1.3–2.1	1.3	0.3–0.4	2.8	5.0	.05–0.3	24–26	High	12
Oxygen transmission rate, cm³-mil/100 in²-day-atm @ 77°F and 0% R.H.	250–840	30–250	250–840	515–645	226–484	5	110	5–1500	100–200	0.08–1.7	2.6	0.01	2
CO₃ permeability, cm³-mil/100 in²-day-atm @ 77°F and 0% R.H.	500–5000	250–645	500–5000	2260–2900	626–1150	N/A	240–285	50–13,500	N/A	0.04–10	4.7	N/A	N/A
Resistance to grease & oil	Varies	Good	Good	Varies	Good	Good	Good	Good	Good	Good	Good	Good	Good
Dimensional change to high R.H., %	0	0	0	0	0	0	0	0	0	0	1.3	0	0
Haze, %	4–10	25–50	6–20	2–10	1–15	4	3	1–2	0–1	2	2	N/A	1–2
Light transmission, %	65	N/A	N/A	55–75	85	88	80	90	90	80–88	N/A	N/A	N/A
Heat seal temperature range, °F	250–350	275–310	250–350	150–300	225–300	275	200–300 (coated)	280–340	194–212	250–300	350–500	300–385	250–300
Service temperature range, °F	–70–180	–60–250	–60–180	–60–140	–150–150	–100–400	–60–250	–20–200	–80–175	0–275	–75–400	N/A	–100–400
Tensile modulus, 1% secant, kpsi	20–40	125	25	8–20	10–50	700	350	350–600	330–475	50–150	N/A	300–385	250–300

a Same PVC data depend on plasticizer content.
b Same EVOH data depend on ethylene content.

purity, and total barrier properties. The inertness and purity of glass assure taste and flavor retention, while the barrier properties provide a safeguard against spoilage related to water vapor and oxygen ingression, as well as microbial contamination. However, glass containers are heavy and brittle. Consumers continue to perceive glass as a wholesome container and it will continue to be used for selected products. Thinner container walls and material recycling are the industries response to the declining use of glass as a packaging material. Recycling of glass is essential if this container form is to maintain a market share.

Metal

Enameled sanitary tin cans have long been used in the vacuum packaging process to preserve food for extended periods of time. They are strong and have high resistance to heat, cold, moisture and oxygen permeation, and other environmental stresses. To maintain its market position, the industry has encouraged recycling, adjusted steel and aluminum sheet stock (thinner wall), improved inner protective coatings, graphics, and closure systems,[6] and developed the two-piece draw–redraw can-making process. Table 2 lists the different kinds of steel can coatings and their application to the packaging of food products.

A listing of selected properties of glass, steel, aluminum, and plastics that are relevant to the use of these materials for packaging applications is given in Table 3.

Storage stability of packaged, moisture-sensitive food products

The availability of free water can be a significant factor in the effectiveness of food processing and preservation and in the potential for biochemical reactions that affect the subsequent stability of the product. Water in foods exerts a vapor pressure, the magnitude of which depends upon the amount that is free to vaporize. Water activity (A_w) is the quotient of the water vapor pressure of the substance divided by the vapor pressure of pure water at the same temperature. Maximum storage stability of many dehydrated substances occurs at moisture contents close to the BET monolayer values[7], corresponding to $A_w = 0.2$ to 0.4. Salwin also suggested that the water molecules covering the active sites of the dry solids form a protective film against oxygen.[8]

Most of the unit operations used in food processing have as a goal (1) the removal of water to stabilize the material, as in drying and concentration; (2) the transformation of water into a nonactive component, as in freezing; or (3) the immobilization of water in gels, structured foods, and low- and intermediate-moisture foods. The main and essential way in which the immobilization of water is measured is through the consideration of A_w and its relationship to moisture content. Based on the thermodynamic concept of chemical potential of water in solutions, A_w has served as an index of how successful we are at controlling water behavior in food systems. A_w is also the parameter that controls the driving force in water removal operations, and is therefore essential for process design purposes.[9]

Water may also change the mobility of the reactants by affecting the viscosity of food systems. It may form hydrogen bonds or complexes with the reacting components. For example, lipid oxidation rates may be affected by hydration of trace metal catalysts or by hydrogen bonding of hydroperoxides to water. The structure of a solid matrix may also change substantially with changes in moisture content, thus indirectly influencing reaction rates. In addition, water influences protein conformation, and the transition of amorphous–crystalline states of sugar and starch.[10] To obtain desirable textures, it is usually necessary to have a high moisture content, which means the A_w is high enough to support

Table 2   Steel Can Coatings

Coating & Products	Can Type	Comments
Acrylic		
Applied as liner for cherries, pears, pie fillings, single-serve puddings, some soups and vegetables Acceptable for aseptic packs	For all exteriors and interiors where a clean, white look is desired As a high-solids side seam stripe for roller and spray application to wire-welded cans Clear wash coat for D&I cans Water-borne spray liner for 3-piece beverage cans	Expensive, but takes heat processing well and is suited to water-borne and high-solids coatings. Employed primarily on can exteriors because of flavor problems with some products. Makes an excellent white coat and assures color retention when pigmented.
Alkyd		
(Exterior use)	Quart oil can ends in clear gold or aluminum pigment Aerosol domes in white enamel Flat sheet-roll-coat decorating as a high-solids white base coat and varnish	Low cost, used mostly as an exterior varnish over inks, because it would present flavor and color problems inside the can. Trend is toward supplanting conventional alkyds with polyesters, which are oil-free alkyds.
Epoxy amine		
Beer and soft drinks Dairy products Fish, ham and sauerkraut Nonfoods such as furniture polish, hair spray, and paint	D&I as beer and beverage base coat Draw/Redraw Can ends Over-varnish on aerosol cans and domes	Costly, but has excellent adhesion, color, flexibility, imparts no off-flavor, scorch resistance and abrasion resistance. Used in interior or as a varnish and size coat. Employed now in water-borne coatings and, with polyamide, as a side seam stripe in high-solids form for welded cans.
Epoxy-phenolic		
Beer and soft drinks (as a base coat) The coating for foods, including fish, fruit, infant formula, juice, meat, olives, pie fillings, ravioli, soups, spaghetti and meat balls, tomato products, vegetables	All steels and cans	Big in volume for can interiors. Used in Europe as a universal coating. Main attributes are steam processing resistance, adhesion, flexibility and imparts no off-flavor. Especially suited for acid-aggressive products. Has excellent properties as a base coat under acrylic and vinyl.
Epoxy-Phenolic with zinc oxide or metallic aluminum powder		
For sulfur-containing foods: fish, meats, soups and vegetables	3-piece tin plate can ends 1-piece single draw cans (307 x 113)	Used primarily to prevent tin sulfide staining, flexibility and clean appearance are the main attributes of these "C" enamels.

Table 2   Steel Can Coatings (continued)

Coating & Products	Can Type	Comments
Oleoresinous		
Beer and soft drinks (as a base coat) Fruit drinks A wide variety of fruits and vegetables, including acid fruits	All except draw/redraw	A general purpose, gold-colored coating least expensive of all. When additional protection is required, it can serve as the undercoat for another lacquer. Can be used in both high-solids and water-borne systems.
Phenolic		
Acid fruits, fish, meats, pet food, soups and vegetables	All steel cans where coating can be flat-applied	Low-cost, exceptional acid resistance and good sulfur resistance. Film thickness restricted by its inflexibility. Tendency to impart off-flavor and odor to same foods.
Polybutadiene		
Citrus plus soups and vegetables (if zinc oxide is added to the coating) Beer and soft drinks	Restricted to 3-piece can bodies because of its poor fabricability	Good adhesion, chemical/resistance and ability to undergo processing are its chief characteristics. Resistance properties and cost comparable to the oleoresinous class.
Sulfide stain-resistant oleoresinous - "C" enamel (has zinc oxide)		
Asparagus, beans, beef stew, beets, chicken broth, chocolate syrup, corn, peas, potatoes, spinach, soups, tomato products and sulfur-containing foods	Soldered and welded tin plate Can ends	Low cost, flexible, often used as a top coat over epoxy-phenolic. Not for use with acid fruits.
Vinyl		
Beer and soft drinks Fruit drinks Tomato juice Nonfoods	Draw/redraw, even in triple draw cans, for coating is very formable Roller-applied topcoat for beer and beverage ends	Flexible and off-flavor-free (used for years with beer). Resistant to acid and alkaline products. Not suitable for high-temperature, long-processed meats and foods. Big as a clear exterior coating.
Vinyl organosol		
Beer and soft drinks Foods	Draw/Redraw Topcoat on beer and beverage can ends 3-piece food can ends Thin tin plate which has to be beaded	Especially suited for thicker film application (9-10 mg per square inch). Good fabricability, superior, corrosion resistance. May prove ideal for welded cans, which allow more severe beading. Still used in low-solids, solvent-borne manner.

Table 3   Selected Properties of Noncellulosic Packaging Materials

Property	Glass	Steel	Aluminum	Plastic
Density (10^3 kg/m^3)	2.5	7.8	2.2	0.9–1.7
Stiffness (10^8 N/M^2)	700	1800	200	0.7–42
Strength (10^7 N/M^2)	1.4–24	14–30	7–21	0.7–10
Thermal conductivity (WM^{-1} °C^{-1})	0.1	35	15	0.1–0.4
Specific heat (J/Kg°C)	775	420	963	1200–1700
Gas and vapor barrier	Impermeable	Impermeable	Impermeable	Permeability range
Resistance to chemical attack	Inert	Reactive	Stable in oxidizing medium	Inert to reactive
Interaction with product	Inert	Corrosion	Inert	Several types
Performance temperature (°C)	400	400	260	80–250
Thermal behavior	Thermal shock	Stable	Stable	Softens
Ovenability (conventional, microwave)	Dual	Conventional	Dual	Dual
Light transmission	Transparent	Opaque	Opaque	Transparent
Shape formability	Limited	Limited	Flexible and rigid	Wide range
Rigidity	Rigid	Rigid	Flexible	Flexible to rigid
Biodegradability	No	No	No	Yes–No
Recyclable	Yes	Yes	Yes	Yes–No
Linear coefficient of thermal expansion (10^{-5} °C)	0.5	0.12	0.23	1–20

microbial growth. When dried to an A_w level that will not support microbial growth, the texture usually becomes too hard, dry, tough, or crumbly.[11]

Shelf life evaluation

The shelf life of a packaged product will depend upon a series of variables associated with the product's composition and processing, its packaging material, and the storage conditions. Since the commercial value of a packaged product, including its quality, consumer safety, and convenience, is affected by the product's shelf life, there is an obvious interest in understanding and correlating the variables affecting the shelf life of a product/package system. The food industry professional can then select the package system that would maintain the desired product quality for the required shelf life period. Once the product formulation is established, the appropriate package system is selected, taking into account the required package composition and retention of package integrity. The package as a functional barrier thus regulates the transfer of gases, vapors, water vapor, or other low molecular weight compounds, as well as the transfer of heat and penetration of electromagnetic radiation of varying wavelengths.

There are three methods of shelf life evaluation: (1) actual storage testing, (2) accelerated tests, and (3) simulation and model-based estimations. Actual storage testing by a long-term stability study involves storing a packaged product under typical storage conditions of temperature and relative humidity. Samples are examined at regular time intervals and the degradation factor is recorded. Such testing is both expensive and time consuming. However, where complex relationships between the product, package, and environment exist, this method provides a very viable alternative.

Accelerated test techniques measure the products stability by conditioning the packaged product for a period of time at rigorous or abusive storage conditions (i.e., high temperature and relative humidity). Stability data obtained from the abusive storage conditions are then extrapolated to actual storage conditions to estimate a shelf life value. This evaluation technique assumes a known relationship between the product's shelf life at accelerated and ambient conditions, and is valid only when the mathematical relationship used in the extrapolation procedure accurately relates product characteristics to storage conditions or package properties for a standard or known product–package system to the product–package system of test. Mistaken assumptions and inherent errors in the procedure are the disadvantages to this type of estimation technique.

Simulation techniques for evaluating or estimating shelf life involve combining expressions for product sensitivity, package effectiveness, and environmental severity into a mathematical model. The solution of such expressions allows one to predict the time over which the product retains the required characteristic quality.

Mathematical models have been used in both the food and pharmaceutical sciences to describe the effect of temperature on the rate of reaction, including high abuse temperatures. The continuously changing conditions in the distribution system could lead to a more rapid rate of product deterioration, if the product is subjected to high abuse temperatures. Also, the water activity of dry foods can increase with temperature increase.[12] If the temperature acceleration factor is given, then extrapolation to low temperatures could be used to predict shelf life of the product. This acceleration factor, which is the Arrhenius Q_{10} factor, is the ratio of shelf life at two temperatures differing by 10°C. It would be used in connection with rates for deteriorative reactions that often can be expressed as a zero-order or a first-order process, giving an equation such as

$$-dA/dt = kA^n, \tag{1}$$

where A is the quality factor measured in some units of amount, n is the reaction order, k is the rate constant obtained from the slope of the plot of appropriate reaction extent (A) versus time (t).[12] However, upon freezing, food product reactants are concentrated in the unfrozen liquid, creating a higher rate of quality loss at certain temperatures, which is not accounted for by the Q_{10} value, and will cause prediction errors.[13]

Chemical reactions can also cause a loss of vitamins such as vitamin C, where oxygen is limiting.[14] Since vitamin C is quite unstable at pH values above 5.0, its degradation is generally considered to reflect a loss in product quality. Labuza also showed that for frozen vegetables, when 15–20% of the vitamin C is lost, they also become unacceptable from a sensory standpoint.[15] This correspondence is due to similar reaction Q^{10} values for sensory characteristics and nutrient levels.

Mizrahi and Karel studied shelf life simulation models extensively by applying computer-aided mathematical solutions to predict chemical deterioration caused by moisture and oxygen interaction with food products.[16,17] Karel and Labuza developed shelf life prediction models for dehydrated space foods.[18] Studies were also conducted on lipid oxidation in potato chips.[3,19,20] Degradation of ascorbic acid in tomato juice was studied independently by two groups, Wanninger[21] and Rimer and Karel.[22] David[23] worked on the evaluation and selection of flexible film for food packaging.

Concepts of mass transfer to study shelf life were introduced by Heiss,[24] who developed a mathematical model that considered the moisture absorption properties of food, the water vapor permeability of the packaging material, and the surrounding atmospheric conditions. Salwin and Slawson also developed a procedure to predict moisture transfer in combinations of dehydrated foods, from knowledge of the sorption isotherm of the individual components.[25] Iglesias et al. extended Salwin and Slawson's work and predicted the moisture transfer in a mixture of packaged dehydrated foods.[26]

Shelf life simulation of a moisture-sensitive product

The moisture content of a packaged product at any time (t) under constant external conditions of temperature and relative humidity depends upon the equilibrium moisture content of the product and the permeability of the package.

$$M_t = f(M_{eq}, P),   (2)$$

where M_t is the moisture content of a packaged product at any time (t), M_{eq} is the equilibrium moisture content of the product ($M_{eq} = f(RH_{eq})$), and P is the permeability constant of the package. Assuming that the shelf life of the product depends solely on the physical and chemical factors that are related to moisture content, it can be estimated. The other assumptions for predicting the shelf life by simulation include the following: (1) the moisture content of the packaged product will come to equilibrium quickly with the internal relative humidity within the package; (2) the relative humidity inside the package is determined by the permeability of the package; and (3) the relationship between the moisture content of the product and relative humidity within the package can be represented by an isotherm equilibrium curve.

The problem of shelf life prediction for a moisture-sensitive food product becomes crucially dependent on packaging, when distribution channels and extended storage possibilities are considered. The shelf life of a packaged moisture-sensitive food product can be predicted by describing two phenomena pertaining to the package–product system. First, transport of water vapor through the package and, second, the uptake of water by the product. The permeability constant P for water vapor through a polymeric structure is composed of a mobility term, described by the diffusion coefficient (D) and a solubility term described by the solubility constant (S). Both D and S, and therefore P, are functions of temperature, and the dependency of each can be described by an Arrhenius-type relationship. The sorption of water vapor by the food product can be described by the equilibrium moisture isotherm. The concept of equilibrium moisture content (M_{eq}) is important in the study of moisture change of packaged moisture-sensitive products and is an integral component of the simulation models developed for estimation of shelf life. The M_{eq} is defined as the moisture content of a product after it has been exposed to a particular environment. Based on the water activity of the product at various moisture contents and temperatures, product with a particular A_w will either gain or lose moisture when exposed to a surrounding environment.

A mathematical relationship between water vapor sorption by the product and package permeability was developed by Wang, which allows calculation of relative humidity within the package at time (t).[27]

Lee developed a more general simulation model that considers the change in product moisture content as a function of storage time over a range of temperature and relative humidity values.[28] The model combines the moisture sorption characteristics of the product and the permeability of the package system, both as a function of temperature. The Brunauer, Emmett, and Teller (BET) equation[7] was applied to describe the experimental sorption isotherm of the product. A cubic polynomial expression was used to fit the data and was substituted into the shelf life estimation model. The agreement between experimental and calculated shelf life stability data obtained from the model was considered to be within acceptable limits. These findings showed that the relationship between product

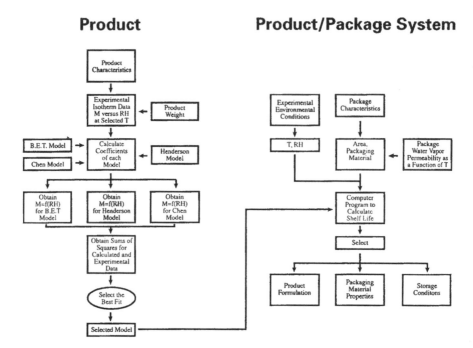

Figure 1 A diagram outlining the simulation model for predicting shelf life of moisture sensitive products.

shelf life and storage environment can be modeled within a practical range of temperature and relative humidity values, based on permeability data of the package and the equilibrium sorption isotherm of the product obtained at three temperatures.

In a study by Kirloskar, a computer simulation model was described for predicting the change in moisture content of a packaged moisture sensitive product over time, as a function of temperature and relative humidity.[29] According to this model, experimental isotherm data for three temperatures are required to predict the shelf life at any temperature within the range. The computer program selects the best-fit equation for the isotherm from among three models provided, based on sums of squares and correlation coefficients. The effect of temperature on the coefficients of the isotherm equation selected is described by a quadratic equation. This allows calculation of the moisture content of the product as a function of relative humidity, at the desired temperature. The shelf life equation is then solved for the required temperature, without actually having to obtain experimental sorption isotherms for that temperature. A model that best fits the experimental data is selected. This model may be modified for isocratic conditions, or by incorporating fluctuating temperature and humidity environments. Various moisture-sensitive products may be considered and fitted to other moisture isotherm equations.

A diagram outlining this simulation modeling approach is shown in Figure 1. The expression that describes the time needed for a product packaged in a moisture-semipermeable material to change its moisture content from an initial to a final equilibrium value, when the environmental humidity condition changes, is given by Equation (3).

$$t = \frac{W}{P(T) \cdot P_s} \cdot \int_{M_o}^{M_t} \frac{dM}{\left[A_w(e) - A_w(M,T) \right]} \tag{3}$$

where t = time, W = dry weight of product (g), $P(T)$ = permeability constant of the package at temperature, T (g water/day · mm Hg · pkg), P_s = saturated vapor pressure of water at temperature T (mmHg), dM = instant moisture content differential, M_o = initial moisture content, Mt = moisture content at time t, $A_w(e)$ = water activity external to the package, $A_w(M,T)$ = internal water activity given as a function of temperature and moisture, content of the product, and T = temperature. The internal water activity, $A_w(M,T)$, can be obtained from the sorption equilibrium isotherm. To solve Equation (3), a trapezoidal numerical integration procedure was used. The value of time (t) obtained would indicate the time required for the initial moisture content (M_o) of the product at temperature T_o and the initial humidity RH_o, to change to a final value at T_t and RH_t.

Storage stability of packaged, oxygen-sensitive food products

The shelf life of food products that deteriorate through an oxidative mechanism is a major consideration of the food industry, and parameters affecting such oxidative processes have been the subject of a number of studies. Quast and Karel determined the rate of oxygen uptake and the effective diffusivity of oxygen in dehydrated food products.[30] Maloney et al. studied the autoxidation of methyl linoleate as a function of water activity.[31] Labuza reported on the effect of both water activity and glycerol content on the oxidation of lipids.[32] Zirlin and Karel described the oxidation of a freeze-dried model food system consisting of methyl linoleate and gelatin.[33] Cabral et al. determined the performance of various packages for potato chips in terms of selected chemical and physical properties, such as peroxide value, hexanal formation, and texture evaluation.[34] Martinez and Labuza studied several deteriorative mechanisms affecting the quality of freeze-dried salmon, such as lipid oxidation and nonenzymatic browning.[35] Tuomy and Walker dealt with the effects of storage time, moisture level, and headspace oxygen concentration on the quality of a dehydrated egg mix.[36]

Storage stability studies have also been extended to include the development of mathematical models for product storage prediction. Karel,[37,38] Karel et al.,[39] Quast and Karel,[19,20] Quast et al.,[40] and Labuza et al.[41] described mathematical models for the evaluation of package requirements for food products that undergo oxidation due to atmospheric oxygen. These models were based on a combination of a kinetic equation of deterioration[42,43] and the permeability characteristics of the package. The computer-aided iteration technique developed enables one to predict the deterioration of a product under defined package requirements and storage conditions. The procedure was applied to several oxidative reactions associated with food systems, such as lipid oxidation in dehydrated shrimp[44] and nonenzymatic browning in dehydrated cabbage[45,46] and tomatoes.[47] The procedure was also applied to potato chips that deteriorate by two mechanisms simultaneously: by oxidation and by textural changes due to moisture pickup.[3,40] In this study, to better simulate actual storage conditions, the rate of lipid oxidation was determined as a function of oxygen partial pressure, extent of oxidation, and equilibrium relative humidity. Singh et al. applied a similar procedure to predict vitamin C degradation in a liquid food product.[14]

For the studies described above, the predictions were made under constant storage conditions. Kwolek and Boukwalter[48] successfully predicted the extent of product deterioration as a function of storage temperature. Labuza reviewed the mathematical models for fluctuating temperature sequences.[49] Further, the author developed equations to predict the quality change in a product undergoing either random, sine, or square wave time–temperature distribution for both zero- and first-order reactions.

Successful simulation of the change in food quality was reported by Lee and Heldman,[50] who developed a mathematical model based on empirical equations to predict the extent of destruction of vitamin C in tomato juice as a function of temperature, pH, and copper concentration.

Sadler described an analytical solution to the diffusion equation to predict vitamin C degradation in orange juice and lycopene changes in tomato juice for a semi-infinite medium with constant surface concentration.[51] A weighted oxygen diffusion coefficient for the liquid food was determined to take into consideration the oxygen diffusion resistance at the polymer membrane. Sadler was able to use an analytical solution without actually employing the variable permeable surface information in direct form. The solution included simultaneous oxygen diffusion and a first-order chemical reaction.

In recent studies Barron et al. applied the finite element method (FEM) to modeling the simultaneous oxygen diffusion and chemical reaction in a packaged liquid food.[52,53] Cylindrical high-density polyethylene (HDPE) packages, top and bottom insulated, were designed to contain apple juice or water containing vitamin C. Oxygen consumption was monitored by a micro-oxygen electrode. A first-order chemical reaction and its reaction rate constant 7.3×10^{-3} per minute (in terms of O_2 consumption) was determined in apple juice with added vitamin C. The simultaneous permeation–diffusion reaction process was modeled by finite element block types developed by Swanson Analysis Systems, Inc., Houston, PA (Engineering Analysis Systems User's and Theoretical Manuals).

The computer model was found to provide acceptable predictions of oxygen levels, as a result of diffusion and reaction with vitamin C. The model can be modified and applied to various liquid-food packaging situations, for investigations of the effects caused by parameters such as oxygen permeability of packages, reaction kinetics of a quality factor, shape and dimensions of a package, and oxygen diffusion coefficients.

Acceptable predictions were obtained for two systems: a cylindrical permeable package and a semi-infinite impermeable container. The model can be modified to simulate other geometrical configurations, packaging materials, and liquid food systems.

Kim described one- and three-dimensional mathematical models for predicting loss of quality of packaged apple juice and a model liquid system (mixture of water and ascorbic acid).[54] The models were based on the mass transfer of oxygen through the package and into the liquid, mass transfer of ascorbic acid (indicator of shelf life) through the product with simultaneous oxidation of ascorbic acid in the packaged liquid-food product. Three computer programs were developed to solve the mathematical model, using the finite difference method, one-dimensional open system (without packaging material), and three-dimensional closed system (with packaging material). Variables such as the permeability of the packaging material, diffusion coefficient of oxygen and ascorbic acid, and the kinetic reaction rate constant of oxidation of ascorbic acid were included. The distinctive characteristic of the mathematical models described by Kim is that they take into account the three-dimensional model of oxygen, ascorbic acid diffusion, with simultaneous oxidation of ascorbic acid in a packaged liquid-food product. Based on experimental values, Kim validated the one- and three-dimensional mathematical models. The computer programs developed were also used to simulate the effect of physical parameters of a packaged liquid model food system such as the diffusion coefficient of oxygen and ascorbic acid, oxygen permeability of the packaging material, and the kinetic reaction rate constant of ascorbic acid oxidation on shelf life.

Developing technologies

Silica-coated packaging material

Considered to be retortable and recyclable–microwaveable, silicon oxide (SiO_x) deposition on polymeric flexible substrates have been applied to packaging to provide a high oxygen barrier to plastic packaging films and bottles that were transparent to microwave energy.[55–58] Felts has reviewed the three vacuum deposition processes—evaporation, sputtering, and plasma-based—that have been developed to deposit SiO_x coatings and the benefits and disadvantages of each. Results showed that the SiO_x coatings from plasma-enhanced chemical vapor deposition (PECVD) yielded the best oxygen barrier properties. Recently, Izu described a proprietary microwave PECVD thin film deposition technology to manufacture high-performance transparent barrier film.[60] Excellent barrier properties of 0.5 cc/m²·day and 0.2 g/m²·day were demonstrated for oxygen and water vapor, respectively.

At present, only one product is commercially available, which is based on silicon oxide deposition onto polyethylene terephthalate (SiO_x PET) for incorporation into high oxygen barrier packaging systems to allow microwaveability in food packaging.[59] In addition to silicon oxide deposition to polyester substrates, Felts reported significant progress on plasma deposition onto low-density polyethylene (LDPE), biaxially oriented polypropylene (BOPP), and biaxially oriented nylon (BON) films.[61] The application of silicon-oxide-coated flexible films for packaging has been reviewed by Brody.[62]

The commercially available glass-coated film, which is based on silicon oxide deposition onto polyethylene terephthalate (SiO_x PET),[63] exhibits a number of important characteristics, to including

- Excellent and stable barrier properties: it has very high barrier properties for water vapor and oxygen.
- Barrier properties that are only slightly influenced by variation in temperature or relative humidity.
- Transparency
- Retortability
- Microwaveability

This silica-deposited film was commercialized as a retort pouch material in lamination with PET and polypropylene in Japan in 1988.[64]

The structure of silica-deposited film (GL film) consists of a base film and the SiO_x coating. GL film is a trademark of the silica-deposited PET film produced by the Toppan Printing Co., Ltd. The oxygen barrier properties of this laminated silica-deposited film are better than that of polyvinylidene chloride (PVDC) or polyamide, and slightly higher than that of EVOH, at dry conditions (Table 4). As a transparent, high-barrier packaging material, the barrier level of 0.5 cc/m²·day·atm allows a long shelf life for most processed foods. Another important property of the silica-deposited film is the limited effect of temperature and humidity on the film's barrier properties.

In comparison to other commonly used barrier films, the silica deposited film's WVTR (0.5 g/m²·day) is equivalent to that of a laminated aluminum metallized film and to that of 500 CPP (Table 5). In production, the 1.0 g/m²·day WVTR is a difficult target for transparent flexible films of less than 100 μ thickness, even though PVDC-coated oriented polypropylene (OPP) or high-density polyethylene (HDPE) are used.[65]

Table 6 lists the oxygen permeance of several laminated barrier films, before and after retort processing (125°C for 20 min). As shown, laminated GL film experienced minimum change in oxygen permeance following retort processing, maintaining a permeance of less than 1 cc/m²·day·atm.

A comparison among silica-deposited film and other barrier materials used in food packaging in terms of oxygen barrier, water vapor barrier, retortability, microwaveability, and transparency is presented in Table 7.[66]

A new deposition technology (QLF coating technology) has been reported for the silicon oxide coating to PET containers that resulted in enhanced oxygen barrier properties.[67] Vacuum-deposited QLF-coated bottles offer a two- to three-fold improvement in gas barrier.

Table 4 Oxygen Permeance of SiO$_x$ Laminated Film and Other Laminated Films

Structure	Oxygen Permeance
PET 12 μ/GL film/CPP 60 μ[a]	0.5
ONy 25 μ/PVDC[a] 15 μ/CPP 60 μ	10.5
PET 12 μ/PVDC[b] 15 μ/ONy 15 μ/CPP 60 μ	1.8
PET 12 μ/PVDC[b] 25 μ/ONy 15 μ/CPP 60 μ	1.0
PET 12 μ/EVOH[c] 15 μ/CPP 60 μ	0.3
OSM[d] 15 μ/ONy 15 μ/CPP 60 μ	3.8
ONy 25 μ/CPP 60 μ	15.8

Note: Units of O$_2$ permeance = cc/m^2 day atm at 25°C, 70% RH.

[a] Kureha Co., "K-Flex".

[b] Asahi Kasei Co., "SARAN UB".

[c] Kurare Co., "EVAL".

[d] Toyobo, "OSM": metaxylene diamine-adipic acid copolymer.

From Sakamaki, C., Kano, M. presented at Barrier Pack '89, Chicago, IL, May 1989. With permission.

Table 5 Water Vapor Transmission Rate of SiO$_x$ Laminated Film and Other Laminated Films

Structure	WVTR
PET 12 μ/GL film/CPP 60 μ[a]	0.5
KOP 25 μ/CPP 25 μ	4.0
OPP 25 μ/EVOH 15 μ/CPP 25 μ	2.5
ONy 25 μ/CPP 60 μ	7.8
PET 12 μ/AL metallizing/CPP 25 μ	0.6
CPP 500 μ	0.5

Note: Unit of WVTR = g/m^2 day at 40°C, 100 RH.

From Sakamaki, C., Kano, M. presented at Barrier Pack '89, Chicago, IL, May 1989. With permission.

Table 6 Oxygen Permeability of SiO$_x$ Laminated Film and Other Films Before and After Retort Process

Structure	O$_2$ Permeability Before Retorting	O$_2$ Permeability After Retorting
PET 12 μ/GL Film/CPP 60 μ[a]	0.5	0.7
ONy 25 μ/PVDC[a] 15 μ/CPP 60 μ	10.5	8.2
PET 12 μ/PVDC[b] 15 μ/ONy 15 μ/CPP 60 μ	1.8	2.6
PET 12 μ/PVDC[b] 25 μ/ONy 15 μ/CPP 60 μ	1.0	1.6
PET 12 μ/EVOH[c] 15 μ/CPP 60 μ	0.3	18.8
OSM[d] 15 μ/ONy 15 μ/CPP 60 μ	3.8	6.0
ONy 25 μ/CPP 60 μ	15.8	21.5

[a] Kureha Co., "K-Flex".

[b] Asahi Kasei Co., "SARAN UB".

[c] Kurare Co., "EVAL".

[d] Toyobo, "OSM": metaxylene diamine-adipic acid copolymer.

Note: Retort condition: 125°C, 20 min. Unit of O$_2$ permeability = cc/m^2 day atm at 25°C, 70% RH. Pouched material: water/oil = 1/1 mixed material.

From Sakamaki, C., Kano, M. presented at Barrier Pack '89, Chicago, IL, May 1989. With permission.

Table 7 Comparison among Silica-Deposited Film and Other Barrier Materials

	Silica-Deposited Film	Aluminum Foil	Aluminum Metallized Film	EVOH	PVDC
Oxygen Barrier 25°C—Dry	Good	Good	Good	Good	Good
Hi Temp.—Hi RH	Good	Good	Good	Low–Med	Medium
W.V. Barrier	Good	Good	Good	Poor	Medium
Retortability	Good	Good	Poor[a]	Poor–Good	Medium
Microwaveability	Good	Poor[b]	Poor[c]	Good	Good
Transparency	Good[d]	No	No	Good	Good

[a] Delamination.

[b] Reflection.

[c] Spark.

[d] Slightly yellowish.

From Toppan Printing, Technical Information Sheet, Tokyo, Japan, 1989. With permission.

Aside from multilayer barrier plastic bottles, target markets are containers for oxygen-sensitive food products now packaged in glass or metal. Multilayer lidding and pouches to replace foil for microwaveability also present opportunities for this developing technology.

Naphthalate-based homopolymer and copolyesters

Among the more exciting developments in the packaging arena recently has been the emergence of high-performance polyethylene naphthalate (PEN homopolymer) and naphthalate-based copolyesters that share some of PEN's performance properties.[68]

Pure PEN homopolymer performs significantly better than PET in key areas, including[68]

- Oxygen and moisture barriers of PEN are four to five times that of PET.
- Glass transition temperature of PEN is 43°C higher than PET. That means PEN containers can be hot-filled without sidewall deformation.
- PEN has high chemical resistance that combined with its ability to withstand high temperatures, makes it a suitable material for returnable/refillable bottles that must be washed in hot caustic solutions.
- PEN has superior resistance to ultraviolet light.
- PEN's superior mechanical properties, including 35% higher tensile strength and 50% greater flexural modulus than PET, should allow some downgauging compared to PET bottles.
- Molding and blow cycles are shorter than those of PET.

With respect to its commercial application, the U.S. Food and Drug Administration has recently amended its regulations to permit PEN homopolymers to be used for food packaging.[69]

Active packaging

The conditions that extend the shelf life of a packaged food or horticultural product can be optimized by maintaining the internal levels of oxygen, carbon dioxide, and water activity within established limits. The control of gas composition within a package system has typically been approached by judicious selection of films with appropriate permeability characteristics. Controlled and modified atmosphere packaging technologies have been developed based on such an approach and have been the subject of numerous studies. In these examples, the packaging film functions as a passive barrier, whose permeability characteristics cannot respond to metabolic or chemical changes associated with the product that may alter the package internal environment. The result is that the internal environment of the package system can change with potentially damaging results. The problem is particularly acute when the packaged product is metabolically active or when there is a chemical change in food.

Recently, active package systems have been developed that are designed to interact with a product or its surrounding environment, resulting in alteration of the in-package environment. Here, *Active packaging* is defined as the inclusion of subsidiary constituents into the packaging material or the package headspace, with the intent of enhancing the performance of the package system.

Two general approaches have been followed that resulted in a packaging film being an active material. In one approach, an active ingredient is incorporated within the polymer matrix, resulting in an improvement of the properties of the material or the development of properties not initially characteristic of the material used to produce the package. The second approach is to modify the structure of the film such that the film performs in a different manner. As indicated by Bigg,[70] the most prominent active packaging materials include antibacterial films, corrosion-inhibiting films, CO_2-permeable films, and oxygen-absorbing films. Each of these is discussed briefly.

Alternatively, the package headspace environment can be modified by inclusion of a reagent directly into the package headspace.

Antibacterial films

A film incorporating a sterilizing agent at a concentration of 0.7–1% has been developed in Japan, in which silver and zinc cations have been bonded to the surface of a synthetic microporous zeolite. This allows generation of active oxygen to halt cell metabolism in most bacterial, including *Staphylococcus aureus*, *Salmonella typimunium*, and *Escherichia coli*.[71,72] The active zeolite has been found to be effective at concentrations as low as 2 ppb (wt/wt) and is usually incorporated into a polymer that functions as an inner layer in a multilayer packaging film.

Other antibacterial additives include a 5% concentration of 10,10'-oxybisophenoxarsine in a 35 mesh polystyrene powder that is intended for incorporation with low-density polyethylene.[73]

CO_2-Permeable films

Porous polypropylene films are available that allow CO_2 and oxygen permeation, while retarding moisture permeation.[74–77] As indicated by Drennan,[78] in most cases the pores, which are on the order of 0.4 μm in diameter, are not straight through the film, but follow a tortuous path. Alternatively, a nonwoven fabric of polymer fibers has been described to provide high levels of oxygen and CO_2 permeability.[79] By producing the nonwoven fabric with a high enough density, water vapor permeability will be restricted, while allowing oxygen and CO_2 to permeate.

Oxygen-absorbing systems

Oxygen-absorbing systems or scavengers used in packaging can be divided into five major categories, which include (1) metal-complex scavengers, (2) nonmetal chemical-complex scavengers, (3) photosensitive dye scavengers, (4) enzyme scavengers, and (5) synthetic heme-complex scavengers.

Iron is the main active component in most metal-complex oxygen scavengers. Iron is relatively inexpensive, safe, has FDA clearance, has a reaction rate with oxygen that can be regulated, and has a much greater affinity for oxygen than most food products. The common iron complexes are generally contained in sachets, which are placed into the food package. "Ageless" is the trade name of the oxygen scavenger system distributed by Mitsubishi, which controls about 70% of the market share.[80] These sachets are used in Japan and Europe for numerous products, such as bakery goods, precooked pasta, cured or smoked meats, dried foods, nuts, coffee, cheese, and chocolates.[81]

The amount of iron needed in a sachet depends upon the initial oxygen level in the headspace, the amount of dissolved oxygen in the food, and the package permeation rate. In general, 1 g of iron can react with 0.0136 mol of oxygen (STP), which is equal to approximately 300 cc. The chemical reaction is

$$4 \, Fe + 3 \, O_2 \rightarrow 2 \, Fe_2O_3.$$

Headspace oxygen concentrations have been maintained at less than 0.01% using metal-complex oxygen scavengers. The rate of oxygen absorption is dependent on oxygen concentration and humidity.

A number of different complexes have been designed to serve the specific needs of food products. For example, Multiform Desiccants, Inc. has specifically designed absorbers for use in moist foods ($A_w > 0.65$), dry foods ($0.0 < A_w < 0.7$), and refrigerated foods (0–5°C). Multiform has also introduced a "FreshMax" oxygen-absorbing label that adheres to a package inner surface. "Ageless" absorbers have also been developed to scavenge carbon dioxide as well as oxygen for use in coffee packages.

Studies have shown that these metal-complex oxygen scavengers have been successful in significantly extending the shelf life of many foods including white bread,[82] pizza crust, snack foods, fried rice cakes, cheese-filled pasta products,[81] and some intermediate moisture foods for the space program.[83]

An active film that absorbs oxygen has been developed using meta-xylene diamine-adipic acid (MXD-6) nylon to which 50–200 ppm (wt/wt) of a cobalt–carboxylic acid salt has been incorporated.[84] Cobalt catalyzes the reaction between oxygen and the nylon, such that the oxygen cannot permeate through the film. The modified nylon is used as either a middle layer in a multilayer barrier film or the exterior layer in a two-layer coextrusion. The presence of the cobalt salt has been shown by Folland to reduce the oxygen transmission rate through a PET/MXD-6 multilayer blow-molded, one-liter bottle from 3.5 cc/m²A·day·atm at 23°C and 55% RH to <0.04 cc/m²·day·atm.[84] In both cases, the nylon layer constituted only 4% of the total bottle wall thickness. The effectiveness of the oxygen barrier persisted for 2 years.

Other tests with this system on beer, juice, and wine have been successful, but the system is not yet being used commercially, although it has FDA approval for use as an additive or blend with PET at levels up to 30%, at temperatures below 49°C.[85]

Nonmetal chemical-complex scavengers

A series of nonmetal oxygen absorbers have also been developed to eliminate the problem of setting off metal detectors on processing lines. They are formed from organometallic molecules that have an affinity for oxygen. Oxygen does not react with these molecules, but is irreversibly bound to them, so there are no harmful byproducts. They are useful with liquid and semiliquid products, and can be blended into polymers as a barrier/scavenging layer.[86]

Photosensitive dye scavengers

It has been proposed[87] that the reaction between ground state oxygen and iron complexes is too slow, and that oxygen should be excited to its singlet state for faster scavenging. Photosensitive dyes impregnated on ethyl cellulose film, when illuminated, activate oxygen to its singlet state, which then can react with an acceptor to form an oxide. Some tested singlet oxygen acceptors are difurylidene erythritol (DFE), difurfurylidene-pentaerythritol (PEF), tetraphenylporphine (TPP), dioctyl phthalate (DOP), and dimethyl anthracine (DMA). None of these are approved for food contact in the U.S. The proposed reaction scheme is as follows[88]:

$$DYE + PHOTON \rightarrow DYE(\text{excited state})$$

$$DYE(\text{excited state}) + OXYGEN \rightarrow DYE + OXYGEN(\text{excited single state})$$

$$OXYGEN(\text{excited single state}) + ACCEPTOR \rightarrow ACCEPTOR\ OXIDE$$

$$OXYGEN(\text{excited single state}) \rightarrow OXYGEN$$

Rubber has been studied, not only as a matrix for holding acceptors, but also as a highly concentrated acceptor in itself. If acceptors approved for food contact can be developed, the advantages of this system would be that no sachets would need to be added to the packages and the scavenger system would not become active prior to use, as long as the packages are stored in the dark.

Enzyme scavenger systems

Glucose oxidase is a known oxidoreductase, transferring two hydrogens from the –CHOH group of glucose to oxygen forming glucono-delta-lactone and hydrogen peroxide. One mole of glucose reacts with one mole of oxygen. However, catalase is a normal contaminant in glucose oxidase, and it decreases the effectiveness of glucose oxidase and is very expensive, when pure by half. Pure glucose oxidase is also very expensive. Glucose oxidase (with catalase) has GRAS status and can be added to food products such as beer or wine to eliminate dissolved oxygen in the product.[81] However, the oxidation reaction forms off-flavor byproducts that are detectable in beer.[89] Scott and Hammer suggested using the glucose oxidase in sachets in dried foods with a humidity factor within the sachet to drive the reaction.[90]

Ethanol oxidase is another enzyme with oxygen scavenging potential. It oxidizes ethanol to acetaldehyde. It has been in use as a breath alcohol analyzer test, but there is no known application for food[81].

Synthetic "heme" scavenger

The Aquanautics Corp. has developed oxygen-binding "heme" complexes that function well with high-water-activity food products and in CO_2 environments.[89] The complexes are called LONGLIFE® and their chemical structures mimic that of a heme molecule. The complexes are water-soluble and so were necessarily immobilized on silica and other supports. As a fixture of the crown closure, they have been successful in reducing oxygen concentration in a package of aqueous solution from over 2000 parts per billion (ppb) to less than 50 ppb within 24 h. The absorber need not be in contact with the liquid to be effective.

Moisture absorbers

Moisture-absorbing desiccants can be incorporated as a packaging component of a broad range of moisture-sensitive foods to include snack chips, candies, and cereals, as well as powdered or granulated products and bulk ingredients. Consumers have become increasingly familiar with the presence and purpose of desiccant packets in packaged goods—first in pharmaceuticals, and more recently in foods.

Most desiccant sachets contain either clay, minerals, or silica gel. Sachets also may contain other sorbant substances, such as activated carbon for odor absorption or iron for oxygen absorption. These are all safe for food contact.

In food applications, the sorbants are typically contained in Tyvek® (DuPont) bags or other spun-bonded bags that are nondusting and tear-resistant. Bag sizes, shapes, and colors can be designed to satisfy the parameters of the particular application. Key parameters include the nature of the food product, the volume of space (consumer-size or foodservice package or bulk bin) being protected, volume of product, initial humidity, moisture content in the product postpackaging, the moisture vapor transmission rate of the package system, and anticipated storage time. In general, 16 g of desiccant can adsorb at least 3 g of moisture vapor at 20% relative humidity, and at least 6 g of moisture at 40% RH at 25°C. Desiccant usage life is about 2 to 4 weeks, depending on the application and humidity levels. With a proper moisture barrier package system, desiccants can provide protection of finished foods, additives, or in-process products for extended periods of time, including short-term storage and overseas shipment.[91]

Product–package interactions and compatibility

Although there are many advantages that plastic materials provide when used for food and beverage packaging, there are also some concerns related to product–package interactions.[92] This broad-based topic is associated with the mass transport of gases, water vapor, and low molecular mass organic compounds between product, packaging material, and storage environment. Mass transport in package systems encompasses a number of phenomena referred to as permeability, sorption, or migration. *Permeability* includes the transfer through the package of molecules from the product to the external storage environment or from the environment to the product. *Sorption* involves the take-up of molecules from the product into, but not through, the package. *Migration* is the passage of molecules originally contained in the packaging material into the product. The mass transfer process provides the basis for further molecular activities within the package system, which can result in changes in the product, as well as physical damage to the package, or both.[93] In a package–product system, we refer to the mass transfer processes, and to the physicochemical activities associated with them, as *product–package interactions*. New packaging barrier polymers reaching the market are designed to minimize such interactions and have been shown to protect foods and beverages for long periods of time.

There are various terms found in the literature to describe the steady-state permeation of molecules through a polymer film of area A and thickness ℓ (see Figure 2). The following terms are defined to characterize the permeation process.

- Permeant transmission rate, $F = \dfrac{q}{A \cdot t}$

- Permeance, $R = \dfrac{q}{A \cdot t \cdot \Delta p}$

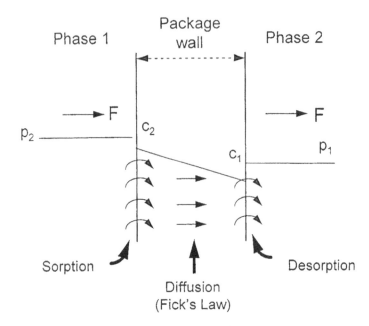

Figure 2 Permeability mechanism of a molecule through a plastic film or sheet. The subscripted parameters p and c correspond to the partial pressure and concentration, respectively, of the permeating molecule in each phase.

- Thickness normalized flow, $N = \dfrac{q \cdot \ell}{A \cdot t}$

- Permeability constant, $P = \dfrac{q \cdot \ell}{A \cdot t \cdot \Delta p}$

where q = is the quantity permeated during time, t; $\Delta p = p_2 - p_1$, pressure drop across the film.

Permeation through a polymer film or sheet is a measure of the steady-state transfer rate of the permeant, and is normally expressed as the permeability coefficient, P. This coefficient can be described in terms of two fundamental parameters, namely, the diffusion and solubility coefficients. The diffusion coefficient, D, is a measure of how rapidly penetrant molecules can advance through the barrier in the direction of lower concentration or partial pressure. The solubility coefficient, S, reflects the amount of the transferred molecules dissolved in the film or slab at equilibrium conditions.

The simplest and most common expression relating P, D, and S is given by Equation 4.

$$P = D \cdot S. \tag{4}$$

Equation (4) is applicable only for situations where D is independent of permeant concentration and S follows Henry's Law of Solubility. However, when the permeation process involves highly interactive organic penetrants such as aroma, flavor, or solvent molecules, the diffusion process is more complex than the diffusion of simple gases, and the diffusion coefficient may vary as a function of penetrant concentration and time.[94-97] When Fick's Second Law takes into account the concentration dependency of D, it is written

$$\frac{dc}{dt} = \frac{d}{dc}\left[D(c)\frac{dc}{dx}\right] \tag{5}$$

for diffusion in a single direction x. Here, c is the concentration of the diffusate in the polymer.[98] Where D varies with t, the diffusion is often called non-Fickian.

Mears derived the following expressions for cases where the diffusion coefficient is concentration dependent.[98]

$$D = D_o (1 + ac) \tag{6}$$

or

$$D = D_o \exp (bc). \tag{7}$$

Equation (7) is more suitable in cases where the diffusion coefficient is strongly concentration dependent. Here D is the differential diffusion coefficient, D_o is the limiting diffusion coefficient, and (a) and (b) are constants.

The loss of volatile low molecular mass organic compounds from a food into polymeric packaging materials, based on a sorption mechanism, has been and continues to be the subject of considerable attention and concern. Several authors have recently reported the loss of aroma components to packaging, and its effect on the change in quality detectable by consumers.[99–106] Sorption may also result in increased permeability to other permeants and lowered chemical and mechanical resistance of the packaging material; it may also affect the kinetics of the migration process. The overall effect during storage could lead to the loss of aroma and flavor volatiles associated with product quality, as well as of other volatile organic food components. In food product–package systems, sorption behavior needs to be characterized for quality control and for predicting the change in product quality, as related to the loss of components associated with product shelf life. Sorption is described as a function of the sorbate concentration by a equilibrium sorption isotherm that can be represented by Henry's Law or other mathematical models.

Sorption, also called scalping, is the uptake of food components, such as flavor, aroma or colorant compounds (called sorbates) by the package material. Many sorption studies of food components have been reported.[107–109] Similarity in chemical structure between the sorbate and polymers enhances sorption.[110] As the molecular weight distribution (MWD) of the polymer increases, sorption also increases. Metallocenes, or in-site polyolefins, which are characterized by narrow MWD are intended to reduce the sorption of volatile compounds compared to LDPE. Sorption studies of benzaldehyde, citral, and ethyl butyrate showed that sorption levels by ionomers are similar to the sorption of LDPE.[111] As in the case of migration, the extent to which the sorption process occurs depends on the initial concentration of the sorbate in the food and the migrant partition coefficient between the plastic and the food. The dynamics of the diffusion process for permeation, migration, and sorption can be calculated from the corresponding solutions to Fick's Second Law.

The fundamental driving force that prompts a molecule to diffuse within the polymer or between a polymer and a surrounding phase is, according to the solution theory, the tendency to equilibrate the specie's activity. In packaging, polymers are generally contacted by gas (air) and liquid phases. In the case of multilayer structures, where several layers of polymers are in direct contact, a layer is contacted by at least one solid phase. Therefore, any mobile molecular species that is not in thermodynamic equilibrium within the phase will tend to equilibrate its activity value. In a packaging system, mass transfer in sorption and migration processes involves two adjacent phases: a polymer phase and a surrounding liquid or gas phase. The mobile substance must diffuse in each phase and move across the interphase. In a permeation process, the permeant needs to move across the two film or sheet interphases and diffuse within the polymer. The maximum concentration of a substance in the polymer, from a contacting gas or liquid phase, is given by the solubility. The solubility is controlled by the equilibrium thermodynamics of the system. From solution theory the activity of a specie, a, can be represented by the activity coefficient γ and concentration c:

$$a = \gamma\, c. \tag{8}$$

In most packaging situations γ is approximately 1, and concentration replaces activity.

Fick's laws of diffusion quantitatively describe permeation, migration, and sorption processes in packaging systems.[112] In anisotropic phases, the diffusion theory states that the rate of transfer of a diffusing substance per unit of area, F, is given by Fick's First Law,

$$F = \pm D \frac{\partial c}{\partial x}, \tag{9}$$

where c is the penetrant concentration in the polymer, x is the direction of the diffusion, and D is the molecular diffusion coefficient. Quantity in F and c are both expressed in the same quantity unit, e.g., gram or cubic centimeter of gas at standard temperature and pressure. The fundamental equation for unsteady state, one-dimensional diffusion in an isotropic phase is Fick's Second Law,

$$\frac{\partial c}{\partial t} = D \frac{\partial^2 c}{\partial x^2}, \tag{10}$$

where t is time. In systems where the diffusant concentration is relatively low, the diffusion coefficient in Equations (9) and (10) is assumed to be independent of both penetrant concentration and polymer relaxations. Diffusion processes in packaging generally involve low values of diffusant concentration. Also, the diffusion is perpendicular to the flat surface of the package with a negligible amount diffusing through the edges. Solutions to the diffusion Equations (9) and (10), together with the initial and boundary conditions, are unidimensional.

For a molecule to diffuse in the polymer it must be dissolved in it; this is the solubility, or concentration c. The solubility of penetrants in polymers (especially polymers above their glass transition temperature and penetrants at low pressure) is, in many cases, well described by the linear isotherm Henry's Law of Solubility

$$c = kp, \tag{11}$$

where p is the partial pressure of penetrant, and k is the Henry's Law constant, which is the solubility coefficient and commonly represented by S. The solubility coefficient S is constant with the pressure p for gases such as O_2 and CO_2, up to 1 atm.

For glassy polymers, and high-pressure penetrants, such as in the cases of CO_2 in PET, a nonlinear Langmuir–Henry's Law model must be followed:

$$c = kp + \frac{C'_H bp}{1 + bp}, \tag{12}$$

where C'_H is the Langmuir capacity constant and b is the Langmuir affinity constant.[113]

In closed systems and when there is an equilibrium in the penetrant activity in the two phases, the partition coefficient K_p is defined as

$$K_p = \frac{c_f^*}{c_p^*}, \tag{13}$$

where c_f^* and c_p^* are the sorbate or migrant equilibrium concentrations in the packaged product and polymer, respectively.

For a specific value of concentration, K_p provides a practical way to calculate the change in a component concentration, either in the food or packaging material, from the time that the food product and packaging material are contacted up to the moment they reach equilibrium, provided the initial concentrations are known.

For organic vapors, Henry's Law is valid, in most cases only at very low pressure, penetrant concentrations in the order of milligrams of penetrant per liter of air (ppm). However, the applicability range of Henry's Law depends on the particular organic penetrant/polymer under consideration. Flory–Huggins equation applies to high solvent activity in polymers.[114]

Since sorption and migration correspond essentially to the same mass transport phenomenon, migration can also be described by partition and diffusion coefficients. The diffusion coefficient determines the dynamics of the sorption process; the larger the value of D, the shorter the time to reach equilibrium.

The focus here is to discuss the principal variables affecting the behavior of the permeation, sorption, and migration processes, specifically in relation to the basic mass transfer parameters such as the diffusion, solubility, and partition coefficients. The variables affecting permeation, sorption, and migration can be grouped as follows:

Compositional variables:

- Chemical composition of the packaging material and penetrant
- Morphology of the polymer
- Concentration of the penetrant
- Presence of co-permeant

Environmental and geometric variables:

- Temperature
- Relative humidity
- Packaging geometry

While an in-depth treatment at each of the above factors is beyond the scope of this review, selected examples are discussed to illustrate their role in the sorption and transport of organic penetrants in barrier polymers. For a more in-depth treatment, the reader is referred to the references listed, and references cited therein.

Diffusion (D) and solubility coefficient (S) values are usually determined by observing the change in weight (i.e., increase or decrease) of a polymer sample during a sorption process. Diffusion (D) and permeability coefficient (P) values are obtained from permeability studies, where the transport of a permeant through a polymer membrane is continually monitored (i.e., isostatic procedure), or by quantifying the amount of the penetrant that has passed through the film and accumulated as a function of time (i.e., quasi-isostatic procedure).

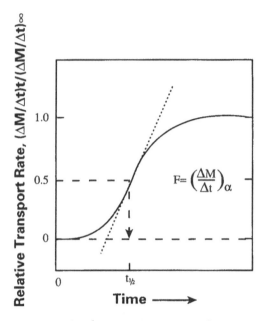

Figure 3   Relative transmission rate for the isostatic test procedure.

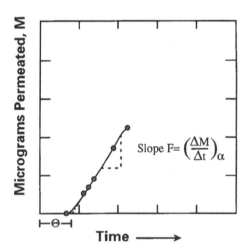

Figure 4   Generalized transmission rate for the quasi-isostatic test procedure.

Typical graphical presentation of data analysis for permeability and sorption experiments are shown in Figures 3–5, respectively. A detailed discussion of permeation and sorption measurements is included in a review by Hernandez et al. and references cited therein.[115]

The relationship between penetrant transfer characteristics and the basic molecular structure and chemical composition of a polymer is rather complex, and a number of factors contribute to the permeability and diffusion processes, among the most important being

- Cohesive-energy density, which produces strong intermolecular bonds, Van der Waals or hydrogen bonds, and regular, periodic arrangement of such bonding groups.

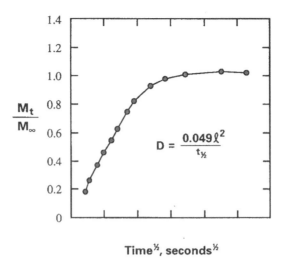

Figure 5 Generalized sorption curve.

- The glass transition temperature (T_g) of the polymer, above which free vibrational and rotational motion of polymer chains occur, so that different chain conformations can be assumed.

As expected, the solubility and diffusivity of liquids and gases in polymers are dependent upon molecular structure, chemical composition, and polymer morphology. Permeability, which depends on the values of the solubility and diffusion coefficient, also behaves in a similar fashion.

According to solubility theory developed by Hildenbrand and Scott,[116] the enthalpy of mixing is given by

$$\Delta H_m = \phi_s \phi_p (\delta_s - \delta_p), \tag{14}$$

where ΔH_m = enthalpy of mixing per unit volume, ϕ_s and ϕ_p = volume fraction of components (i.e., solvent and polymer), and δ_s and δ_p = solubility parameters of the components. According to Hildenbrand, ΔH_m has to be equal to or near 0 for solubility to take place.

Equation (14) indicates that $\Delta H_m = 0$, if $\delta_s = \delta_p$. This is equivalent to saying that if the polymer and solvent have equal solubility parameters they should be mutually soluble or that substances with similar chemical structures are quite likely to dissolve in each other. However, Equation (14) takes into consideration only the chemical structure or dispersion forces between molecules, and assumes that no additional noncovalent interactions occur or physicochemical bonds are formed between the two types of molecules. These include van der Waals forces, hydrogen bonds, and dipole interactions. The existence of intermolecular forces between polymer and solvent may result in a lower value of ΔH_m than that calculated using Equation (14), and as a result solubility between the two species may occur,[110] even when the values of δ_s and δ_p differ. If only one of the substances contains polar groups or is capable of forming hydrogen bonds, ΔH_m may actually be higher than that calculated and the solubility may be low, even if δ_s and δ_p are similar or equal.

For situations where there is interaction between polymer and solvent, a more refined treatment of the solubility parameter concept is necessary. The component group contributions method of Hoftyzer and VanKrevelen[117] was used to estimate the solubility param-

Table 8 Predicted Solubility Parameter Values and Interaction-Type Contribution

Polymer	Solubility Parameter (δ)	Dispersion Contribution (δ_D)	Polar Contribution (δ_p)	Hydrogen Bonding Contribution (δ_H)
MXD-6 nylon	25	17.0	13.7	12.2
PET	22	17.9	7.3	10.5
HDPE	18	18.0	0	0
Acetone	20.5	15.5	10.4	7

Note: Units = $(J/cm^3)^{1/2}$.

Table 9 $\overline{\Delta\delta}$ Values for Polymer/Acetone Systems

Solvent Polymer System	$\overline{\Delta\delta}$ $(J/cm^3)^{1/2}$
Acetone/MXD-6 nylon	6.3
Acetone/PET	4.6
Acetone/HDPE	13.0

eter for the polymers meta-xylene diamine/adipic acid (MXD-6 nylon); polyethylene terephthalate (PET); and high-density polyethylene (HDPE). Table 8 gives the solubility parameter (δ) values and the dispersion (δ_d), polar (δ_p), and hydrogen bonding (δ_H) contributions to the solubility parameter. Values for the penetrant acetone are also given.

Equation (15) determines the solubility of a polymer or organic liquid, when dispersion, polar, and hydrogen bonding contribution are considered.[110]

$$\overline{\Delta\delta} = \left[\left(\delta_{d,p} \pm \delta_{d,s} \right)^2 + \left(\delta_{p,p} \pm \delta_{p,s} \right)^2 + \left(\delta_{H,p} \pm \delta_{H,s} \right)^2 \right]^{1/2} \tag{15}$$

For a good solubility, $\overline{\Delta\delta}$ must be less than 5. The $\overline{\Delta\delta}$ values calculated for the respective polymer/acetone systems are summarized in Table 9. These values predict that acetone would have the highest solubility in PET, and a much lower solubility in both MXD-6 nylon and HDPE. Experimental sorption equilibrium and diffusion coefficient values for acetone vapor and the polymers, MXD-6 nylon, PET, and HDPE are presented in Table 10. Although the method of Hoftyzer and Van Krevelen[117] was initially developed for amorphous polymers, the solubility data for acetone agrees well with the trend predicted from solubility parameter value estimations.

This method may therefore be used to predict qualitative changes in polymer/sorbate systems, based purely on the chemical structure of the polymer. The method of Hoftyzer and Van Krevelen does not, however, take into consideration the morphology of the polymer, which affects the sorption and diffusion characteristics of the polymer. The morphology of polymeric materials is influenced by a number of factors, including

1. Structural regularity or chain symmetry, which can readily lead to a three-dimensional order of crystallinity. This feature is determined by the type of monomer(s) and the conditions of the polymerization reaction.
2. Intermolecular forces such as van der Waals or hydrogen bonds, and the periodic arrangement of such interacting groups.
3. Chain alignment or orientation, which allows laterally bonding groups to approach each other to the distance of best interaction, enhancing the tendency to form more orderly materials.

Table 10 Solubility and Diffusion Coefficient Values for Acetone Vapor in Barrier Films

Film Sample	Solubility $(kg/kg \times 10^2)$	Diffusion Coefficient $(cm^2/sec \times 10^{12})$
MXD-6 Nylon	1.4	5.1
PET[a]	8.7	80
PET[b]	8.4	84
High-density	1.3	210

Note: Values determined at 22 ± 1°C and 250 gm/m³ acetone vapor concentration.

[a] 400% orientation, with an orientation temperature of 115°C.

[b] 400% orientation, with an orientation temperature of 90°C.

With respect to the glass transition temperature of barrier polymer structures, DeLassus reported that glassy polymers have very low diffusion coefficients for flavor, aroma, and solvent molecules at low concentrations. Typically, these values are too low to measure by standard analytical procedures. The diffusion coefficient determines the dynamics of the permeation process and thus the time to reach steady state, which accounts for glassy polymers exhibiting high barrier characteristics to organic permeants. Polyolefins, being well above their glass transition temperature, are nonglassy polymers and have high diffusion coefficients for organic permeants, and steady-state permeation is established quickly in such structures.

Polymer free volume is also a function of structural regularity, orientation, and cohesive energy density. The aforementioned structure–property relationships all contribute to a decrease in solubility and diffusivity, and thus permeability.

Salame has proposed a relationship between polymer molecular structure and permeability based on an empirical constant (π), or "Permachor" constant which, when substituted into the Permachor equation, predicts the gas permeability of polymer structures.[92] The correlation parameter or the Permachor constant is based on the cohesive energy density and free volume of the polymer, two major properties of the polymer. Agreement between the Permachor constant and film permeability has been shown to be quite good for oxygen, CO_2, and nitrogen, but not for water vapor and organic vapors.

The equation for relating gas permeability to the Permachor constant is as follows:

$$P = A \exp(-s\pi), \tag{16}$$

where A and s are constants for any given gas at temperature (T) and π is the Permachor constant of the polymer.

Morphology of polymer

Solid-state polymer chains can be found in a random arrangement associated with an amorphous structure or in a highly ordered, crystalline phase arrangement. Morphology in a polymer refers to the physical state by which amorphous and semicrystalline regions coexist and relate to each other and depends not only on its stereochemistry, but also on whether the polymer has been oriented or not, and at which conditions of temperature, strain rate, and cooling temperature the film has been oriented, as well as the melt cooling temperature.

Fundmental properties that are associated with polymer morphology and will therefore influence the permeability and diffusivity characteristics of the polymer include

- structural regularity or chain symmetry, which can readily lead to a three-dimensional order of crystallinity. This is determined by the type of monomer(s) and the conditions of the polymerization reaction.
- chain alignment or orientation that allows laterally bonding groups to approach each other to the distance of the best interaction, enhancing the tendency to form crystalline materials.

In the development of the diffusion, sorption, and permeation theory, it is assumed that the polymer phase is a homogeneous and isotropic phase, i.e., an isotropic amorphous polymer. The presence of a crystalline microphase complicates this assumption considerably and makes the diffusion process in semicrystalline polymers a complex phenomenon. Semicrystalline polymers consist of a microcrystalline phase dispersed in an amorphous phase. The dispersed crystalline phase decreases the sorption of penetrants, whenever the crystal conformations produce regions of higher density than the amorphous polymer. A closer atomic packing tends to exclude relatively large molecules such as organic permeants. For this reason it is generally accepted that gases and vapors are normally sorbed, and therefore able to diffuse, only in the polymer's rubbery or amorphous phase. The dispersed microcrystals are impermeable to penetrant diffusion and create a more tortuous path for the diffusing molecule. Additionally, the microcrystalline phase also acts as a tridimensional crosslinking agent, increasing the nonisotropism of the polymer. The combined decrease in sorption and diffusion contributes then to a lower permeability.

Sorbate solubility (S) in a semicrystalline polymer, having a crystalline volume fraction of α, and an amorphous phase solubility S_a is given by

$$S = (1 - \alpha)S_a. \tag{17}$$

One of the earliest examples in support of Equation (17) was reported by Van Amerongen.[119] The work of Michaels and coworkers on polyethylene terephthlate (PET) and PE also supported the model presented in Equation (17).[120] Puleo et al. reported the gas sorption and permeation in a semicrystalline polymer, for which the crystal phase has a higher density than the amorphous phase.[121] At 100% crystallinity, the sorption of CO_2 and CH_4 was about 25–30% of the solubility in the amorphous phase.

The diffusion coefficient in the amorphous phase, D_a, has been shown to be related to the diffusion coefficient in the semicrystalline phase D by

$$D_a = D\beta\tau, \tag{18}$$

where β is a "chain immobilization factor" and τ is a "geometric impedance factor." Both β and $\tau > 1$. Michael et al. applied Equation (18) to semicrystalline polyethylene terephthalate (PET).[122] Application of Equation (18) does not work well, however, especially for annealed polymers.[123]

Morphology is thus important in determining the barrier properties of semicrystalline polymers. This importance is illustrated by the results of permeability studies carried out on biaxially oriented polyethylene terephthalate (PET) films of varying thermomechanical history.[124] Film samples were oriented biaxially at a strain of 350%/s based on the initial dimension of 4 × 4 in., which corresponded to an orientation rate of 14 in./s, biaxially.

The degree of orientation was 400% based on the initial dimensions. The orientation temperatures were 90°C, 100°C, and 115°C, respectively. Table 11 lists the crystallinity percentage values calculated from density values of the films, at the respective orientation temperatures.

Table 11   Density, and Mass-Fraction Crystallinity of the PET Sample Films

Orientation Temperature (°C)	Density (g/cc)	Mass Fraction Crystallinity Percentage
90	1.360	24
100	1.366	30
115	1.371	33

Table 12   Permeability of Ethyl Acetate through PET Film Biaxially Oriented at 90°C and 115°C

Orientation Temperature (°C)	Vapor Activity (a)	Run Temperature (°C)	P Permeability Coefficient[a] $\times 10^{20}$	D Diffusion Coefficient[b] $\times 10^{12}$
90	0.59	30	2.6	1.8
	0.43	37	4.8	2.9
	0.21	54	15.4	11.0
115	0.59	30	0.014[c]	—
	0.21	54	3.6	5.3

[a] Permeability coefficient units are kg m/m²·s·Pa.

[b] Diffusion coefficient units are cm²/s.

[c] No permeation after 550 h. Value of P reported represents an upper bound.

Table 12 summarizes the results of permeability studies carried out with ethyl acetate in PET film biaxially oriented at 90°C and 115°C, respectively, and serves to illustrate the effect of thermomechanical history (crystallinity percentage) on the relative barrier properties of PET for the permeation of ethyl acetate. The crystallinity percentage of PET film oriented at 90°C was 22%, while the crystallinity percentage of the film sample oriented at 115°C was 31%. As shown, ethyl acetate permeability values decreased by approximately four times by increasing the film orientation temperature from 90°C to 115°C.

As indicated above, the solubility of vapors and gases in polymers is also strongly dependent on crystallinity percentage, since solubility is usually confined to the amorphous regions. This is shown by the results of sorption studies carried out on the oriented PET film samples. A representative plot of M_t/M_∞ vs. $t^{1/2}$ for sorption of ethyl acetate in the PET film oriented at 90°C, obtained at an ethyl acetate vapor concentration of 300 g/m³ (ppm) and a temperature of 37°C, is presented in Figure 6. The calculated curve is also presented for comparison to the Fickian model. A similar plot was obtained for the PET film sample oriented at 115°C.

The equilibrium solubility and solubility coefficient values obtained for ethyl acetate in the PET films are presented in Table 13.

For the conditions of test (i.e., 37°C, 300 g/m³ vapor concentration) the specific concentration, C_a,[125] of ethyl acetate in PET film oriented at 90°C and 115°C is 0.021 and 0.016 g ethyl acetate per gram amorphous component, respectively. These findings indicate that the difference between the solubility of ethyl acetate vapor in PET film oriented at 90°C and 115°C cannot be attributed solely to the difference in crystallinity percentage, but may be the result of penetrant-induced swelling of the polymer structure, which permits additional sorption. The solubility parameter for ethyl acetate is 9.1 cal$^{1/2}$/cm$^{3/2}$ and that for PET is 10.7 cal$^{1/2}$/cm$^{3/2}$. Since the polymer and penetrant are of similar polarity, penetrant–polymer interaction that results in the swelling of the polymer structure is not unexpected.[94-96]

Table 14 contains the diffusion and solubility coefficient values for toluene in PET film oriented at 100°C, as a function of toluene vapor concentration. Sorption studies were carried out at 23°C.[126]

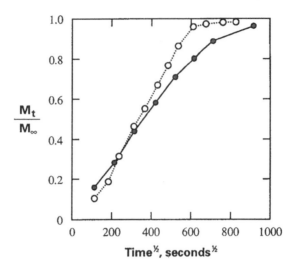

Figure 6 Sorption of ethyl acetate vapor in oriented PET (orientation achieved at 90°C), 300 ppm (µg ethyl acetate/ml N_2). Symbols: ○ - experimental; ● - calculated.

Table 13 Solubility of Ethyl Acetate Vapor in PET Film at 37°C

Orientation Temperature (°C)	S_e Solubility (kg vapor/kg polymer)	S Solubility Coefficient (kg/kg·Pa × 10⁷
90	0.016	8
115	0.011	5

Table 14 Diffusion and Solubility of Toluene Vapor in PET Oriented at 100°C

Toluene Vapor Concentration (g/m³)	Vapor Activity	Diffusion Coefficient $D \times 10^{12}$ (cm²/sec)	S_e Solubility (kg vapor/kg polymer × 10⁷)
76	0.50	—	1.0
80	0.53	2.6	2.7
85	0.57	3.5	3.7
90	0.60	4.1	4.7
95	0.63	4.4	6.0
102	0.68	4.6	8.0

Concentration dependence of the transport process

The results of studies on the permeance of limonene vapor through (1) oriented polypropylene; (2) Saran-coated, oriented polypropylene; (3) two-sided, acrylic (heat seal)-collated, biaxially oriented polypropylene; and (4) one-side Saran coated, one-side acrylic coated, polypropylene film samples, as a function of penetrant concentration, are presented graphically in Figure 7, where permeance (R) is plotted as function of penetrant concentration.[127] The observed concentration dependency of the permeance may be the result of penetrant–polymer interaction, resulting in configurational changes and alteration of polymer chain conformational mobility. Zobel reported similar findings for the transport of the penetrant benzyl acetate through coextruded, oriented polypropylene and Saran-coated, oriented polypropylene, at various penetrant concentrations.[128]

Figure 8 provides a further example of the concentration dependency of the mass transport process for the permeance of limonene through two typical cereal package liners,

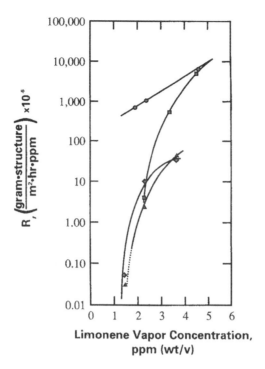

Figure 7 Effect of limonene vapor concentration on permeability of oriented polypropylene and coated oriented polypropylene structures (22° ± 1°C). Symbols: ● - oriented polypropylene; ■ - two-side, heat-seal-coated oriented polypropylene; ♦ - one-side coated Saran and one-side acrylic-coated oriented polypropylene; ▲ - Saran-coated oriented polypropylene.

namely, a high density polyethylene (HDPE) sealant laminate and a glassine-based structure (wax/polyvinyl alcohol/glassine/polyvinyl alcohol/wax).[129] As shown for both structures, the permeance increases exponentially with increased limonene vapor concentration over the entire vapor concentration range studied. Assuming these relationships can be extrapolated to vapor concentration levels below those evaluated experimentally, an expression derived to describe such a relationship could be used to estimate the permeance at *d*-limonene levels associated with a product system, which would be impractical to determine experimentally.

To provide a relative evaluation of the aroma barrier properties of a polymer film or slab, one or more probe compounds characteristic of the flavor profile are selected and their permeability through the structure determined. Such probe compounds are selected based on their relative concentration in the aroma profile, contribution to the aroma, and ease of analysis.

Presence of co-permeant

As shown above, organic vapors are capable of exhibiting concentration-dependent mass transport processes. Therefore, the type and/or mixture of organic vapors permeating will determine the magnitude of sorption and permeation, as well as the effect of a co-permeant on penetrant permeability. The synergistic effect of a co-permeant is illustrated by the results of permeability studies carried out on a biaxially oriented polypropylene film. The degree of film orientation was 430% (machine direction) and 800% (cross machine/direction), based on the initial dimensions. Binary mixtures of ethyl acetate and limonene of varying concentration were evaluated as the organic penetrants.[130]

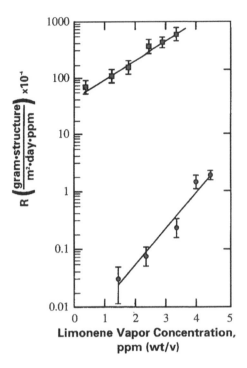

Figure 8 Effect of limonene vapor concentration on permeance constant (*R*) for HDPE and glassine-based structures. Symbols: ■ - HDPE/sealant laminate structure; ● - wax/PVOH/glassine/PVOH/wax structure. Adapted from Reference 129.

Results of permeation studies for selected ethyl acetate–limonene binary vapor mixtures are summarized in Tables 15 and 16, respectively. Permeability values for the pure penetrants are also listed for comparison. As shown in Table 15, at the lowest vapor activity levels evaluated (ethyl acetate $a_v = 0.10$ and limonene $a_v = 0.18$), limonene vapor had a significant effect on the transport properties of the co-permeant. A 500% increase in the permeability coefficient of ethyl acetate was obtained when compared to ethyl acetate vapor permeability alone, at similar test conditions. However, at this concentration level, ethyl acetate did not appear to influence the permeation of the limonene vapor. Selected results are presented graphically in Figure 9, where the transmission curves for ethyl acetate in the binary mixture and for ethyl acetate vapor alone are shown. The transmission rate profile curve for limonene vapor in the binary mixture is superimposed in Figure 9 to provide a complete description of the transmission characteristics of the mixed vapor system.

For the ethyl acetate $a_v = 0.1$/limonene $a_v = 0.29$ binary mixture, a permeation rate 40 times greater than the transmission rate of pure ethyl acetate vapor, at an equivalent concentration, was obtained (see Table 15). This is illustrated in Figure 10, where the transmission profile plot of the binary mixture is presented, and compared to the transmission rate profile curve for ethyl acetate vapor alone. Again, at this concentration level, ethyl acetate did not appear to effect the permeability characteristics of limonene vapor (see Table 16).

For studies carried out with the binary mixture of ethyl acetate $a = 0.48$ and limonene $a = 0.18$, the individual components of the mixture were found to have a significant effect on the permeation rates of the co-penetrant. As shown in Tables 15 and 16, the permeability of both ethyl acetate and limonene through the oriented polypropylene film increased by an order of magnitude, when compared to the permeability of the individual components

Table 15 Effect of Limonene Vapor on the
Permeability of Ethyl Acetate in Binary Mixtures
through Polypropylene

Ethyl Acetate Vapor Activity	Limonene Vapor Activity	Ethyl Acetate Permeability Coefficient[a]
0.05	0	0.35
0.10	0.18	2.30
0.10	0.29	15.0
0.12	0	0.40
0.18	0	0.42
0.30	0	0.5
0.48	0	1.5
0.48	0.18	8.5

[a] Permeability constant values expressed as $kg \cdot m / m^2 \cdot s \cdot Pa$. Average of replicate runs, with a confidence limit of 10% (maximum). Values obtained at $22 \pm 1°C$.

Table 16 Effect of Ethyl Acetate Vapor on the Permeability of Limonene in Binary Mixtures through Polypropylene

Limonene Vapor Activity	Ethyl Acetate Vapor Activity	Limonene Permeability Coefficient[a]
0.18	0.1	49
0.18	0.48	540
0.21	0	37
0.29	0.1	420
0.29	0	57
0.42	0	64
0.50	0	66

[a] Permeability constant values expressed as $kg \cdot m / m^2 \cdot s \cdot Pa$. Average of replicate runs, with a confidence limit of 10% (maximum). Values obtained at $22 \pm 1°C$.

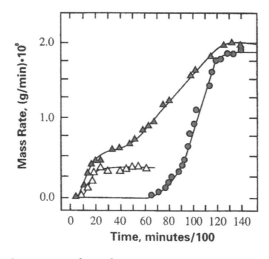

Figure 9 Comparison of mass rates through oriented polypropylene for the binary mixture, ethyl acetate ($a_v = 0.1$)/limonene ($a_v = 0.18$), with pure ethyl acetate ($a_v = 0.12$). Symbols: △ - ethyl acetate; ▲ - ethyl acetate as co-permeant; ● - limonene as co-permeant.

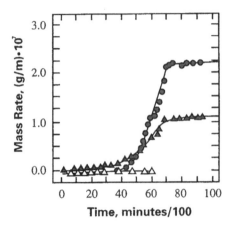

Figure 10 Comparison of mass rates through oriented polypropylene for the binary mixture, ethyl acetate (a_v = 0.1)/limonene (a_v 0.29), with pure ethyl acetate (a_v = 0.12). Symbols: △ - ethyl acetate; ▲ - ethyl acetate as co-permeant; ● - limonene as co-permeant.

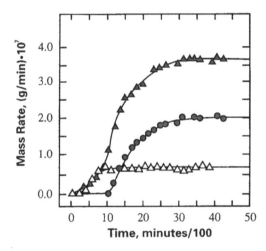

Figure 11 Comparison of mass rates through oriented polypropylene for the binary mixture, ethyl acetate (a_v = 0.48)/limonene (a_v = 0.18), with pure ethyl acetate (a_v = 0.5). Symbols: △ - ethyl acetate; ▲ - ethyl acetate as co-permeant; ● - limonene as co-permeant.

of the mixture, at equivalent concentration levels. Typical results are presented in Figure 11, where the transmission profile plot of the binary mixture is compared to the transmission rate profile curve for ethyl acetate vapor alone. It should be noted that, for each binary mixture investigated, the collective permeation rate for the mixture was significantly higher than the transmission rates for the pure components of the mixture.

Although sorption equilibrium values were not determined in the present study, it appears that by comparing permeability and diffusivity data, the solubility coefficient of limonene is much higher (10–100 times) than that of ethyl acetate. This difference is supported by the numerical values of the solubility parameter of the components of this system, 9.1 $(cal/cm^3)^{1/2}$ for ethyl acetate, 7.8 $(cal/cm^3)^{1/2}$ for limonene, and 8.1 $(cal/cm^3)^{1/2}$ for polypropylene. The difference between the solubility parameters values of limonene and polypropylene is less than 0.5, while the difference for ethyl acetate and polypropylene is 1.0. Accordingly, from solubility theory it is expected that the value of solubility for

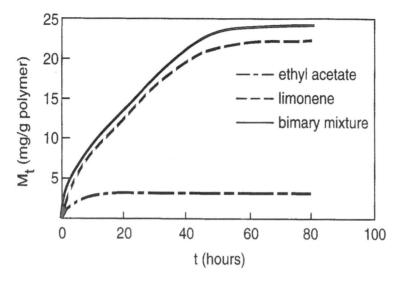

Figure 12 Sorbed amounts of ethyl acetate ($a = 0.10$), limonene ($a = 0.21$), and a binary mixture of ethyl acetate ($a = 0.10$) and limonene ($a = 0.21$) by oriented polypropylene. Adapted from Reference 131.

limonene in polypropylene should be higher than for ethyl acetate. This difference may explain the fact that only at the highest ethyl acetate activity levels is the permeability of limonene affected. These co-penetrant considerations are reasonable since other interactions such as hydrogen bonding are not expected.

Assuming Equation (4) is applicable to the sorbate/polymer systems studied, the diffusion coefficient and/or the solubility coefficient must also vary in the presence of a co-penetrant to account for the observed increases in the transmission rates for the components of ethyl acetate/limonene binary mixtures. By direct measurement of the equilibrium solubility, Nielsen and Giacin found that the solubility coefficient values were independent of sorbate vapor activity, over the range of activity levels studied, and were not affected by the presence of a co-penetrant.[131] This is illustrated graphically in Figure 12, where sorption profiles for ethyl acetate ($\alpha = 0.10$) and limonene ($\alpha = 0.21$) determined as pure single component vapors and as a binary mixture (ethyl acetate, $\alpha = 0.10$/limonene, $\alpha = 0.21$) are presented. As shown, the total level of the binary mixture sorbed (mg uptake/g polymer) is equal to the sum of the quantity of the individual constituents sorbed.

For the ethyl acetate ($a_v = 0.10$)/limonene ($a_v = 0.21$) binary mixture, Hensley et al. reported a permeation rate of ethyl acetate 40 times greater than the transmission rate for pure ethyl acetate vapor of an equivalent concentration,[130] while sorption studies showed the solubility coefficient of ethyl acetate to be constant and did not deviate from Henry's law in the presence of a co-penetrant. Assuming the relationship given in Equation (4) is valid, a possible explanation for the dramatic increase in the permeability coefficients for ethyl acetate in the presence of limonene as a co-permeant lies with the high co-permeant dependency of the diffusion coefficient.

For ethyl acetate/limonene binary mixtures, limonene as a co-penetrant appears to have little or no effect on the solubility of ethyl acetate in oriented polypropylene film, while significantly changing the inherent mobility of ethyl acetate within the polymer bulk phase. This accounts for the observed increase in the transmission rates for ethyl acetate through the OPP film in the presence of limonene.

While not fully understood, the proposed co-penetrant dependency of the diffusion coefficient may be due in part to co-penetrant-induced relaxation effects occurring within the polymer matrix. The absorption of organic vapors can result in polymer swelling and thus change the conformation of the polymer chains. These conformational changes are controlled by the retardation times of polymer chains. If these times are long, stresses may be set up which relax slowly. Thus, the absorption and diffusion of organic vapors can be accompanied by concentration as well as time-dependent processes within the polymer bulk phase, which are slower than the micro-Brownian motion of polymer chain segments which promote diffusion.[132]

There is precedence in the literature in support of such long time period relaxation effects occurring in polymer films above their glass transition temperature.[133,134] Thus, there may be co-penetrant-induced relaxation effects occurring during the diffusion of ethyl acetate/limonene binary mixtures in the oriented polypropylene film investigated. Such relaxation processes, which occur over a longer time-scale than diffusion, may be related to a structural reordering of the free volume elements in the polymer, thus providing additional sites of appropriate size and frequency of formation, which promote diffusion and account for the observed increase in the permeation rate of ethyl acetate in the presence of limonene as a co-permeant.

DeLassus et al. have alluded briefly to the permeation of multicomponent mixtures of organic vapors in their discussion of the transport or apple aroma in polymer films, where permeation of organic vapor mixtures in low-density polyethylene was studied.[135]

Effect of relative humidity

The effect of water activity or moisture content on the diffusion of toluene vapor through a multilayer coextrusion film structure containing moisture sensitive hydrophilic barrier layers (i.e., nylon and EVAL) is presented in Figure 13, where the permeance is plotted as a function of water activity.[136] In this study, the effect of sorbed water on the diffusion of toluene vapor was evaluated, where the test film was preconditioned to a fixed water activity prior to test. From Figure 13 it becomes evident that the permeance increased exponentially with an increase in water activity, at a constant penetrant level.

When dry, there is little permeation through the multilayer structure, since interchain hydrogen bonding reduces polymer chain segmental mobility, and thus restricts the diffusion of penetrant molecules. When moisture is present, however, sorbed water acts to plasticize the hydrophilic barrier layers, which decreases the cohesive forces between the polymer chains, resulting in an increase in polymer chain segmental mobility.[137] This plasticization of the hydrophilic barrier layers will result in a significant increase in toluene vapor diffusivity, and accounts for the accelerating effect of water activity.

A further illustration of the effect of water activity and penetrant vapor concentration on the barrier properties of a packaging film is presented in Figure 14, where the permeance is plotted as a function of water activity for the permeability of toluene vapor through a two-sided, Saran-coated, oriented polypropylene film.[138] As shown, at a constant penetrant level of 60 g/m^3, the permeance is dependent upon water activity and increased exponentially with an increase in sorbed water. The plot of log permeance vs. water activity for toluene vapor concentrations of 40 and 81 g/m^3 are superimposed in Figure 14 to illustrate the relationship among penetrant vapor concentration, water activity, and the barrier properties of the test film. The permeance showed a marked dependency upon sorbed water at toluene vapor concentration levels of 40 and 60 g/m^3, while the diffusion process appeared to be independent of relative humidity at 81 g/m^3. DeLassus et. al. evaluated the effect of relative humidity on the permeation of *trans*-2-hexanal through a vinylidene chloride copolymer (CoVDC) and observed no difference. While precedence for the accelerating effect of relative humidity on the barrier properties of

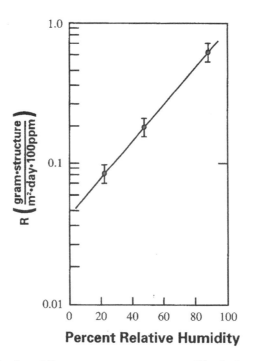

Figure 13 Effect of relative humidity on permeance constant (*R*) of toluene vapor through a coextrusion film structure (polyethylene/nylon/EVAL/nylon/polyethylene) at 81 ppm (μg toluene/ml of N_2) and 23°C. Adapted from Reference 136.

polymers with an affinity for water is found in the literature,[139-144] Landois-Garza and Hotchkiss reported an example of an increase in the effectiveness as a barrier of polyvinyl alcohol (PVOH) with increasing relative humidity for the penetrant, ethyl propionate.[145] The permeability coefficients, *P*, as a function of relative humidity, are presented graphically in Figure 15, and show a trend toward lower permeability values as the relative humidity increases.

The authors explained the observed decrease in permeation as relative humidity increases in terms of the change in diffusion (*D*) and solubility (*S*) coefficient values as a function of relative humidity. As shown in Figure 16, diffusion coefficient values remained relatively constant, but at high relative humidity showed a slightly downward trend. Solubility coefficients, however, showed a clear trend toward lower solubility at high relative humidity. The authors concluded that the decrease in *P* as relative humidity increases was mainly the result of the decrease in penetrant solubility.

As discussed above, studies on the permeability of acetone vapor through amorphous polyamide (nylon 6I/6T) were conducted under both dry and humid conditions. The temperatures were 60°C, 75°C, 85°C, and 95°C and the penetrant partial pressure was constant at 92 mmHg (290 g/m²). Water activity (A_w) of the penetrant steam was maintained at 0.7 (70% RH) when measured at 23°C. Each experiment was run for a period of 8 to 14 days after attaining steady state to ensure the system was at equilibrium.[146]

A summary of the respective permeability coefficient values is presented in Table 17. As shown, sorption of water vapor resulted in an increase in acetone permeability, as compared to dry conditions, with an increase of approximately 1.5 times being observed.

An additional example of the effect of water activity on the barrier properties of polymer films is presented in Figure 17, where the total quantity of ethyl acetate permeated is plotted as a function of time, for the permeability of ethyl acetate through SiO_x PET

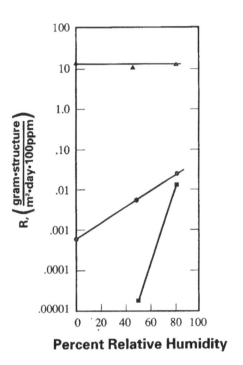

Figure 14 Effect of relative humidity and penetrant concentration on the permeance constant (*R*) of toluene vapor through two-sided, PVDC-coated opaque white OPP film (21°C). Symbols: ■ - 40 ppm (μg/ml of N_2); ● - 60 (μg/ml of N_2); ▲ - 81 (μg/ml of N_2). Adapted from Reference 138.

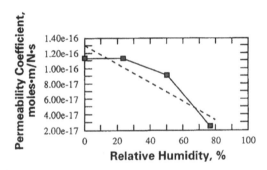

Figure 15 Permeability of ethyl propionate vapor (353 μmol/l of N_2) (875 Pa) through PVOH at 25°C as a function of relative humidity. Adapted from Reference 145.

and EVAL-F films.[147] The test conditions were as follows: temperature, 22°C; concentration of ethyl acetate vapor, 190 g/m³, at 87% RH and 56% RH. Over the course of the entire run, fluctuation of relative humidity was ± 2% and of concentration was ± 5%.

In over 500 hours of continuous testing, there was no measurable permeation of ethyl acetate through either SiO_x PET or EVAL-F at 56% RH. However, at 87% RH there was a significant permeation rate of ethyl acetate vapor through the EVAL-F film.

Table 18 summarizes the permeance data. Also presented in Table 18 are upper-limit value estimations for permeance. As shown, permeation of ethyl acetate vapor through SiO_x PET was not influenced by relative humidity. On the other hand, the permeation of ethyl acetate vapor through EVAL-F film was strongly affected by relative humidity. When the EVAL-F film is in a dry condition, there is little permeation through the amorphous regions, since hydrogen bonding reduces polymer chain segmental mobility, and thus

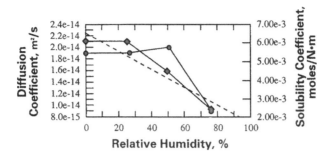

Figure 16 Diffusivity and solubility of ethyl propionate vapor (353 μmol/l of N_2) in PVOH at 25°C as a function of relative humidity (● - diffusion coefficient values, ◆ - solubility coefficient values). Adapted from Reference 145.

Table 17 Summary of Permeability Coefficient Values of Acetone through Nylon 6I/6T

Temperature (°C)	$P \times 10^{19}$ Dry Condition	$P \times 10^{19}$ Humidified Conditions
60	3.7	4.9
75	6.5	11.2
85	9.8	17.6
95	11.8	—

Note: Permeability constant expressed in kg·m/m²·s·Pa.

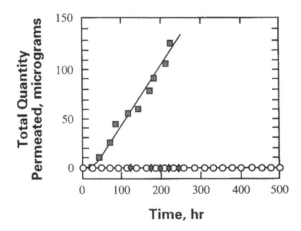

Figure 17 Quantity of permeated ethyl acetate vapor through SiO_x PET(◆ - 87% RH) and EVAL-F (○ - 56% RH; ■ - 87% RH) films at 22°C, and 190 ppm (μg toluene/ml of N_2) quasi-isostatic method.

restricts the diffusion of ethyl acetate vapor. When moisture is present, sorbed water acts to plasticize the hydrophilic barrier polymer, which decreases the cohesive forces between the polymer chains and increases segmental mobility. Plasticization of the barrier layer will result in a significant increase in ethyl acetate vapor diffusivity, and accounts for the accelerating effect of relative humidity.

Temperature dependency

Permeability, diffusion, and solubility coefficients follow a van't Hoff–Arrhenius relationship as given in Equations (19), (20), and (21).

Table 18 Effect of Relative Humidity on the Permeation
of Ethyl Acetate Vapor through SiO$_x$ PET and EVAL-F Film

RH	Sample	Permeance (kg/m^2·sec·Pa) × 10^{17}
56%	SiO$_x$ PET	< 1.1
	EVAL-F	< 2.2
87%	SiO$_x$ PET	< 2.2
	EVAL-F	840 ± 40

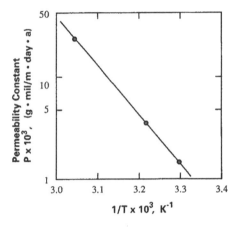

Figure 18 Temperature dependence of the permeability constant for ethyl acetate vapor in PET film oriented at 90°, 300 ppm (μg ethyl acetate/ml of air).

$$P(T) = P_o \exp(-E_P/RT) \tag{19}$$

$$D(T) = D_o \exp(-E_D/RT) \tag{20}$$

$$S(T) = S_o \exp(\Delta H_s/RT) \tag{21}$$

where E_P is the activation energy for permeation, E_D is the activation energy for diffusion, ΔH_S is the molar heat of sorption. P_o, D_o, and S_o are constants, R is the gas constant, and T is the absolute temperature.

The above expressions are valid over a relativey small range of temperatures, which should not include the polymer's glass transition temperature. In particular for P, polymers show values of E_P larger at temperatures above T_g than below T_g. Equation (19) can be used to estimate the permeability coefficient at a desired temperature from a known value.

Figure 18 shows the temperature dependency of the permeation of ethyl acetate through biaxially oriented polyethylene terephthalate (PET) film, where P is plotted vs. $1/T$ (in Kelvin). As can be seen, the temperature dependency of the permeability constant (P) over the temperature range studied (30–54°C) follows the Arrhenius relationship. From the slope of Figure 18, the activation energy of the permeation process (E_P) was determined to be 23.3 kcal/mol. An Arrhenius plot of the diffusion coefficient (D) is shown in Figure 19. The activation energy of diffusion (E_D) was determined to be 15 kcal/mol.[124]

Figures 20 and 21 present the results obtained by DeLassus et al. on the permeation of *trans*-2-hexenal in a vinylidene chloride copolymer film (CoVDC).[135] As shown, the temperature dependency of the transport process for this penetrant/polymer system can be described by an Arrhenius relationship, and gave activation energies of 17.1 kcal/mole

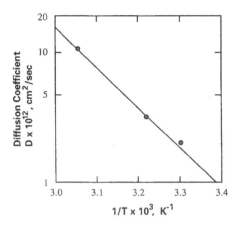

Figure 19 Temperature dependence of the diffusion coefficient for ethyl acetate vapor in PET film oriented at 90°, 300 ppm (μg ethyl acetate/ml of N_2).

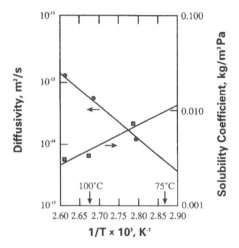

Figure 20 Diffusivity and solubility coefficients of *trans*-2-hexenal vapor in a polyvinylidene chloride (PVDC) copolymer film (● -diffusion coefficient values; ■ - solubility coefficient values). Adapted from Reference 135.

and 26.1 kcal/mole for E_P and E_D, respectively. As shown, although the solubility of *trans*-2-hexenal decreased with temperature, the diffusion coefficient value more than proportionately increased. This disproportionate effect on the latter resulted in the observed increase in the *trans*-2-hexenal permeability value as a function of temperature.

Figure 22 shows the results of permeability experiments on acetone in amorphous polyamide (nylon 6I/6T) film that were carried out under dry and humidified conditions.[146] The activation energy (E_P) for the studies conducted under dry conditions was 9.3 kcal/mole, while E_P for studies carried out under humidified conditions was 12.5 kcal/mole. Statistical analysis showed that the slopes of the respective Arrhenius plots were different, within a confidence level of 99.5%, indicating a statistically significant difference in the permeability of this penetrant/polymer system in the absence and presence of sorbed water vapor. A similar increase in the activation energy was observed in the case of oxygen permeability when comparing dry and humid conditions.

The results of studies to evaluate the temperature dependency of the transport of ethyl acetate vapor through PET, silica-coated PET (SiO_x PET), and an ethylene-vinyl alcohol

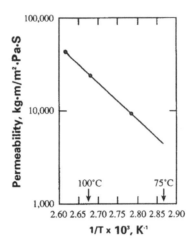

Figure 21 Permeability of *trans*-2-hexenal vapor in a polyvinylidene chloride (CoVDC) copolymer film as a function of temperature. Adapted from Reference 135.

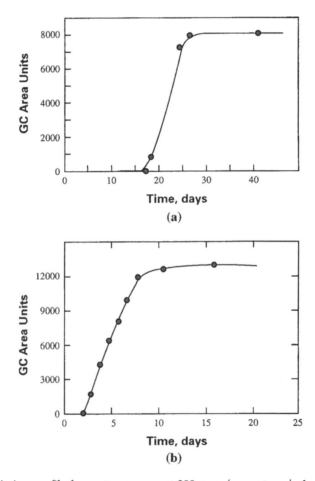

Figure 22 Transmission profile for acetone vapor at 290 ppm (μg acetone/ml of N_2): upper frame, at 60°C through amorphous nylon film; lower frame, at 75°C through amorphous nylon film sorbed by water vapor.

Table 19 Effect of Temperature on the Permeability of
Ethyl Acetate Vapor (300 g/m³) through Barrier Films

| | Permeability Rate[a] × 10⁶ | | |
| | Temperature (°C) | | |
Sample	50	65	76
PET	5.8 ± 5	440 ± 16	820 ± 20
SiO$_x$/PET	2.1 ± 0.6	6.4 ± 0.9	2.9 ± 3
EVAL-F	< 0.3[b]	11.2 ± 1.3	114 ± 15

[a] Transmission rate values expressed as kg/m²·s.

[b] Value reported represents an upper bound based on estimation.

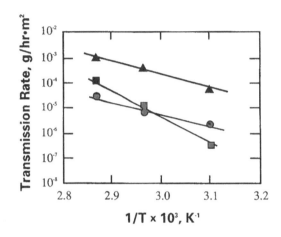

Figure 23 Temperature dependence of the transmission rate for ethyl acetate vapor through PET(▲), SiO$_x$ PET(●), and EVOH (■) film structures.

Table 20 Activation Energy for Permeation of Ethyl
Acetate Vapor through SiO$_x$ PET, EVAL-F and PET

Sample	Activation Energy
SiO$_x$ PET	22.1 Kcal/mol
EVOH	48.8 Kcal/mol
PET	23.2 Kcal/mol

copolymer (EVAL-F) are summarized in Table 19.[147] An Arrhenius plot of the permeability rates vs. $1/T$ for the respective barrier films is shown in Figure 23. Activation energy values for the respective test films are summarized in Table 20.

For comparison, Shirakura reported an activation energy value of 23.3 kcal/mole for the ethyl acetate/PET system,[124] in good agreement with the value obtained by Sajiki.[147] The activation energy seemed to be high when compared with the activation energy of oxygen gas through PET below T_g ($E_P = 8$ kcal/mol).[148] A possible explanation for this is that the molecular size of ethyl acetate is larger than oxygen and requires a higher activation energy for diffusion through the PET film sample.

In a recent study Lin determined the permeability of ethyl acetate and toluene for the following commodity films: (1) oriented polypropylene, (2) high-density polyethylene, (3) glassine, (4) Saran-coated oriented polypropylene, (5) acrylic-coated oriented polypropylene, and (6) metallized polyethylene terephthalate/oriented polypropylene laminate, as a function of temperature and vapor activity.

Table 21 Permeance Constants of Ethyl Acetate Through Oriented Polypropylene (OPP) as a Function of Vapor Activity and Temperature

Vapor Activity[a]	Temperature (°C)	Permeance[b] $R \times 10^{12}$
0.095	30	0.6
	40	1.7
	50	4.4
0.21	30	0.7
	40	1.9
	50	4.4
0.41	30	0.9
	40	2.1
	50	4.8

Note: The results reported are the average of duplicate analyses.

[a] Vapor activity values were determined at ambient temperature (24°C).

[b] Permeance units are expressed in kg/m²·s·Pa.

Table 22 Permeance Constants of Ethyl Acetate through High-Density Polyethylene (HDPE) as a Function of Vapor Activity and Temperature

Vapor Activity[a]	Temperature (°C)	Permeance[b] $R \times 10^{11}$
0.095	30	1.0
	40	1.7
	50	2.7
0.21	30	2.4
	40	2.8
	50	3.4
0.41	30	2.9
	40	3.8
	50	4.8

Note: The results reported are the average of duplicate analyses.

[a] Vapor activity values were determined at ambient temperature (24°C).

[b] Permeance units are expressed in kg/m²·s·Pa.

Permeability studies were carried out at three temperatures to allow evaluation of the Arrhenius relationship. For each temperature, three vapor activity levels were evaluated. Since several films investigated are multilayer or barrier coated structures, permeance values are presented to describe the barrier properties of the total structure.

Permeance values determined at ethyl acetate vapor activity levels of $a_v = 0.095$, 0.21, and 0.41 for the respective barrier structures are summarized in Tables 21–25. For the metallized polyethylene terephthalate/OPP laminate, no measurable rate of diffusion was detected, following continuous testing for 44 h at 70°C and a vapor activity of $a_v = 0.41$.

Determined permeance values for toluene activity levels of $a_v = 0.067$, 0.22, and 0.44, for the respective barrier structures, are summarized in Tables 26–30. There was no mea-

Table 23 Permeance Constants of Ethyl Acetate through
Glassine as a Function of Vapor Activity and Temperature

Vapor Activity[a]	Temperature (°C)	Permeance[b] $R \times 10^{12}$
0.095	23	1.1
	30	2.0
	40	4.2
0.21	23	2.3
	30	2.9
	40	5.2

Note: The results reported are the average of duplicate analyses.

[a] Vapor activity values were determined at ambient temperature (24°C).

[b] Permeance units are expressed in kg/m²·s·Pa.

Table 24 Permeance Constants of Ethyl Acetate through Saran-
Coated Oriented Polypropylene as a Function of Vapor Activity
and Temperature

Vapor Activity[a]	Temperature (°C)	Permeance[b] $R \times 10^{13}$
0.095	40	1.0
	50	5.0
	60	15.1
0.21	40	1.6
	50	5.2
	60	14.9
0.41	40	2.8
	50	7.7
	60	26.5

Note: The results reported are the average of duplicate analyses.

[a] Vapor activity values were determined at ambient temperature (24°C).

[b] Permeance units are expressed in kg/m²·s·Pa.

surable rate of diffusion for the polyethylene terephthalate/OPP laminate structure following continuous testing for 44 h at 70°C and a vapor activity of 0.44.

From Equation (19) the temperature dependency of the transport process associated with the respective barrier membranes, over the temperature range studied, was found to follow the Arrhenius relationship. From the slopes of the Arrhenius plots, the activation energy for the permeation process (E_P) was determined for the respective film structures, as a function of vapor activity. The determined activation energy values for both ethyl acetate and toluene are summarized in Tables 31 and 32, respectively.

Numerical consistency of permeability data

In addition to determining the permeability constants of organic penetrants, it is also important to determine the diffusion coefficients and to evaluate the consistency of the experimental data obtained. The numerical consistency of the permeability data will affect the values of both the diffusion coefficient and permeability constant and would indicate

Table 25 Permeance Constants of Ethyl Acetate through Acrylic-
Coated Oriented Polypropylene as a Function of Vapor Activity
and Temperature

Vapor Activity[a]	Temperature (°C)	Permeance[b] $R \times 10^{12}$
0.095	50	0.25
	60	1.0
	70	2.6
0.21	50	0.3
	60	1.2
	70	3.6
0.41	50	0.3
	60	1.2
	70	3.3

Note: The results reported are the average of duplicate analyses.
[a] Vapor activity values were determined at ambient temperature (24°C).
[b] Permeance units are expressed in kg/m²·s·Pa.

Table 26 Permeance Constants of Toluene through Oriented
Polypropylene (OPP) as a Function of Vapor Activity and
Temperature

Vapor Activity[a]	Temperature (°C)	Permeance[b] $R \times 10^{11}$
0.067	30	0.5
	40	1.15
	50	1.8
0.22	30	0.75
	40	1.4
	50	2.7
0.44	30	1.4
	40	2.0
	50	3.1

Note: The results reported are the average of duplicate analyses.
[a] Vapor activity values were determined at ambient temperature (24°C).
[b] Permeance units are expressed in kg/m²·s·Pa.

any variations of the system parameters, such as temperature, or permeant concentration changes during the course of the permeability experiment. Gavara and Hernandez have described a simple procedure for determining the diffusion coefficient and for performing a consistency analysis on a set of experimental permeability data from a continuous-flow permeation study.[150] This procedure was applied to the continuous-flow permeation data obtained by Huang,[151] to provide a better understanding of the mechanism of the diffusion and sorption processes associated with the permeation of organic penetrants. The consistency test for continuous flow permeability experimental data has been described in detail by Gavara and Hernandez[150] and is summarized briefly below.

The value of the permeation rate at any time F_t, during the unsteady-state portion of the permeability experiment varies from 0, at time equal to 0, up to the transmis-

Table 27   Permeance Constants of Toluene through-High Density Polyethylene (HDPE) as a Function of Vapor Activity and Temperature

Vapor Activity[a]	Temperature (°C)	Permeance[b] $R \times 10^{11}$
0.067	30	2.1
	40	3.4
	50	6.5
0.22	30	4.0
	40	5.5
	50	8.0
0.44	30	7.7
	40	8.6
	50	9.2

Note: The results reported are the average of duplicate analyses.
[a] Vapor activity values were determined at ambient temperature (24°C).
[b] Permeance units are expressed in kg/m²·s·Pa.

Table 28   Permeance Constants of Toluene through Glassine as a Function of Vapor Activity and Temperature

Vapor Activity[a]	Temperature (°C)	Permeance[b] $R \times 10^{12}$
0.067	30	2.8
	40	4.4
	50	6.0
0.22	30	3.5
	40	5.3
	50	6.6
0.44	30	4.1
	40	5.5
	50	6.8

Note: The results reported are the average of duplicate analyses.
[a] Vapor activity values were determined at ambient temperature (24°C).
[b] Permeance units are expressed in kg/m²·s·Pa.

sion rate value (F_∞) reached at the steady state. This is described by the following expression.[152]

$$F_t/F_\infty = \left(4/\pi^{1/2}\right)\left(1^2/4Dt\right)^{1/2} \sum_{n=1,3,5}^{\infty} \exp\left(-n^2 1^2/4Dt\right). \tag{22}$$

Equation (22) is simplified to the following form:

$$\gamma = \left(4/\pi^{1/2}\right)(x)^{1/2} \exp(\pm x), \tag{23}$$

Table 29   Permeance Constants of Toluene through Saran-Coated
Oriented Polyethylene (Saran OPP) as a Function of Vapor
Activity and Temperature

Vapor Activity[a]	Temperature (°C)	Permeance[b] $R \times 10^{13}$
0.067	40	1.0
	50	5.8
	60	21.9
0.22	40	2.5
	50	9.7
	60	25.0
0.44	40	3.1
	50	12.8
	60	28.9

Note: The results reported are the average of duplicate analyses.

[a] Vapor activity values were determined at ambient temperature (24°C).

[b] Permeance units are expressed in kg/m²·s·Pa.

Table 30   Permeance Constants of Toluene through Acrylic-
Coated Polypropylene as a Function of Vapor Activity and
Temperature

Vapor Activity[a]	Temperature (°C)	Permeance[b] $R \times 10^{13}$
0.067	50	—[c]
	60	—[c]
	70	—[c]
0.22	50	—[c]
	60	—[c]
	70	—[c]
0.44	50	1.4
	60	2.0
	70	3.1

Note: The results reported are the average of duplicate analyses.

[a] Vapor activity values were determined at ambient temperature (24°C).

[b] Permeance units are expressed in kg/m²·s·Pa.

[c] Without detectable response after 44 h test.

where γ is equal to the transmission rate ratio F_t/F_∞ and $x = \ell^2/4Dt$. In Equation (22), D is assumed to be independent of permeant concentration and time.

For each value of F_t/F_∞ (γ), a value of x can be calculated, and from a plot of $1/x$ versus (t), the diffusion coefficient (D) can be determined. The authors further described two dimensionless constants, k_1 and k_2;

$$k_1 = t_{1/4}/t_{3/4} = x_{1/4}/x_{3/4} = 0.4405 \tag{24}$$

$$k_2 = t_{1/4}/t_{1/2} = x_{1/4}/x_{1/2} = 0.6681, \tag{25}$$

Table 31 Activation Energy Values for the Permeation
of Ethyl Acetate through Polymer Membranes

Polymer	E_p (kcal/mol)		
Membranes	$a_v = 0.095$	$a_v = 0.21$	$a_v = 0.41$
OPP	17.7	14.6	14.5
HDPE	9.3	4.4	4.9
Glassine	14.6	9.6	—
Saran OPP	24.9	23.0	24.5
Acrylic OPP	25.9	27.9	25.0

Table 32 Activation Energy Values for the Permeation
of Toluene through Polymer Membranes

Polymer	E_p (kcal/mol)		
Membranes	$a_v = 0.095$	$a_v = 0.21$	$a_v = 0.41$
OPP	12.63	12.53	7.79
HDPE	9.45	6.90	1.90
Glassine	5.78	6.21	5.02
Saran OPP	34.00	34.10	35.41
Acrylic OPP	—	—	30.25

where $x_{1/4}$, $x_{1/2}$, and $x_{3/4}$ denote the numerical values of x when the permeability experiment has reached values of 0.25, 0.5, and 0.75, respectively for F_t/F_∞, the transmission rate ratio.

The numerical values of the constants, k_1 and k_2, as given in Equations (24) and (25), together with the linear relationship of $1/x$ versus (t), will provide values of the diffusion coefficient and a criteria to evaluate the consistency of the experimental data.

The results of a series of permeability studies carried out with limonene vapor are presented graphically in Figures 24 and 25, where the transmission rate is plotted as a function of time, and serve to illustrate the applicability of the consistency test to permeability data obtained for polyethylene and polypropylene.[151] The values of k_1 and k_2 calculated from the experimental data and the associated diffusion coefficients, $D_{1/4}$, $D_{1/2}$, and $D_{3/4}$, obtained by substitution into the expressions

$$D_{1/4} = \ell^2/(4 \, X_{1/4} t_{1/4}) \tag{26}$$

$$D_{1/2} = \ell^2/(4 \, X_{1/2} t_{1/2}) \tag{27}$$

$$D_{3/4} = \ell^2/(4 \, X_{3/4} t_{3/4}), \tag{28}$$

are also summarized in the respective transmission rate profile plots. The values of k_1 and k_2 calculated from the experimental data ranged between 4% and 14% of the theoretical values given by Gavara and Hernandez.[150]

Other factors affecting permeation

Chain orientation, permeant concentration, plasticizers, and fillers affect permeability: Chain orientation in general decreases the permeability to gases (see Table 33). Permeant concentration of gases below 1 atm of pressure in general does not effect the permeability coefficient. However, concentration dependent effects has been observed in the permeability of organic compounds. For instance, as described above, the permeability of organic

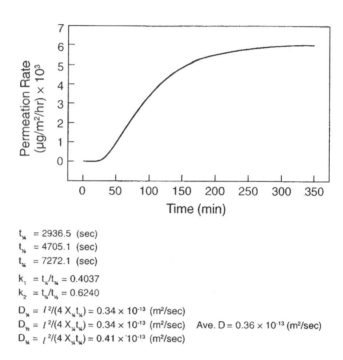

$t_{1/4}$ = 2936.5 (sec)
$t_{1/2}$ = 4705.1 (sec)
$t_{3/4}$ = 7272.1 (sec)

k_1 = $t_{1/4}/t_{3/4}$ = 0.4037
k_2 = $t_{1/4}/t_{1/2}$ = 0.6240

$D_{1/4}$ = $l^2/(4 X_{1/4} t_{1/4})$ = 0.34 × 10^{-13} (m²/sec)
$D_{1/2}$ = $l^2/(4 X_{1/2} t_{1/2})$ = 0.34 × 10^{-13} (m²/sec)   Ave. D = 0.36 × 10^{-13} (m²/sec)
$D_{3/4}$ = $l^2/(4 X_{3/4} t_{3/4})$ = 0.41 × 10^{-13} (m²/sec)

Figure 24 Transmission rate consistency test for the permeability of limonene vapor (a_v = 0.4) through high-density polyethylene at 50°C.

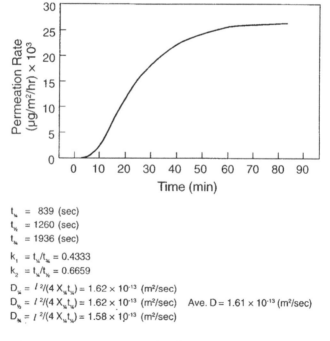

$t_{1/4}$ = 839 (sec)
$t_{1/2}$ = 1260 (sec)
$t_{3/4}$ = 1936 (sec)

k_1 = $t_{1/4}/t_{3/4}$ = 0.4333
k_2 = $t_{1/4}/t_{1/2}$ = 0.6659

$D_{1/4}$ = $l^2/(4 X_{1/4} t_{1/4})$ = 1.62 × 10^{-13} (m²/sec)
$D_{1/2}$ = $l^2/(4 X_{1/2} t_{1/2})$ = 1.62 × 10^{-13} (m²/sec)   Ave. D = 1.61 × 10^{-13} (m²/sec)
$D_{3/4}$ = $l^2/(4 X_{3/4} t_{3/4})$ = 1.58 × 10^{-13} (m²/sec)

Figure 25 Transmission rate consistency test for the permeability of limonene vapor (a_v = 0.2) through oriented polypropylene at 50°C.

Table 33 Effect of Orientation on PP Properties

	Nonoriented PP	Oriented PP
WVTR g mil/m² day at 90% RH and 100°F	15	6
Stiffness	Very low	High, similar to cellophane
Propagated tear strength	High	Very low CD
		Very high MD
Heat sealability	Yes, 350–450°F	No, film distorts
Density	0.902	No change
Optics	Very good	Excellent
Surface adhesivity to inks, etc.	Low	Low
Oxygen permeability cc mil/m² day atm	3700	2500

Table 34 Oxygen Permeability of LDPE with a Calcium Carbonate Filler

Filler % by Volume	Surface Tested	Oxygen Permeability (cc·m/m²·day·atm)
0	None	0.189
15	Yes	0.098
25	Yes	0.059
15	None	0.394
25	None	0.787

vapors such as aromas, flavors, and solvents in general are strongly depend on concentration.[98] The addition of plasticizers usually but not always increases the permeability. Film thickness per se does not affect permeability, diffusion, or solubility coefficients of a penetrant, provided the polymer morphology is not affected. However, polymer film produced in different thickness may have different morphology due to different cooling conditions during processing. The molecular weight of a polymer has been found to have little effect on the permeability of polymer, except in the very low range of molecular weight. Inorganic mineral fillers such as talc, $CaCO_3$ or TiO_2, used as much as 40%, affect the permeability of a film. When coupling agents, such as titanates, are used to improve the interfacial bond between polymer and filler, the permeability to gases and vapors decreases. The absence of bonding agents may result in an increase in permeability.

Permeability of a polymer containing inorganic fillers can be estimated by Equation (29),[153] provided good adhesion and wettability exist between the polymer and the filler:

$$P_{eff} = P_p \phi_p \left[1 + (L/2W) \phi_f \right] \tag{29}$$

where P_p and Φ_p are the permeability and volume fractions of the unfilled polymer, Φ_f is the volume fraction of the filler, and W/L is the aspect ratio. The aspect ratio is the average dimension of the dispersed filler particles parallel to the plane of the film, L, divided by the average dimension perpendicular to the film, W. When there is not good adhesion between the filler and polymer, permeability may increase in a less predictable form, by diffusion through interphase microvoids. Table 34 summarizes values of oxygen permeability of LDPE filled with calcium carbonate at different loading levels.[154]

Packaging geometry

The sorption or "scalping" of intrinsic food flavors by food contact polymers has been reported by a number of authors and has involved exposure of solutions of either flavor extracts or pure compounds to a polymer film or slab.[155-161] Goto monitored the sorption of limonene, cineole, and menthol by a series of polymer structures, and evaluated their relative sorption capacity, the time to reach equilibrium, and the effect of temperature on sorption levels.[162]

In order to estimate the changes in sorbate concentration associated with the sorption process and with the migration of low molecular weight organic compounds from the package to a contacting product phase, the partition coefficient (K) is required. K is defined as

$$K = \frac{C_f^*}{C_p^*},\tag{30}$$

where C_f^* and C_p^* are the equilibrium concentrations of the sorbate in the contacting product phase and in the packaging material, respectively. In addition, the volume (or mass) ratio of the food phase and packaging material is also required for quantification of the equilibrium concentration for a specific product/package/sorbate system.

Equations (31) and (32) describe the equilibrium concentration levels in the case of migration ($^mC_f^*$) and sorption ($^sC_f^*$) respectively, as a function of initial sorbate concentration C^o:

$$^mC_f^* = \frac{C_p^*}{(K \pm 1) + \left(V_f/V_p\right)}\tag{31}$$

$$^sC_f^* = \frac{K\,C_f^o}{\dfrac{V_p}{V_f} + K}\tag{32}$$

For geometrically simple forms, Table 35 shows the relationship between the volume of the food V_f contained in packaging of varying shapes, when the volume of the packaging material is V_p. Table 35 compares all packages with the same thickness.

Tables 36 and 37 give numerical values for the relative concentration of sorbate in a contacting product phase resulting from transport involving sorption and migration processes, respectively. The equilibrium concentration levels presented were obtained by solution of Equations (31) or (32) as a function of the partition coefficient (K) and the volume ratio (V_f/V_p). In the case of sorption, the effect of geometry, as defined by V_f/V_p, is significant for values of K between 0.01 to 1 (see Table 36). Similarly, in the case of migration the geometric volume ratio is important for values of K above 0.1 (see Table 37).

Partition coefficient values of 7.1×10^{-3} for styrene monomer distribution between polystyrene and 50% aqueous ethanol[163] and of 1.0×10^{-4} for BHT distribution between HDPE and water[164] serve to illustrate typical K values obtained experimentally.

Table 38 presents estimated values for the equilibrium distribution of dimethyl disulfide and dipropyl disulfide, representative flavor components from an onion/garlic-flavored sour cream, between the sour cream and a high-impact polystyrene container.[165] As shown, partition coefficient values indicate that the equilibrium distribution strongly favors the sorption of the flavor compounds by the polymeric container material.

Table 35 Relationship between Package
Shape and Product–Package Volume Ratio

Package Shape	V_f/V_p
Rectangular	$0.150\ V_f^{1/3}$
Cubic	$0.167\ V_f^{1/3}$
Cylindrical	$0.181\ V_f^{1/3}$
Sphere	$0.207\ V_f^{1/3}$

Table 36 Values of ${}^sC_f^*/C_f^*$ as a
Function of K and Volume Ratio (V_f/V_p)

$K = \dfrac{C_f^*}{C_p^*}$	V_f/V_p		
	10	20	30
10^{-4}	10^{-3}	-10^{-3}	10^{-3}
10^{-4}	0.01	0.02	0.03
10^{-2}	0.09	0.17	0.23
10^{-1}	0.50	0.67	0.75
1	0.99	0.95	0.97
10	0.99	0.98	0.99
10^2	0.999	0.99	1.00
10^3	0.999	1.00	1.00

Table 37 Values of ${}^sC_f^*/C_f^*$ as a Function of K
and Volume Ratio (V_f/V_p)

$K = \dfrac{C_f^*}{C_p^*}$	V_f/V_p		
	10	20	30
10^{-4}	10^{-4}	10^{-4}	10^{-4}
10^{-3}	10^{-3}	10^{-3}	0.97×10^{-3}
10^{-2}	0.009	.0083	0.0077
10^{-1}	0.05	0.033	0.025
1	0.09	0.047	0.033
10	0.099	0.050	0.033
10^2	~0.10	0.050	0.033
10^3	~0.10	0.050	0.033

Table 38 Partition Coefficient Values for Dimethyl-
Disulfide and Dipropyl-Disulfide Distributed between
Sour Cream and High-Impact Polystyrene at 5°C

Compound	C_f (kg/m)	C_p (kg/m)	K
Dimethyl disulfide	8×10^{-2}	40	2×10^{-3}
Dipropyl disulfide	1×10^{-2}	220	4.5×10^{-5}

In this study, the ratios of availability to capacity were found to be only 0.05 and 0.001, for dimethyl disulfide and dipropyl disulfide, respectively. From these values, it is clear that the product flavor characteristic will deteriorate rapidly, if relatively high levels of flavor compounds are needed to maintain a minimum acceptability of the product. The depletion of flavor compounds from the product will occur over time, with the sorption of flavor compounds by the polymer being the major mechanism of interaction with flavor moieties. Knowledge of the partition distribution of flavor volatiles between product–headspace–packaging material is therefore of value in selecting and designing a package for a product of this nature.

References

1. Risch, S. J. Migration of toxicants, flavors, and odor-active substances from flexible packaging materials to food, *Food Technol.*, 7, 95, 1988.
2. Harte, B. R. Packaging of aseptic products, in *Principles of Aseptic Processing and Packaging*, Nelson, P. E., Chambers, J. V. and Rodriquez, J. H., Eds., The Food Processing Institute, Washington, D.C., 1987, p. 63.
3. Quast, D. G., Karel, M. Effect of environmental factors on the oxidation of potato chips, *J. Food Sci.*, 37, 584, 1972.
4. Jenkins, W. A., Harrington, J. P. Packaging foods with plastics, in *Packaging Foods with Plastics*, Technomics Publishing, Lancaster, PA, 1991.
5. Brown, W. E. Plastics in food packaging: properties, design, and fabrication, in *Plastics in Food Packaging*, Marcel Dekker, New York, NY, 1992.
6. Rauch Assoc., Inc., *The Rauch Guide to the U.S. Packaging Industry*, Rausch Assoc., Inc., P.O., Box 6802, Bridgewater, NJ, 08807, 1986.
7. Brunauer, S., Emmett, P. H., Teller, E. Adsorption of gases in multimolecular layers, *J. Am. Chem. Soc.*, 60, 309, 1938.
8. Salwin, H. Defining minimum moisture contents for dehydrated foods, *Food Technol.*, 594, 1959.
9. Le Maguer, M. Mechanics and influence of water binding on water activity, in *Proc. 10th Basic Symposium*, Institute of Food Technologists; Dallas, TX, 1986, p. 1-25.
10. Leung, H. K. Influence of water activity on chemical reactivity, in *Proc. 10th Basic Symposium*, Institute of Food Technologists; Dallas, TX, 1986, p. 27-54.
11. Bourne, M. C. Effects of water activity on textural properties of food, in *Proc. 10th Basic Symposium*, Institute of Food Technologists; Dallas, TX, 1986, p. 75-99.
12. Labuza, T. P., Riboh, D. Theory and application of Arrhenius kinetics to the prediction of nutrient losses in foods, *Food Technol.*, 36, 66, 1982.
13. Fennema, O. Water and ice, in *Principles of Food Science*, 2nd ed., Marcel Dekker, New York, NY, 1985, chap. 2.
14. Singh, R. P., Heldman, D. R., Kirk, J. R. Computer simulation of quality degradation in liquid foods during storage, in *Proceedings 4th Int. Congress Food Sci. & Technol.*, 1974, p. 19-21.
15. Labuza, T. P. Open shelf life dating of foods, in *Food and Nutrition Press*, Westport, CT, 1982.
16. Mizrahi, S., Karel, M. Accelerated stability tests of moisture sensitive products in permeable packages by programming rate of moisture content increase, *J. Food Sci.*, 42, 958, 1977.
17. Mizrahi, S., Karel, M. Accelerated stability tests of moisture sensitive products in permeable packages at high rates of moisture gain and elevated temperatures, *J. Food Sci.*, 42, 1575-1579, 1977.
18. Karel, M., Labuza, T. P. *Optimization of Protective Packaging of Space Foods*, U.S. Air Force Contract, 1969.
19. Quast, D. G., Karel, M., Rand, W. M. Development of a mathematical model for oxidation of potato chips as a function of oxygen pressure, extent of oxidation and equilibrium relative humidity, *J. Food Sci*, 37, 673, 1972.
20. Quast, D. G., Karel, M. Computer simulation of foods undergoing spoilage by two interacting mechanisms, *J. Food Sci.*, 37, 679, 1972.
21. Wanniger, L. A. Mathematical model predicts stability of ascorbic acid in food products, *Food Technol.*, 26, 42, 1972.
22. Rimer, J., Karel, M. Shelf life studies of vitamin C during food storage. Prediction of L-ascorbic acid retention in dehydrated tomato juice, *J. Food Proc. Pres.*, 1, 293, 1978.
23. Davis, E. G. Evaluation and selection of flexible films for food packaging, *Food Technol.*, 22, 62, 1970.

24. Heiss, R. Shelf life determinations, *Modern Packaging*, 8, 119, 1958.
25. Salwin, H., Slawson, V. Moisture transfer in combination of dehydrated foods, *Food Technol.*, 13, 715, 1959.
26. Iglesias, H. A., Viollaz, P., Chirife, J. Technical Note: A technique for predicting moisture transfer in mixtures of packaged dehydrated foods, *J. Food Sci.*, 14, 89, 1979.
27. Wang, M. J. M. M.S. Thesis, Michigan State University, East Lansing, MI, 1985.
28. Lee, C. H. M.S. Thesis, Michigan State University, East Lansing, MI, 1987.
29. Kirloskar, M. M.S. Thesis, Michigan State University, East Lansing, MI, 1991.
30. Quast, D. G., Karel, M. Effect of oxygen diffusion on oxidation of some dry foods, *J. Food Technol.*, 6, 95, 1971.
31. Maloney, J. F., Labuza, T. P., Wallace, D. H., Karel, M. Autooxidation of methyl linoleate in freeze-dried model systems, *J. Food Sci.*, 31, 878, 1966.
32. Labuza, T. P. Kinetics of lipid oxidation in foods (a review), *CRC Crit. Rev. Food Technol.*, 2, 355, 1971.
33. Zirlin, A., Karel, M. Oxidation effects in a freeze-dried gelatin-methyl linoleate system, *J. Food Sci.*, 34, 160, 1969.
34. Cabral, A. C. D., Orr, A., Stier, E. F., Gilbert, S.G. Performance of metallized polyester as a fatty food packaging material, *Package Develop. Systems*, 9, 18, 1979.
35. Martinez, F., Labuza, T. P. Rate of deterioration of freeze-dried salmon as a function of relative humidity, *J. Food Sci.*, 33, 241, 1968.
36. Tuomy, J. M., Walker, G. C. Effect of storage time, moisture level and headspace oxygen on the quality of dehydrated egg mix, *Food Technol.*, 24, 1287, 1970.
37. Karel, M. Use tests only way to determine effect of package on food quality, *Food Canada*, 4, 43, 1967.
38. Karel, M. Packaging protection for oxygen-sensitive products, *Food Technol.*, 28, 50, 1974.
39. Karel, M., Mizrahi, S., Labuza, T. P. Computer prediction of food storage, *Mod. Packaging*, 44, 54, 1971.
40. Quast, D. G., Karel, M., Rand, W.H. Development of a mathematical model for oxidation of potato chips as a function of oxygen pressure, extent of oxidation and equilibrium relative humidity, *J. Food Sci.*, 37, 673, 1972.
41. Labuza, T. P., Mizrahi, S., Karel, M. Mathematical model for optimization of flexible film packaging of foods for storage. Transactions, *Am. Soc. Agric. Eng.*, 15, 150, 1972.
42. Marcuse, R., Fredriksson, P. Fat oxidation at low oxygen pressure. Kinetic studies on the rate of fat oxidation in emulsions, *J. Am. Oil Chem. Soc.*, 45, 400, 1968.
43. Labuza, T. P., Tsuyuki, H., Karel, M. Kinetics of linoleate oxidation in model systems, *J. Am. Oil Chem. Soc.*, 45, 409, 1969.
44. Simon, I. B., Labuza, T. P., Karel, M. Computer-aided prediction of food storage stability: oxidative deterioration of a shrimp product, *J. Food Sci.*, 36, 280, 1971.
45. Mizrahi, S., Labuza, T. P., Karel, M. Computer-aided predictions of extent of browning in dehydrated cabbage, *J. Food Sci.*, 35, 799, 1970.
46. Mizrahi, S., Labuza, T. P., and Karel, M. Feasibility of accelerated tests for browning in dehydrated cabbage, *J. Food Sci.*, 35, 804, 1970.
47. Mizrahi, S., Karel, M. Accelerated stability tests of moisture-sensitive products in permeable packages by programming rate of moisture content increase, *J. Food Sci.*, 42, 958, 1977.
48. Kwolek, W. F., Bookwalter, G. N. Predicting storage stability from time-temperature data, *Food Technol.*, 25, 1025, 1971.
49. Labuza, T. P. A theoretical comparison of losses in food under fluctuating temperature sequences, *J. Food Sci.*, 44, 1162, 1979.
50. Lee, Y.C., Kirk, J.R., Bradford, L.L., and Heldman, D.R. Kintetics and computer simulation of ascorbic acid stability in tomato juice as a function of temperature, pH, and metal catalyst, *J. Food Sci.*, 42, 640, 1977.
51. Sadler, G. D. A mathematical prediction and experimental confirmation of food quality loss for products stored in oxygen permeable polymers, Ph.D. Thesis, Purdue University, 1984.
52. Barron, F. H., Harte, B., Giacin, J., Hernandez, R. Modeling of oxygen diffusion through a model package and simultaneous degradation of vitamin C in apple juice, *Packaging Technol. Sci.*, 6, 301, 1993.

53. Barron, F. H., Harte, B., Giacin, J., Hernandez, R., Segerlind, L. Finite element computer simulation of oxygen diffusion in packaged liquids, *Packaging Technol. Sci.*, 6, 311, 1993.
54. Kim, J. H. Ph.D. Thesis, Michigan State University, East Lansing, MI, 1995.
55. Felts, J. presented at 36th Society of Vacuum Coaters Conference, Dallas, TX, 1993.
56. Allison, H. in *Proc. Pack Alimentaire '93*, Session A-1, 1993.
57. Krug, T., Marcantonio, J. in *Proc. Pack Alimentaire '93*, Session A-1, 1993.
58. Schwier, R. W. in *Proc. Pack Alimentaire '93*, Session A-1, 1993.
59. Felts, J. T. Transparent gas barrier technologies, *J. Plastic Film Sheeting*, 9, 201, 1993.
60. Izu, M. High performance organic clear coat barrier film technology, presented at Future-Pak 93, October, 1993.
61. Felts, J. T. Transparent barrier coatings update: flexible substrates, *J. Plastic Film Sheeting*, 9, 139, 1993.
62. Brody, A. L. Glass-coated flexible films for packaging: an overview, *Packaging Technol. Eng.*, February, 44, 1994. ·
63. Brody, A. L. Silica super barrier coated films for the next generation of microwaveable food pouches, presented at the Micro-Ready Foods '88, October 1988.
64. Sasaki, H. Development of silica deposited retort pouch, presented at Japan Packaging Institute, Fall 1988.
65. Sakamaki, C., Kano, M. Application study of high barrier ceramics deposited film, presented at Barrier Pack 89, Chicago, IL, May, 1989.
66. Toppan Printing, Technical Information Sheet on Ceramic Vacuum Coating Film - GL, 1, Kanda-Izumicho, Chiyoda-ku, Tokyo, Japan, 1989.
67. Felts, J. E. QLS - a newly developed transparent barrier, presented at Barrier Pack, London, England, May 1990.
68. Reynolds, P. *Packaging World*, July, 26, 1995.
69. Anonymous, *Packaging Strategies*, 14, 7, April 15, 1996.
70. Bigg, D. M. The newest developments in polymeric packaging materials, *IoPP Tech. J.*, Fall, 33, 1992.
71. Sacharow, A. Packaging meets 1990s needs through active technology, *Paper Film Foil Converters*, 65, 52-53, 1991.
72. Anonymous. *Packaging Strategies*, 3, 1988.
73. Anonymous. Adhesives & coatings, *Modern Plastics*, January, 121, 1990.
74. Anonymous. *Packaging Strategies*, 3, 1988.
75. Anonymous. *Packaging Strategies*, October 20, 1988
76. Varriano-Marston, R. in *Proc. Pack Alimentaire '89*, 1989.
77. Anonymous. *Plastics Packaging*, 2, 11, 1989.
78. Drennan, W. C. *Plastics Packaging*, 2, 16, 1989.
79. Anonymous. Japanese Packaging Report No. 10., Packaging Planning Services, 1988.
80. Sacharow, S. Packaging meets 1990s needs through active technology, *Paper, Film Foil Converter*, July, 52-53, 1991.
81. Labuza, T. P., Breene, W. M. Application of "active packaging" for improvement of shelf-life and nutritional quality of fresh and extended shelf-life foods, *J. Food Proc. Preser.*, 13, 1, 1989.
82. Nakamura, H., Hoshino, J. Techniques for the preservation of food by employment of an oxygen absorber, in *Sanitation Control for Food Sterilizing Techniques*, Sanyo Publ., Tokyo, Japan, 1983, chap. 12.
83. Waletzko, P., Labuza, T. P. Accelerated shelf-life testing of an intermediate moisture food, *J. Food Sci.*, 41, 1338, 1976.
84. Holland, R. Oxbar: A total oxygen barrier system for PET packaging, *Pack Alimentaire '90*, Session B-2, Oxfordshire, England, 1990.
85. Rice, J. Polymeric oxygen scavenger system, *Food Process.*, July, 44, 1990.
86. Anonymous. The packaging activists, *Prepared Foods*, August, 172, 1990.
87. Rooney, M. Oxygen scavenging from air in package headspaces by singlet oxygen reaction sin polymer media, *J. Food Sci.*, 47, 291, 1981.
88. Rooney, M. Photosensitive oxygen scavenger films: an alternative to vacuum packaging, *CSIRO Food Res. Q.*, 43, 9, 1981.

89. Zenner, B. D., Salame, M. A new oxygen absorbing system to extend the shelf-life of oxygen sensitive beverages, presented at the BEV-PAK '89 13th International Ryder Conference on Beverage Packaging, 1989.

90. Scott, D., Hammer, F. Oxygen scavenging packet for in package deoxygenation, *Food Technol.*, 15, 99, 1961.

91. Rice, J. Fighting moisture: desiccant sachets poised for expanded usage, *Food Proc.*, 65, 46, 1994.

92. Salame, M. The use of barrier polymers in food and beverage packaging, *J. Plastic Film Sheeting*, 2, 321, 1986.

93. Harte, B. R., Gray, J. R. The influence of packaging on product quality, in *Proc. Food Product-Package Compatibility*, J. I., Harte, B. R. and Miltz, J., Eds., Technomic Publishing, Lancaster, PA, 1987.

94. Bagley, E., Long, F. A. Two-stage sorption and desorption of organic vapor in cellulose acetate, *J. Am. Chem. Soc.*, 77, 2172, 1958.

95. Fujita, J. Diffusion in polymer-diluent systems, *Fortsch-Hochpolym-Forsch*, 3, 1, 1961.

96. Crank, J. *The Mathematics of Diffusion*, 2nd ed., Clarendon Press, Oxford, England, 1975.

97. Berens, A. R. Diffusion and relaxation in glassy polymer powders: 1. Fickian diffusion of vinyl chloride in poly(vinyl chloride), *Polymer*, 18, 697, 1977.

98. Mears, P. Transient permeation of organic vapors through polymer membranes, *J. Appl. Polymer Sci.*, 9, 917, 1965.

99. Gilbert, S. G., Hatzidimitriu, E., Lai, C., Passy, N. Studies on barrier properties of polymeric films to various organic vapors, *Instrumental Anal. Food*, 1, 405, 1983.

100. Mannheim, C. H., Miltz, J., Letzter, A. Interaction between polyethylene laminated cartons and aseptically packed citrus juices, *J. Food Science*, 52, 737, 1987.

101. Ikegami, T., Shimoda, M., Koyama, M., Osajima, Y. Sorption of volatile compounds by plastic polymers for food packaging, *Nippon Shokuhin Kogyo Gakkaishi*, 34, 267, 1987.

102. Kwapong, O. Y., Hotchkiss, J. H. Comparative sorption of aroma compounds by polyethylene and ionomer food-contact plastics, *J. Food Sci.*, 52, 761, 1987.

103. Durr, P., Schobinger, U., Waldvogel, R. Aroma quality of orange juice after filling and storage in soft packages and glass bottles, *Lebensmittel-Verpackung*, 20, 91, 1981.

104. Marshall, M. R., Adams, J. P., Williams, J. W. Flavor absorption by aseptic packaging materials, in *Proc. Aseptipak '85*, Princeton, NJ, 1985.

105. DeLassus, P. T., Hilker, B. L. Interaction of high barrier plastics with food: permeation and sorption, in *Proc. Food Product-Package Compatibility*, Gray, J. I., Harte, B. R. and Miltz, J., Eds., Technomic Publ., Lancaster, PA, 1987.

106. Gilbert, S. G. Food/package compatibility, *Food Technol.*, 39, 54, 1985.

107. Ikegami, T., Nagashima, M., Shimoda, Y., Tanaka, and Y. Osajima. *J. Food Sci.*, 56(2), 500, 1991.

108. Nielsen, T. J. *J. Food Sci.*, 59(1), 227, 1994.

109. Toebe, J., Hoojjat, H., Hernandez, R. J., Giacin, J., Harte, B. *Packaging Technol. Sci.*, 3, 133, 1990.

110. Van Krevelen, D. W. *Properties of Polymers*, 3rd ed., Elsevier, New York, 1990.

111. Kwapong, O. Y., Hotchkiss, J. H. *J. Food Sci.*, 52, 761, 1987.

112. Crank, J. in *The Mathematics of Diffusion*, 2nd ed., Clarendon Press, Oxford, 1975.

113. Michaels A. S., Wieth, W. R., Barrie, J. A. *J. Appl. Phys.*, 34, 13, 1963.

114. Berens, A.R. *Angew. Makromol. Chem.*, 47, 85, 1985.

115. Hernandez, R. J., Giacin, J. R., Baner, A. L. The evaluation of the aroma barrier properties of polymer films, *J. Plastic Film Sheet*, 2, 187, 1986.

116. Hildebrand, J. H., Scott, R. I. *The Solubility of Non-Electrolytes*, 3rd ed., Reinhold Publishing, New York, 1950.

117. Hoftyzer, P. J., VanKrevelen, D. W. presented at the International Symposium on Macromolecules (IUPAC) Leyden, Paper (No. IIIa-15), 1990.

118. DeLassus, P. T. Permeation of flavors and aromas through glassy polymers, in *Proc., TAPPI, 1993 Polymers, Laminations and Coatings Conference*, Chicago, IL, p. 263, 1993.

119. Van Amerongen, G. J. *J. Polym. Sci.*, 2, 318, 1947.

120. Michaels, A. S., Vieth, W. R., Barrier, J. A. *J. Appl. Phys.*, 34, 1, 1963.

121. Puleo, A. C., Paul, D. R., and Wong, P. K. *Polymer*, 30, 13, 1989.

122. Michaels, A. S., Vieth, W. R., Barrier, J. A. *J. Appl. Phys.*, 34, 13, 1963.
123. Veith W. R. *Diffusion In and Through Polymers*, Hanser Publishers, New York, 1991.
124. Shirakura, A. M.S. Thesis, Michigan State University, East Lansing, MI, 1987.
125. Choy, C. L., Leung, W. P., Ma, T. L. *J. Polymer Sci. Polymer Phys. Ed.*, 22, 707, 1984.
126. Hernandez, R. J., Giacin, J. R., Jayaraman, K., Shirakura, A. Diffusion and sorption of organic vapors through biaxially oriented poly(ethylene) terephthalate films of varying thermomechanical history, *Proc. ANTEC '90*, Dallas, TX, May 7-11, 1990.
127. Giacin, J. R., Hernandez, R. J. Evaluation of aroma barrier properties of polymer films by sorption and permeation methods, *Proc. Activities Report of the R & D Associates, Fall 1986 Meeting of Research and Development Associates for Military Food and Packaging Systems, Inc.*, 39, 79, 1987.
128. Zobel, M. G. R. Measurement of odour permeability of polypropylene packaging films at low odourant levels, *Polymer Testing*, 3, 133, 1982.
129. Mohney, S., Hernandez, R. J., Giacin, J. R., Harte, B. R., Miltz, J. The permeability and solubility of d-limonene vapor in cereal package liners, *J. Food Sci.*, 53, 253, 1988.
130. Hensley, T. M., Giacin, J. R., Hernandez, R. J. Permeation of multi-component organic vapor mixtures through polymeric barrier films, in *Proc. IAPRI 7th World Conference on Packaging*, Utrecht, The Netherlands, April 14-17, 1991,
131. Nielsen, T. J., Giacin, J. R. The sorption of limonene/ethyl acetate binary vapour mixtures by a biaxially oriented polypropylene film, *Packaging Technol. Sci.*, 7, 247, 1994.
132. Mears, P. Transient permeation of organic vapors through polymer membranes, *J. Appl. Polymer Sci.*, 9, 917, 1965.
133. Berens, A. R. Diffusion and relaxation in glassy polymer powders: 1. Fickian diffusion of vinyl chloride in poly(vinyl chloride), *Polymer*, 18, 697, 1977.
134. Blackadder, D. A., Keniry, J. S. Difficulties associated with the measurement of the diffusion coefficient of solvating liquid or vapor in semicrystalline polymer. I. Permeation methods, *J. Appl. Polymer Sci.*, 17, 351, 1973.
135. DeLassus, P. T., Tou, J. C., Babinec, M. A., Rulf, D. C., Karp, B. K., Howell, B. A. Transport of apple aromas in polymer films, in *Food and Packaging Interactions*, Hotchkiss, J. H. Ed., ACS Symposium Series 365, American Chemical Society, Washington, DC, 1988, chap. 2, p. 11.
136. Liu, K., Giacin, J. R., Hernandez, R. J. Evaluation of the effect of relative humidity on the permeability toluene vapor through a multi-layer coextrusion film, *Packaging Technol. Sci.*, 1, 57, 1988.
137. Barrie, J. A. Water in polymers, in *Diffusion in Polymers*, Crank, J., Park, G. S., Eds., Academic Press, New York, 1968, p. 259.
138. Liu, K. J., Hernandez, R. J., Giacin, J. R. The effect of sorbed water and penetrant vapor activity on the permeation of toluene vapor through a two-sided PVDC coated opaque white oriented polypropylene film, *J. Plastic Film & Sheeting*, 52, 67, 1991.
139. Hatzidimitriu, E., Gilbert, S. G., Loukakis, G. Odor barrier properties of multi-layer packaging films at different relative humidities, *J. Food Sci.*, 52, 472, 1987.
140. Meyer, J. A., Rogers, C., Stannett, V., Szwarc, M. Studies in the gas and vapor permeability of plastic films and coated papers part III. The permeation of mixed gases and vapors, *TAPPI*, 40, 142, 1957.
141. Ito, Y. Effect of water vapor on permeation of gases through hydrophilic polymers, *Chem. High Polym.*, January 18, 158, 1961.
142. Petrak, K., Pitts, E. Permeability of oxygen through polymer II, the effect of humidity and film thickness on the permeation and diffusion coefficients, *J. Appl. Polym. Sci.*, 25, 879, 1980.
143. Long, F. A., Thompson, L. J. Diffusion of water vapor in polymers, *J. Polym. Sci.*, 8, 321, 1953.
144. Long, F. A. Diffusion of water vapor in polymers, *J. Polym. Sci.*, 15, 413, 1960.
145. Landois-Garza, J., Hotchkiss, J. H. Permeation of high-barrier films by ethyl esters: Effect of Permeant molecular weight, relative humidity, and concentration, in *Food and Packaging Interactions*, Hotchkiss, J. H., Ed., ACS Symposia Series 365, American Chemical Society, Washington, DC, 1988, Chapter 4, p. 59.
146. Nagaraj, S. M.S. Thesis, Michigan State University, East Lansing, MI, 1991.
147. Sajiki, T. M.S. Thesis, Michigan State University, East Lansing, MI, 1991.

148. Michaels, A. S., Vieth, W. R., Barrier, J. A. Diffusion of gases in polyethylene terephthalate, *J. Appl. Phys.*, 34, 13, 1963.
149. Lin, C. H. M.S. Thesis, Michigan State University, 1996.
150. Gavara, R., Hernandez, R. J. Consistency test for continuous flow permeability experimental data, *J. Plastic Film Sheet.*, 9, 126, 1993.
151. Huang, S. J. M.S. Thesis, Michigan State University, 1996.
152. Pasgternak, R. A., Schimscheimer, J. F., Heller, J. A dynamic approach to diffusion and permeation measurements, *J. Polymer Sci.*, Part A-2, 8, 467, 1970.
153. Nielsen, L. E. *J. Macromol. Sci. (Chem.)*, A1, 929, 1967.
154. Steingiser, S., Rubb, G. J., Salame, M. *Encyclopedia of Chemical Technology*, Vol. 3, 3rd ed., 1980, p. 480.
155. Shimoda, M., Ikegami, T., Osajema, Y. Sorption of flavour compounds in aqueous solution into polyethylene film, *J. Sci. Food Agric.*, 42, 157, 1988.
156. Meyers, M. A. Ph.D. Thesis, Rutgers - The State University, New Jersey, 1988.
157. Hirose, K., Harte, B. R., Giacin, J. R., Miltz, J., Stine, C. The influence of aroma compound absorption on mechanical and barrier properties of sealant films used in aseptic packaging, in *Food and Packaging Interactions*, Hotchkiss, J. H., Ed., ACS Symposium Series 365, American Chemical Society, Washington, DC, 1988, chap. 3, p. 28.
158. Mannheim, C. H., Miltz, J., Passy, N. Interaction between aseptically filled citrus products and laminated structures, in *Food and Packaging Interactions*, Hotchkiss, J. H., Ed., ACS Symposium Series 365, American Chemical Society, Washington, DC, 1988, chap. 6, p. 68.
159. Schimoda, M., Nitanad, Kodota, T., Ohta, H., Suetsana, K., Osajima, Y. Adsorption of satsuma mandarin orange juice aroma on plastic films, *Nippon Shokuhin Kogyo Gakkaishi*, 31, 697, 1984.
160. Schimoda, M., Matsui, T., Osajima, Y. Effects of the number of carbon atoms of flavor compounds on diffusion, permeation, and sorption with polyethylene films, *Nippon Shokuhin Kogyo Gakkaishi*, 34, 535, 1987.
161. Shimoda, M., Matsui, T., Osajima, Y. Behavior of diffusion, permeation and sorption of flavor compounds in vapor phase with polyethylene films, *Nippon Shokuhin Kogyo Gakkaishi*, 34, 402, 1987.
162. Goto, A. Packaging for flavor-contained products effect in use of polyacylonitrile-based films, *Packaging Japan*, September, 25, 1988.
163. Summary Report, Migration from styrene-based polymeric styrene from crystal polystyrene and BHT from impact polystyrene, Arthur D. Little, Inc., Project No. 81166, FDA Contract Number 223-77-2360, 1981.
164. Gavara, R., Hernandez, R. J., Giacin, J. R. Methods to determine partition coefficient of organic compounds in water/polystyrene systems, *J. Food Sci.*, 61, 947, 1996.
165. Toebe, J. M., Hoojjat, H., Hernandez, R. J., Giacin, J. R., Harte, B. R. Interaction of flavour components from an onion/garlic flavoured sour cream with high impact polystyrene, *Packaging Tech. Sci.*, 3, 133, 1990.

chapter eleven

Mathematical modeling of quality loss

Edward W. Ross

Contents

Introduction

The purpose of this chapter is to present and discuss some of the most important examples of mathematical modeling as they relate to changes in quality during food storage. No universal mathematical theory exists that can be used to model all quality changes in foods. The variety of mechanisms and phenomena is too great to be encompassed within any single, usable set of equations. Although we describe a general framework for such a theory in the next section, it is too general and vague to be of much immediate help in practical situations, and we are forced to specify the theory more narrowly in order to make it applicable.

The word "quality," as used in descriptions of food, can be interpreted in many ways. A consumer may perceive the food as of high quality for a number of reasons, and a food scientist may make a similar (or contradictory) judgment for different reasons.In this chapter we do not define quality precisely but leave it as a somewhat intuitive concept. Further discussion of the definition can be found in some of the other sections of this book, especially Chapters 2 and 12.

The quality of foods, however defined, is assumed to be capable of change with the passage of time. The scale of the time change may be short (e.g., minutes, during processing) or long (e.g., months, during storage). The agents causing the quality change are usually physical, chemical, or biological and may act in extremely complicated combinations. Often, temperature is the single, dominant factor, but the change may also be affected by the presence of moisture, oxygen, and bacteria of many kinds. The main part of the review deals with the situation where there is a single, dominant variable (say, temperature) whose time variation is known. The problem of finding the time variation of the dominant variable is not discussed here.

There is an intimate relationship between the mathematics involved in estimating quality loss and that involved in the study of food-processing methods. The quality loss during storage may be regarded as a form of processing at relatively low temperatures but going on for rather long times. It is natural that many of the concepts developed in connection with food processing will play similar roles in a quality-loss context, and so quantities such as z, Q_{10} and F_o are important to us.

The literature dealing with this subject is large, and we make no attempt to cite all or even most of it. Some references that the author has found valuable are given in the bibliography at the end of this chapter. Any review of a technical field is bound to reflect the viewpoint of the author, and the present one is no exception. The main intention here is to describe a version of the existing theory which (the author believes) offers the best basis for future work in the field. Undoubtedly this slights many worthwhile contributions of others in this field, a flaw which the author regrets but cannot correct within existing limitations of space.

General theory

The quality is assumed to be defined numerically in some way as the quantity A, a function of time, t, and possibly also of spatial position, x. The latter dependence will not play a major role in what follows, but, if food is stored in a large mass, it may be necessary to consider it. We define also $v_1(t,x), v_2(t,x), \ldots, v_m(t,x)$ as the variable agents of change. For example, $v_1(t,x)$ might be the temperature, $v_2(t,x)$ moisture content, etc. Any of these quantities may also involve constants or parameters, which we collectively denote by P. Then it is reasonable to think that the quality change is governed by some mathematical model of the general form

$$dv_1/dt = \Psi_1\left(A,\, v_1,\, v_2,\, \ldots,\, v_m,\, P\right)$$

$$dv_2/dt = \Psi_2\left(A,\, v_1,\, v_2,\, \ldots,\, v_m,\, P\right)$$

$$\vdots \qquad \vdots$$

$$dv_m/dt = \Psi_m\left(A,\, v_1,\, v_2,\, \ldots,\, v_m,\, P\right).$$

This system of ordinary differential equations is to be solved with initial conditions at time t_o:

$$A\left(t_o\right) = A_o,\; v_1\left(t_o\right) = v_{1o},$$

$$v_2\left(t_o\right) = v_{2o},\; \ldots,\; v_m\left(t_o\right) = v_{mo}.$$

In the differential equation system we usually assume that the quantities on the left and right sides of the equality signs are to be evaluated at the same time, t. However, it may occasionally be desirable to evaluate some of the quantities on the right side at an earlier time, in which case the model is said to be of differential-delay or differential-difference form. We do not pursue that case in the present chapter.

This system is too general to be useful. It must be specialized by choosing the functions $\Psi_o, \Psi_1, \Psi_2, \ldots, \Psi_m$ and the parameters, P. An example of a system which is more specific but still general enough to contain many commonly used models is obtained by assuming that $m = 2$ and making the definitions

$$\Psi_o = v_1\alpha(A, P)$$

$$\Psi_1 = \Psi_2 \ d\beta/dv_2$$

and $v_1(t_o) = \beta(t_o)$. Then the model equations become

$$dA/dt = v_1\alpha(A, P)$$

$$dv_1/dt = \Psi_2(v_2) \ d\beta/dv_2$$

$$dv_2/dt = \Psi_2(v_2).$$

The function α may be any reasonable function of A containing parameters P, and β is any reasonable function of v_2. If Ψ_2 is eliminated between the last two equations, and the chain-rule used, we obtain in place of the second equation

$$dv_1/dt = d\beta/dt.$$

If this is integrated and the initial condition used, we see that the above differential equation for v_1 can be replaced by

$$v_1 = \beta = \beta[v_2(t)].$$

This system may be regarded as a prototype for the models used in estimating shelf life and quality loss.

To see how this model reduces to a familiar form, define

$$\pm v_1 = \pm\beta = k = \text{the rate constant}$$

$$v_2 = T = \text{absolute temperature, degrees Kelvin}$$

$$\alpha(A, P) = A^n.$$

and the symbol P is taken as the lone parameter n, the order of the process. Then the model becomes

$$dA/dt = \pm k[T(t)] A^n$$

$$dT/dt = \Psi_2(T).$$

The latter of these equations simply determines the function $T(t)$, the temperature history which the food experiences. Henceforth, we assume that the temperature history, $T(t)$, is known, and so it is unnecessary to consider the last equation further. The form of the function $k[T]$ specifies the kinetics of the process, which will be discussed later.

In the following derivation, the quality A is assumed to obey the above differential equation,

$$dA/dt = \pm k\, A^n \qquad (k \geq 0,\ n \geq 0),$$

and the initial condition $A(t_o) = A_o$. That is, A obeys an n-th order process. The rate function, k, has dimension (dim) given by

$$\dim k = (\text{time})^{-1}\left(\dim A^{1-n}\right).$$

The rate function k may depend on many variables, e.g., temperature, moisture, oxygen level, and any of these may itself depend on time. We assume that one of these variables is dominant, that the dominant one is temperature, $T(t)$, and that $T(t)$ is known. However, the derivation is quite general and applicable if the dominant variable is not temperature but some other variable or combination of variables, as long as the time dependence of that variable is known. It is easily verified that, if $k = k[T(t)]$, the time dependence of the solution to the differential equation can be expressed in terms of

$$\phi(t) = \int_{s=t_o}^{s=t} k\big[T(s)\big]\, ds, \tag{1}$$

and the solution can be written

$$A(t)/A_0 = \left[1 - cA_o^{-c}\phi(t)\right]^{1/c} \qquad \text{for } c \neq 0 \tag{2}$$

$$= e^{[-\phi(t)]} \qquad \text{for } c = 0. \tag{3}$$

Here $c = 1 - n$, and so $c \leq 1$.

The time dependence of A is entirely governed by the behavior of $\phi(t)$, which is a nondecreasing function of t with $\phi(t_o) = 0$.

The above solutions may be expressed in terms of the quantities

$$R(t) = \text{fraction of quality remaining at time } t$$

$$= A(t)/A_o$$

$$D(t) = \text{fraction of quality lost by time } t$$

$$= 1 - R(t) = \left[A_o - A(t)\right]/A_o$$

as

$$R(t) = 1 \pm D(t) = G\big[\phi(t);\ c,\ A_o\big]$$

in which

$$G\big[\phi(t);\ 0,\ A_o\big] = e^{\pm\phi(t)} \qquad\qquad \text{for } c = 0$$

$$G\big[\phi(t);\ c,\ A_o\big] = \big[1 \pm cA_o{}^{\pm c}\phi(t)\big]^{1/c} \qquad \text{for } c \neq 0.$$

The simplest situation is that of constant temperature, say $T = T_o$, and $k = k_o = k(T_o)$, for which case

$$\phi(t) = k_o\big(t \pm t_o\big)$$

is to be substituted in the preceding equations. We may then write the solutions in several important circumstances as follows: For $n = 0$, $c = 1$, a zero-order process,

$$R(t) = A(t)/A_o = 1 \pm \phi/A_o = 1 \pm k\big(t \pm t_o\big)/A_o$$

$$A(t) = A_o \pm \phi = A_o \pm k\big(t \pm t_o\big).$$

For $n = 1$, $c = 0$, a first-order process,

$$R(t) = \exp\big[\pm\phi\big] = e^{\big[\pm k(t \pm t_o)\big]}$$

$$A(t) = A_o \exp\big(\pm\phi\big) = A_o\, e^{\big[\pm k(t \pm t_o)\big]}$$

The change in form between the cases $c = 0$ and $c \neq 0$ is sometimes inconvenient, and it is useful to have a formula that displays explicitly the behavior of A in the case where $|c|$ is small. One such formula can be found by expanding ϕ in powers of c, equating coefficients and neglecting all terms after the first correction in the general solutions. The result is

$$R(t) = A(t)/A_o \approx e^{\big[\pm\phi(t)\{1+c\tau(t)\}\big]}$$

$$\tau(t) = (1/2)\phi(t) \pm \ln A_o. \tag{4}$$

The formula is accurate only when $|c|$ is small but shows how the transition occurs between the cases $c = 0$ and $c \neq 0$.

The behaviors of these solutions in more-or-less typical situations are displayed in Figures 1 through 3. In these graphs the remaining quality, R, is shown as a function of t (hours) for various values of the parameters. The temperature, T, is assumed constant, and A is taken to have the initial value $A = A_o = 100$ at time $t = t_o = 0$. For this special case the parameters are c (i.e., the order) and k_o. Figure 1 shows the effect on $R(t)$ of varying c for a fixed $k_o = k = 0.1$. The effect of changes in k for fixed c are illustrated in Figure 2, and Figure 3 displays a comparison between exact solutions, Equations (2) and (3), and approximate solutions, Equation (4), for small c.

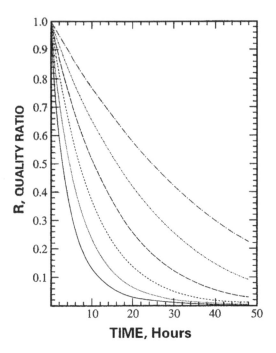

Figure 1 R as a function of time, t, in hours for k = 0.1 and various values of c. (--.--.--), c = 0.3; (-.-.-.), c = 0.2; (-- -- --), c = 0.1; (- - - -), c = 0; (.....), c = -0.1; (-------), c = -0.2.

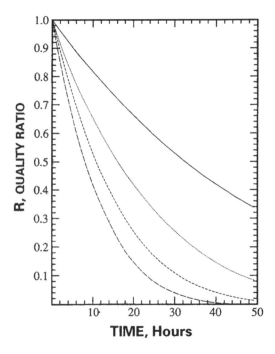

Figure 2 R as a function of time, t, in hours for c = 0.2 and various values of k. (-- -- --), k = 0.2; (- - - -), k = 0.15; (........), k = 0.1; (-------), k = 0.05.

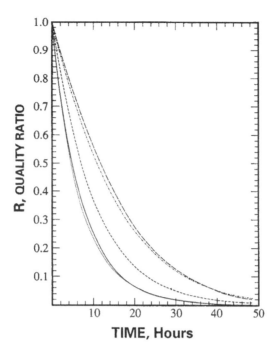

Figure 3 R as a function of time, t, in hours for $k = 0.1$ and small values of c. Comparison of exact solution, Equation (2), with approximate solution, Equation (4). (-.-.-.), $c = 0.1$, exact; (-- -- --), $c = 0.1$, approx.; (- - - -), $c = 0$, exact and approx.; (........), $c = -0.1$, exact; (-------), $c = -0.1$, approx.

Shelf life

We define the lowest acceptable value of A as A_b. If A declines to or below A_b, the quality is assumed to be unacceptable. The time at which this happens is t_b, and the shelf life is

$$L = t_b \pm t_o.$$

If $\phi(t)$ is a strictly increasing function ($d\phi/dt > 0$), t_b and L are unique. If $R_b \equiv R(t_b)$, the value of t_b is found by solving for t the equation

$$R(t) = R_b.$$

This condition can also be expressed as

$$G\left[\phi(t_b); \ c, \ A_o\right] = R_b,$$

or, solving for $\phi(t_b)$,

$$\phi_b = \phi(t_b) = W(R_b, \ c, \ A_o), \tag{5}$$

provided we define the function W to be

$$W = A_o^c\left(1 \pm R_b^c\right)/c \qquad \text{for } c \neq 0$$

$$= \pm \ln R_b \qquad\qquad \text{for } c = 0.$$

The values of t_b and L depend on c, R_b, A_o, (if $c \neq 0$) and very much on the form of $\phi(t)$. In the special but important case where T is constant, and so k is constant, $k = k_o$, and $c \neq 0$ we have to solve

$$k_o\left(t_b \pm t_o\right) = W = A_o^c\left(1 \pm R_b^c\right)/c, \tag{6}$$

which leads to

$$L = A_o^c\left(1 \pm R_b^c\right)/\left(ck_o\right). \tag{7}$$

When $c = 0$, we obtain

$$L = \pm\left(1/k_o\right)\ln\left(R_b\right). \tag{8}$$

Similarly, an approximate formula for the shelf-life when $|c|$ is small and T constant can be found from Equation (4) by solving

$$\pm \ln R_b = \phi_b\left(1 \pm c\ln A_o\right) + c\,\phi_b^2/2$$

for $\phi_b = k_o L$. After some simplification the formulas

$$L = \phi_b/k_o$$

$$\phi_b \approx \pm\left(\ln R_b\right)\left[1 + c\ln\left(A_o R_b^{1/2}\right)\right]$$

are found.

When T is constant, it is clear that the shelf life, L, depends inversely on k_o in all cases considered here. The dependence on c and R_b is illustrated in Figure 4 for the case where $A_o = 100$. In general, L increases as c increases and R_b decreases.

Arrhenius kinetics

In this section we discuss the dependence of k on $T(t)$, i.e., the kinetics, an aspect of the problem that we have largely ignored up till now. It is assumed in what follows that $T(t)$, the absolute temperature history of the food, is known. In practice it is often difficult to find or calculate this function, but for the purposes of the present exposition these obstacles are ignored.

The most common assumption about the behavior of k is embodied in Arrhenius kinetics, i.e., in the assumption that

$$k = k_1\, e^{\left[\pm E_a/\left\{R_g T(t)\right\}\right]}$$

where k_1 = the pre-exponential constant, E_a = the activation energy, and R_g = the universal gas constant.

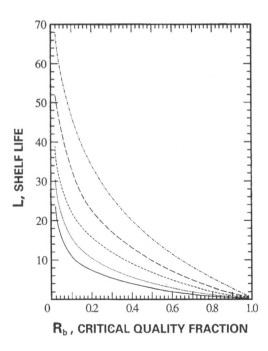

Figure 4 Dependence of shelf life, *L*, on critical *R*, R_b, for *k* = 0.1 and various values of *c*. (-.-.-.), *c* = 0.2; (-- -- --), *c* = 0.1; (- - - -), *c* = 0; (......), *c* = −0.1; (--------), *c* = −0.2.

This can be expressed in a slightly different form as follows:

$$k = k_r \, e^{\left[\pm E\left\{(T_r/T)\pm 1\right\}\right]}$$

$$E = E_a \big/ \left(R_g T_r\right)$$

$$k_r = k_1 \, e^{\left[\pm E_a \big/ \left(R_g T_r\right)\right]}$$

T_r is assumed to be a "standard" or "reference" temperature at which the rate is known to be k_r. It is clear that $k = k_r$ when $T = T_r$. In a food-processing context T_r is typically 394.1 K, corresponding to 250°F. In almost all food-storage situations, reasonable values for T_r would lie in the range $250 < T_r < 350$. The parameter *E* may be regarded as a dimensionless version of the activation energy. Since $R_g \approx 2$, and usually $E_a > 10{,}000$ cal/mol, the values of *E* commonly exceed 15 and may reach 100 or more.

For many purposes it is convenient to define

$$x(t) = \left[T_r/T(t)\right] \pm 1 \quad \text{or} \quad T(t) = T_r \big/ \left[1 + x(t)\right],$$

in which case we can express the Arrhenius Law as

$$k = k_r e^{\pm E x(t)} \tag{9}$$

In this form $x(t)$ is a dimensionless function of temperature. Notice, however, that $x(t)$ is, loosely speaking, "inverse" to temperature: $x(t)$ increases when *T* decreases and conversely. Figure 5 shows the relation between *x* and T/T_r, the absolute temperature ratio. As *T*

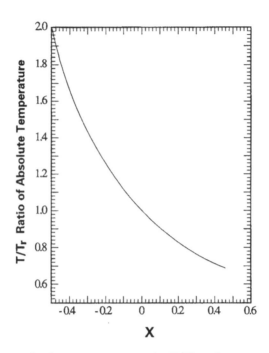

Figure 5 Relation between absolute temperature ratio, T/T_r and x.

decreases, x increases, and $-Ex$ becomes large and negative, implying that k gets smaller, and the process slows down. When $x = 0$, $T = T_r$ and $k = k_r$. If logarithms are taken of both sides in the above equation, we obtain

$$\ln k = \ln k_r \pm Ex$$

Thus, a plot of $\ln k$ versus x will be a straight line with slope $= -E$ and y-intercept $= \ln k_r$.

If T changes over a narrow enough range in the vicinity of T_r, then $|x| \ll 1$, and a binomial approximation shows that

$$T \approx T_r(1 \pm x).$$

Therefore, T and x are approximately related in linear fashion as long as the temperature range is not too great, as can be seen in Figure 5. It is also true that, if $|x|$ is sufficiently small, Equation (9) can be approximated by using a Taylor series to obtain

$$k \approx k_r\left[1 \pm Ex\right].$$

However, this is inaccurate unless $Ex \ll 1$; since E is usually quite large, this approximation is good only when x is extremely small, i.e., T is very near T_r.

After making the above definitions, the important quantity $\phi(t)$ can be expressed as

$$\phi(t) = k_r F(t)$$

$$F(t) = \int_{s=t_0}^{t} e^{\pm Ex(s)}\, ds = \int_{s=t_0}^{t} 10^{\pm Ex(s)/B}\, ds$$

$$B = \ln 10 \approx 2.303.$$

The function $F(t)$ always has the dimension of time. If T is constant and equal to T_r between times t_o and t, then $x = 0$ over the entire range of integration, and

$$F(t) = t \pm t_o.$$

$F(t)$ is therefore the elapsed time at temperature T_r. The time dependence of the quality loss is governed only by $\phi(t)$ or $F(t)$. If $T(t)$ and $x(t)$ are variable, $F(t)$ is the time at temperature T_r that produces the quality loss caused by the changing temperature. If the storage or process starts at time t_o and ends at time t_e, then it is usual to call $F(t_e) = F_o$, the equivalent processing time at the reference temperature. Commonly the term F_o is used in a food-processing context only when the reference temperature is 250°F and refers to a complete process. Here, in a food-storage context we shall use $F(t)$ and $F_o(t)$ interchangeably to mean the processing time at the reference temperature (whatever that temperature may be) that is equal to the integral from t_o to t of $e^{[-Ex(s)]}$.

With these definitions the shelf life can be expressed as follows: the time, t_b, at which the quality becomes unacceptable is found by solving Equation (5),

$$\phi(t_b) = k_r F(t_b) = W,$$

which leads to

$$t_b = \phi^{\pm 1}(W) = F^{\pm 1}(W/k_r). \tag{10}$$

Then

$$L = t_b - t_o.$$

Frequently, it is difficult to solve the above equation and find t_b explicitly even if we know the temperature history, $T(t)$. However, it is possible to do this in a few cases, and an example is given later in this chapter.

In dimensionless form the Arrhenius Law can be written

$$k = k_r 10^{\pm Ex/B}$$

A quantity that is often used in defining food processing is the z-value, defined with reference to T_r as follows: $-z$ is the decrease in temperature required to reduce k_r to $k_r/10$. It is easily seen that the x which satisfies that condition is

$$x = B/E$$

corresponding to the temperature

$$T = T_r E/(E + B) = T_z.$$

The definition of z implies

$$T_z = T_r \pm z,$$

which combines with the preceding equation to give

$$z = T_r B / (E + B).$$

If $E \gg B$, as is often the case, this simplifies to the familiar form

$$z \approx T_r B / E = B R_g T_r^2 / E_a.$$

This relationship states that as activation energy increases, smaller reductions in temperature are needed to reduce the rate to (1/10)th of its original value. Notice that z is approximately, but not exactly, proportional to $1/E_a$.

The Q_{10} value

Another quantity that has been defined and studied in connection with food processing is Q_{10}, defined as

$$Q_{10} = \frac{\left[\text{Rate at Celsius temperature } (h+10)\right]}{\left[\text{Rate at Celsius temperature } h\right]}$$

where h is the temperature in °Celsius. Q_{10} is usually described as a measure of the sensitivity of the reaction to temperature change. The activation energy, E_a, and its dimensionless counterpart, E, are also measures of the same sensitivity, so it is reasonable that there is a relation between Q_{10} and E_a.

Instead of examining this relationship for Q_{10}, we generalize the definition slightly by defining

$$Q_\delta = \frac{\left[\text{Rate at Celsius temperature } (h+\delta)\right]}{\left[\text{Rate at Celsius temperature } h\right]} = \frac{k(h+\delta)}{k(h)}$$

δ is the temperature difference in degree Celsius. Since T is the temperature in degrees Kelvin,

$$h = T \pm S, \qquad S = 273°.$$

The relation between Q_δ and E_a may be found by using

$$Q_\delta = k(h+\delta)/k(h) = k(T \pm S + \delta)/k(T \pm S)$$

$$= \frac{k_1 \exp\left[\pm(E_a/R_g)(T \pm S + \delta)^{\pm 1}\right]}{k_1 \exp\left[\pm(E_a/R_g)(T \pm S)^{\pm 1}\right]},$$

$$\ln Q_\delta = E_a \delta / \left[R_g (T \pm S)(T \pm S + \delta)\right]$$

$$= E_a \delta / \left[R_g h(h+\delta)\right],$$

$$\log_{10} Q_\delta = E_a \delta / \left[R_g B h(h+\delta)\right].$$

If $\delta = 10$, we obtain

$$\log_{10} Q_\delta = \left\{ 10 / \left(R_g B \right) \right\} E_a / \left\{ h(h + \delta) \right\}$$

$$\approx 2.19 E_a / \left\{ h(h + \delta) \right\},$$

If, instead, it is true that $\delta \ll h$, then

$$\log_{10} Q_{10} \approx E_a \, \delta / \left[R_g B h^2 \right].$$

If the temperature is constant, say h, and the shelf life and rate are $L(h)$ and $k(h)$, respectively, Equation (6) implies

$$k(h) = W / L(h).$$

Similarly, at temperature $h + \delta$,

$$k(h + \delta) = W / L(h + \delta),$$

and so the definition of Q_δ implies

$$Q_\delta = L(h) / L(h + \delta).$$

Statistical determination of shelf life

Usually, two kinds of information are needed in order to estimate the shelf life of a food, namely, the temperature history and the food parameters. Here we shall focus on the problem of determining the food parameters by experimental and statistical means. We assume throughout this section that the kinetics are of Arrhenius form, as is usually done. Then the parameters that we must determine are c (or n), E (or E_a), k_r and possibly A_o the initial quality.

The experiments commonly consist of measurements of quality at various times and constant temperatures. If the times are designated as $t_1, t_2, \ldots, t_N$ and the temperatures as $T_1, T_2, \ldots, T_M$, the data can be arranged in an array or spreadsheet as follows:

Time | Temp =	T_1	T_2	$\ldots$	T_M
t_1	A_{11}	A_{12}		
t_2	A_{21}	A_{22}		
.	.	.		
.	.	.		
t_N	A_{N1}	A_{N2}		

The entry in the i-th row and j-th column is the quality measurement at time t_i and absolute temperature T_j. The entries in the first column, A_{1i}, describe the experimentally obtained time-variation of quality at the constant temperature T_1, and so on. From these measurements we want to find estimates of c, E, k_r and A_o based on the formulas described in previous sections.

Since we assume Arrhenius kinetics and constant temperature in each column of the spreadsheet, the function $\phi(t)$ can be expressed for the j-th column (j-th temperature, T_j, or x_j), and i-th time, t_i, as

$$\phi_j(t_i) = k_j(t_i \pm t_o)$$

$$k_j = k_r e^{(\pm E x_j)}.$$

If we also assume $t_o = 0$, the entry A_{ij} in the spreadsheet should be of the form

$$A_{ij} = \left(A_o^c - c\, k_j t_i\right)^{1/c} \qquad \text{if } c \neq 0$$

$$= A_o\, e^{-k_j t_i} \qquad \text{if } c = 0$$

Determination of the parameters is usually carried out by doing statistical regression calculations based on these models.

The procedures for estimating the parameters are affected by what is assumed to be known. The simplest situation is that in which the order, i.e., c, is taken as given. In that case only linear regressions have to be done; otherwise, nonlinear regression calculations are usually required, and the computational burden becomes considerable. In the next paragraph we summarize the regression procedures when c is known. In all these cases a two-stage regression procedure is described, in which each temperature is first treated separately to find A_o and k_j, and then the results for k_j at the different temperatures are combined to estimate k_r and E. This appears to be the most commonly used method.

If $c = 0$ (a first-order process), the regression is based on the model found by taking the natural logarithm of the above model for $c = 0$, which gives the model

$$\ln(A_{ij}) = \ln(A_o) - k_j t_i.$$

If a general, linear regression for U as a function of t has model

$$U_i = B_o + B_1 t_i,$$

then the regression for $c = 0$ and $T = T_j$ consists of choosing

$$U_i = \ln(A_{ij})$$

and finding B_o and B_1 by the usual linear-regression procedure. Then

$$B_o = \ln A_o \qquad \text{or} \qquad A_o = e^{B_o}$$

$$B_1 = -k_j \qquad \text{or} \qquad k_j = -B_1$$

The usual regression procedure also furnishes estimates of σ, the error standard deviation, and the standard errors in B_0 and B_1.

If $c \neq 0$, the model can be found by taking the c-th power of each side in Equation (2) to obtain

$$A_{ij}^{\ c} = A_o^{\ c} \pm c k_j t_i.$$

The usual regression method is performed with

$$U_i = A_{ij}^{\ c}$$

and then

$$B_o = A_o^{\ c} \quad \text{or} \quad A_o = B_o^{\ 1/c}$$
$$B_1 = \pm c\, k_j \quad \text{or} \quad k_j = \pm B_1/c.$$

Again the various standard deviations are also estimated. The simplest such case is $c = 1$ (a process of order zero), for which the regression consists merely of fitting straight lines to the plots of the data, A_{ij} versus t_i, for each j.

Having found estimates of k_j for each j, the quantities k_r and E are found by doing a linear regression based on the model

$$\ln k_j = \ln(k_r) \pm E x_j$$

which results from taking the natural logarithm of Equation (9). The general linear regression model has

$$U_j = B_o + B_1 x_j,$$

where

$$U_j = \ln(k_j),$$

After finding B_0 and B_1 by regression, k_r and E are found from

$$B_o = \ln k_r \quad \text{or} \quad k_r = e^{B_o}$$
$$B_1 = \pm E \quad \text{or} \quad E = \pm B_1$$

The usual regression method once more gives estimates of σ and the standard deviations of B_0 and B_1. The above descriptions are all based on the assumption that c is known in advance. In practice the data are usually subjected to preliminary visual checks, to see whether the graphs support the claim that c has the hypothesized value. These "eyeball" checks are valuable, but it is questionable whether they can detect modest departures from the assumed value of c. It may be desirable, therefore, to subject the data to a further analysis by nonlinear regression, in which the four parameters c, A_o, k_r and E are all sought at the same time. An example of such a calculation has been presented by Haralampu et al.[2] Also, a recent paper by Rhim et al. takes a step in this direction and is a departure from the usual approach in assuming that tests are conducted with linearly varying temperature.[12]

The conventional two-stage process described above is slightly open to questions of interpretation since the intermediate quantities k_j are treated as data in the second stage but are really quantities derived from the data of the first stage. It is not likely that much error in E or k_r is caused by this, but it appears to miscalculate some of the correlations among the parameters and hence some of the error estimates. Again, more study is needed on the question, and this also suggests that nonlinear regression be undertaken in spite of its computational demands.

Shelf life at variable temperature

An important problem in shelf life estimation is that of finding the effect of temperature variation. In principle, if the temperature history of the food is known, the quantity

$$\phi(t_e) = k_r \int_{t_o}^{t_e} e^{\pm Ex(t)} \, dt$$

can always be calculated, using numerical integration if necessary. The shelf life, L, is then determinable by solving the equation $\phi(t_e) = k_r F(t_e) = W$ for t_e, provided the parameters of the food are known. If Arrhenius kinetics are assumed, and E is known, $F(t_e)$ becomes

$$F(t_e) = \int_{t_o}^{t_e} e^{\pm Ex(t)} \, dt$$

The process of determining L is relatively easy when the temperature is constant during storage (i.e., x is constant) but is somewhat more taxing if x varies. While the solution can usually be found by numerical approximations, these are not always helpful in understanding the effects of changes in the parameters or the temperature history. For the latter purpose it would be useful to have exact or accurate approximate solutions, from which the influence of these changes could be seen.

A small number of viable formulas are known, some of which have been obtained by assuming that the temperature range is narrow enough so that T and x are approximately related in linear fashion:

$$x = 1 - (T/T_r).$$

Some results that are accurate over a wider temperature range are given by Rhim et al.[12] Rather than describing these solutions, we shall present a different example of an exact formula that is valid over any physically meaningful temperature range, provided the pulse has the assumed shape. This solution will be used later in the present chapter to illustrate the calculation of shelf life.

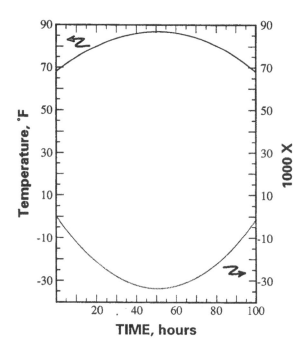

Figure 6 Temperature pulse shape as a function of time, t, in hours. Lower curve is $1000x$; upper curve is temperature in degrees Fahrenheit.

In an recent paper,[14] the author considered a temperature pulse of the form

$$x = x_m \left[1 \pm \left\{ (t \pm m)/(t_s \pm m) \right\}^2 \right]$$

The pulse starts from $x = 0$ at time t_s, attains $x = x_m$ at time $t = m$ and returns to $x = 0$ at $t = t_e = 2m - t_s$. The pulse is symmetric about $t = m$. Figure 6 displays the dependence of both x and temperature on t. We assume $x_m < 0$, so the pulse in x is negative, corresponding to a positive pulse in T. Then a result was found that was equivalent to

$$F(t) \pm F(t_s) = C \left\{ q(aV) + q(a) \pm 1 \right\}$$

$$u = \pm E x_m, \quad a = (2u)^{1/2}, \quad V = (t \pm m)/(m \pm t_s)$$

$$C = (m \pm t_s)(\pi/u)^{1/2} e^u,$$

where q is the standard, normal cumulative function, provided $t_s < t < t_e$. In particular at the end of the pulse we have

$$F(t_e) - F(t_e) = C\{2q(a) - 1\}$$

This formula shows that the F accumulated during the pulse is simply proportional to $m - t_s$, i.e., to the pulse width. Moreover, since E and x_m influence F only through the combination u, changes in E and x_m that lead to the same u also lead to the same accumulated F. Although these results are specific to the temperature pulse assumed here, they are useful as a guide to what we might expect for pulses of generally similar shape. The author described formulas for an entire family of such pulses.[14]

An example

This fictitious example is about finding the shelf life, L, of a food item whose quality, A, varies with time according to the model of Equation (2) with $n = 0.2$, $c = 1 - n = 0.8$. The initial quality, at time $t_o = 0$, is taken to be $A = A_o = 100$ on some scale whose true meaning is irrelevant to the example. The item is expected to be stored at 68°F = 20°C = 293 K for 500 h or about 21 days. The item obeys Arrhenius kinetics with activation energy, E_a, of 20 kcal/mol. Then

$$T_r = 293$$

$$E = 20,000 / 293 / 1.986 = 34.37.$$

The quality is assumed to have $A_b = 40$, i.e., the item is unacceptable when A deteriorates to or below 40. Then

$$R_b = 0.4.$$

The rate constant, $k_r = k_o$, is found by solving Equation (6) in the form

$$k_r = A_o^{\ c}\left(1 \pm R_b^{\ c}\right)\big/(cL)$$

$$= 100^{0.8}\left(1 \pm .4^{0.8}\right)\big/\left[0.8(500)\right] = 0.05171.$$

Thus at the reference temperature, 68°F ($T = T_r = 293$), we have

$$x = 0, \ k = k_r = 0.05171, \ L = 500 \text{ h}.$$

The values of the rate constants and shelf lives for storage at other constant temperatures are found by applying the formula

$$k = k_r e^{\pm Ex}$$

to obtain k, and substituting in Equation (6)

$$L = A_o^{\ c}\left(1 \pm R_b^{\ c}\right)\big/(ck)$$

to find L. At 77°F = 25°C = 298 K we obtain

$$x = (293 / 298) \pm 1 = \pm.01678$$

$$k = (0.05171)\, e^{(34.37)(0.01678)} = 0.09205$$

$$L = 100^{0.8}\left(1 \pm 0.4^{0.8}\right)\big/\left[.8(0.09205)\right] = 281 \text{ h}.$$

Similarly at 86°F = 30°C = 303 K we obtain

$$x = \pm 0.03300, \ k = 0.16075, \ L = 161 \text{ h}.$$

Now suppose that there has been a failure of the cooling system that controls the storage area, leading to a temperature pulse of the type considered at the end of the preceding section. We assume that the pulse commences at $t_s = 100$ h, reaches its maximum, 86°F or $x_m = -0.0330$, at $m = 150$ h, and returns to 68°F or $x = 0$ at $t = t_e = 200$ h. We want to determine the decreased shelf life resulting from this cooling failure. The formulas of the preceding section give

$$u = \pm E x_m = \pm(34.37)(\pm 0.03300) = 1.1342$$

$$C = 50(\pi/1.1342)^{1/2} e^{1.1342} = 258.7$$

$$a = \left[2(1.1342)\right]^{1/2} = 1.506$$

$$q(a) = 0.934$$

$$F(t_e) \pm F(t_s) = 258.7\left[2(0.934) \pm 1\right] = 224.6 \text{ h}.$$

Then the shelf life is found by solving

$$F_n = t_s + \left(L \pm t_e\right) + \left[F(t_e) \pm F(t_s)\right],$$

where F_n is the nominal or intended shelf life at $x = 0$, i.e., $F_n = 500$. This leads to

$$L = F_n + t_e \pm t_s \pm \left[F(t_e) \pm F(t_s)\right]$$

$$L = 500 + 200 \pm 100 \pm 224.6 = 375.4 \text{ h}.$$

The effect of the pulse has therefore been to reduce L from its nominal value, 21 days, to about 16 days.

We might also ask about the effects that errors in estimating E_a have on the shelf life. This could be investigated by simply making several different calculations for various values of E. It might also be studied by carrying out a Taylor series expansion of Equation (8) in powers of $(E - E_n)$, where E_n is the nominal or intended value of E. Omitting the details, we find that

$$(\delta L) \approx \pm 5.6 \, (\delta E),$$

where (δL) is the change in L caused by the change (δE) in E. As expected, an increase in E causes a loss of shelf life.

Conclusions

The theory given above is reasonably complete as to the implications of the simplest and most used mathematical model. The "General Theory" Section of this chapter makes an attempt to describe a framework into which a more complete theory might fit, but there seems not to have been much effort to go beyond the simple theory. Certainly, the situation where two or more interactions affect the quality is of importance, and assumptions other than Arrhenius kinetics may sometimes be appropriate.

As mentioned in the chapter's Introduction, this model must usually be combined with equations that determine the temperature in the food. This adds a formidable layer of difficulty to the entire problem. The temperature most often depends on both time and position within the food and is governed by one or more partial differential equations. These have to be solved to determine the time variation of the highest temperature in the food, which is then taken as $T(t)$. In many cases the solutions of the partial differential equations have to be carried out by numerical means and may be the most time-consuming part of the problem. The present review has completely ignored this aspect of quality loss estimation and focused entirely on the narrower and more manageable aspects that arise when $T(t)$ is known. However, if significant energy is absorbed or liberated during the chemical reactions that affect quality, it may be impossible to treat separately the quality loss and thermal parts of the problem.

It is clear that much remains to be done. Many of the recent attempts to deal with quality loss are computer-based, and this is to some degree inevitable. However, there are large bodies of mathematical theory dealing separately with quality loss and thermal variation, some portions of which may help to reduce the difficulty of these problems.

Bibliography

1. Ball, C. O. and Olson, F. C. W., 1957: Sterilization in Food Technology, McGraw-Hill, New York.
2. Haralampu, S. G., Saguy, I., and Karel, M., 1985: Estimation of Arrhenius model parameters using three least squares methods, J Food Process Preserv 9, p. 129.
3. Hayakawa, K. I., 1978: A critical review of mathematical procedures for determining proper heat-sterilization processes, Food Technol 32 (3), p. 59.
4. Hayakawa, K. I. and Wong, Y., 1974: Performance of frozen-food indicators subjected to time-variable temperatures, ASHRAE J 16 (4), p. 44.
5. Karel, M., 1983: Quantitative analysis and simulation of food quality losses during processing and storage, in Computer-Aided Techniques in Food Technology (ed. I. Saguy), Marcel Dekker, New York, p. 117.
6. Labuza, T. P., 1982: Shelf-Life Dating of Foods, Food and Nutrition Press, Westport, CT.
7. Labuza, T. P., 1984: Application of chemical-kinetics to deterioration of foods, J Chem Edu 61, p. 348, (State-of-the-Art Symposium, Chemistry of the Food Cycle).
8. Labuza, T. P. and Kamman, J., 1983: Reaction kinetics and accelerated tests simulation as a function of temperature, in Computer-Aided Techniques in Food Technology (ed. I. Saguy), Marcel Dekker, New York, p. 71.
9. Lenz, M. K. and Lund, D. B., 1977: The lethality-Fourier number method. Experimental verification of a model for calculating temperature-profiles and lethality in conduction-heating canned foods, J Food Sci 42, p. 989.
10. Lund, D. B., 1978: Statistical analysis of thermal process calculations, Food Technol 32 (3), p. 76.
11. Rand, W. M., 1983: Development and analysis of empirical mathematical, kinetic models pertinent to food-processing and storage, in Computer-Aided Techniques in Food Technology (ed. I. Saguy), Marcel Dekker, New York, p. 49.
12. Rhim, J. W., Nunes, R. V., Jones, V. A., and Swartzel, K. R., 1989: Determination of kinetic parameters using linearly increasing temperature, J Food Sci 54 (2), p. 446.
13. Rogers, A. R., 1963: An accelerated storage test with programmed temperature rise, J Pharm Pharmacol 15, p. 101.

14. Ross, E. W.,1993: Relation of bacterial destruction to chemical marker formation during processing by thermal pulses, J Food Process Eng 16, p. 247.
15. Simon, J. and Debreczeney, E., 1971: 1/T heating program in reaction kinetic studies, J Thermal Anal 3, P. 301.
16. Singh, R. P. and Wells, J. H., 1988: A kinetic approach to food-quality prediction, using full-history time-temperature Indicators, J Food Sci, 53 (6), p. 1866.
17. Stumbo, C. R., 1965: Thermobacteriology in Food Processing, Academic Press, New York.

chapter twelve

Stored product quality: open dating and temperature monitoring

Bruce B. Wright and Irwin A. Taub

Contents

Introduction

The quality of a product held in storage will degrade over time, the rate of degradation being dependent on temperature. If the process associated with such degradation is highly sensitive to temperature, reflecting a high activation energy, the rate increases substantially as the temperature rises. Assuming that throughout the storage duration the temperature is maintained constant, a point in time will be reached when the product quality degrades, perceptively or imperceptively, to a level taken as a cutoff limit. That point in time corresponds to the product shelf life. In situations where the temperature is not constant, either because the temperature of the storage facility is uncontrolled and subject to fluctuations or because of malfunctioning temperature controllers, the cutoff quality will be reached in shorter or longer times. The dynamics underlying the quality degradation process and the modeling of the relationship to temperature and time are described in Chapters 11, 13, and 15 and are covered in detail in the book by Labuza[1] and in various journal articles.[2-5]

The quality cutoff limit and the shelf life under specific storage conditions vary over a wide range and depend on the nature of the product, the predominant degradation process, and the mode of storage. Roughly speaking, products can be classified as either fresh or processed, recognizing that there are gradations within and between these classifications. Minimal processing falls between the two, while commercial sterilization represents the extreme in processing. Roughly speaking again, the degradation processes can be classified as microbial, biochemical, and physicochemical. Growth of spoilage and pathogenic microorganisms, first and foremost, sets the limit on storage of fresh or minimally processed products. Biochemical processes, such as enzymatic browning and rancidity development, if occurring on a time scale shorter than microbial growth, can determine the cutoff. Physicochemical processes, including flavorant loss, moisture redistribution, and starch crystallization, some of which significantly affect texture, can determine the shelf life of products in which microbial growth is not a factor. Such biochemical and physicochemical instabilities are discussed in Chapters 3–6. Depending then on the susceptibility of the product to one or more of these degradation processes, it would be stored at freezing, chilled, or ambient temperatures to maintain a marketable quality for as long as is needed to move through the food chain from producer to consumer.

Even in a class of products made commercially sterile and storable at ambient temperature by thermal processing, there can be a wide range of shelf lives, as illustrated by studies on the storage stability of the prepackaged combat ration, Meal, Ready to Eat (MRE). In studies done in the 1980s,[6,7] the quality of MRE components stored at different temperatures from 2 to over 5 years depending on temperature was assessed by a consumer panel on the basis of a nine-point hedonic rating for overall acceptance and other sensory attributes. The data were analyzed by Ross et al.[7] in terms of the time it took (or would take) for the quality score to drop to a rating of 5 ("neither like nor disklike"). Although many components were used, the results were either pooled altogether or grouped into five categories: entrees, pastries, vegetables, fruits, and miscellaneous. Accordingly, the shelf life defined in this way for each group as a function of constant storage temperature can be plotted semilogarithmically,[8] as shown in Figure 1 for some of these categories, including the combination of all.

The distinctions among the products relative to their susceptibility to degrade and their sensitivity to temperature are clearly evident. Starch components, the pastries, have relatively short shelf lives that are almost independent of temperature. Fruits (i.e., dehydrated items) have relatively longer shelf lives at moderate storage temperatures, but much shorter shelf lives at higher temperatures, because of a high temperature sensitivity. The entrees are longer lived than either starches or fruits at moderate storage temperatures, but show shelf lives longer than fruits but shorter than starches at higher storage temperatures. These results emphasize the importance of knowing product sensitivity to storage temperature and time and the importance of being able to properly specify the cutoff limits in terms of end of storage times.

Product dating

Providing consumers with information on a package that indicates how long it has been in storage or how soon it should be used is now an accepted practice. Such *product dating*

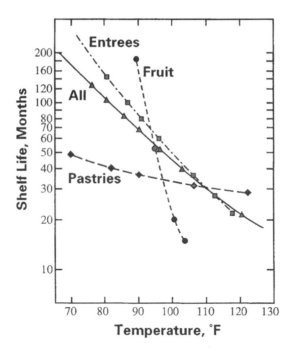

Figure 1 Shelf life of MRE components as a function of storage temperature. "Shelf life" is defined here as the time in months for the rating to decrease to 5 on the hedonic scale. The line for "All" (▲) corresponds to the pooling of data for all components.

or *open dating*, however, varies in terminology and is not always easily understood. Some of the terms used include "Pull Date," "Use by Date," "Sell by Date," "Best if Sold by Date," "Better if Used by Date," "Best When Used by Date," and "Date of Pack." Sometimes just the date corresponding to 3 years in the future will be stamped on the package or can. Clearly, terms that indicate the date to sell the product are aimed at the retailer, but the consumer relies on them as well.

Product dating, if properly understood, minimizes but does not eliminate the possibility of purchasing and consuming a spoiled or significantly deteriorated product. The *sell by* terms tend to imply that the food suddenly becomes unacceptable by that date and should be discarded. The *best if used by* terms avoid this perception and tend to imply a more gradual deterioration. In either case, such date designations cannot guarantee that a spoiled product will not be bought or consumed, because the reference temperature is not provided and the temperature history is not known. Their limitations notwithstanding, they do serve as a useful guide to the consumer when purchasing the product from the retailer and when using it while stored at home.

In practice, the best way to assure that a product will be safe and of an acceptable quality when consumed is to control temperature during storage and distribution and to integrate the temperature exposure over time.

Temperature monitoring

Monitoring devices

The devices and procedures used to monitor or control the temperature in a storage or transportation facility depend on the product and the location of the product in the distribution chain. Thermometers such as the mercury-in-glass type can provide an accurate measurement of the temperature, but only on an instantaneous basis. Generally, one would need to take thermometer readings on a regular basis to obtain meaningful data on the temperature history. This approach would involve considerable labor, so in most cases a recording thermometer, or thermograph, is used. Such an analog recorder can be useful both in determining if a certain temperature had been exceeded and in providing feedback for regulating the temperature. Analog recorders now are being supplanted by digital recorders, including battery-powered portable devices, that not only sample the temperature at adjustable intervals, but store the information for subsequent downloading to a computer. Through available programs, the sampled data can be printed out in various formats or utilized for further analysis, including integrating temperature over time and correlating product quality with such an integration.

Monitor placement

The distribution of food products occurs over a wide geographic area and can involve multiple modes of transportation and diverse warehousing in urban as well as in rural areas. Storage of military rations, in particular, includes facilities in remote and sometimes forbidding regions. The complex nature of the distribution system and the wide range of daily and seasonal temperatures impose significant stresses on the products that need to be monitored.

Transportation to and from warehouses in the chain often involves a combination of land, sea, or air transport. Each loading and unloading operation at a transfer point or interim storage facility introduces a thermal stress or an excursion from an otherwise controlled temperature environment. Moreover, each mode of transport would not usually be designed for extremes in temperature, which put a greater load than expected on the temperature control systems. Often the amount of product that is transported comes close to the volumetric limit of the transport, especially in the case of trucks. Although their cooling systems work more efficiently when near design capacity, the temperature of the product when loaded onto such transporters can introduce a heat load that causes the capacity to be exceeded. Even if the equipment continues to function properly, the added heat load requires a longer time for the product to reach its set temperature and, in some cases, the product could reach its destination without having attained the proper temperature. This excursion from the desired temperature can easily happen aboard ships where the chill boxes are filled to capacity and the destination is a hot climate.

Under such circumstances, the placement of the monitoring/recording devices becomes critical if the record is to be representative of the thermal history of products being transported and stored. Since the cost of the portable recorders is significant, the number that can be used in an individual shipment is limited. Ideally, they should be packed in cases placed in strategic locations on pallets placed strategically within a van to capture both the most stable environment as well as the most fluctuating environment (e.g., near the door of the container van).

All products will be stressed in some way during such distribution, but they will not necessarily be abused. There will always be normal variations in temperature during the life cycle of a product. Even in a freezer or refrigerator, there is an expected fluctuation in temperature above or below which a deteriorative effect can occur. When this upper temperature limit is exceeded or when, in the case where freezing is to be avoided, the lower limit is exceeded, then the product might be abused. The degree of abuse would depend on the length of time the product is beyond the limit and the tolerance used in setting the limit.

Time–temperature indicators

Time–temperature indicators (TTIs) can be either full history or partial history temperature monitors that integrate the exposure to a temperature over time by accumulating the effect of such exposures. TTIs of either type usually require activation at start-up time. The full-history TTI will integrate continuously until it has expired. The partial-history TTI will integrate only when the temperature exceeds a predetermined set point, e.g., 90°F; moreover, when the temperature returns to or passes through the set point, the integrating ceases. TTIs have different, sometimes adjustable, sensitivities to temperature, which correspond to different reaction rate constants and activation energies. A discussion of the kinetics as well as other aspects of using TTIs is provided in a paper by Grabiner and Prusik.[9] More details on TTI devices are in a paper by LeBlanc,[10] especially related to their application with frozen foods. A TTI standard has been developed by ASTM.[11]

Principle of operation

TTIs can be classified on the basis of the phenomenological processes that governs their principle of operation. These phenomena include enzymatic or nonenzymatic chemical reactions and various physical transformations. With respect to chemical reactions, the original I-Point TTI was based on the enzymatic hydrolysis of a fatty acid ester, which leads to a change in pH that would be discernible through a color change in a pH indicator. The LifeLines TTI, to be discussed in more detail below, is based on the thermally induced polymerization of a substituted diacetylene monomer, which leads to a change in color and optical reflectivity. With respect to physical processes, the 3M TTI is based on diffusion of a dye in a material whose viscosity decreases with increasing temperature, which leads to a migration of the dye front along a linear strip. In all cases, the monitoring and integration relies on a temperature-dependent phenomenon that can be made to give a visual indication by a color change or color migration.

The review by Taoukis et al.[2] includes a discussion of TTI classification and principles. It also summarizes patents in the field. Depending on the principle, the TTI label is configured so that the integrated effect can be assessed. With some labels, an instrumental reading of change in color intensity can be made that gives a quantitative value of the time–temperature integration. Some labels are designed so additional information about the product to which it is attached can be obtained in order to facilitate a computer-based product management system.

Integration concept

The concept of an integrating indicator label, though based directly on kinetic principles, can be pictorally illustrated using a ticking clock analogy (Figure 2). The ticking rate depends on temperature and the accumulated number of ticks corresponds to the integrated effect. In this illustration, frame A shows an arbitrary ticking rate at 60°F of 1 tick/min that doubles every 10°F, frame B shows two temperature profiles, 1 and 2, over a 2-hour period, and frame C shows the accumulation or integration of ticks for each profile. Note that the time-averaged temperature for both profiles is 80°F. Nevertheless, the total number of ticks for profile 2 is greater than for profile 1, because of the larger residence time at 100°F. This simple depiction merely follows the very detailed and often sophisticated treatments of nonisothermal kinetics obeying the Arrhenius relation, but it emphasizes pictorially the influence of high temperature exposure time on the overall thermal effect.

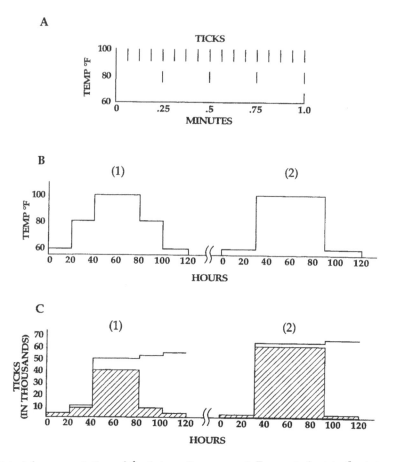

Figure 2 Pictorial representation of the integration concept. Frame A depicts the interval between ticks for each temperature, and indicates that the ticking rate at 100°F is 16 times the rate at 60°F. Frame B depicts two different temperature profiles; the mean temperature is 80°F for both profiles 1 and 2. Frame C depicts the number of ticks for each time interval (the shaded area) and the accumulated number of ticks (the upper line) at the end of each time interval. The total number of ticks is 50,400 and 61,200 for profiles 1 and 2, respectively. This difference in tick count corresponds to a decrease in shelf life of about 18%.

Illustrative applications

TTIs have been used in diverse applications to track thermal exposure and to signal the associated degradation in the product. A common feature is that there is at least one attribute of the product that undergoes a temperature-dependent change responsible for, or related to, a decrease in product quality with time. In the applications to be described, a TTI label that indicates visually was used.

The label (Figure 3), produced by LifeLines, consists of three main sections: a bar code both for identifying the label (a serial number in this instance) and for providing additional information (including product code), a bar code to represent the thermally sensitive material used as the indicator (MC38 in this instance), and a stripe or patch containing the indicator material which is applied in a printing process. A computer-based scanner is used to obtain the digital signals from the bar codes and the analog signal from the stripe on each label. Since the scanner is responding to the light intensity reflected by the stripe, this analog signal is transformed into *percent reflectance*. As the stripe darkens with

Figure 3   An example of a TTI label showing the three main sections. The third section simulates the decreased reflectance of the active indicator material over time.

time (the rate of darkening increasing with increasing temperature), the percent reflectance value decreases and the permanent change in percent reflectance corresponds to the integrated thermal effect.

Transcontinental and transoceanic shipment

Temperature history

MRE rations were monitored under realistic transportation and storage conditions involving shipment from the ration assembler in McAllen, Texas to facilities in California, in Massachusetts, and in Turkey. Special MREs were obtained for this study in order to monitor ration components that are relatively sensitive to heat stress. They corresponded to 4 of the 12 available menus, each having one of the following entree items: omelet with ham, chicken-a-la-king, beef and rice meatballs in spicy tomato sauce, and escalloped potatoes with ham. Consequently, each case contained 12 meals, with the four menus in triplicate.

Two TTI models (MC11 and MC18) were placed in duplicate inside and outside the outer carton, providing a total of eight TTI labels for each case. The labels were scanned prior to applying them to the cases to obtain the initial reflectance measurements.

A total of 64 cases were shipped from the assembler: 24 to George Air Force Base (AFB) in California; 24 to an AFB in Incirlik, Turkey; and 16 to the Natick RD&E Center in Massachusetts for use as controls. Temperature recorders were included in some of the cases to obtain a history of temperature changes, based on readings taken every 2 h. The two AFB locations were very similar in average temperature, being 68.9 and 68.6°F for California and Turkey, respectively, but they had very different temperature profiles during the 31 Aug 1988 to 7 Sep 1989 time period.

The results of the long-term temperature histories are shown in Figure 4 for cases shipped to California. The rations were shipped by the assembler at the end of the summertime, Day 0 corresponding to 31 Aug 1988. Figures 5a and 5b show a more detailed rendering of the early temperature profile for these shipments, including the high tem-

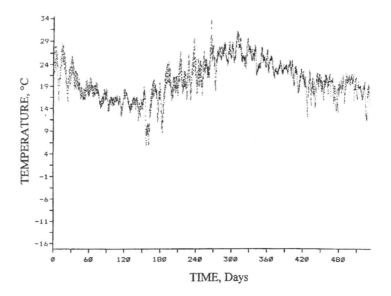

Figure 4 Temperature profile recorded for MREs stored at George AFB, CA, for over a year showing the seasonal variations.

peratures encountered during transport by ship to Turkey. The daily fluctuations (i.e., highs and lows) are evident in these printouts.

Some cases of rations stored in California and Turkey were returned at selected time intervals to Natick for scanning of the TTI labels. After 1 year in storage, the remainder of the rations were returned to Natick for scanning and sensory evaluation. For the rations shipped to and stored at George AFB, Figure 6 shows both the temperature profile from the digital recorder and the reflectance measurements from the TTIs. Although only four readings of the TTIs were made, they are consistent with the thermal history. The lines connecting the readings are steepest during the hotter months and least steep during the colder season of the year, corresponding to a slower rate of reaction for the polymerization. Table 1 gives the results of the TTI measurements for samples stored in California and Turkey for up to 13 months and then returned to Natick.

Product quality correlation

A limited attempt at correlating MRE product quality with the temperature integration, expressed as percent reflectance, was successful and is instructive. The correlations involved MREs stored under constant temperature conditions at Natick where the TTIs could be scanned monthly and quality evaluations could be made by a sensory panel.

Sensory panel evaluations were conducted on products stored at 70°F and 100°F. Products included omelet with ham, chicken-a-la-king, beef and rice meatballs in spicy tomato sauce, escalloped potatoes with ham, potato au gratin, pears, and peaches. It was not expected that all items would show effects that could be correlated with the percent reflectance measurements or would change at rates that could be observed either because of the temperatures involved or because of the elapsed time at which the evaluations were made. Storage data at 70°F did not prove to be useful because of the short storage time. The data at 100°F, however, was useful for assessing the omelet, chicken, meatballs, escalloped potatoes, and pears.

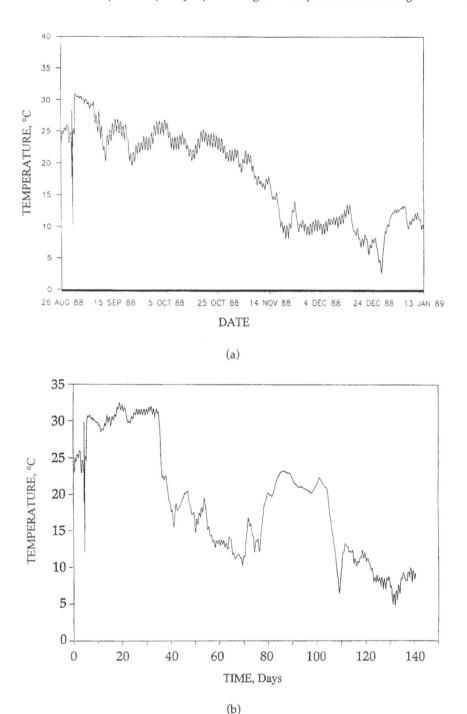

(a)

(b)

Figure 5   (a) Temperature profile recorded for MREs stored at George AFB, CA, showing on an expanded scale the daily cycling and the general decrease during the fall and early winter seasons. (b) Temperature profile recorded for MREs shipped to Incirlik, Turkey. Day 0 corresponds to 26 August 1988. It shows the general decrease during the fall season along with a higher temperature period at about 80–100 days when the rations were aboard ship.

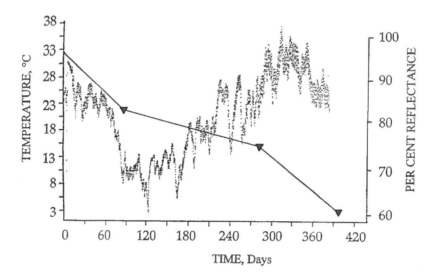

Figure 6 The temperature profile for MREs stored at George AFB, CA, and the percent reflectance of the TTIs both plotted as a function of time. The slope of the lines connecting the percent reflectance measurements clearly changes in accordance with the seasonal variation in temperature.

Table 1   Percent Reflectances for California (CA) and Turkey (T) Shipments

Time (Months)	MC11 CA	MC11 T	MC18 CA	MC18 T
0	96	96	92	92
2	84	—	48	—
9	74	67	22	15
13	59	61	2	7

The results for pears from MRE Menu #9 and for omelet with ham from Menu #4 are illustrative. These were stored at 100°F for over 2 years and rated for color and overall acceptance, respectively, after withdrawals at 2, 14, and 26 months. Panelists evaluated these using an open-ended magnitude estimation scale in which high scores are favorable. As the comparisons of percent reflectance with sensory scores over time in Figures 7 and 8 show, the correlations are qualitatively good. Since the scales for sensory scores were "stretched" so the 2-month value would coincide with the reflectance curve, the other values indicate a relative, not a quantitative, match. Moreover, reflectance readings below 20% (corresponding to extensive polymerization) do not provide an adequate basis for assessing the strength of the correlation. The correspondence in these plots, nevertheless, is clear and indicates that the TTIs can be matched to either subjective or objective assessments of quality.

Container vans in a desert environment

Temperature history

Three container vans, each with a different type of packaged ration, were transported to the U.S. Army Yuma Proving Ground for use in studying ration storage in a high-temperature desert environment. One van contained cases of MREs; another contained cases of T-rations, which are meal components thermoprocessed in metal trays suitable

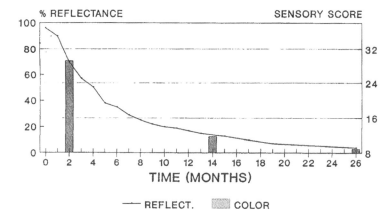

Figure 7 Percent reflectance readings and sensory scores for color plotted as a function of time for MRE pears stored at 100°F. The sensory scale has been adjusted to make the 2-month score coincident with the line for the reflectance readings in order to show the close correspondence.

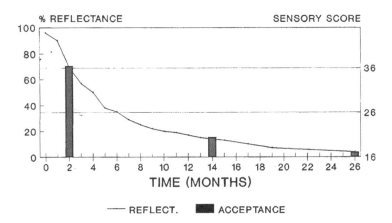

Figure 8 Percent reflectance readings and sensory scores for overall acceptance plotted as a function of time for MRE omelet with ham stored at 100°F. The sensory scale has been adjusted to make the 2-month score coincident with the line for the reflectance readings in order to show the close correspondence.

for serving a group of 18 individuals; and the third contained cases of B-rations, which are a mix of dry and canned items, also suitable for group serving. These vans were 8' × 8' × 40' in size, with the capacity to hold a large quantity of rations for transportation and storage. It was possible to equip a small number of cases within a van for temperature monitoring and still have a van with the mass and other characteristics typical of a loaded van.

A total of 64 thermocouples and TTI labels were placed at selected locations in the vans. Most of the thermocouples were in the central van, containing the MREs; two thermocouples were placed on the outside of the van for comparison with the inside van temperatures and with local weather records. The daily temperature variations for a 2-week period that includes the hottest day of the year are shown in Figure 9.

Some of the TTI labels were placed near thermocouples for comparison. Table 2 shows the TTI percent reflectance measurements for labels with both MC11 and MC18 temperature sensitivities attached to cases placed in similar locations (southwest corner of the

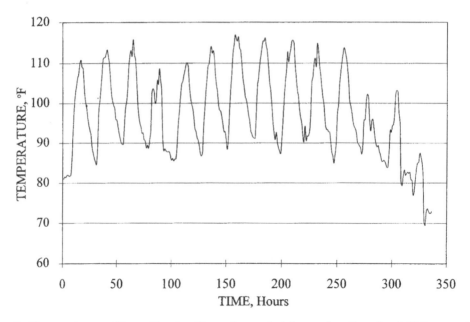

Figure 9 Temperatures at Yuma Proving Ground recorded from a location about 20 ft away from the vans during a 2-week period in August 1992 showing the cyclic nature of the variations.

Table 2 Comparison of Mean Reflectances for Similar Locations in Three Vans

	B-Ration Van		T-Ration Van		MRE Van	
Indicator	MC18	MC11	MC18	MC11	MC18	MC-11
June	94	95	96	94	95	95
July	5	40	19	54	9	52
August	NA	NA	NA	NA	1	38
September	3	12	3	24	5	31

Table 3 Percent Reflectances from MC11 in the MRE Van

	Van Top						Van Bottom	
	South Side			North Side			South	North
AT[a]	27	31	46	43	50	52	45	51
TIME[b]								
0	95	95	95	95	95	95	96	95
1	52	51	46	49	58	47	59	—
2	38	34	27	36	—	29	48	—
3	31	22	19	22	37	19	45	43
10	19	16	10	—	21	15	37	—
11	15	11	9	—	18	12	32	32
13	9	8	7	7	11	7	23	27
16	8	3	2	5	4	4	19	16
17	4	3	2	3	3	1	16	—

[a] The number of the associated thermocouple for the TTI labels whose average percent reflectances are given.

[b] Time in months.

van) in the B, T, and MRE vans. The rate of change in reflectance is faster for the MC18 label than it is for the MC11. After one summer of exposure, the MC18 label had expired, the residual reflectance being about 10% or less. (Although it was not suitable here, this label can still be a useful indicator for short monitoring periods that might include the initial transportation of items to a warehouse.) The MC11 July and September readings are both lower for the B van than for the other two vans, which can be attributed in part to the differences in the color of the vans. Since the B van was darker, a higher average temperature was attained, resulting in lower reflectance readings.

Table 3 summarizes the percent reflectance measurements of the MC11 TTIs in the MRE van taken over 17 months. As expected, TTI label readings in the top sections of the van decreased more rapidly than those in the two bottom sections, and generally the TTI labels picked up differences throughout the van. These results show that it is possible to use TTI labels where it would be impractical to use a large number of thermocouples in order to monitor the storage environment and thus to reflect the condition of the products in all parts of a van or a similar storage space.

Product quality correlation

Another attempt was made to correlate product quality with reflectance readings, this time using TTI data from the uncontrolled storage of MREs in the vans and using both instrumental and sensory evaluations of various quality attributes. MRE applesauce, cheese spread, escalloped potatoes with ham, grape jelly, and peanut butter were selected for comparison. All were subjected to the 3 months of summer stress at Yuma Proving Ground. The results on correlating color change in the applesauce and cheese spread with TTI reflectance are illustrative and instructive.

The basic approach is based on the premise that both the degradation of a particular quality attribute and the decrease in TTI reflectance essentially follow first-order kinetics and that both the label and the product experience essentially the same temperature history. Consequently, at any point in time, the measurements of TTI reflectance and quality attribute should correspond to the integrated effect of that temperature history and should bear a clear relationship to each other. Since the activation energies for both processes could be different, the correspondence is not expected to be one-to-one. Nevertheless, in the simplest case, the fractional change in a particular attribute should be proportional to the fractional change in TTI reflectance. This concept is similar to the concept used in relating bacterial destruction to the formation of chemical markers in thermoprocessed foods.[12-13]

In applying this concept to the degradation of color of MRE applesauce and cheese spread, certain simplifications and estimations had to be made, because measurements of TTI reflectance and product color were not always made at identical times. Reflectance readings were made initially and at 1, 2, and 3 months afterwards, whereas instrumental color measurements, a^* and L values, were made initially and at 3, 12, 24, and 36 months afterwards. To obtain color measurements comparable in time to the TTI reflectance measurements, plots of a^* and L vs. time were made for both applesauce and cheese spread, and then fitted with curves that accommodated well the more significant changes that took place over the first 12 months. The a^* and L values for each product at months 1, 2, and 3 were then calculated from the fit and used in making the correlation with TTI reflection.

As Figures 10 and 11 indicate, the correlation of instrumentally determined color change in applesauce and cheese spread with reflectance is relatively good. For both products, the a^* values, which measure the degree of redness, for both products linearly correlate with the TTI reflectance, expressed as the ratio of reflectance at any time to the

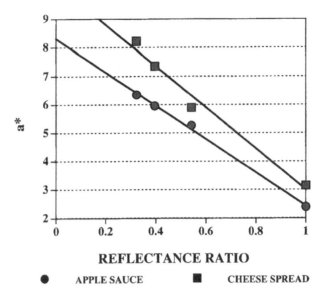

Figure 10   Measured *a** values from stored samples of applesauce and cheese spread plotted as a function of the ratio of the TTI reflectance at time of withdrawal to its initial reflectance.

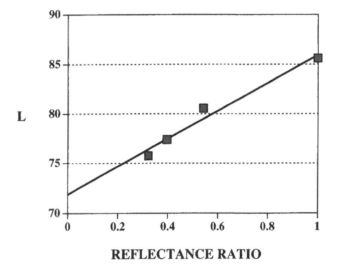

Figure 11   Measured or extrapolated *L* value for cheese spread plotted as a function of the ratio of the TTI reflectance at time of withdrawal to its initial reflectance.

initial reflectance. The redness intensity increases with time and the reflectance decreases with time, so the correlations shown in Figure 10 are negative. The separation of the two lines is primarily a consequence of different absolute values of *a**. The *L* value for cheese spread (Figure 11) also correlates well with the ratio of TTI reflectances. Since it is a measure of lightness and darkness and the product became darker with time, the correlation of *L* with the ratio of TTI reflectances is positive. It must be acknowledged that the correlations here are rough, in part because the values of *a** and *L* for 1, 2, and 3 months were obtained by fitting data, most of which pertain to times up to 36 months. Despite this necessity, the inference is clear that the TTIs, if properly chosen and calibrated for a particular attribute, give useful information about product quality.

References

1. Labuza, T.P. Shelf-Life Dating of Foods, Food & Nutrition Press, Westport, CT (1982).
2. Taoukis, P.S., Fu, B. and Labuza, T.P. Time-Temperature Indicators, *Food Technol.* 45, 70 (1991).
3. Wells, J.H., Singh, R.P. and Noble, A.C. A Graphical Interpretation of Time-Temperature Related Quality Changes in Frozen Food, *J. Food Sci.* 52, 436 (1987).
4. Wells, J. H. and Singh, R.P. A Kinetic Approach to Food Quality Prediction Using Full-History Time-Temperature Indicators, *J. Food Sci.* 53, 1866 (1988).
5. Taoukis, P.S. and Labuza, T.P. Applicability of Time-Temperature Indicators as Shelf Life Monitors of Food Products, *J. Food Sci.* 54, 783 (1989).
6. Ross, E. W., *et. al.*, Acceptance of a Military Ration after 24 Month Storage, *J. Food Sci.*, 50, 178 (1985).
7. Ross, E. W., *et. al.*, A Time-Temperature Model for Sensory Acceptance of a Military Ration, *J. Food Science*, 52, 1712 (1987).
8. Ross, E. W., private communication.
9. Grabiner, F.R. and Prusik, T. The Application of Time-Temperature Indicators for Monitoring Temperature Abuse and/or Shelf Life of Foods, Food Preservation 2000 Conference Proceedings, Vol. 2, 589, 19-21 October 1993.
10. LeBlanc, D.I. Time-Temperature Indicating Devices for Frozen Foods, *J. Inst. Can. Sci. Technol. Aliment.* 21(3) 236 (1988).
11. American Society for Testing and Materials. ASTM F 1416, Standard Guide for the Selection of Time-Temperature Indicators, ASTM.
12. Ross, E.W. Relation of Bacterial Destruction to Chemical Marker Formation During Processing by Thermal Pulses. *J. Food Process. Eng.* 16, 247, 1993.
13. Kim, H.-J. and Taub, I.A. Intrinsic Chemical Markers for Aseptic Processing of Particulate Foods, *Food Technol.* 47(1), 91, 1993.

chapter thirteen

Quality management during storage and distribution

John Henry Wells and R. Paul Singh

Contents

Introduction

All food products, regardless of preservation technique, will eventually deteriorate. Moreover, the keeping quality of perishable foods, those that are preserved either by freezing or by storing at refrigeration temperature, is particularly sensitive to the environmental conditions in which they are stored. Fruits and vegetables that are marketed

0-8493-2646-X/97/$0.00+$.50
© 1998 by CRC Press LLC

as fresh products, for example, require refrigerated conditions to limit the biological functions of respiration (Wills et al., 1981). Meats, fish, and poultry products preserved by freezing need controlled temperature conditions to avoid a proliferation of resident microorganisms and to retard biochemical changes that result from enzymatic activity (Desrosier and Tressler, 1977). Other primary factors contributing to quality maintenance are initial product composition and quality, processing techniques, and the packaging materials and processes (Fennema et al., 1973; Goodenough and Atkin, 1981). Inpackage gas composition and addition of preservatives are additional environmental factors that affect the keeping quality of perishable foods. No single factor, however, has a more pronounced impact on the quality of stored perishables than does temperature history. Consequently, maintenance of proper temperature throughout the entire food distribution chain is essential in order to deliver the highest quality product possible to the consumer.

Considerable research has been reported in the literature on the keeping quality and shelf life of perishable foods. The books by Van Arsdel et al. (1969) and Jul (1984) review the results of storage investigations on the keeping quality of frozen foods. These document the influence of storage temperature on the length of time that frozen fruits, vegetables, and meats may be stored. A comprehensive review of the keeping quality of fresh fruits and vegetables, dairy products, and other refrigerated foods is available in a book by Labuza (1982). All these reviews emphasize that cumulative storage time and temperature (i.e., the temperature history) is the single most important factor affecting keeping quality. Temperature history also influences the beneficial effects of a secondary treatment such as modified atmosphere storage in extending shelf life (Marshall et al., 1991). For most perishable foods, storage temperatures higher than recommended adversely reduce the length of time that these products can be held in storage.

Perishable food items that are exposed to variable temperature conditions during storage will experience deterioration rates dissimilar to those of items stored at constant temperature. The relationships between temperature history and the rate of food quality loss have been described with various mathematical expressions. Earliest interest in the mathematical modeling of food quality loss was motivated by the observation that frozen foods stored at fluctuating temperatures did not have the same shelf life as products stored at constant temperature conditions, even though the two storage environments had the same average temperature (Hicks, 1944; Schwimmer et al., 1955). Recent efforts in relating food quality loss to temperature history have been discussed under the heading of "shelf life kinetics." Researchers have adapted the Arrhenius equation (Lai and Heldman, 1982), an analogous approach to thermal death time (Labuza, 1979), and the Q-value technique (Schubert, 1977) toward estimating the food quality changes in well-defined chemical reactions such as vitamin loss or browning. Additionally, a wide variety of empirical relationships specific to a given set of product preparation and processing parameters have been suggested.

Food quality modeling is typically conducted at known temperature conditions, but can be extended to variable temperature conditions if used in conjunction with a digital data acquisition system to precisely record temperature history. However, computer-based data acquisition systems are difficult to utilize for in-transit monitoring of temperature, since electronic components are prone to failure under the extremes of temperature encountered as food products move through the distribution system. An alternative to utilizing real-time temperature monitoring systems is the use of time–temperature indicators.

Time–temperature indicators are not precise temperature recorders, but are monitors that exhibit a change in color (or another physical characteristic) in response to temperature history. Wells and Singh (1985) classified time–temperature indicators as either partial- or

Table 1 Studies Correlating Food Products and TTI Response

Product	Reference
Pasteurized whole milk	Mistry and Kosikowski (1983)
	Cherng and Zall (1989)
	Grisius et al. (1987)
Ice cream	Dolan et al. (1985)
Frozen hamburger	Singh and Wells (1986)
Chilled cod fillets	Tinker et al. (1985)
Refrigerated ready-to-eat salads	Campbell (1986)
Frozen bologna	Singh and Wells (1986)
Whole milk, UHT	Zall et al. (1986)
Refrigerated orange juice	Chen and Zall (1987a)
Pasteurized milk	Chen and Zall (1987b)
Pasteurized cream	Chen and Zall (1987b)
Cottage cheese	Chen and Zall (1987b)
Pasteurized whole milk	Grisius et al. (1987)
Frozen strawberries	Singh and Wells (1987)
Tomatoes, lettuce	Wells and Singh (1988c)
UHT-sterilized milk	Wells and Singh (1988b)
Cottage cheese	Shellhammer and Singh (1991)

From Taoukis, P.S., et al. 1991. Food Technol. 45(10):70-82. With permission

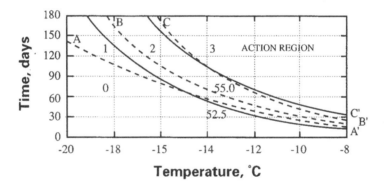

Figure 1 Full-history time–temperature indicator action diagram for frozen hamburger rancidity and I-POINT time temperature monitor model 3015. Color change ranges are denoted by 0, 1, 2, and 3; they are separated by dotten lines A–A', B–B', and C–C'. Rancidity contours are shown as solid lines and are given for rancidity values of 52.5 and 55.0 (From Wells, J.H., et al. 1987. J. Food Sci. 52(2):436-439, 444. With permission.)

full-history indicators, depending upon their response mechanism. Full-history indicators respond independently of a temperature threshold, whereas partial-history indicators are temperature dependent and do not respond unless a temperature threshold is exceeded. A full-history time–temperature indicator can be used to monitor the cumulative temperature exposure during storage and distribution, and can provide a means for comparing items that have been exposed to different temperature histories.

The responses of time–temperature indicators have been related to quality changes in several frozen and refrigerated foods (Table 1). Wells et al. (1987) depicted the time–temperature-quality relationship as a plot superimposed on a scaled response of a full-history time–temperature indicator at known time–temperature exposures. This depiction was the first attempt to utilize time–temperature indicators for assisting in the management of perishable food inventories. The correspondence between rancidity and monitor response in an indicator action diagram, as shown in Figure 1, suggested that the response of time–temperature indicators could be used in the management of perishable inventory. Wells (1985) discussed how a partial-history time–temperature indicator could be used in the context of inventory management.

Inventory management policies that schedule the issue priority for a stockpile are traditionally based on the elapsed storage time of a product. These policies do not account for nonuniform deterioration within a perishable stockpile. Typically, such a perishable stockpile would contain items of different ages and in various stages of deterioration, since varying temperature conditions may have been encountered during storage and distribution. The time-based inventory issue policies, such as first-in first-out (FIFO) and last-in first-out (LIFO), will not produce an optimal stock issue priority for products in inventory that have different states of deterioration. Alternative issue criteria, based on a quality criterion such as the estimated quality level or the remaining shelf-life of a product, are feasible if there exists a means to monitor temperature history. Utilizing time–temperature indicators to predict the extent of quality change in a product based on temperature history provides the link for implementing a quality-based criterion for inventory management. An appropriate criterion for issue of perishable products from an inventory stockpile and the utilization of time–temperature indicators in such an inventory management strategy are described in this chapter.

Criteria for evaluating food quality

Characteristics of perishable food quality

Natural deterioration processes eventually render all food products unsuitable for their intended purpose. The processing and handling techniques used in the food industry are designed to retard product deterioration by stabilizing product composition and reducing the chemical, enzymatic, and microbial changes that normally occur in unprocessed commodities. These common processing techniques can extend the life of perishable products, but cannot fully arrest the physiological and biochemical degradation within a commodity nor eliminate microbial contaminants that reside on a host product. Any discussion of perishable food quality must include the considerations of consumptive safety, product composition and physical properties, chemical and enzymatic activity, and microbiological interaction and growth.

The term "quality" is used as a gross measure of the state of deterioration that has occurred in a food. While "quality" has no scientific meaning of its own, from the perspective of the consumer, "quality" can be referenced to specific desirable characteristics or attributes that are inherent to a food (Schutz et al., 1972). That is, the sensory expectations established by an individual for a food product can be expressed in terms of the degree of the presence (or absence) of desirable (or undesirable) characteristics within the product. Thus, an item with a greater amount of a desirable characteristic would be perceived to be a higher-quality product, whereas an item with a lesser amount of that same characteristic would be considered a lower-quality product.

There are strong arguments that the aggregate of food quality is most appropriately defined by sensory perception, since perishable food is destined for human consumption. Furthermore, each physical, chemical, enzymatic, and microbial change that is important in the aggregate of perishable food quality should find an expression in terms of changes in a single (or multiple) sensory characteristic(s). For example, lipid oxidation in meats and fish (a change in the chemical and physical properties of a product) is expressed in the development of rancid off-flavors as determined by sensory analysis. Since many of the component changes that contribute to an overall sensorial quality of food are temperature dependent, examination of changes in specific attributes of food quality with respect to temperature should correlate with overall sensory changes with respect to the same temperature exposure. Sensory techniques, however, cannot be used as the sole source of food quality monitoring; they must be augmented with an objective means of microbiological and toxicity testing for proof of food safety.

The quality of a perishable food (the criterion by which consumers judge food to be acceptable or not) may be considered as a combination of many distinctive sensory attributes, one or more of which may change during storage. The resulting changes in prominent sensory attributes eventually lead to consumer rejection of the product. Identification of the sensory attributes that change and the quantitative definition of these attributes provide a means for monitoring time–temperature-related quality changes. The changes in defined sensory characteristics that are important to consumer acceptance must be identified and monitored. An appropriate methodology would include identification of desirable attributes of food quality, the quantification of the magnitude of a characteristic, and the scoring of the relative importance of an attribute to acceptance of the product by consumers. Various specific sensory evaluation methodologies could be used to study the extent of time-dependent changes in an attribute of food quality (Amerine et al., 1965). Additionally, studies of time-dependent quality changes in foods (i.e., food storage studies) have formed the basis for establishing the shelf life of a food product (Dethmers, 1979).

Shelf life of perishable foods

Much of the research undertaken to evaluate the quality of perishable food has been aimed at identifying the length of time necessary for overall quality changes to result in an unacceptable product: the product shelf life. For frozen foods, the definition of shelf life suggested by the International Institute of Refrigeration is practical storage life (PSL), "the period of frozen storage after freezing of an initially high quality product during which the organoleptic quality remains suitable for consumption or the process intended" (International Institute of Refrigeration, 1972). This definition is generally considered to mean the period of time during which sensory changes within a product cannot be detected by an untrained consumer panel. A host of specific sensory procedures and statistical techniques have been developed to define product shelf life in term of overall quality differences. These definitions employ sensory difference tests (discrimination methods) and strict statistical criteria for shelf life failure (Amerine et al., 1965).

One popular shelf life failure procedure is just noticeable difference (JND). Shelf life, as defined by JND procedures, is the earliest time that a difference in quality between experimental and control samples can be detected by a predetermined number of trained sensory panelists (Van Arsdel et al., 1969). In this type of test, the noticeable difference between experimental and control samples is determined for the product considering all things and not just a specific sensory attribute. This type of difference testing is usually terminated and the product is declared to be at the end of its shelf life when a difference between control and experimental samples is observed by a sensory panel at a predetermined level of statistical significance. That level can be specified so as to call attention to a noticeable difference that may be of commercial significance if the product had been introduced in commerce (Dethmers, 1979).

The results of storage studies that use difference techniques provide useful information on the expected length of time a product can be stored (i.e., an end-point measure), but they are of limited value in determining the way a quality change takes place. Difference testing procedures do not provide information about either the extent to which the quality change occurs or the rate at which it occurs. Such data can be obtained through periodic measurement of selected sensory attributes over the storage life of a product. The aim of a sensory testing methodology for monitoring changes in perishable food quality should include procedures to measure the rate and extent of changes while remaining sensitive to noticeable changes that could have an economic impact on the industry.

Predicting changes in food quality

A chemical kinetics approach to food quality modeling has been advocated by Kwolek and Brookwalter (1971) and Labuza (1984) as the most general and widely applicable mathematical modeling techniques to describe the influence of temperature on the rate of quality loss. With the kinetic approach, the rate of quality loss is expressed as an exponential function of the reciprocal of absolute temperature (i.e., the Arrhenius relationship). An Arrhenius relationship was recommended by Saguy and Karel (1980) for food-quality modeling; however, other researchers (Moreno, 1984) have preferred modified forms of the model with additional parameters. As researchers gain a further understanding of the pathological, biochemical, and other mechanisms that contribute to food deterioration, further attention will likely be given to increasingly complex mathematical models that describe changes in food quality.

The Arrhenius relationship is a two-parameter mathematical expression that describes the rate of a chemical reaction as a function of absolute temperature. The parameters of the model are referred to as the preexponential factor and the activation energy. The preexponential factor is the magnitude of the reaction rate independent of temperature, and the activation energy describes the temperature sensitivity of a reaction (Wells et al., 1987; Wells and Singh, 1988b). Lai and Heldman (1982) derived a methodology to determine the value of the activation energy of food quality losses from shelf life data at known storage temperatures. Lai and Heldman (1982) documented the values of activation energy for the quality changes in a number of frozen foods. The activation energies for quality changes in various refrigerated and semiperishable foods have also been published (Labuza, 1982).

Wells and Singh (1988c) discussed the need for obtaining kinetic information about changes in frozen and refrigerated food quality. They proposed the use of sensory rating methods, specifically the *deviation-from-reference* technique, as a possible method of gathering kinetic information. Sensory rating method can be utilized for estimating the kinetic parameters for food quality changes observed during isothermal storage. In turn, these kinetic models can be used to predict food quality changes under varying temperature histories (Wells and Singh, 1988b). The use of Arrhenius models in conjunction with computer simulation to predict food quality changes during frozen storage has been demonstrated by Singh (1976) and Wells and Singh (1989).

Monitoring food quality with time–temperature indicators

A review of several storage study investigations using refrigerated, frozen, and shelf-stable foods have confirmed the use of full-history time–temperature indicators as food quality monitors (Taoukis et al., 1991). Research has provided evidence that time–temperature indicators can be used with various perishable and semiperishable foods, including meat, fish, dairy, and bakery products (Mistry and Kosikowski, 1983; Singh et al., 1984; Singh et al., 1986). In these investigations, the responses of several indicator models were correlated with sensory and other measures of food quality, when both indicator and food were exposed to the same temperature conditions. Grisius et al. (1987) reported statistical correlations between the microbial changes in pasteurized milk and full-history

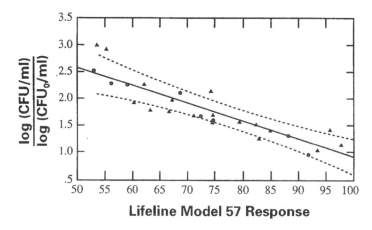

Figure 2 Plot of normalized total count enumeration and response of Life Lines indicator model 57 for constant (▲) and variable (●) temperature treatments. The relationship between normalized count and indicator response is given by: Y = –0.033X + 4.249. (From Grisius, R., et al. 1987. J. Food Process. Preserv. 11:309-324. With permission.)

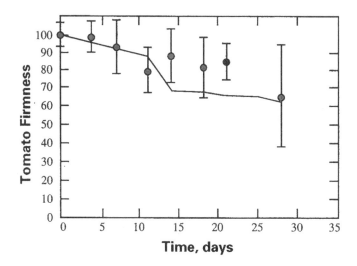

Figure 3 Comparison of the sensory (●) scores of tomato firmness to the scores predicted (solid line) from the mean response of the Life Lines Indicator Model 57 during variable temperature storage. (From Wells, and Singh, 1988).

indicator response (Figure 2). Wells and Singh (1988b) compared the observed changes in sensorial qualities of fresh tomatoes over time with those predicted by indicator response (Figure 3).

Stock management concepts for perishable inventories

Inventory management strategies for perishable foods

Consumer demands require management strategies that promote delivery of perishable products with consistently high quality. Important management strategies that have been adopted to fulfill this goal include (1) temperature control; (2) modified atomsphere storage; (3) grading, packaging, and other quality assurance standards; and (4) inventory management and stock rotations.

Inventory management and stock rotation typically rely on time-based criterion such that items within an inventory stockpile are scheduled for distribution according to the length of time that an item has been in storage. As a demand arises for items to be distributed from inventory, the stockpile is issued according to a ranked priority. Examples of such procedures include (1) shipment or disposal of inventoried product held in storage longer than some specified time (i.e., a stock rotation policy), or (2) shipment of inventory items in order of priority beginning with the oldest (or youngest) items in storage (i.e., an inventory issue policy). The two most common inventory issue policies used to establish the priority in which a stockpile will be distributed are:

1. first-in, first-out (FIFO) policy, and
2. last-in, first-out (LIFO) policy.

The FIFO policy requires the oldest item within a stockpile to be issued first, and the LIFO policy allows the youngest item on hand to take highest priority. Both policies are based on the age of an item (e.g., the time that a product is retained in storage) regardless of conditions that might render a product otherwise unsuitable for distribution. An alternative management strategy would be to implement a quality-based policy with the management objective of issuing consistently high quality products. For such a consideration, one must suppose that the shelf life of perishable food can be defined in terms of some threshold level of quality beyond which a consumer would no longer have a preference based on the perceived quality of the product.

Classic inventory depletion problem

The operations research literature related to perishable inventory management has addressed two fundamental problems: the inventory depletion problem and the inventory replenishment problem. The order in which items are dispersed from an inventory stockpile (i.e., the decision on which items are to be issued from storage) is considered in the inventory depletion problem, and the procedure by which the issue decision is made is referred to as the inventory issue policy (Bomberger, 1961). The inventory depletion problem is generally treated independent of other inventory management concerns, such as the regular balancing of stock levels to minimize system operating costs. These concerns are addressed in the inventory replenishment problem (Silver, 1981; Nahmias, 1982), which is beyond the scope of the chapter.

The inventory depletion problem was formulated by Derman and Klein (1958) as a way of determining an optimal sequence for removing items from a stockpile with units of varying ages. It was assumed that an item issued from the stockpile had a field life that was a known function of the age of the item, and that an item was issued in response to a specific demand when previously issued items had expired or been consumed. The assumptions in the problem formulation meant that the total field life of the stockpile was dependent upon the sequence in which items were removed from the stockpile. An inventory issue policy was considered an optimal policy when the total field life of the entire stockpile was maximized.

The classic inventory depletion problem was formulated based on the assumption that field life of the stockpile was a known function of age. When applied to perishable food products that have been exposed to differing temperature histories, this assumption is invalid and, therefore, so are the resulting management strategies for the inventory depletion problem. Wells and Singh (1989) advocated that quality management should be the aim of an inventory issue policy for perishable foods. Inventory issue policies based on the quality of items within the stockpile—rather than the age of stockpile items—is the most appropriate issue criterion for perishable foods.

Time-based criterion for perishable inventory management

Perishable inventories may be considered to exhibit a utility function that describes the deterioration of an item as a function storage time (i.e., a deterioration function) (Derman and Klein, 1958). The specific mathematical form of a deterioration function constrains the type of age-based (or time-based) issue criterion that is optimal for the inventory depletion problem. The FIFO and LIFO issue policies were devised as the optimal solution for the inventory depletion problem based on the uniform deterioration function of the perishable inventory under examination (Derman and Klein, 1959; Bomberger, 1961; Pierskalla, 1967a and 1967b; Nahmias, 1974; Albright, 1976). Perishable inventories that exhibit concave deterioration functions are optimally issued under a LIFO policy, and stockpiles that deteriorate with convex functions are optimally issued with the FIFO policy (Pierskalla and Roach, 1972). The optimal use of these policies hold only for stockpiles where all items deteriorate with the same uniform deterioration function. The FIFO and LIFO inventory issue policies are time-based strategies, prioritizing inventory items for issue, based on the total elapsed time that an item has been in storage (i.e., product age). A time-based issue criterion does not compensate for items with an inventory stockpile that undergo differing deterioration functions based on outside, uncontrolled factors, such as variations in temperature history.

First-in first-out rationale

Traditionally, the FIFO issue policy is used when issuing perishable foods from frozen or refrigerated storage. However, if zero- or first-order kinetic models are used to describe food quality change, the deterioration function is either a decreasing linear or decreasing concave decay function, respectively. Viewing this in terms of the classic inventory depletion problem, the optimal policy for maximizing the total life of the stockpile would be the LIFO policy. The widely used FIFO policy for perishable food suggests that there is an alternative interpretation for the inventory depletion problem, since the LIFO policy is considered the optimal solution for maximizing the total life of the stockpile with items that have concave deterioration functions.

The conflict between the theoretically optimal LIFO policy and the widely used FIFO policy can be understood by examining how each policy affects the age of an issued item. Implementation of the FIFO policy gives highest issue priority to the oldest items within a stockpile, while the LIFO policy gives priority to the most recently processed items. If a LIFO issue policy (the theoretically optimal policy) was placed into effect, it is likely that a portion of the accumulated product within the stockpile (the perpetual inventory) would never be distributed, because any recently manufactured product would preempt the issue of any older item. In this case, perishable foods in the perpetual inventory would eventually become unsuitable for consumption and require disposal (stockpile obsolescence). On the other hand, the FIFO issue policy does not have the problem of stockpile obsolescence, because the oldest items within the inventory always have the highest priority for issue.

When all items within an inventory stockpile have the same temperature history, and thus deteriorate with the same uniform deterioration function, use of the FIFO issue policy will outperform the LIFO issue policy with respect to the issuance of products with uniform quality. That follows, because when all items within the inventory stockpile deteriorate in the same manner, the use of a FIFO issue policy gives rise to the shipment of items that have the most consistent quality.

As stated previously, research into the keeping quality of perishable foods has firmly established a direct relationship between the rate of quality change and the storage temperature. Thus, when elevated or variable temperature exposures occur during the history of an item prior to being placed in the stockpile, a time-based issue policy (such as FIFO or LIFO) is unable to compensate for the nonuniformity in deterioration functions of the stockpile items. As a result, the consistency in the quality of the product distributed from the stockpile may be compromised. This inconsistency is a serious drawback, especially for foodstuffs that undergo indirect movement from manufacture to consumer and therefore risk possible exposure to uncontrolled or irregular temperature conditions.

Using an Arrhenius model to predict the quality changes in frozen broccoli stored at fluctuating temperatures, Wells and Singh (1989) demonstrated that items issued under a FIFO policy exhibit a more consistent level of quality at time of issue than do items stored under the same conditions that were issued under a LIFO policy. It was hypothesized that a quality-based criterion for determining the issue priority for perishable food should include the objective of minimizing variations in product quality. With food products, a consumer likely would indicate a preference for items of consistent quality from one purchase to the next over a product that has a history of inconsistent quality. Additionally, consumers will only purchase items that have not deteriorated beyond a level of quality that is considered unacceptable.

For refrigerated and frozen foods, efforts to maintain consistent quality products include careful raw material selection and strict quality control procedures during processing. However, there is no analogous procedural framework in inventory management systems that assist in ensuring that consistent quality products are issued from perishable stockpiles. A FIFO issue policy could inadvertently retain an item that, during an additional period of storage, would become grossly inferior in quality to the issued item if the issued item had remained in storage for the additional period of time. Under such a scenario the variation in product quality at time of issue would be increased. In effect, the use of an inappropriate inventory issue policy could negate measures taken during processing to improve the consistency of quality.

In contrast to the classic inventory depletion problem, an appropriate inventory issue policy for perishable food would seek to manage the stockpile quality in such a way that items would be distributed with the most consistent level of quality. Perishable foods that are exposed to variable temperature conditions will have deterioration functions that are dissimilar to those of items stored at a constant temperature. Only in situations where items within an inventory stockpile have completely uniform deterioration functions is the use of time-based issue policies appropriate. An alternative inventory management criterion would be to determine issue priority based on *observed (or estimated) food quality rather than elapsed time in storage.*

Quality criterion for perishable inventory management

Quality-based interpretation of shelf life

The shelf life of a perishable food is defined by the period of storage after processing that an initially high-quality product remains suitable for consumption (International Institute of Refrigeration, 1972). This implies that there is some limiting threshold of quality (or deteriorative change) corresponding to the end of a product's shelf life. Assuming that the time- and temperature-dependent changes in quality characteristics can be satisfactorily predicted by a deterioration function, it would follow that shelf life could be expressed mathematically in a quality-based interpretation.

For the quality-based interpretation, the elapsed storage time that defines an item's shelf life, $t_{Q,ref}$, occurs when the limiting threshold quality, Q_{th}, is reached. Implied in the shelf life data reported in the literature is that a product is stored at a recommended reference temperature. Thus, for any level of quality, Q_n, between the initial and threshold levels, shelf life may be expressed as the sum of an equivalent age and remaining shelf life (Wells and Singh, 1989.) This relationship is denoted as $t_{Q,ref} = A_{e,n} + A_{r,n}$, where $A_{e,n}$ and $A_{r,n}$ are the equivalent age and remaining shelf life, respectively.

The equivalent age, $A_{e,n}$, represents the length of time necessary to bring about the same level of quality, Q_n, if the product had been stored at a reference temperature; while the remaining shelf life, $A_{r,n}$, represents the length of time for food quality to change from the observed level, Q_n, to the threshold level, Q_{th}, if the product is stored at the same reference temperature. The remaining shelf life and equivalent age are complementary functions of quality and are related to temperature history (time and temperature) by the deterioration function of the perishable product.

Estimation of remaining shelf life

A generalized mathematical model to predict food quality and remaining shelf life from the response of a full-history time–temperature indicator was presented by Wells and Singh (1988b). In the derivation, it was presupposed that an Arrhenius model would adequately describe both the indicator response and the changes in food quality attribute within the range of storage temperatures to which indicator and product were exposed. Also, the prediction of remaining shelf life was considered to hold true only for situations of continuous quality deterioration. That is, the quality prediction model (and remaining shelf life calculation) would not be valid in situations where temperature (or other conditions) cause a discontinuity in quality deterioration. Examples of a discontinuity in quality deterioration include thawing of frozen products, excessive proliferation or sporulation of microbial contaminants, changes in product composition caused by protein denaturation, and mechanical injury due to product damage or loss of package integrity.

Shortest remaining shelf life inventory issue policy

Wells and Singh (1989) and Taoukis et al. (1991) discussed an alternative to the time-based FIFO issue policy based on the maximum expected remaining shelf life of a product. The remaining shelf life was calculated from quality predictions estimated by time–temperature indicator response during storage. Variously called the shortest remaining shelf life (SRSL) or the least shelf life first-out (LSFO) inventory issue policy, the object of the quality-based inventory issue policy is to prioritize stockpile distribution beginning with the items that have the least quality reserve. Since food quality continually changes throughout storage and distribution, the maximum expected remaining shelf life will also continue to change largely as a consequence of the temperature history that the item has experienced.

Scheduling stockpile issue priority in the sequence from shortest to longest remaining shelf life, the shortest remaining shelf-life (SRSL) issue policy would retain items with the greatest amount of quality reserve within the stockpile and expedite issue of items that would be less tolerant (or perhaps become unacceptable) if they remained in storage for an additional period of time. In effect, the SRSL issue policy would allow inventory items that have undergone the greatest amount of quality change to move most rapidly through the remainder of the food distribution system. The SRSL and LSFO issue policies are identical, because items having the least (remaining) shelf-life would be the first out of storage. A comparison of the SRSL inventory issue policy to the FIFO and LIFO policies was given by Wells and Singh (1989).

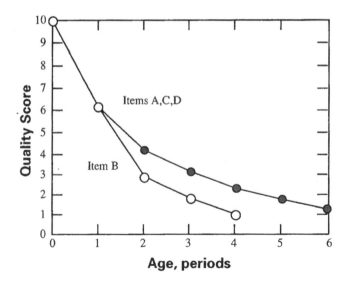

Figure 4   Example of a nonuniform quality deterioration function for items (○ — A, C, D; ● — B) in a perishable inventory stockpile. (From Wells, J.H. and Singh, R.P. 1989. J. Food Process. Preserv. 12:271-292. With permission.)

The utility of the SRSL issue policy is seen by considering the deterioration functions of several items that are of the same age (Figure 4). One item (Item B) has undergone some heat abuse during storage or distribution, while a second item (Items A, C, and D) has been stored at a recommended isothermal condition. Without a means of detecting temperature history, and subsequently predicting the product quality (or remaining shelf life), neither item would be given higher priority for issue than the other. However, the remaining shelf life of the heat-abused product (Item B) is greatly reduced as compared to the remaining shelf life of product items stored at the recommended reference temperature (Items A, C, and D). Prioritizing inventory issue based on the estimated remaining shelf life, predicted from time–temperature indicator response, would expedite movement of the heat-abused product (Item B) from the stockpile. In turn, this would reduce the variation in the product quality issued from the stockpile even when remaining stockpile items are issued at a later point in time (compare quality levels for Items A, C, and D subsequent to issue of Item B from the stockpile).

SRSL issue policy and shelf life dating

Shelf life dating remains a crucial decision in the management of perishable food inventory. Manufactured products are stamped with a *use by* or *pull date* to reflect the estimated time when the shelf life of a product will be reached. These dates are stated, given some conservative assumption of storage temperature and handling procedure. Management of perishable inventories then rely on high rates of inventory turnover, assisted by price discounting if necessary, to ensure that items are not stored beyond their pull dates. The use of conservative pull-dating likely results in the loss, disposal, or deeply discounted sales of significant amounts of high quality product. Such disposal practices are conducted without the benefit of a means of estimating if quality changes are sufficient to deem that product performance would warrant removal from inventory.

Inasmuch as management of perishable inventories with the SRSL policy establishes issue priority on the remaining shelf life, the SRSL policy could be used as an objective means for establishing a product pull date. Pull dates are established on the basis of an assumed storage condition and a stated limiting performance threshold for the product.

Since the remaining shelf life is calculated as the difference between a current estimate of quality and a limiting threshold, such a procedure could be used to determine the time remaining to reach the threshold that defines the pull date. A remaining shelf life of zero (or a negative remaining shelf life) would mean that the objective quality threshold had been reached (or exceeded). A remaining shelf life of zero would imply that there is no difference between the quality of a product in its current state and the threshold quality level requiring a product to be pulled from stock. A negative remaining shelf life would mean that the product quality has already deteriorated below that considered tolerable.

Quality-based management of perishable inventories

Food storage and distribution systems

The typical storage and distribution system for movement of perishable food from manufacture to consumption was detailed by Wells and Singh (1992). The perishable distribution chain encompasses the movement of food from manufacturer (the point in the processing of a food such that the item is suitable for retail purchase) to consumer (a retail outlet where products are delivered to the final user/consumer). Little control can be exerted by processors and suppliers over the destiny of perishable foods beyond the retail outlet other than point-of-purchase information and other consumer education measures. Within a distribution center and other storage locations throughout the distribution chain, inventory stockpiles accumulate as product "lots" are received in excess of the number of "case lots" or stock keeping units (SKUs) that are distributed. Also, quantities of inventory are held in reserve to supply multiple locations either on demand or because of seasonable manufacturing. Transaction records of the inventory stockpile form the primary source of information on which to base scheduling of inventory distribution. Since perishable food inventories are destined for retail consumption, a quality-based management strategy must be oriented to deliver the highest, most consistent quality products possible.

Use of time–temperature indicators to manage food quality

The use of time–temperature indicators as food quality monitors has been demonstrated experimentally by several researchers. Furthermore, the mathematical relationship between time–temperature indicator response and food quality change has been detailed in Chapter 11 and provides the foundation on which product inventory could actively be managed with the use of time–temperature indicators rather than the elapsed storage time. Such a scheme would establish a framework to justify taking action to expedite the shipment of heat-abused products from the warehouse and would call attention to any segment of the distribution chain that was deficient in temperature maintenance procedures.

Distribution decision support systems for stockpile management

The mathematical nature of the SRSL inventory issue policy lends itself to the implementation of additional logistical and quality constraints that could be used in the context of a distribution decision support system (DDSS) (Wells, 1987). Inventory distribution from a perishable stockpile is based on an inventory issue policy in conjunction with stockpile transaction records or from physical observation of the stockpile to determine the quantity of each product lot on hand. The information is then synthesized into a form in which items in the stockpile can be prioritized for shipment.

Supplemental decision criteria regarding inventory quality can also be included in a DDSS. Constraints imposed by the logistics of distribution and/or premium quality standards, in addition to ranking the stockpile items for shipment priority, could be accommodated within the framework of the SRSL issue policy. For instance, in situations where logistics of delivery require lengthy travel time, stockpile items could be selected such that $A_{r,n} > A_{r,min}$, where $A_{r,n}$ is the remaining shelf life of the product to be issued and $A_{r,min}$ is the length of time that is required to deliver the shipment (plus any required length of storage at the destination). This constraint would ensure that the remaining shelf life of a product be greater than the length of time it takes to ship the inventory to its destination, plus the expected time prior to consumption of the product. Such a constraint, of course, assumes that no temperature mishandling will occur and that recommended reference temperature is maintained. Remaining shelf life predictions at the time of shipment would be based on the temperature history of the product as monitored by the time–temperature indicator; the remaining shelf life after shipment would be conditional on maintenance at the recommended reference temperature throughout shipment and storage. That is, the remaining shelf life depends on the actual storage temperature maintained subsequent to shipment and not on the temperature upon which the SRSL issue decision is based.

The SRSL inventory issue policy could also be used in conjunction with constraints placed on a threshold of premium quality. Since items with a longer remaining shelf life—and hence a higher quality level—may command a premium price at a retail outlet compared to items of a lesser quality, an inventory stockpile could be segregated into groupings of premium and nonpremium items by comparing the estimated quality to the limiting value of a premium quality threshold. Items then that met premium quality requirements could be distributed to preferred retail locations or otherwise marketed separately.

Implementation of an inventory management system to provide distribution decision support can be facilitated with advanced microcomputer systems (Wells, 1987; Singh and Wells, 1987). The microcomputer provides both the environment to conduct the calculation of remaining shelf life from time–temperature indicator response and the means to structure the information relevant to inventory management. Additionally, telecommunication links to send time–temperature indicator responses from remote locations to a central base can be accomplished via a computer modem (Kral et al., 1988).

Menu-driven software has been developed by Singh and Wells (1987) for demonstrating a computer-based inventory management system, and the development of commercial systems were reported by Taoukis et al. (1991). Figure 5 shows a screen display by Wells and Singh (1987) indicating the provision for choice of inventory issue policy and supplemental distribution decision support criteria. Other features of this software include a means to calculate quality predictions based on time–temperature indicator response and the ability to predict the effect of known (or simulated) temperature histories on food quality changes.

Development of user-friendly software, designed to function as an add-on to existing transaction recording and inventory management software, should remain a priority. Such systems must incorporate the SRSL issue policy as well as conventional issue policies. The requirement of computer-based inventory management systems includes (1) industry standard data and file transfer protocol, (2) a high level of compatibility with commercial databases used in tracking inventory transactions, and (3) a telecommunications interface for remote access and data transfer. Software packages for perishable inventory management using a hand-held scanner to monitor time–temperature indicator response must also be examined with respect to statistical sampling plans and normal variations within food products (Wells, 1987). A quality-based inventory management system promises to improve the quality and consistency of products delivered to the consumer, and such a system has been demonstrated to be feasible using time–temperature indicators.

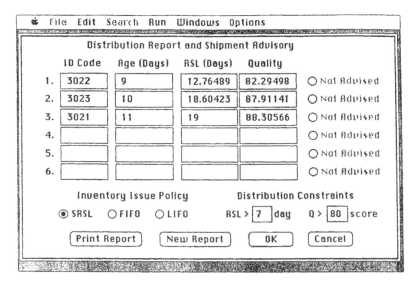

Figure 5 Screen graphic of the results of a decision support calculation showing the SRSL issue priority with no shipment advisory.

Conclusions

Full-history time–temperature indicators can be used to provide important information regarding unusual or fluctuating temperature conditions during storage and distribution of perishable foods. Methods have been established that utilize indicator responses to predict changes in food quality, thus allowing inventory management decisions to be based on food quality rather than elapsed storage time. Introduction of a quality-based inventory management strategy has the potential to enhance the consistency in the quality of perishable foods delivered to the consumer without compromising the overall level of quality acceptable to the consumer.

An important characteristic of the SRSL policy is that it retains the same issue order as the FIFO policy when all items deteriorate in the same manner, but compensates for nonuniform deterioration when evidenced by the time–temperature indicator. Using SRSL as a criterion for inventory management has the potential to improve the quality of perishable foods delivered to a consumer. Improvements in consistency of perishable food quality by implementing a quality-based issue policy could heighten the reputation of products that have established consistent quality standards in raw material selection and process quality control. Additionally, the SRSL policy could be used to establish objective criteria for shelf life dating that does not lead to inventory removal unless warranted by the magnitude of the estimated quality change.

References

1. Albright, S.C. 1976. Optimal stock depletion policies with stochastic lives. Manage. Sci. 22(8):852-857.
2. Amerine, M.A., Pangborn, R.M. and Roessler, E.B. 1965. Principles of Sensory Evaluation of Food. Academic Press, New York.
3. Bomberger, E.E. 1961. Optimal inventory depletion policies. Manage. Sci. 7(3):294-303.
4. Cambell, L.A. 1986. Use of a time-temperature indicator in monitoring quality of refrigerated salads. M.S. Thesis. Michigan State Univ., East Lansing.

5. Chen, J.H. and Zall, R.R. 1987a. Refrigerated orange juice can be monitored for freshness using a polymer indicator label. Dairy Food Sanit. 7(6):280-282.

6. Chen, J.H. and Zall, R.R. 1987b. Packaged milk, cream and cottage cheese can be monitored for freshness using polymer indicator labels. Dairy Food Sanit. 7(8):404-404.

7. Cherng, Y.S. and Zall, R.R. 1989. Use of time temperature indicators to monitor fluid milk movement in commercial practice. Dairy, Food Environ. Sanit. 9(8):493-443.

8. Derman, C. and Klein, M. 1958. Inventory depletion management. Manage. Sci. 4(4):450-456.

9. Derman, C. and Klein, M. 1959. A note on the optimal depletion of inventory. Manage. Sci. 5(2):210-213.

10. Desrosier, N.W. and Tressler, D. K. 1977. Fundamentals of Food Freezing. AVI Publishing, Westport, CT.

11. Dethmers, A.E. 1979. Utilizing sensory evaluation to determine product shelf life. Food Technol. 33(9): 40-42.

12. Dolan, K.D, Singh, R.P. and Wells, J.H. 1985. Evaluation of time temperature related quality changes in ice cream during storage. J. Food Process. Preserv. 9:253-271.

13. Fennema, O.R., Powrie, W.D. and Marth, E.H. 1973. Low Temperature Preservation of Foods and Living Matter. Marcel Dekker, New York.

14. Goodenough, P.W. and Atkin, R.K. 1981. Quality in Stored and Processed Vegetables and Fruit. Academic Press, London.

15. Grisius, R., Wells, J.H., Barrett, E.L. and Singh, R.P. 1987. Correlation of full-history time-temperature indicator response with microbial growth in pasteurized milk. J. Food Process. Preserv. 11:309-324.

16. Hicks, E.W. 1944. Note on the estimation of the effect of diurnal temperature fluctuations on reaction rates in stored foodstuffs and other materials. J. Counc. Sci. Indust. Res., Austr. 17:111-114.

17. International Institute of Refrigeration. 1972. Recommendations for the Procession and Handling of Frozen Foods, 2nd ed. IIR, Paris.

18. Jul, M. 1984. The Quality of Frozen Foods. Academic Press, New York.

19. Kral, A.H., Zall, R.R. and Prusik, T. 1988. Use of remote communications to transmit product quality information from polymer-based, time-temperature indicators. Dairy Food Sanit. 8(4):74-178.

20. Kwolek, W F. and Bookwater, G.N. 1971. Predicting storage stability from time-temperature data. Food Technol. 25(10):1025, 1026, 1028, 1029, 1031, 1037.

21. Labuza, T.P. 1979. A theoretical comparison of losses in food under fluctuating temperature sequences. J. Food Science 44(4):1162-1168.

22. Labuza, T.P. 1982. Shelf-Life Dating of Foods. Food and Nutrition Press, Westport, CT.

23. Labuza, T.P. 1984. Application of chemical kinetics to deterioration of food. J. Chem. Ed. 6(14):348-358.

24. Lai, D. and Heldman, D.R. 1982. Analysis of kinetics of quality change in frozen foods. J. Food Process Eng. 6:179200.

25. Marshall, D.L., Wiese-Lehigh, P.L., Wells, J.H. and Farr, A.J. 1991. Comparative growth of *Listeria monocytogenes* and *Pseudomonas ~fluorescens* on precooked chicken nuggets stored under modified atmospheres. J. Food Prot. 54(11):841-843, 851.

26. Mistry, V.V. and Kosikowski, F.V. 1983. Use of time-temperature indicators as quality control devices for market milk. J. Food Prot. 46(1):52-57.

27. Moreno, J. 1984. Quality deterioration of refrigerated foods and its time-temperature mathematical relationships. Int. J. Refrig. 7(6):371-376.

28. Nahmias, S. 1974. Inventory depletion management when the field is random. Manage. Sci. 20(9):1276-1283.

29. Nahmias, S. 1982. Perishable inventory theory: a review. Operations Res. 30(4):680-708.

30. Pierskalla, W.P. 1967a. Optimal issuing policies in inventory management -I. Manage. Sci. 13(5):395-412.

31. Pierskalla, W.P. 1967b. Inventory depletion management with stochastic field life functions. Manage. Sci. 13(11):8778.

32. Pierskalla, W.P. and Roach, C.D. 1972. Optimal issuing policies for perishable inventory. Manage. Sci. 18(11):603-614.
33. Saguy, I. and M. Karel. 1980. Modeling of quality deterioration during food processing and storage. Food Technol. 34(2):78-85.
34. Schubert, H. 1977. Criteria for application of T-TI indicators to quality control of deep frozen products. Science et Technique Froid IIF-IIR 1977-1:407-423.
35. Schutz, H.C., Damrell, J.D. and Locke, B.H. 1972. Predicting hedonic ratings of raw carrot texture by sensory analysis. J. Texture Stud. 3: 227-232.
36. Schwimmer, S., Ingraham, L.L. and Hughes, H.M. 1955. Temperature tolerance in frozen food processing: Effective temperatures in thermal fluctuating systems. Indust. Eng. Chem. 47(6):1149-1151.
37. Shellhammer, T.H. and Singh, R.P. 1991. Monitoring chemical and microbial changes of cottage cheese using a full-history time-temperature indicator. J. Food Sci. 56(2):402-405, 410.
38. Silver, E.A. 1981. Operations research in inventory management: A review and critique. Operations Res. 29(4):628-645.
39. Singh, R.P. 1976. Computer simulation of food quality during frozen food storage. Int. Inst. Refrig. Bull. Supp. 1976-1:197-204.
40. Singh, R.P. and Wells, J.H. 1986. Keeping track of time and temperature: New improved indicators may become industry standards. Meat Process. 25(5):41-42, 46-47.
41. Singh, R.P. and Wells, J.H. 1987. Development, evaluation and simulation of an automated stock management system utilizing time temperature indicators. Report prepared for U.S. Army Natick Research & Development Center, Natick, MA. May. (Contract No. DAAK60-85-C-0111.)
42. Singh, R.P., Barrett, E.L., Wells, J.H., Grisius, R.C. and Marum, W. 1986. Critical evaluation of time-temperature indicators for monitoring quality changes in perishable and semi-perishable foods. Report prepared for U.S. Army Natick Research & Development Center, Natick, MA. January. (Contract No. DAAK6084-C-0076.)
43. Singh, R.P., Wells, J.H., Dolan, K.D., Gonnet, E.J. and Munoz, A.M. 1984. Critical evaluation of time-temperature indicators for monitoring quality changes in stored subsistence. Report prepared for U.S. Army Natick Research & Development Center, Natick, MA. September. (Contract No. DAAK60-83-C-0100.)
44. Taoukis, P.S., Fu, B. and Labuza, T.P. 1991. Time-temperature indicators. Food Technol. 45(10):70-82.
45. Tinker, J.H., Salvin, J.W., Learson, R.J. and Empola, V.B. 1985. Evaluation of automated time-temperature monitoring system in measuring the freshness of chilled foods. IIF-IIR Commission C2, Do 4:286. Intl. Inst. of Refrig., Paris.
46. Van Arsdel, W.B, Coply, M.J. and Olson, R.L. 1969. Quality and Stability of Frozen Foods. Wiley-Interscience, New York.
47. Wells, J.H. 1985. Performance evaluation and application of time-temperature indicators to frozen food storage. M.S. Thesis. University of California, Davis.
48. Wells, J.H. 1987. Computer-based inventory management system for perishable foods. D.E. Dissertation. University of California, Davis.
49. Wells, J.H. and Singh, R.P. 1985. Performance evaluation of time-temperature indicators for frozen food transport. J. Food Sci. 50(2):369-371, 378.
50. Wells, J.H. and Singh, R.P. 1988a. Application of time temperature indicators in monitoring changes in quality attributes of perishable and semi-perishable foods. J. Food Sci. 53(1): 148-152, 156.
51. Wells, J.H. and Singh, R.P. 1988b. Kinetic approach to food quality prediction using full-history time-temperature indicators. J. Food Sci. 53(6):1866-1871, 1893.
52. Wells, J.H. and Singh, R.P. 1988c. Response characteristics of full-history time-temperature indicators for perishable food handling. J. Food Process. Preserv. 12:207-218.
53. Wells, J.H. and Singh, R.P. 1989. A quality-based inventory issue policy for perishable foods. J. Food Process. Preserv. 12:271-292.

54. Wells, J.H. and Singh, R.P. 1992. Application of time temperature indicators to food quality monitoring and perishable inventory management. in Mathematical Modelling of Food Processing Operations, S. Thorne, ed. Elsevier Science Publishers, Essex, England.

55. Wells, J.H., Singh, R.P. and Noble, A.C. 1987. A graphical interpretation of time-temperature related quality changes in frozen food. J. Food Sci. 52(2):436-439,444.

56. Wills, R.B.H., Lee, T.H., Graham, D., McGlasson, W.B. and Hall, E.G, 1981. Postharvest. AVI Publishing, Westport, CT.

57. Zall, R.R. Chen, J. and Field, S.C. 1987. Evaluation of automated time-temperature monitoring systems in measuring freshness of UHT milk. Dairy Food Sanit. 6(7):285-290.

List of Abbreviations and Acronyms

DDSS Distribution Decision Support System
FIFO   First-In First-Out
JND   Just Noticeable Difference
LIFO   Last-In First-Out
LSFO   Least Shelf-life First-Out (same as SRSL)
PSL   Practical Storage Life
SKU   Stock Keeping Unit
SRSL   Shortest Remaining Shelf Life (same as LSFO)

chapter fourteen

Freezing preservation of fresh foods: quality aspects

David Reid

Contents

Introduction

The freezing process is of considerable importance to the food preservation industry. Freezing preservation is based on the concept that for most chemical reactions, the rate of change is slower at lower temperatures. Indeed, in general terms, the Arrhenius relationship indicates the dependence of reaction rates upon temperature.[1,2] For this reason, refrigeration is useful in preserving the freshness of food in storage, by reducing the rate of chemical reactions and reducing the rate of growth of microorganisms. Since this relationship appears to have general validity, rates would be expected to be even slower in frozen storage, and one might ask the question why frozen storage has not been employed for long-term storage of all foods. This chapter will attempt to answer this question, and will indicate the scientific basis for the preservation of fresh tissues by freezing. The main factors that influence the long-term changes in quality during frozen storage will be discussed.

In order to better appreciate the factors that are important for the long-term frozen storage of food tissues, it is necessary to consider the structural features of both plant and animal systems, since these behave differently in the freezing process. It may be necessary to include a prefreezing step, which might stabilize a tissue material prior to freezing. An example of a preprocessing step would be blanching, which inactivates enzymes and is necessary to stabilize some vegetables for frozen storage.

Technological basis of freezing

As has already been indicated, the lowering of the temperature to the freezing point of a food tissue tends to extend shelf life, with some exceptions. The exceptions are primarily associated with the potential for chill damage, whereby the metabolic processes of cells are disrupted by low temperature and abnormal metabolism ensues. The proposed mechanisms of chilling damage often invoke phase change in the lipids of the cell membranes and conformational change in the cell biopolymers as factors that contribute to the disruption of metabolism.[3,4] These physical changes within the cell are believed to lead to changes in the cell's chemistry and biochemistry.

Since there appears to be an adverse effect resulting from physical change in certain components (e.g., lipids) of many cellular systems at temperatures above the freezing point of water, we might expect a similar potential for adverse change below this temperature. Indeed, there are clear signs of damage (similar to chill damage) at such temperatures, perhaps as a consequence of a phase change in some membrane lipids.[5] The phase change of water from liquid to ice, as expected, has important consequences, as can be seen by a comparison of the effects of low temperature on both supercooled (i.e., with no ice formation) tissues and the same tissues frozen (i.e., with ice formation) at the same temperatures. In the supercooled state, food systems may be chill damaged. The frozen systems frequently exhibit much more in the way of disruption of metabolic processes.[6]

In the field of cryobiology, it is often found that cryoprotective compounds have to be added to tissue systems to enable them to retain viability through a freeze–thaw cycle. In contrast, supercooled products stored at the same temperature do not require the addition of cryoprotectants.[7]

As has been discussed elsewhere,[8] the requirements for food freezing are different from those of cryobiology in that high-quality attributes must be maintained. This does not necessarily equate with viability. The mechanisms of change during food freezing need to be considered in order to understand the constraints on maintaining acceptable quality in frozen foods.

The first important factor in food freezing is the rate of change of temperature. It is dependent in part upon external conditions, such as the equipment used, and upon the size of the object to be frozen. There has long been controversy as to how important the rate of freezing is to product quality. Some researchers have indicated that for vegetables the freezing rate is very important for maintaining acceptable quality, while others point out that the freezing rate of animal tissue is of little importance. Jul considers freezing rate to be of minor importance, drawing most of his examples from animal tissue systems.[9]

When considering freezing rate, most investigators focus on the rate of change of temperature at some selected locations, and define a *fast rate* and a *slow rate* of freezing. Insofar as the rates of heat removal, and advance of the 0°C isotherm (or freezing interface at the actual freezing temperature) are linked to rate of change of temperature through the relationships of heat transfer for any particular geometry this procedure would appear to be appropriate. However, it neglects an important aspect of the freezing process. To understand this aspect, it is necessary to consider the freezing of tissues in more detail.

An important characteristic of food tissue systems is the existence of cell walls and cell membranes as barriers that separate the interior of the cell from the external environment. It has been shown in many studies that this barrier can block the propagation of ice crystals.[11] The consequences of this barrier are of great importance to the freezing process for food.

Consider the situation when the cooling process has produced ice crystals in the external environment of a cell. As the temperature continues to decrease, the concentration of the fluid medium between the ice crystals increases. If the ice fails to penetrate into the interior of the cell, then an osmotic driving force will now exist between the aqueous interior phase of the cell and the unfrozen aqueous component of the exterior medium. Depending upon the permeability properties of the cell wall, water will transfer from the interior of the cell into the extracellular medium, there to form ice crystals. The rate of transfer of this water may be either sufficiently rapid to maintain the internal cell contents close to equilibrium with the unfrozen extracellular medium or much slower than this limiting rate.[12] Figure 1 attempts to illustrate the possibilities. If the osmotic transfer is slower than that required to maintain approximate balance, the internal aqueous medium will become increasingly supercooled. When a critical level of supercooling is exceeded, the ice will nucleate within these internal solutions and freezing will occur inside the cell.. Once this internal freezing has occurred, there no longer is any driving force to export water and to dehydrate the cell, since osmotic balance is now automatically maintained, with the "surplus" water within the cell transferred to the ice.

This description of the process by which a cell can seek to maintain a balance with the external medium leads to an operational definition of *slow* and *fast freezing. Slow freezing* is defined as taking place under conditions where the rate of cooling does not exceed the capability of the cells to export water. Only extracellular ice is formed. *Fast freezing* describes conditions for which the ability of the cell to export water has been exceeded. As a consequence, at some point, critical internal supercooling has also been exceeded and water has been nucleated within the cells. Slow freezing leads only to extracellular ice crystals, whereas fast freezing leads to both intracellular and extracellular ice. Slow freezing is indicative of potential cell dehydration. Since the permeability of cell walls and membranes and the potential magnitude of critical supercooling are properties of individual cell types, the actual conditions that correspond to either slow freezing or to fast freezing will differ from cell to cell. It should be noted that the ability of animal cells to resist internal ice formation is much less than that of many plant cells. Thus, the rates of change of temperature that produce only extracellular ice in many plant systems may still result in the formation of intracellular ice in animal tissue systems.

On the basis of this description, the source of some of the disagreements as to whether fast freezing leads to quality maintenance as compared to slow freezing (freezing rate being defined as rate of change of temperature) can be more easily understood. In plant tissue systems, the range of rates of change of temperature attained in freezing processes might lead to products with only extracellular ice, or to products with both intracellular and extracellular ice. For muscle tissue systems, similar freezing conditions more often would result in products with a significant proportion of intracellular ice. Another factor that has to be taken into consideration is the difference between the cells. Plant cells, in general, are characterized by moderately rigid to nonrigid cellulose cell walls with associated membranes. Shrinkage within the cells might be expected to result in major tissue damage as the interrelationship between cell wall and cell membrane alters. Animal cells, in particular muscle cells, do not have cell walls, but have flexible membranes as barriers.

The effects of freezing rate per se have to be evaluated immediately after freezing, as further change during frozen storage might obscure the effects of freezing rate. Studies have shown that, even where there is a clear effect of freezing rate on frozen product quality, this effect is not necessarily sustained through frozen storage.[13,14] The effect of

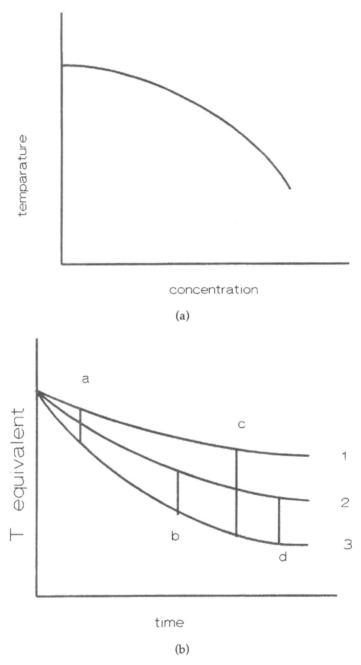

Figure 1 (a) A typical freezing curve, showing equilibrium freezing temperature as a function of a solute concentration in the aqueous phase (schematic). (b) Curve 3 illustrates the temperature–time history of a cooling experience. Curves 1 and 2 show the change in concentration of cell contents as water is removed by osmosis. This is indicated by plotting the equilibrium freezing temperature of the liquid aqueous phase (obtained from 1a). The distance between Curves 1 and 3 or Curves 2 and 3 illustrates the extent of supercooling for each system at any time. Vertical lines indicate freezing. At this point the composition of the unfrozen liquid phase becomes equal to that of the external unfrozen phase, illustrated by Curve 3. Curve 1 is typical of a system that loses water more slowly. In this system, the supercooling of the contents increases rapidly. The supercooling increases less rapidly for Curve 2.

Table 1 Heat Transfer Coefficients ($W·m^{-2}·K^{-1}$)

Air (convective)	5.5–17
Air blast	17–55
Plate freezer	35–120
Liquid nitrogen	115–570

storage temperature and of storage conditions of food tissues can overwhelm the differences in quality that initially are a result of differing freezing rates.

The principles and technology of the freezing process for food have been described elsewhere.[15,16] The product to be frozen is brought into contact with a medium that can remove heat. This medium, for example, might be cold air in an air-blast freezer. The efficiency of heat removal will depend upon air temperature, air velocity, and also upon the effectiveness of contact (for heat transfer) between the air and the product. The air will then be recooled by passing over the refrigeration coils. The design of the coils, the moisture load of the air, and other factors will all affect the efficiency of this process. Too high a moisture load can lead to icing up of the cooling coils, and a need for frequent defrosting of the coils.

Instead of cold air produced through a mechanical refrigeration system, a cryogenic fluid such as CO_2 or N_2 could be used as the heat transfer medium. These cryogenic refrigerants are at lower temperatures than the temperatures typically attained by the circulation of mechanically refrigerated air. A variety of systems exist to control the rate of delivery of cryogen and, hence, to control the cooling effect. Due to the lower temperature of the cooling medium, the rates of surface heat transfer can be higher. Direct contact of the product with refrigerated surfaces may also be employed for freezing (e.g., plate freezer). The plates are internally cooled by either a direct or an indirect refrigeration system. Unless the product has a geometry suited to intimate contact with a solid surf ace, this method can be inefficient for heat transfer. Table 1 lists typical heat-transfer coefficients for the different types of systems.[17] The higher the heat-transfer coefficient, the more rapid the removal of heat from the product, and hence the more rapidly the product may be frozen.

In addition to the freezing system, the size of the product will affect the conditions of freezing. The amount of heat that has to be removed depends upon the volume of the product, while the rate at which heat can be removed is a function both of the surface area of the product and of the heat transfer characteristics of the freezer. The freezing characteristics of small products are controlled by the surface heat transfer characteristics, while those of large products are controlled by the surface and internal heat transfer characteristics. (It should be noted that the larger the product, the longer the time for the inner portions of the food to become frozen. It should also be noted that the thermal conductivity of ice at about –20°C is about four times greater than that for water at 0°C). Individual strawberries, for example, can be rapidly frozen (e.g., IQF freezing), whereas bulk packed strawberries in a large 55-gal drum can take several days for the center to be fully frozen.

Aside from the freezing process, there are other aspects of frozen storage that need to be considered before discussing the stability of particular tissue types in frozen storage. Several factors influence the rates of change in frozen systems.

The actual temperature of storage is important. As indicated, the Arrhenius relationship can often describe the change in the rate of a process as temperature is altered, providing that the mechanism of the process remains the same. In frozen systems, this assumption is not always valid. First, due to the formation of ice, the unfrozen medium becomes more concentrated. As discussed by Fennema,[18] this concentrating effect can result in a change in reaction rate in addition to the simple effect of temperature. Second, other components of the aqueous medium might crystallize out. As a result, physical and chemical properties could change markedly. For example, selective precipitation of salts can lead to changes in pH and, hence, affect the mechanisms of reactions.

An additional factor to be considered is the physical state of the medium between ice crystals. It is now realized that this physical state may be very important to product stability.[19,20] In many situations, the *maximally freeze-concentrated* material between ice crystals may enter the glassy state at low temperatures. Molecular mobility of solute molecules as in the glassy state should be markedly reduced, so stability would be enhanced. Thus, there might be a critical temperature below which change will markedly slow down. It has also been argued that above this temperature, while the product is still frozen, Arrhenius kinetics might not be the most appropriate description of the temperature dependence of product stability. The Williams–Landel–Ferry relationship has been proposed as a better description of the kinetics in the region above the glass temperature and below the final melting temperature.[22] Product stability might therefore be manipulated by altering this glass temperature. There is still very active discussion on this point. It has been shown, however, that the rate of change of a reaction in a food under frozen storage at a particular temperature can be reduced if the temperature at which the system attains the maximally freeze concentrated glass is raised by addition of appropriate polymers.[23]

Two categories of change should be considered when assessing the effects of temperature of storage and the temperature history of storage. These are (1) chemical (or biochemical) change and (2) physical change. It has often been demonstrated that the rate of chemical change in a fluctuating temperature environment is close to that which would be expected if a constant temperature, close to the time average, were to be maintained. This is a key conclusion of the comprehensive studies on product stability carried out by the U.S. Department of Agriculture in the late 1950s.[24-28] In contrast, physical change (e.g., ice crystal size) is much more sensitive to temperature fluctuation. Temperature gradients, in particular, provide a driving force for moisture migration. The effect of temperature fluctuation on physical change can be clearly seen in many products. In some cases, physical change can be so extensive as to influence the characteristics of chemical change. Freezer burn, the oxidation of surface regions where sublimation of ice has occurred, is an example of this influence.

Practical applications

Vegetables

Vegetables are important frozen commodities. The freezing of untreated vegetables can cause significant tissue damage. Microscopic studies show widespread microstructural disruption of many cells.[29] Upon thawing, many vegetables exhibit significant off-flavors, off-odors, and discoloration. These detrimental effects have been shown to be a consequence of the disruption of membranes and the activation of enyzme systems within the cells. Consequently, many vegetables must be blanched prior to freezing. Blanching is a thermal process designed, in part, to inactivate the enzymes responsible for generating the off-flavors and off-odors. Blanching of vegetables has been an integral part of freezing processing since Joslyn and Cruess first recommended the procedure.[30] The ideal conditions for blanching to optimize frozen product quality and stability are still not settled. Many processors utilize a heat treatment sufficient to inactivate peroxidase, one of the more stable enzymes present, and not incidentally, one of the enzymes whose activity is relatively easy to measure. Recent research has suggested that targeting peroxidase leads to a more severe heat treatment than is required for many vegetables and that the enzymes responsible for quality loss, which have been identified, have a lower stability than peroxidase.[31] A less severe heat treatment, then, is possible and should be used, assuming that a satisfactory test for enzyme inactivation exists for these critical enzymes. The commercial availability of controllable and thermally efficient blanchers makes targeting these enzymes a more attractive approach to take. Recommendations of appropriate blanching conditions continue to be published. It is important to note that blanching conditions used prior to freezing can affect the stability of the frozen vegetable while in storage.

Not all vegetables require a blanching pretreatment to stabilize them for frozen processing and storage. Koslowski has described some that can be frozen without blanching.[32] Often the decision whether to blanch can be made by considering the required shelf life. If an adequate shelf life can be achieved without blanching, then freezing by itself may be the preferred processing route. If blanching is required to inactivate the enzymes, then there may be a cost incurred in terms of decreased overall product quality.

Since blanching is a heat treatment, changes associated with mild thermal processing can be expected. These include loss of turgor in cells, due to thermal destruction of membrane integrity, and partial degradation of cell wall polymers. Such destruction of cell membranes and cell walls would not necessarily promote the propagation of ice, since many polymer systems can markedly slow down the rate of ice propagation at high concentrations. (A barrier to ice propagation can still exist, so the distinction between fast and slow freezing remains.)

A further effect of thermal processing is the degradation of chromophores such as chlorophyll, resulting in color change. Pigment degradation will continue to take place through frozen storage. The extent of the color change and the rate of the reactions could be affected by the severity of heat treatment.[33,34] For example, the greater the extent of conversion of chlorophyll in the initial product, the more rapid will be the deterioration during the frozen storage. Blanching can lead to thermally induced degradation of nutrients such as ascorbic acid. It can also lead to leaching of nutrients. Hence, a determination must be made as to whether or not the unblanched product has adequate stability to retain structural integrity and quality attributes upon freezing and during frozen storage, and a decision must be reached as to the extent of blanching needed to ensure optimum product quality.

Storage stability of frozen vegetables

Table 2 illustrates the storage lives of some typical vegetables at a variety of temperatures in terms of color and flavor changes. Clearly, the effect of increasing temperature is to reduce the storage life for all products. The practical storage life of different vegetables is controlled by different factors. For example, the table shows that for green beans color is a limiting factor. Overblanching of green beans can further reduce the shelf life. The flavor of blanched green beans, however, tends to be maintained well in storage. In contrast, unblanched frozen green beans develop off-flavor very rapidly, but maintain good color and texture. Cauliflower can be seen to be very sensitive to storage temperature, as is broccoli. The results given in the table are typical averages.

It should be noted that there can be considerable variation among cultivars. For example, some varieties of green bean have much higher lipoxygenase levels, requiring a longer blanch, which further destabilizes the chlorophyll. Some varieties of carrot can be frozen without blanching, and are stable for 6–8 months at about –20°C, whereas others will undergo unacceptable changes within a matter of weeks at this temperature. For some vegetables, such as brussels sprouts, the size can be very important to the stability, since size determines the required heat treatment for enzyme inactivation. The data in Tables 2 through 6 have been compiled from a wide variety of sources, including many of the

Table 2   Storage Stability of Frozen Vegetables: Typical Storage Life, in Days, Based on Flavor/Color

	–18°C (0°F)	–12°C (10°F)	–7°C (20°F)
Green beans	290/100	150/28	28/8
Peas	260/200	120/50	30/11
Spinach	270/350	130/140	35/35
Cauliflower	350/60	60/18	17/6
Broccoli	350/120	80/40	—
Carrots	380	190	—
Corn	300	100	—
Peppers	150	50	—
Asparagus	300	70	—

Note: These results are mean values. Data show considerable variation. A slight change in process conditions can markedly change storage life. A change in maturity can also affect storage life. Where only one figure is given, it is an estimated storage life based on overall consumer acceptibility.

references cited elsewhere in this chapter. Several useful compilations of stability behavior have been utilized in producing the estimates shown.[35-38]

Fruits

Many fruits are consumed raw. This constrains us as to suitable methods of freeze preservation. Unless a fruit tissue has significant mechanical integrity, blanching results in an unsuitable product. In general, two freezing methods are used on fruits, which result in products with different stabilities. The first method is to freeze the individual fruits, using either blast freezing or cryogenic freezing. It produces a free-flowing product, generally referred to as IQF fruit. The second method is to pack the fruit into containers, with added sugar, and to freeze the product. The containers range from retail packages to large drums of 30-lb size or even 55-gal size. Fruit frozen in this manner is often destined for further processing. The freezing rate in the large packages can be quite slow, and there is a significant osmotic dehydration in the individual fruits. The primary factors that limit the storage life of frozen fruits are color and flavor change. Oxidative color changes are common. Treatment of fruit with ascorbic acid will often slow down this deteriorative process. Textural damage is also important to the final quality of frozen fruit, but this effect is influenced more by the method of freezing and by the addition of sugar than by storage conditions.

The stabilities of a variety of frozen fruits are given in Table 3. Estimates of storage life of frozen fruits vary markedly, due in part to differing opinions as to the quality cutoff for acceptibility. The values given in the table are estimated from a variety of sources. For IQF (raw) fruits, the storage life can be affected by the packaging. The estimates in the table are for storage where access of oxygen to the product is minimized.

Meats

The conditions for freezing meat appear to have little effect on the quality of the frozen product immediately after freezing. Jul has provided examples of the changes that can take place.[9] During frozen storage, these changes include oxidation of pigments and of lipids. There is also evidence for an insolubilization of proteins, which may contribute the textural change. Table 4 lists some typical storage lives. Clearly, frozen pork is a relatively unstable product. Another factor that will shorten the storage life is the degree of com-

Table 3   Storage Stability of Frozen Fruits (in Days)

	–23°C (–10°F)	–18°C (0°F)	–12°C (10°F)
Strawberry	>700	600	150
Strawberry in sugar	—	600	100
Raspberry	>700	600	150
Raspberry in sugar	—	600	100
Peach	>700	500	120
Peach in sugar	—	500	100

Table 4   Storage Stability of Frozen Meat (in Days)

	–29°C (–20°F)	–18°C (0°F)	–12°C (10°F)
Beef	540	350	130
Lamb	540	350	130
Veal	480	300	100
Pork	300	230	70
Ground beef	300	230	110

Table 5   Storage Stability of Frozen Poultry (in Days)

	–23°C (–10°F)	–18°C (0°F)	–12°C (10°F)
Chicken	750	540	260
Chicken parts	750	540	260
Turkey	700	450	230
Turkey parts	700	450	230
Duck	540	350	180
Goose	540	350	180

minution, which influences the access of oxygen. It accounts for the lower stability of ground beef. The effect of comminution on stability is described by Berry and Leddy.[39,40]

Poultry

Regulations require the temperature of frozen poultry to be down to –18°C within 72 h of being chilled and packaged. Freezing is commonly carried out by blast freezing, immersion freezing, or cryogenic freezing. The resulting product is relatively stable. Of particular concern is the growth of microbiological contaminants, which could be a consequence of too long a time in prefreeze handling or too long a freezing process. Poor packaging can also lead to freezer burn. The stability of frozen poultry can be very dependent upon the initial condition of the raw product. Table 5 indicates how the relative stabilities of poultry products can vary.

Fish

Fish is particularly unstable during frozen storage. There are several reasons for this instability, some of which are more evident in certain fish species than in others. High-fat fish tend to have a short shelf life in frozen storage due to lipid oxidation, which produces a variety of off-flavors.

With other fish, protein insolubilization can be a problem, leading to tough, inedible fish. Various mechanisms have been proposed for this insolubilization. It has been linked to the degradation of trimethylamine oxide in gadoid fish. Lipid oxidation has also been

Table 6 Storage Stability of Frozen Fish (in Days)

	−23°C (−10°F)	−18°C (0°F)	−12°C (10°F)
Trout	300	260	110
Cod fillet	300	210	90
Ocean perch	300	220	120
Mackeral	110	80	50
Halibut	350	260	170

implicated. The handling of fish prior to freezing also affects shelf life. Even a short period of time in storage above the desired freezing temperature can lead to a major reduction in shelf life. Microbial growth and biochemical degradation are two factors contributing to this loss in potential shelf life. Another important factor is the nutritional status of the fish, so the fishing harvest can be of variable quality.

Because of the instability of frozen fish, it is often recommended that very low storage temperatures be employed. Table 6 illustrates the relative frozen storage stability of several species of fish, typical of lean fish, fatty fish, and gadoid fish. A categorization of frozen fish products as high, moderate, or low stability is given by Slavin.[41]

Conclusion

From Tables 2–6 for storage stabilities, it can be seen that different tissue types exhibit both different absolute stabilities toward frozen storage and different temperature sensitivities. It is also clear that for many products, −18°C is too high a temperature for effective long-term frozen storage. There is a trend within the frozen food industry to employ lower temperatures for the storage of many products. Most refrigerated warehouses operate at temperatures below −18°C, and indeed −30°C is not uncommon. Only at the retail end of the chain should products experience temperatures of −18°C or higher for any period, and it is to be hoped that product turnover will be sufficiently rapid to ensure that a product when purchased is within the useful shelf life. The effect of storage temperature is most marked for fish; though well-controlled low-temperature storage could contribute to a greatly improved product, there is growing evidence that the effects of temperature mishandling of products may be more solid than previously believed. A simple trial with frozen green beans[42] in which a one-time temperature fluctuation was imposed has shown that even a rise in temperature from −18°C to −15°C for a short time can result in a detectable color change in the thawed product after only 3 more months of storage.

References

1. Atkins, P.W. *Physical Chemistry*. Oxford University Press.
2. Villota , R. and J.G. Hawkins (1992). "Reaction kinetics in food systems" in *Handbook of Food Engineering*, eds. D.R. Heldman and D.B. Lund, Marcel Dekker, New York, p 56.
3. Graham, D. and B.D. Patterson (1982). Responses of plants to low, non-freezing temperatures. *Ann Rev Pl Physiol* 33, 347-72.
4. Morris,G.J. and A. Clarke (1987). "Cells at low temperatures" in *The Effects of Low Temperatures in Biological Systems*, eds. B.W.W. Grout and G.J. Morris, Edward Arnold, London, Chapter 2.
5. Lyons, J.M., J.K. Raison and P.L. Steponkus (1979). "The plant membrane in response to low temperature" in *Low Temperature Stress in Crop Plants: The Role of the Membrane*, eds. J.M. Lyons, D. Graham and J.K. Raison, Academic Press, New York, p 1-24.
6. Mazur, P. (1970). Cryobiology, the freezing of biological systems. *Science* 168, 939-49.

7. Franks, F. (1988). "Storage in the undercooled state" in *Low Temperature Biotechnology, Emerging Applications and Engineering Contributions*, ed. J.J. McGrath and K.R. Diller, ASME, New York, pp 107-113.
8. Reid, D.S. (1987). "The freezing of food tissues" in *The Effects of Low Temperatures in Biological systems*, eds. B.W.W. Grout and G.J. Morris, Edward Arnold, London, Chapter 15.
9. Jul, M. (1984). *The Quality of Frozen Foods*, Academic Press, London.
11. Mazur, P. (1966). "Physical and chemical basis of injury in single celled organisms subjected to freezing and thawing" in *Cryobiology*, ed. H. Meryman, Academic Press, New York, Chapter 6.
12. Reid, D.S. and S. Charoenrein (1988). "Studies on the freezing of plant cells under a cryomicroscope" in *Proceedings 17th International Congress of Refrigeration, Vol E.*, IIR, Paris.
13. Reid, D.S. (1990). "Freezing" in *Vegetable Processing*, eds. D. Arthey and C. Dennis Blackie, London, Chapter 5.
14. Singh, R.P. and C.Y. Wang (1977). The quality of frozen foods: a review. *J Food Process Engin* 1, 97-127.
15. Holdsworth, S.D. (1987). "Physical and engineering aspects of food freezing" in *Developments in Food Preservation - 4*, ed. S. Thorne, Elsevier Applied Science, London, Chapter 5.
16. Londahl, G and P.O. Persson (1993). "Freezing methods" in *Frozen Food Technology*, ed. C.P. Mallett, Blackie, London, Chapter 2.
17. Heldman, D.R. and R.P. Singh (1981). *Food Process Engineering, 2nd Edition*, AVI, Westport CT.
18. Fennema, O. (1975). "Reaction kinetics in partially frozen systems" in *Water Relations of Foods*, ed. R.B. Duckworth, Academic Press, London, pp 539-58.
19. Slade, L. and H. Levine (1991). "A food polymer science approach to structure property relationships in aqueous food systems," in *Water Relationships in Foods*, eds. H. Levine and L. Slade, Plenum, New York.
20. Slade, L. and H. Levine (1993). "The glassy state phenomenon in food molecules" in *The Glassy State in Foods*, eds. J.M.V. Blanshard and P.J. Lillford, Nottingham University Press, Loughborough, Chapter 3.
21. Williams, M.L., R.F. Landel and J.D. Ferry (1955). The temperature dependence of relaxation mechanisms in amorphous polymers and other glass forming liquids. *J Am Chem Soc* 77, 3701-7.
22. Reid, D.S. (1993). "Basic physical phenomena in freezing and thawing plant and animal tissues" in *Frozen Food Technology*, ed. C.P. Mallett, Blackie, London, Chapter 1.
23. Lim, M.H. and D.S. Reid (1991). "Studies of reaction kinetics in relation to the Tg' of polymers in frozen model systems," in *Water Relationships in Foods*, eds. H. Levine and L. Slade, Plenum, New York, pp 103-22.
24. Van Arsdel, W.B. (1957). Time temperature tolerance of frozen foods, I, introduction—the problem and the attack, *Food Technol* 11, 28-33.
25. Dietrich, W.C., F.E. Lindquist, J.C. Miers, G.S. Bohart, H.J. Neumann and W.F. Talburt (1957). The time-temperature tolerance of frozen foods, IV, objective tests to measure adverse changes in frozen vegetables. *Food Technol* 11, 109-13.
26. Hanson, H.L., L.R. Fletcher and A.A. Campbell (1957). The time temperature tolerance of frozen foods. V. Texture stability of thickened precooked frozen foods. *Food Technol* 11, 339.
27. Klose, A.A., M.F. Pool and H. Lineweaver (1955). Effect of fluctuating temperatures on frozen turkey. *Food Technol* 9, 372.
28. Van Arsdel, W.B. and D.G. Guadagni (1959). Time temperature tolerance of frozen foods, XV, method of using temperature histories to estimate changes in frozen food quality. *Food Technol* 13, 14-19.
29. Brown, M.S. (1979). Frozen fruits and vegetables, their chemistry, physics and cryobiology *Adv Food Res* 25, 181-235.
30. Joslyn, M.A. and W. V. Cruess (1929). Freezing of fruits and vegetables. *Fruit Products J* 8, 9-12.
31. Velasco, P.J., M.H. Lim, R.M. Pangborn and J.R. Whitaker (1989). Enzymes resonsible for off-flavor and off-aroma in blanched and frozen stored vegetables. *Appl Biochem Biotechnol* 11, 118-127.
32. Koslowski, A. Personal communication.

33. Katsaboxakis, K.T. (1984). "The influence of the degree of blanching on the quality of frozen vegetables" in *Thermal Processing and Quality of Foods*, eds. P. Zeuthen, J.C. Cheftel. C. Eriksson, M. Jul, P. Linko, G. Varella and G. Vos, Elsevier Applied Science, London p 559-65.

34. Katsaboxakis, K.T. and P.N. Papenicolaou (1984). "The consequences of varying degrees of blanching on the quality of frozen green beans" in *Thermal Processing and Quality of Foods*, eds. P. Zeuthen, J.C. Cheftel, C. Eriksson, M. Jul, P. Linko, G. Varella and G. Vos, Elsevier Applied Science, London, p 684-90.

35. ASHRAE Handbook (1991). *Refrigeration*, ASHRAE, Atlanta.

36. IIR (1986). *Recommendations for the Processing and Handling of Frozen Foods*, IIR, Paris.

37. TRREF. *Commodity Storage Manual*, TRREF, Bethesda, MD.

38. Van Arsdel, W.B., M.J. Copley and R.L. Olson (1969). *Quality and Stability in Frozen Foods*, Wiley, New York.

39. Berry, B.W. and K.F. Leddy (1989). Effects of freezing rate, frozen storage temperature and storage time on tenderness values of beef patties. *J Food Sci* **54**, 291-6.

40. Berry, B.W. and K.F. Leddy (1990). *Meat Freezing, a Source Book*, Elsevier Applied Science, London.

41. Spink, D. Personal communication.

chapter fifteen

Quality changes during distribution of deep-frozen and chilled foods: distribution chain situation and modeling considerations

W. E. L. Spiess, T. Boehme, and W. Wolf

Contents

Introduction

Fresh and processed foods, being materials of biological origin, undergo reactions that in virtually all cases reduce the quality of such foods during their entire lifespan and

0-8493-2646-X/97/$0.00+$.50
© 1998 by CRC Press LLC

Table 1 Characteristic Quality-Affecting Processes: Examples and Governing Conditioning
Factors

Process	Example	Conditioning Factor
Physical	Evaporation of volatile substances (loss of water and weight loss, loss of cell turgor, freezer burn, loss of aroma); Loss of color; Uptake of off-flavors	Water vapor concentration; Air circulation; Presence of light
(Bio)chemical	Oxidation; Protein-denaturation; Formation of Maillard products; Formation of off-flavors	Temperature; Atmosphere (water vapor and oxygen concentration); Composition of produce/product (pH, redox potential, etc.)
Biological-physiological	Dissimilation; Maturation; Senescence; Loss of aroma; Formation of off-flavors	Temperature; Atmosphere (water vapor and oxygen concentration); Composition of produce/product (pH, redox potential, etc.)
Microbiological	Deterioration; Fermentation; Formation of flavors/off-flavors	Temperature; Water vapor and oxygen concentration; Composition of product/produce (pH, redox potential, etc.)

ultimately lead to their spoilage. The main cause of food spoilage is the presence of microorganisms; fresh foods in particular provide an ideal medium for microbial growth, because of their high water content and nutrient density. Physical and chemical changes also reduce food quality. Possible effects of these deteriorative processes are summarized in Table 1.

The gentlest measures that can be used to retard food spoilage and to maintain product quality are lowering temperature, water content, and, in certain cases, pH. Lowering the temperature even down to the freezing point of water, however, merely reduces the metabolic activity of microorganisms; it does not damage them. The complete retardation of microbial growth is achieved only by lowering the product temperature below –10°C. However, when chilled or frozen products are subsequently exposed to ambient temperatures (~20°C), an increase of metabolic activity will occur and consequent growth and reproduction can be expected. Chemical changes on the other hand are minimized only at temperatures below –18°C to –20°C.[1] Other minimal processing techniques that can potentially extend shelf life without significantly changing the product nature and that can be applied either individually or in combination include irradiation treatment (UV radiation, ionizing radiation); ultrahigh pressure treatment; application of chemicals such as salt, nitrite, or smoke; and packaging with materials having specific barrier properties, with vacuum, with materials having active components (such as modified atmospheres or embedded reactants), or with edible coatings. If food products have to be stored for a longer period of time at ambient temperatures, microorganisms and active chemical constituents such as enzymes have to be inactivated by an adequate heat treatment (i.e., pasteurization or sterilization).

The consumer in the industrialized world has recognized the benefits of minimally processed foods. They are available in a wide variety and, through the application of HACCP concepts, are safely produced and distributed. Their market share is expanding well above average. Among these products, chilled and deep-frozen products with a high degree of convenience and sophistication are finding their way to the table of the consumer. Chilled food products are carefully treated without the addition of chemical preservatives to maintain their "fresh" quality; however, they are storable for only a limited time. Deep-frozen food products in some cases lack the attribute of freshness; however, they can be stored over a longer period of time and are better suited for long distance transportation and distribution.[2]

Because of the limited time period during which the high quality of chilled products can be maintained and because of the complicated and expensive operations involved in storage, transportation, distribution, and retailing, it is important to understand the factors that control quality loss. Such an understanding can be exploited in ways to correct for deficiencies in the overall operations as well as to predict possible quality losses, which would help in specifying deadlines for consumption.

The classical TTT concepts based on a zero-order reaction kinetics and developed for deep-frozen food products, in general, are not suited for such operations, since their underlying principles do not apply to the quality-loss processes associated with a wide variety of chilled and deep-frozen products. Furthermore, these concepts are applicable only to foods within individual lots experiencing a uniform time–temperature history. Consequently, concepts are required that describe the possible fate of a product unit processed, stored, transported, distributed, and retailed over a period of time under nonuniform temperature conditions. In one such concept, individual processing steps can be considered as reactive units (i.e., reactors) that are characterized by distinct residence time and temperature distributions. The quality losses in such units can then be determined by applying specific quality functions with time and temperature as parameters.

Using these assumptions, the overall quality of a product unit can be calculated through a convolution of the quality losses of the individual reactive units. The result will be a distribution function of product quality at the end of the distribution chain. To predict this distribution function at any point in the distribution chain for either deep frozen products or chilled products, four kinds of information are required:

1. Specific product stability (quality function)
2. Initial quality of the product
3. Temperature distribution within reactive units
4. Residence time distribution within reactive units.

Deep-frozen and chilled food chain

The main elements of the deep-frozen food and chilled food chains are production, storage, distribution, and sale (Figure 1). They are only part of a system in which a large number of steps have an influence on the final product quality. The integral processing and distribution chain may comprise the following stages:

- Raw material selection and collection*
- Raw material storage
- Product conditioning
- Primary production*
- Lowering of product temperature
- Advanced production
- Packaging
- Factory storage
- Transport to central storage
- Central storage*

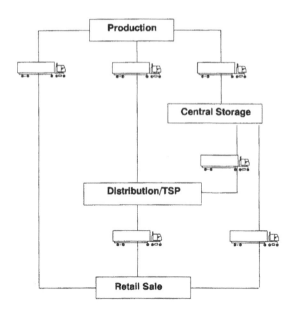

Figure 1 Distribution chains.

- Transport to distribution storage
- Distribution storage (transshipment points)
- Transport to retail sale*
- Retail sale*
- Transport to private homes*
- Storage in private homes (household refrigeration)*
- Preparation for consumption*
- Consumption

Operations with the highest potential for quality losses are indicated by an asterisk (*).

Production and packaging

Material selection and collection contribute to the final quality of deep-frozen and chilled products to a high degree. For a wide variety of products, special breeding programs have made available raw materials especially well suited for both freezing and chilling; they should be used for producing intermediate products and also ready-to-serve products. The processing parameters have to be adapted consistent with product properties and subsequent production steps; the same is true for the addition of spices and processing aids. Processing determines the intensity of chemical modifications in product components, the formation of precursors of Maillard reactions, the uptake of oxygen, etc. Consequently, the rate of quality deterioration is controlled basically by the nature of the raw material and the method and extent of processing. Further operations influencing the specific rate of quality loss in products even before entering the actual chain system are the packaging and the rate at which the temperature is lowered down to the final storage temperature. For deep-frozen products and processed chilled products, optimal packaging involves the use of barrier materials for water vapor and gas (e.g., oxygen and nitrogen), as well as the use of materials sufficient for protection against mishandling.

Living cellular materials show improved shelf-life properties when packed in "intelligent" materials, i.e., materials that have an influence on the metabolic activities of the products and can possibly adapt their permeability/scavenger properties according to special regimes of temperature, gas concentration, or level of ethylene oxide in the atmosphere. In such systems, the quality function is influenced by the product and the packaging situation.

Another important factor influencing the pattern of quality loss, especially in the case of chilled food, is the degree of microbial contamination when a product leaves the processing line. Good manufacturing practices (GMP) require that the microbial load should be as low as possible; however, only a few countries have specified legal upper limits for various food contaminants.[3] It is generally agreed that pathogenic microorganisms such as *Listeria monocytogenes*, *Salmonella* spp., *Shigella* ssp., or *Vibrio cholerae* and similar species should be at a zero-level tolerance (under different definitions); other less dangerous species may be tolerated up to $10^3/10^4/10^5$ cfu/g, depending on the type of food and the prospective group of consumers. Food spoilage bacteria, yeasts, or molds can be tolerated up to 10^5 cfu/g. It must be stressed, however, that low initial counts are important for a long shelf life at a high quality level.[4]

Storage and distribution modules

Factory storage
Immediately after production, deep-frozen/chilled products have to undergo quality-control measures. Deep-frozen products are usually kept under factory storage or in special sections of large central stores during the time those measures are taken, which may last up to 2 weeks. Chilled products with a short shelf life, in general, are distributed after production under conditions that would allow a call-back operation; they are released from this regime once the quality measures taken are completed. From the modeling point of view, the residence time distribution in the factory storage can be associated with plug flow, the time scales differing for deep-frozen and chilled products.

Central storage
Central storage is particularly appropriate for deep-frozen products that are seasonally harvested. These products have to be stored up to several months. Moreover, products to be distributed over large areas or shipped intercontinentally are also stored in central stores. Examples of seasonally harvested and produced products are peas, beans, brussels sprouts, carrots, cauliflower, kohlrabi, and raspberries, with main harvesting periods lasting over 6 to 8 weeks. Spinach is an example of a product with long harvesting periods, but the production is concentrated in certain agricultural centers.

Non seasonal products, in general, are ready-to-serve products, which are usually manufactured corresponding to the requirements of the market in order to minimize capital investments. With the help of computer systems and appropriate labeling, e.g., bar code labeling, it is possible to maintain good control of storage influxes and outfluxes. Before these supporting systems were introduced, random outfluxes were the normal way of operation. The improved control systems allow the realization of *first-in first-out* systems without problems. Under these conditions, the residence time distributions for seasonal products are flat over the entire storage period (a characteristic of fast filling and slow emptying operations). Foods produced over the entire year and stored in central storage systems show binomial residence time distributions. A certain stock has to be maintained to guarantee a sufficient supply. This fact and irregular outgoing fluxes cause deviations from plug flow characteristics. The temperature distributions in central storage are—as in factory storage—narrow, with minor deviations from the required values being around –24°C to –30°C for deep-frozen products and around +2°C to 4°C for chilled products.

Distribution stores

Products in distribution stores (transshipment points, TSP), in general, show rather narrow residence time distributions, which can be approximated by either plug flow characteristics or narrow binomial distributions with defined minimum and maximum residence times. Plug flow is exhibited in the case of products with high turnover, while slower-selling products develop binomial distribution. Deep-frozen products are kept for longer periods of time in the TSPs than chilled products. Temperature control in TSPs is also rather tight, so that narrow temperature distributions have to be expected.

Retail sale

In retail sales the residence times and temperatures are more difficult to control. Depending on the type of operation, rather different characteristic distributions are observed. Retail outlets with high turnover rates or retail outlets specializing in certain products exhibit residence time and temperature distributions that can be well approximated by plug flow characteristics. In the case of retail outlets with lower turnover rates, binomial residence time distributions with tailing characteristics and temperature distributions of a Gaussian nature have to be expected. This observation holds for both deep-frozen and chilled foods, with different time and temperature scales.

In the case of retail outlets with high turnover rates, the individual sales units are picked at random, which means each unit has the same chance of being selected. This selection principle applies also to packages in deeper layers, because many consumers prefer packages from deeper layers rather than from top layers. In the case of slow-selling products, the units in upper/front layers have higher chances of being selected, which results in binomial residence time distributions. Product temperatures are also distributed over wide ranges.

Transport

The transportation steps for deep-frozen and chilled products can be considered as operators corresponding to plug flow with regard to time distributions and having narrow temperature distributions with Gaussian characteristics. In the case of the final distribution step from TSP to retail outlet (carried out in a way so that many shops are served during one distribution journey), the temperature distribution has a more binomial character. In general, the distribution steps have little impact on the product quality.

These characteristic distribution curves have to be obtained experimentally: in the case of residence times, through controlled input and output measurements; and in the case of temperature, through appropriate temperature measurements at standardized locations within the sales units.

Quality functions and time/temperature distribution functions

Specific product quality: the quality function

Changes in the quality of food, corresponding to the kinetics of chemical reactions, may be characterized by a general approach of the form, $dC/dt = -k \cdot C^n$, where dC characterizes the change in the state or concentration C in the time differential dt; k characterizes the rate constant, while n is a coefficient characterizing the order of reaction. Describing the quality by Q, the change in quality can be expressed as

$$\frac{dQ}{dt} = \pm k \cdot Q^n. \tag{1}$$

When Equation (1) is integrated, the quality of a product of initial quality Q_0 after a storage time t for zero-, first-, and second-order reactions is:

$$\text{zero - order } (n = 0) \qquad Q = Q_0 \pm k \cdot t, \tag{2}$$

$$\text{first - order } (n = 1) \qquad Q = Q_0 \cdot \exp(\pm k \cdot t), \tag{3}$$

$$\text{second - order } (n = 2) \qquad \frac{1}{Q} = \frac{1}{Q_0 + kt} \tag{4}$$

For any reaction order except for $n = 1$, Q can be expressed as

$$Q = \left(Q_0^{1\pm n} \pm (1 \pm n) \cdot k \cdot t \right)^{\frac{1}{1\pm n}}. \tag{5}$$

The influence of temperature T on the reaction is described by Arrhenius' description for the rate constant k:

$$\frac{d \ln k}{dT} = \frac{E_a}{R \cdot T^2}, \tag{6}$$

with E_a being activation energy and R being the universal gas constant. After integration it follows

$$\ln k = \ln k^* \pm \frac{E_a}{R} \cdot \frac{1}{T} \tag{7}$$

or

$$k = k^* \cdot \exp\left(\pm \frac{E_a}{R \cdot T} \right). \tag{8}$$

The activation energy, E_a, is assumed to be constant for the temperature range under investigation; k^* is a temperature-independent rate constant.

From Equations (3) and (5), the quality functions

$$Q = \left[Q_0^{1\pm n} \pm (1 \pm n) \cdot k^* \cdot \exp\left(\pm \frac{E_a}{R \cdot T} \right) \cdot t \right]^{\frac{1}{1\pm n}} \tag{9}$$

and

$$Q = Q_0 \cdot \exp\left[\pm k^* \cdot \exp\left(\pm \frac{E_a}{RT} \right) \cdot t \right] \tag{10}$$

are obtained, for $n \neq 1$ and for $n = 1$, respectively. As indicated by these relationships, quality can be described as a function of time and temperature.

The size (i.e., absolute value) and the unit of the quality function Q depend on what quality-controlling parameter is used. When Q is described by a component (e.g., concentration of an ingredient), the quality function is comparable to that of a chemical reaction. When Q is described by a complex state of quality (e.g., taste evaluated by sensory analysis), Equations (7) and (8) must be regarded as pseudo-reaction functions. In any case, quality decreases as a function of time and temperature. If quality loss from an initial quality Q_0 to a quality level corresponding either to the end of High-Quality Life, Q_{HQL}, or to the end of Practical Storage Life, Q_{PSL}, is used, the following considerations may be applied.

Using Equations (2), (3), (4), and (8), storage time can be described by the general relationship

$$\ln t = \pm \frac{E_a}{R} \cdot \frac{1}{T} + \ln Q^*, \tag{11}$$

where the term Q^* depends on the reaction order n and can be expressed as

$$Q^* = \frac{Q_0 \pm Q}{k^*} \tag{12}$$

for zero-order ($n = 0$),

$$Q^* = \frac{\ln Q_0 \pm \ln Q}{k^*} \tag{13}$$

for first-order ($n = 1$), and

$$Q^* = \frac{(Q_0/Q) \pm 1}{k^* \cdot Q_0} \tag{14}$$

for second-order ($n = 2$). Equation (11) shows that, if the results from storage tests in which only the time to reach a certain quality level is considered are then plotted in terms of $\ln t$ vs. $1/T$, a straight line will be obtained so E_a and $\ln Q^*$ can be determined from the slope and intercept, respectively. Using this determination of $\ln Q^*$, k^* can be obtained from Equations (12), (13), and (14).

The values of n, k^*, and E_a for the quality functions (9) and (10) are determined experimentally by measuring changes in quality over defined periods of time, which is normaly done at constant temperature.

Detailed studies on the stability of frozen and chilled foods have shown reaction patterns corresponding to first order reactions ($n = 1$), no matter whether quality losses were characterized by chemical–analytical methods (e.g., decrease in vitamin C) or by sensory analysis.[5,6]

Examples of data obtained in controlled storage tests at various temperatures are shown in Figures 2 and 3. Both sets of data are elaborated through sensory evaluation of the flavor impression. In Figures 4 and 5, flavor-quality scores for green beans and stuffed pasta have been converted into their logarithmic values and then replotted against time. It becomes obvious that both sets of data can be represented by straight lines, which means that both types of quality deteroration can be described by (pseudo) first-order kinetics. The rate constant k^* and the activation energy E_a can be calculated from the set of

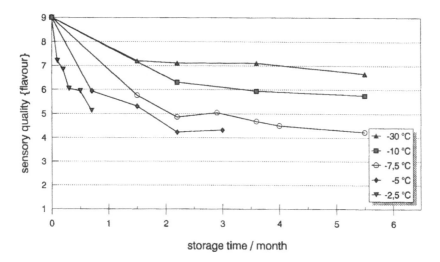

Figure 2 Quality of frozen green beans stored at different temperatures (original data).

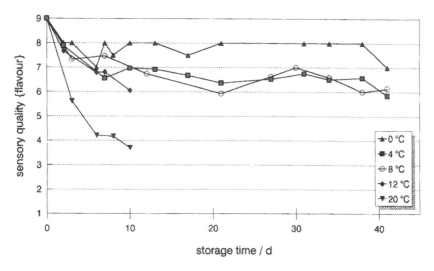

Figure 3 Quality of chilled stuffed pasta stored at different temperatures (original data).

Equations (3) and (7) according to Figures 6 and 7. Data obtained for various products are given in Table 2.

Residence time distribution functions

Distribution functions for residence time and temperature are not predictable on a theoretical basis; they have to be established experimentally.[7] Residence time distributions, however, can be derived from controlled storage input and output measurements. As already explained, residence times in factory storage and in distribution centers are rather evenly distributed; data obtained for deep-frozen products in central stores and retail sale cabinets are shown in Figures 8 and 9.

Residence times of chilled products in the distribution chain are much shorter when the distribution is properly controlled. A pronounced distribution of residence times is to be expected only during retail sale. Typical distributions[8] are shown in Figure 10.

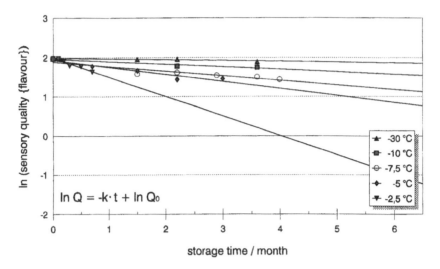

Figure 4 Quality of frozen green beans stored at different temperatures (first-order reaction).

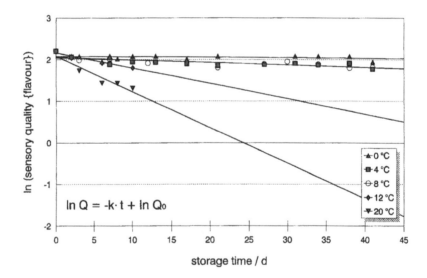

Figure 5 Quality of chilled stuffed pasta stored at different temperatures (first-order reaction).

Distribution curves for the data presented can be mathematically treated based on approaches used in chemical reaction engineering.[9] For skewed, bell-shaped distributions of residence times, the distribution (i.e., distribution density) can be described through a simple equation using the most frequently observed storage time t_{mod} (i.e., modal value) as a parameter (Figure 11):

$$y(t) = \frac{t}{t_{mod}} \cdot \exp\left(\pm \frac{t}{t_{mod}}\right). \tag{15}$$

With the help of these or similar equations, residence time distributions for entire chains can be obtained by means of a mathematical convolution, for example, by a Laplace transformation. Since residence time distributions are frequently irregular and not suited for description by an analytical approach, a stepwise convolution is a more appropriate approach.

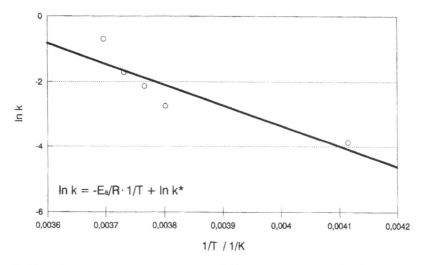

Figure 6 Quality of frozen green beens stored at different temperatures (Arrhenius approach).

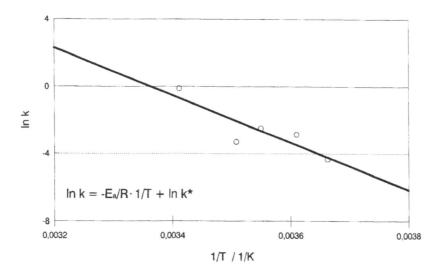

Figure 7 Quality of chilled stuffed pasta stored at different temperatures (Arrhenius approach).

Temperature distribution functions

Temperature measurements within the deep-frozen and chilled food product chains show that temperatures are approximately normally distributed (see Figures 12 and 13.) Hence, the temperature distributions can be analytically described by the normal distribution curve:

$$y(T) = \frac{1}{s \cdot \sqrt{2 \cdot \pi}} \cdot \exp\left(\pm \frac{\left(T \pm T_m\right)^2}{2 \cdot s^2} \right), \tag{16}$$

where T_m is the mean observed temperature, T is the storage temperature, and s is the standard deviation.

Table 2   Reaction Rate Constant, k^* (First Reaction Order, $n = 1$), and Activation Energy for Selected Deep-Frozen and Chilled Food Products

Product	$k^*_{(n=1)}$/month^{-1}	E_a /kJ·mol^{-1}
Frozen products		
Ice cream[a]	$4.370 \cdot 10^{28}$	14529.3
Bakery products[a]	$3.109 \cdot 10^{11}$	6347.5
Pork[a]	$9.161 \cdot 10^{6}$	4509.5
Beef[a]	$1.494 \cdot 10^{10}$	5625.8
Poultry[a]	$1.688 \cdot 10^{8}$	4732.8
Strawberry[a]	$9.757 \cdot 10^{35}$	18321.9
Green beans[a]	$3.310 \cdot 10^{9}$	52.6
Fatty fish[a]	$6.375 \cdot 10^{13}$	7038.2
Lean fish[a]	$4.139 \cdot 10^{8}$	4674.3
Chilled products		
Product	$k^*_{(n=1)}$/day^{-1}	E_a /kJ·mol^{-1}
Yogurt	$2.143 \cdot 10^{5}$	3908.7
Maultaschen (stuffed pasta)	$5.966 \cdot 10^{23}$	13877.3

a = data according to reference 5.

If the residence time distributions are treated by stepwise convolution, it is recommended either that the frequency of temperatures be described in the form of histograms or that categories of prevailing temperatures be defined.

Convolution of distributions

Stepwise convolution: quality distributions at the end of a chain

For calculating quality distributions, it is necessary to have the specific quality function and a precise knowledge of temperature and residence time distributions within the individual links of the chain. Moreover, the probability of encountering an individual product quality is determined by convoluting the distributions for initial quality, temperature, and residence time.

The probability theorem defines the probability of simultaneously occurring independent events as the product of the probabilities of these individual events. Consequently, the probability (i.e., relative frequency) of a quality event as determined by a combination of one initial quality event and one temperature event and one residence time event can be expressed as

$$f(Q) = f\left(Q_{0_i} \text{ AND } T_j \text{ AND } t_k\right) = f\left(Q_{0_i}\right) \cdot f\left(T_j\right) \cdot f\left(t_k\right). \tag{17}$$

In addition, quality Q has to be calculated by the product quality function Equation (9) or Equation (10) by using the same parameters Q_{0_i}, T_j, and t_k. When the quality and its corresponding probability of all possible combinations of all initial qualities Q_{0_i}, temperatures T_j, and residence times t_k have been calculated, the distribution of quality at the end of the chain link under consideration is obtained. In order to obtain a distribution histogram needed for further computations, the calculated quality data have to be grouped into classes. Figure 14 shows schematically the mathematical procedure for calculating the quality distribution at the end of a single chain link.

Product quality at the end of the chain

To predict the final quality distribution at the end of distribution chain, this procedure has to be repeated for each chain link, so that the product quality distribution at the end

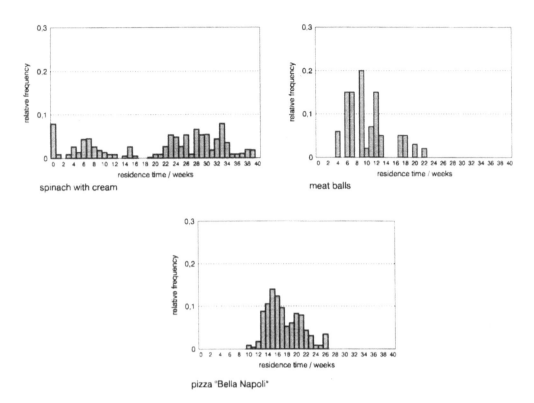

Figure 8 Typical residence time distributions of selected deep-frozen food products in central stores.

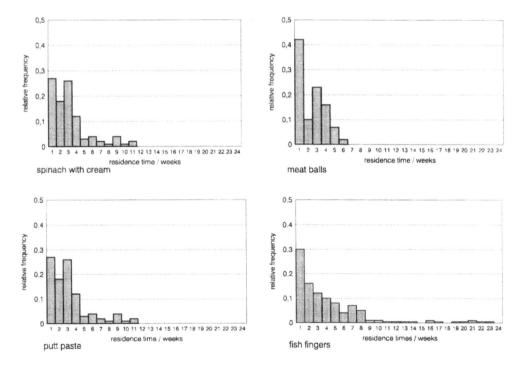

Figure 9 Typical residence time distributions of selected deep-frozen food products in retail freezer cabinets.

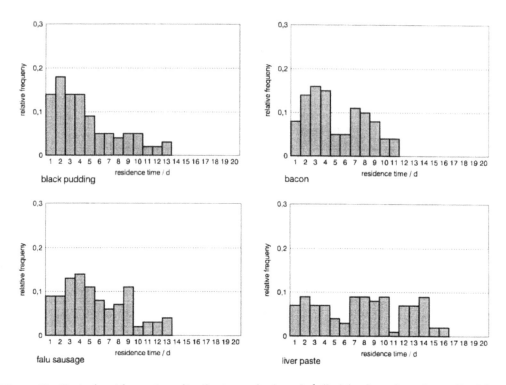

Figure 10 Typical residence time distributions of selected chilled food products in retail cabinets. (From Olsson, P. In: Zeuthen, P. et al. (eds.) Processing and Quality of Foods (3). Elsevier Applied Science Publishers, London, 1990. With permission.)

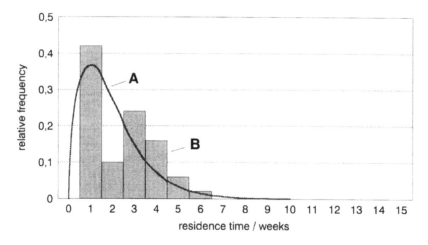

Figure 11 Residence times distribution of meatballs in central store. A, analytical, Equation (15). B, histogram.

of one chain link becomes the initial quality distribution for the following chain link. For the purpose of making the calculation more transparent, links of a similar nature can be regarded as one link. In most cases, it is sufficient to consider two stages: factory/central storage and distribution/retail sale. Results of a complex calculation are shown in Figures 15 to 17, where the quality loss at the end of a deep-frozen food chain for lean fish is described. The product enters the chain with a uniform quality scored as "8". As

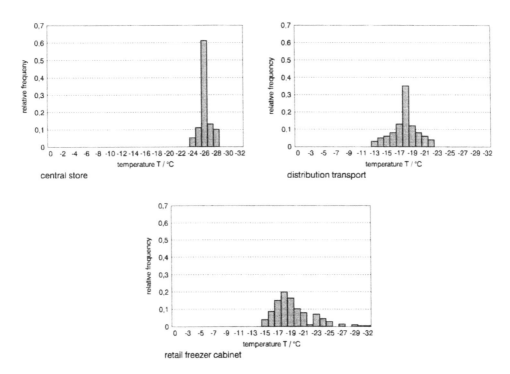

Figure 12   Typical temperature distributions at various links of a deep-frozen food chain.

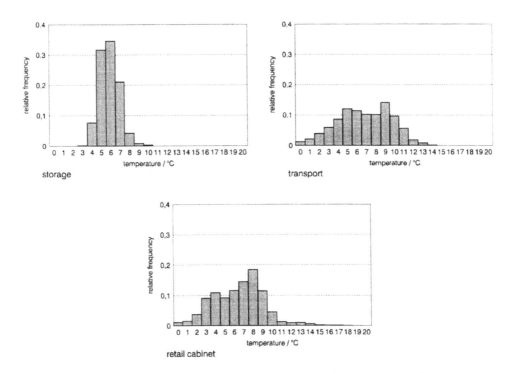

Figure 13   Typical temperature distributions at various links of a chilled food chain.

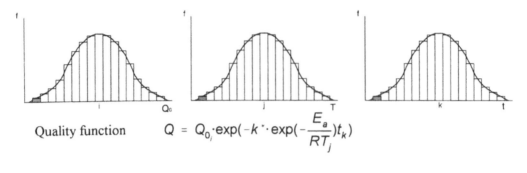

Quality function $Q = Q_{0_i} \cdot \exp(-k^* \cdot \exp(-\dfrac{E_a}{RT_j})t_k)$

Stepwise convolution $f(Q) = f(Q_{0_i} \; AND \; T_j \; AND \; t_k) = f(Q_{0_i}) \cdot f(T_j) \cdot f(t_k)$

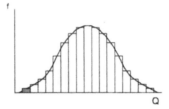

Figure 14 Calculation of product quality distribution at the end of a chain link; the quality function is given for the reaction order $n = 1$.

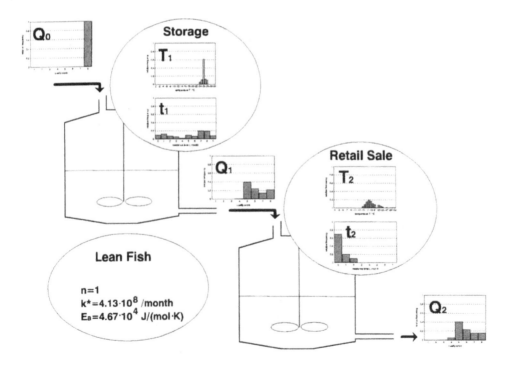

Figure 15 Reactor model for stepwise convolution to calculate the final quality of lean fish.

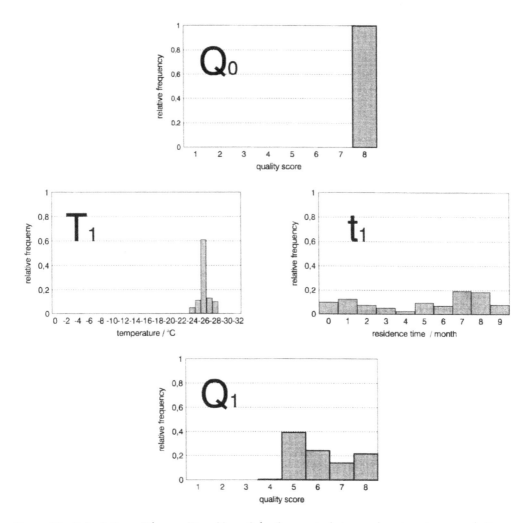

Figure 16   Calculation of the quality of lean fish after central storage by stepwise convolution.

a consequence of exposure to different temperatures for various storage times, the product quality changes thus resulting in a nonuniform quality distribution upon leaving the factory/central stores. Products with different qualities enter retail sale at the same time and are exposed again to different temperature/time events.

Summary

After manufacture, food is subject to changes in quality caused by temperature-dependent and time-dependent reactions that can be described in mathematical terms. Experimental studies provide the information required to establish a product-specific quality function.

When data on temperature, residence times, and initial quality for a particular system are available in the form of distribution functions, the method of stepwise convolution combined with the quality function of the product under consideration allows one to determine the quality distribution at the end of a distribution chain.

For representing distributions of product quality within chains (e.g., chilled and frozen food distribution chain), the quality distribution at the end of one link in the chain is taken as initial quality for the following link.

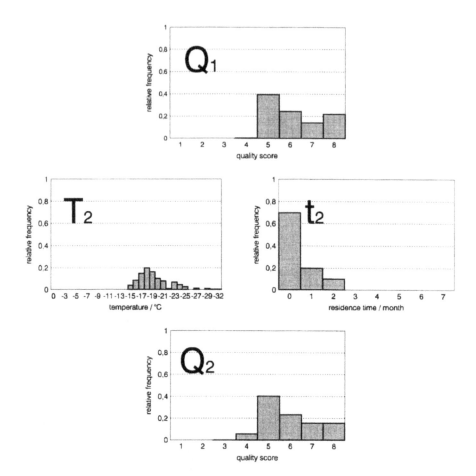

Figure 17 Calculation of the quality of lean fish after retail sale (final quality) by stepwise convolution.

Such representations make it possible to set limits on time in the chain and to manage stocks of deep frozen or chilled products in a systematic manner.

References

1. Poulsen, K.P. and Lindelof, F. Acceleration of chemical reactions due to freezing. Deutscher Kaelte- und Klimatechnischer Verein. Tagungsbericht, Hamburg, 1978.
2. Jul, M. The Quality of Frozen Foods. Academic Press, London, 1984.
3. N. N. Verordnung über die hygienisch-mikrobiologischen Anforderungen an Lebensmittel, Gebrauchsgegenstaende, Raeume, Einrichtungen und Personal (Hygieneverordnung, HgV) vom 26. Juni l996 des Eidgenoessischen Departments des Innern, Bern, Switzerland.
4. Baumgart, J. (Ed.) Mikrobiologische Untersuchung von Lebensmitteln. B. Behr´s Verlag, Hamburg, l994.
5. Labuza, T.P. Shelf-Life-Dating of Foods. Food & Nutrition, Westport, CT, 1982.
6. Spiess, W.E.L. and Gruenewald, T. Das Lagerverhalten tiefgefrorener Lebensmittel. Personal communication.
7. Spiess, W.E.L. Zum Verweilzeitverhalten tiefgefrorener Lebensmittel in der Tiefkuehlkette. KI Klima-Kaelte-Heizung (7:8):319-324, 1988.

8. Olsson, P. Improved economy and better quality in the distribution of chilled foods. In: Zeuthen, P. et al. (eds.) Processing and Quality of Foods (3). Elsevier Applied Science Publishers, London, 1990.

9. Levenspiel, O. Chemical Reaction Engineering. John Wiley and Sons, New York, 1962.

10. Spiess, W.E.L., Wien, K.J., Jung, G. und Wolf, W. Welche Temperaturen werden in Verteiler- fahrzeugen und Kuehltruhen des Einzelhandels in der Bundesrepublik Deutschland bei tiefgefrorenen Lebensmitteln tatsächlich eingehalten? Berichte der Bundesforschungsanstalt für Ernaehrung, Karlsruhe, 1/1975.

11. Spiess, W.E.L. and Folkers, D. Time-temperature surveys in the frozen food chain. In: Zeuthen, P. et al. (eds.) Thermal Processing and Quality of Foods. Elsevier Applied Science Publishers, London, 1984.

chapter sixteen

Technologies to extend the refrigerated shelf life of fresh fruits

Adel A. Kader, R. Paul Singh, and Jatal D. Mannapperuma

Contents

Controlled atmosphere storage

Modified atmospheres (MA) or controlled atmospheres (CA) mean removal or addition of gases resulting in an atmospheric composition around the commodity that is different from that of air (78.08% N_2, 20.95% O_2, 0.03% CO_2). Usually this involves reduction of oxygen (O_2) and/or elevation of carbon dioxide (CO_2) concentrations. MA and CA differ only in the degree of control; CA is more exact. The use of MA or CA should be considered as a supplement to proper temperature and relative humidity management.

Effects of controlled atmospheres

The potential for benefit or hazard from atmospheric modification depends upon the commodity, variety, physiological age, atmospheric composition, and temperature and duration of storage. This helps explain the wide variability in results among published reports for a given commodity.

Potential beneficial effects

Used properly, MA or CA can supplement proper temperature management and can result in one or more of the following benefits, which translate into reduced quantitative and qualitative losses during postharvest handling and storage of some horticultural commodities:

1. Retardation of senescence (ripening) and associated biochemical and physiological changes, i.e., slowing down respiration and ethylene production rates, softening, and compositional changes.
2. Reduction of fruit sensitivity to ethylene action at O_2 levels below about 8% and/or CO_2 levels above 1%.
3. Alleviation of certain physiological disorders such as chilling injury of various commodities, russet spotting in lettuce, and some storage disorders of apples.
4. MA/CA can have a direct or indirect effect on postharvest pathogens and consequently decay incidence and severity. For example, elevated CO_2 levels (10% to 15%) significantly inhibit development of botrytis rot on strawberries, cherries, and other fruits.
5. Low O_2 (0–5% or lower) can be a useful tool for insect control in some commodities.

Potential harmful effects

In most cases, the difference between beneficial and harmful MA combinations is relatively small. Also, MA combinations that are necessary to control decay or insects, for example, cannot always be tolerated by the commodity and may result in faster deterioration. Potential hazards of MA to the commodity include the following:

1. Initiation and/or aggravation of certain physiological disorders such as blackheart in potatoes, brown stain on lettuce, and brown heart in apples and pears.
2. Irregular ripening of fruits such as banana, pear, and tomato can result from O_2 levels below 2% or CO_2 levels above 5%.
3. Development of off-flavors and off-odors at very low O_2 concentrations as a result of anaerobic respiration.
4. Increased susceptibility to decay when the commodity is physiologically injured by too-low O_2 or too-high CO_2 concentrations.
5. Stimulation of sprouting and retardation of periderm development in some root and tuber vegetables such as potatoes.

Requirements and recommendations

During the past 50 years, uses of CA and MA has increased steadily and has contributed significantly to extending the postharvest life and maintaining the quality of several fruits and vegetables. This trend is expected to continue as technological advances are made in attaining and maintaining CA and MA during transport, storage, and marketing of fresh produce. Several refinements in CA storage include low O_2 (1.0–1.5%) storage, low ethylene CA storage, rapid CA (rapid establishment of the optimum levels of O_2 and CO_2),

Table 1 Classification of Fruits and Vegetables According to Their Tolerance of Low O_2 Concentrations

Minimum O_2 Concentration Tolerated (%)	Commodities
0.5	Tree nuts, dried fruits, and vegetables.
1.0	Some cultivars of apples and pears, broccoli, mushroom, garlic, onion, most cut or sliced (minimally processed) fruits and vegetables.
2.0	Most cultivars of apples and pears, kiwifruit, apricot, cherry, nectarine, peach, plum, strawberry, papaya, pineapple, olive, cantaloupe, sweet corn, green bean, celery, lettuce, cabbage, cauliflower, brussels sprouts
3.0	Avocado, persimmon, tomato, pepper, cucumber, artichoke
5.0	Citrus fruits, green pea, asparagus, potato, sweet potato

Table 2 Classification of Fruits and Vegetables According to Their Tolerance of Elevated CO_2 Concentrations

Minimum CO_2 Concentration Tolerated (%)	Commodities
2	Apple (Golden Delicious), Asian pear, European pear, apricot, grape, olive, tomato, pepper (sweet), lettuce, endive, Chinese cabbage, celery, artichoke, sweet potato.
5	Apple (most cultivars), peach, nectarine, plum, orange, avocado, banana, mango, papaya, kiwifruit, cranberry, pea, pepper (chili), eggplant, cauliflower, cabbage, brussels sprouts, radish, carrot.
10	Grapefruit, lemon, lime, persimmon, pineapple, cucumber, summer squash, snap bean, okra, asparagus, broccoli, parsley, leek, green onion, dry onion, garlic, potato.
15	Strawberry, raspberry, blackberry, blueberry, cherry, fig, cantaloupe, sweet corn, mushroom, spinach, kale, Swiss chard.

and programmed or sequential CA storage (e.g., storage in 1% O_2 for 2 to 6 weeks followed by storage in 2–3% O_2 for the remainder of the storage period). Other developments that may expand use of MA during transport and distribution include using edible coatings or polymeric films to create a desired MA within the commodity.

Fresh fruits and vegetables vary greatly in their relative tolerance to low O_2 concentration (Table 1) and elevated CO_2 concentrations (Table 2). These are the levels beyond which physiological damage would be expected. These limits of tolerance can be different at temperatures above or below recommended temperatures for each commodity. Also, a given commodity may tolerate brief exposures to higher levels of CO_2 or lower levels of O_2 than those indicated. The limit of tolerance to low O_2 would be higher as storage temperature or duration increases, because O_2 requirements for aerobic respiration of the tissue increases with higher temperatures. Depending on the commodity, damage associated with CO_2 may either increase or decrease with an increase in temperature. CO_2 production increases with temperature, but its solubility decreases; thus, CO_2 in the tissue can be increased or decreased by an increase in temperature. Further, the physiological effect of CO_2 could be temperature dependent. Tolerance limits to elevated CO_2 decrease with a reduction in O_2 level, and similarly the tolerance limits to reduced O_2 increase with the increase in CO_2 level.

Current MA/CA recommendations are summarized in Table 3 (fruits) and Table 4 (vegetables). Also included is an estimate of potential benefits and extent of current

Table 3 Summary of recommended CA or MA conditions
during transport and/or storage of selected fruits

Commodity	Temp. range (°C)[a]	CA[b] %O$_2$	%CO$_2$	Potential for benefit[c]	Remarks[d]
Deciduous Tree Fruits					
Apple	0–5	1–3	1–5	A	About 50% of production is stored under CA
Apricot	0–5	2–3	2–3	C	No commercial use
Cherry, sweet	0–5	3–10	10–15	B	Some commercial use
Fig	0–5	5–10	15–20	B	Limited commercial use
Grape	0–5	2–5	1–3	C	Incompatible with SO$_2$ fumigation
Kiwifruit	0–5	1–2	3–5	A	Some commercial use; C$_2$H$_4$ must be maintained below 20 ppb
Nectarine	0–5	1–2	3–5	B	Limited commercial use
Peach	0–5	1–2	3–5	B	Limited commercial use
Pear, Asian	0–5	2–4	0–1	B	Limited commercial use
Pear, European	0–5	1–3	0–3	A	Some commercial use
Persimmon	0–5	3–5	5–8	B	Limited commercial use
Plum and prune	0–5	1–2	0–5	B	Limited commercial use
Raspberry and other cane berries	0–5	5–10	15–20	A	Increasing use during transport
Strawberry	0–5	5–10	15–20	A	Increasing use during transport
Nuts and dried fruits	0–25	0–1	0–100	A	Effective insect control method
Subtropical and tropical fruits					
Avocado	5–13	2–5	3–10	B	Limited commercial use
Banana	12–15	2–5	2–5	A	Some commercial use during transport
Grapefruit	10–15	3–10	5–10	C	No commercial use
Lemon	10–15	5–10	0–10	B	No commercial use
Lime	10–15	5–10	0–10	B	No commercial use
Olive	5–10	2–3	0–1	C	No commercial use
Orange	5–10	5–10	0–5	C	No commercial use
Mango	10–15	3–5	5–10	C	Limited commercial use
Papaya	10–15	3–5	5–10	C	No commercial use
Pineapple	8–13	2–5	5–10	C	No commercial use

[a] Usual and/or recommended range. A relative humidity of 90% to 95% is recommended.

[b] Best CA combination may vary among cultivars and according to storage temperature and duration.

[c] A = excellent, B = good, C = fair.

[d] Comments about use refer to domestic marketing only; many of these commodities are shipped under MA for export marketing.

commercial use. There is no doubt that some of these MA combinations will change as more research is completed. The possibility of adding carbon monoxide to MA/CA for some commodities may change its potential for benefit.

Current CA use for long-term storage of fresh fruits and vegetables is summarized in Table 5. Its use on nuts and dried commodities (for insect control and quality maintenance, including prevention of rancidity) is increasing as it provides an excellent substitute for chemical fumigants, (such as methyl bromide) used for insect control. Also the use of CA on commodities listed in Table 5 other than apples and pears is expected to increase as international market demands for year-round availability of various commodities expand.

Table 4 Summary of Recommended CA or MA Conditions during Transport and/or Storage of Selected Vegetables

Commodity	Temp. range (°C)[a]	CA[b] %O$_2$	%CO$_2$	Potential for benefit[c]	Remarks[d]
Artichokes	0–5	2–3	2–3	B	No commercial use
Asparagus	0–5	air	5–10	A	Limited commercial use
Beans, snap	5–10	2–3	4–7	C	Potential for use by processors
Beets	0–5	None		D	98–100% rh is best
Broccoli	0–5	1–2	5–10	A	Limited commercial use
Brussels sprouts	0–5	1–2	5–7	B	No commercial use
Cabbage	0–5	2–3	3–6	A	Some commercial use for long-term storage of certain cultivars
Cantaloupes	3–7	3–5	10–15	B	Limited commercial use
Carrots	0–5	None		D	98–100% rh is best
Cauliflower	0–5	2–3	2–5	C	No commercial use
Celery	0–5	1–4	0–5	B	Limited commercial use in mixed loads with lettuce
Corn, sweet	0–5	2–4	5–10	B	Limited commercial use
Cucumbers	8–12	3–5	0	C	No commercial use
Honeydews	10–12	3–5	0	C	No commercial use
Leeks	0–5	1–2	3–5	B	No commercial use
Lettuce	0–5	1–3	0	B	Some commercial use with 2–3% CO added
Mushrooms	0–5	air	10–15	C	Limited commercial use
Okra	8–12	3–5	0	C	No commercial use; 5–10% CO$_2$ is beneficial at 5–8°C
Onions, dry	0–5	1–2	0–5	B	No commercial use; 7% rh
Onions, green	0–5	1–2	10–20	C	Limited commercial use
Peppers, bell	8–12	3–5	0	C	Limited commercial use
Peppers, chili	8–12	3–5	0	C	No commercial use; 10–15% CO$_2$ is beneficial at 5–8°C
Potatoes	4–12	None		D	No commercial use
Radish	0–5	None		D	98–100% rh
Spinach	0–5	air	0–20	B	No commercial use
Tomatoes, mature	12–20	3–5	0–3	B	Limited commercial use
Tomatoes, green-partially ripe	8–12	3–5	0–5	B	Limited commercial use

[a] Usual and/or recommended range. A relative humidity of 90% to 98% is recommended unless otherwise indicated under "Remarks."

[b] Best CA combination may vary among cultivars and according to storage temperature and duration.

[c] A = excellent, B = good, C = fair, D = slight or none.

[d] Comments about use refer to domestic marketing only; many of these commodities are shipped under MA for export marketing.

CA/MA use for short-term storage and transport of fresh horticultural crops (Table 6) will continue to increase supported by technological developments in transport containers, MA packaging, and edible coatings. Carbon monoxide at 5–10%, when added to O$_2$ levels below 5%, is an effective fungistat that can be used for decay control on commodities that do not tolerate 15% to 20% CO$_2$. However, CO is very toxic to humans, and special precautions must be taken.

CA/MA conditions, including MA packaging (MAP), can replace certain postharvest chemicals used for control of some physiological disorders, such as scald on apples. Proper

Table 5 Summary of CA Use for Long-Term Storage of Fresh Fruits and Vegetables

Storage Duration (months)	Commodities
More than 12	Almond, filbert, macadamia, pecan, pistachio, walnut, dried fruits and vegetables
6–12	Some cultivars of apples and European pears
3–6	Cabbage, Chinese cabbage, kiwifruit, some cultivars of Asian pears
1–3	Avocado; olive; some peach, nectarine, and plum cultivars; persimmon; pomegranate

Table 6 Summary of CA/MA Use for Short-Term Storage and/or Transport of Fresh Horticultural Crops

Primary benefit of CA/MA	Commodities
Delay ripening and avoid chilling temperatures	Avocado, banana, mango, melons, nectarine, papaya, peach, plum, tomato (picked mature-green or partially ripe)
Control decay	Blackberry, blueberry, cherry, fig, grape, raspberry, strawberry
Delay senescence and undesirable compositional changes (including tissue brown discoloration)	Asparagus, broccoli, lettuce, sweet corn, fresh herbs, minimally processed fruits and vegetables

use of CA can also eliminate the need for using daminozide on apples. Furthermore, some postharvest fungicides and insecticides can be reduced or eliminated where CA/MA provides adequate control of postharvest pathogens or insects.

CA/MA may facilitate picking and marketing more mature (better flavor) fruits by slowing their postharvest deterioration to permit transport and distribution. Another potential use for CA/MA is in maintaining quality and safety of lightly processed fruits and vegetables, which are increasingly being marketed as value-added convenience products.

The residual effects of CA/MA on fresh commodities after transfer to air (during marketing) may include reduction of respiration and ethylene production rates, maintenance of color and firmness, and delayed decay. Generally, the lower the concentration of O_2 and the higher the concentration of CO_2 (within the tolerance limits of the commodity), and the longer the exposure to CA/MA conditions, the more prominent are the residual effects.

Modified atmosphere packaging

The positive effects of film packaging, other than creation of MA/CA conditions, can include:

1. maintenance of high relative humidity and reduction of water loss;
2. improved sanitation by reducing contamination of the commodity during handling;
3. minimized surface abrasions by avoiding contact between the commodity and the material of the shipping container;
4. reduced spread of decay from one unit to another;
5. possible exclusion of light, when needed, for commodities such as potato and Belgian endive;
6. use of the film as carrier of fungicides, sprout inhibitors, scald inhibitors, or other chemicals;
7. facilitation of brand identification.

The negative effects include slowing down cooling of the packaged commodity and increased potential for water condensation within the package, which may encourage fungal growth.

Modified atmospheres can be created either passively by the commodity or intentionally. With *commodity-generated* or *passive MA*, if commodity and film permeability characteristics are properly matched, an appropriate atmosphere can passively evolve within a sealed package through consumption of O_2 and production of CO_2 by respiration. The gas permeability of the selected film must allow O_2 to enter the package at a rate offset by the consumption of O_2 by the commodity. Similarly, CO_2 must be vented from the package to offset the production of CO_2 by the commodity. Furthermore, this atmosphere must be established rapidly and without creating anoxic conditions or injuriously high levels of CO_2.

Because of the limited ability to regulate a passively established atmosphere, it is likely that atmospheres within MAP will be actively established and adjusted known as active modified atmosphere. This can be done by pulling a slight vacuum and replacing the package atmosphere with the desired gas mixture. This mixture can be further adjusted through the use of absorbing or adsorbing substances in the package to scavenge O_2, CO_2, or C_2H_4.

Although active modification implies some additional costs, its main advantage is that it ensures the rapid establishment of the desired atmosphere. Ethylene absorbers can help delay the climacteric rise in respiration for some fruits. Carbon dioxide absorbers can prevent the buildup of CO_2 to injurious levels, which can occur for some commodities during passive modification of the package atmosphere.

Many plastic films are available for packaging, but relatively few have been used to wrap fresh produce. Low-density polyethylene and polyvinyl chloride are the main films used in packaging fruits and vegetables. Polystyrene has been used, but Saran™ and polyester have such low gas permeabilities that they would be suitable only for commodities with very low respiration rates.

Design of modified atmosphere packages

The design of modified atmosphere packages involves the use of mass transport properties of packages and respiration rates of produce placed inside the package. In this section we will examine how simple mathematical models may be developed for this purpose. The design methodology presented in this chapter is based largely on a paper presented by Mannapperuma and Singh (1994).

The basic principles involved in the design of modified atmospheric packages can be illustrated by plotting the recommended modified atmospheric compositions, such as those listed in Tables 3 and 4, on a two-dimensional chart with O_2 concentration as the *x* axis and CO_2 concentration as the *y* axis. Figures 1 and 2 are such plots for some common fruits and vegetables (Mannapperuma and Singh, 1994).

Most of the recommended modified atmospheres contain lower O_2 concentrations and higher CO_2 concentrations compared to the ambient. In such cases, these concentration gradients create a flux of O_2 into the package and a flux of CO_2 out of the package. Under steady-state conditions these two fluxes should be equal to the O_2 consumption and the CO_2 generation by the produce in the package. The design parameters of the package are determined using these equalities.

Development of a mathematical model

Using the quantities shown in Figure 3, a mathematical model for the package may be formulated. The gas flow rate through the package is expressed as the product of film

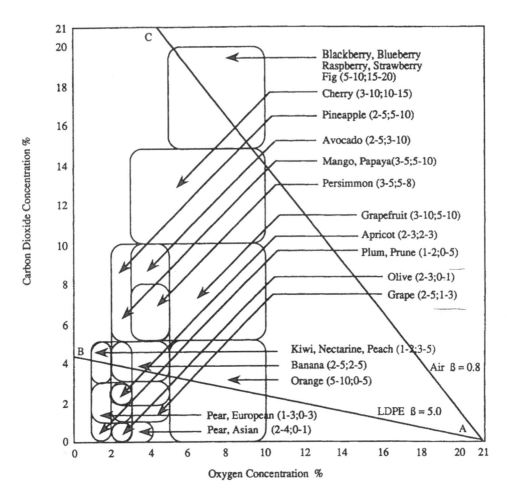

Figure 1 Recommended modified atmospheres for storage of fruits (numbers within parenthesis are oxygen and carbon dioxide ranges).

permeability, film area, and the concentration gradient. Multiplying the weight of the produce by the specific respiration rate, we obtain the respiration rate of the produce. Thus the mathematical model is expressed as follows:

$$WR_1 = P_1 A\{(c_1 \pm x_1)/b\} \tag{1}$$

$$WR_2 = P_2 A\{(x_2 \pm c_2)/b\}, \tag{2}$$

where A = area of the film (m²), b = thickness of the film (m), c = gas concentrations in ambient (mol/m³), P = permeability (m²/s), R = respiration rate (mol/kg·s), W = weight of produce in the package (kg), and x = gas concentrations in package atmosphere (mol/m³); suffixes 1 and 2 denote O_2 and CO_2 gases.

In Equations (1) and (2), there are 11 different variables. By rearranging these equations and introducing relating terms, we can show the effect of these variables on the package design.

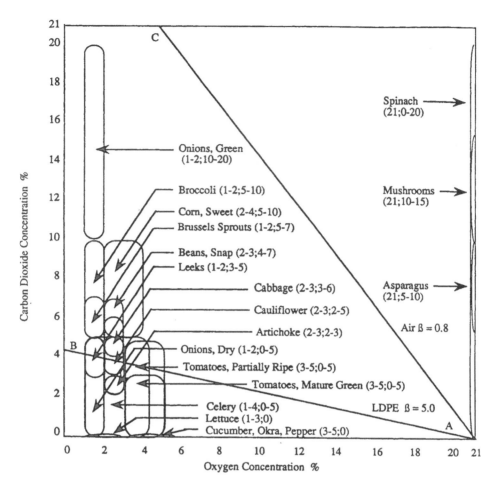

Figure 2 Recommended modified atmospheres for storage of vegetables (numbers within parenthesis are oxygen and carbon dioxide ranges).

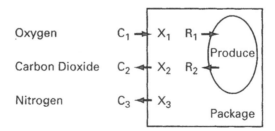

Figure 3 Schematic of a modified atmosphere package.

Influence of permeability ratio

It is well known that the gas permeabilities of commercially available polymers vary over a wide range. However, the ratio of the permeability of any two gases falls within a very narrow range for all the polymers (Stannett, 1968). For most of the polymers used in commercial practice, the ratio of permeability of CO_2 to O_2 (permeability ratio, β) falls between 4 and 6. This narrow range in permeability ratio causes numerous limitations in the design of polymeric packages for fresh produce. We may combine Equations (1) and (2) and the permeability ratio, β, to illustrate this limitation.

$$x_2 = c_2 + \beta^{\pm 1}(c_1 \pm x_1)R_2/R_1. \tag{3}$$

From Equation (3), when the permeability ratio is close to 5, as for low-density polyethylene (LDPE), we obtain a straight line with a slope of 1/5 on the plot of recommended modified atmospheres. In Figures 1 and 2, this line is shown as A-B. This analysis indicates that in a package made of LDPE film, we can generate only those modified atmospheres that are represented by points along the line A-B. For example, we can obtain a modified atmosphere of 2% CO_2 and 11% O_2 or 1% CO_2 and 16% O_2, and so on.

We can then relate the modified atmospheres generated by the LDFE film with products that have recommended requirements of specific modified atmospheres. For example, from Figure 1, the recommended modified atmosphere conditions for cauliflower, celery, and pepper, and those of kiwifruit, nectarine, peach, cranberry, plum, orange, banana, and avocado from Figure 2 lie along the line A-B. Thus for the aforementioned commodities, packages made of LDPE film are suitable, as they can create the recommended modified atmospheres. Conversely, packages made of LDPE film are not appropriate for other commodities that do not lie along the A-B line on Figures 1 and 2.

Influence of respiratory quotient

The slope of line A-B is influenced also by the ratio of carbon dioxide generation to oxygen consumption. This ratio, known as respiratory quotient, α, is close to unity when substrate used in the metabolic process is carbohydrate and sufficient amount of oxygen is available. This ratio is <1 when the substrate is a lipid and >1 when the substrate is an organic acid. When anaerobic conditions prevail, the respiratory quotient is >1 even when the substrate is a carbohydrate.

The influence of respiratory quotient on the package atmosphere can be illustrated by introducing the symbol in Equation (3), as follows:

$$x_2 = c_2 + (\alpha/\beta)(c_1 \pm x_1). \tag{4}$$

In Equation (4), it is obvious that the effect of the permeability ration, β, is opposed by the respiratory quotient, α. Therefore, respiratory quotients greater than unity will result in slopes greater than $1/\beta$ and vice versa.

Influence of package design parameters

When designing packages for modified atmosphere, the values of R_1, R_2, and α are determined by the type of produce placed inside a package. Similarly, the produce also determines the recommended modified atmospheric composition x_1 and x_2, and therefore the value of β. With the preceding quantities, a certain type of film is identified, therefore, fixing the values of P_1 and P_2. As a result of this selection process, the package designer is left with the choice of only three variables: thickness of the film, area of the package, and weight of the produce in the package. The criterion for the selection of these three variables can be illustrated by rewriting Equations (1) and (2), and by introducing a new parameter, ϕ.

$$x_1 = c_1 \pm \left(R_1/P_1\right)\phi \tag{5}$$

$$x_2 = c_2 + \left(R_2/P_2\right)\phi, \tag{6}$$

where

$$\phi = Wb/A. \tag{7}$$

From Equations (6) and (7), we can examine the influence of increasing ϕ. As ϕ increases, the package atmosphere moves toward point B along line A-B in Figures 1 and 2, and vice versa. The quantity ϕ increases by decreasing the area, A, or by increasing the weight of the produce, W, or thickness of the film, b.

The package atmosphere affects the respiration rtes of fresh produce. In most cases, the respiration rate increases with increasing oxygen content. And, the respiration rate decreases with increasing carbon dioxide content or as a package atmosphere moves toward B along the A-B line. This relationship between the package atmosphere and the respiration rate tends to moderate the effect of ϕ. An increase in ϕ increases x_2 and decreases x_1. This will decrease respiration rates R_1 and R_2. Therefore, the final value of x_1 will be higher and x_2 will be lower than those predicted by Equations (5) and (6) where constant values of R_1 and R_2 were used. These relationships are helpful to a package designer. Any errors in values available for respiration rate and film permeability data will result in smaller deviations of package atmospheres. In addition, the effect of temperature on the respiration rate and film permeability will be moderated to a large extent.

Influence of temperature

The respiration rates of produce and the film permeability are influenced by the temperature. Both these effects can be expressed as Arrhenius-type relations in the temperature range encountered in modified atmosphere packages of fresh produce. These relationships are shown as follows:

$$P = P_0 \exp\left[\pm E_p/R'T\right] \tag{8}$$

$$R = R_0 \exp\left[\pm E_R/R'T\right], \tag{9}$$

where R' = universal gas constant (J/mole·K), T = temperature (K), and E = activation energy (J/mol).

The effect of a change in temperature on package atmosphere can be illustrated by combining Equations (8) and (9) and using two temperatures, T_{ref} and T, as follows:

$$R_1/P_1 = \left(R_{1,ref}/P_{1,ref}\right)\exp\left[-\left(\left(E_{R1} - E_{P1}\right)/R'\right)\left(\left(1/T\right) - \left(1/T_{ref}\right)\right)\right]. \tag{10}$$

The difference in activation energy of respiration and permeation governs the change in package atmosphere due to change in temperature. The activation energy of permeability of some films are listed in Table 7. When the activation energy of respiration of some fresh produce in modified atmosphere conditions are within the same range, the

Table 7 Permeability Data of Some Polymeric Films, Air, and Water

	Permeability (std ml·mil/m²·h·atm)		Activation Energy (J/mol)		Permeability Ratio
	O_2	CO_2	O_2	CO_2	
Polyethylene (density 0.914)	310	1360	42,400	38,900	4.44
Polypropylene	180	780	47,700	38,100	4.33
Polyvinyl chloride	4.9	16.9	55,600	56,900	3.46
Polyvinylidene chloride (Saran)	0.57	3.23	66,500	51,400	5.66
Air	2.6×10^9	2.0×10^9	3,600	3,600	0.8
Water	9.3×10^3	2.2×10^5	15,800	15,800	23.3

Note: Data from Mannapperuma, J.D., et al. 1989. In: Proceedings of the Fifth International Controlled Atmosphere Research Conference, vol. 2, Other Commodities and Storage Recommendations, ed. J. K. Fellman, p. 225-233, June 14-16, 1989; Wenatchee, WA.

changes in temperature do not cause any large changes in the package atmosphere (Mannapperuma et al., 1989).

Influence of holes, micropores, and water layers in a package

Packages made of polymeric films are susceptible to puncturing during handling. Sometimes, special microporus films and labels are used in modified atmosphere packaging. The presence of holes in the film allows the flow of gases by diffusion as well as by convection. The analysis of these phenomena is extremely complicated. However, the relative effect of diffusive flow through holes on package atmosphere can be studied by comparing the permeability of gases in air with permeability of the gases in polymers. As seen in Table 7, the air is much more permeable than polymeric films. Therefore, even a very small hole in a polymeric package can affect the package atmosphere very significantly. However, this phenomenon can be used as an advantage by using carefully introduced micropores in the package film.

The permeability ratio β for air may be calculated as 0.8. This represents line A-C in Figures 1 and 2. This implies that an otherwise impermeable package, with a few small holes, can be used to create atmospheres along this line, such as in the requirement for berries and figs in Figure 2. The carbon-dioxide–oxyen atmospheres that lie between lines A-B and A-C may be created by using packages that use LDPE film with pinholes or microporus windows (Mannapperuma and Singh, 1994).

It is common to see an accumulation of water on the interior surface of the film as a thin layer or as droplets due to condensation in fresh produce packages. The effect of such water layers on the package atmosphere may be studied by converting the data on diffusion of gases in water to the same basis as film permeability using the convention proposed by Yasuda (1975). The converted values (Table 7) show that the permeability of gases in water is much higher than in polymers. Therefore, thin layers and droplets of water forming inside polymeric packages do not affect the package atmosphere significantly.

Dynamic behavior of gas concentrations inside a package

In this chapter, the method presented to design a modified atmosphere package described earlier was based entirely on the steady-state or equilibrium behavior of the package and its contents. Normally, when produce is packaged, the internal gas atmosphere may not reach the design conditions for a certain period of time which may be a few hours or a few days. The gas composition inside the package during this initial period may be studied using a dynamic model. Of course, if the gas atmosphere is flushed with the design conditions at the time of packaging, then the following analysis is not necessary.

In order to study the dynamic behavior of the package, the free volume inside the package must be known. The time required by the package to reach the equilibrium conditions is closely related to the free volume inside the package, V. Three mathematical equations of the dynamic model consist of three equations describing the rate of change of the concentrations of the three gases, oxygen, carbon dioxide, and nitrogen. For oxygen,

$$V\left(dx_1/dt\right) = \left(P_1 A/b\right)\left(c_1 \pm x_1\right) \pm WR_1, \tag{11}$$

for carbon dioxide,

$$V\left(dx_2/dt\right) = \left(P_2 A/b\right)\left(c_2 \pm x_2\right) + WR_2, \tag{12}$$

and for nitrogen,

$$V\left(dx_3/dt\right) = \left(P_3 A/b\right)\left(c_3 \pm x_3\right). \tag{13}$$

In addition to the above equations, the ideal gas equation relation has to be obeyed by the gases inside the package. This provides a restriction on total pressure, p, of the package, as follows:

$$p = \left(x_1 + x_2 + x_3\right)R'T. \tag{14}$$

The solution to this set of equations depends on the nature of the dependence of R_1 and R_2 on x_1 and x_2. When the relationship is simple, an analytical solution can be obtained. A numerical technique such as Runge–Kutta or predictor–corrector method may be employed when the relationship is complicated.

Analytical solutions of Equations (11) and (12) were obtained by Deily and Rizvi (1981), assuming that R_1 and R_2 were constant in the case of peaches. Hayakawa et al. (1975) assumed R_1 and R_2 to be piecewise multilinear functions of x_1 and x_2 and obtained analytical solutions using the Laplace transformation method. However, both of these studies do not include Equations (13) and (14). A number of models for packages with constraints on volume and pressure were presented by Mannapperuma and Singh (1987). In these models, the gas concentrations inside a product were considered as a separate set of variables.

According to Equation (14), the sum of three gas concentrations remains invariant in spite of unequal changes in x_1 and x_2. This equality will hold only by changing the volume of the package. In fact, this phenomenon is often observed as a contraction of polymeric packages when a modified atmosphere is established inside a package. Because Equation (14) imposes a change in volume, the system of equations becomes nonlinear, therefore requiring numerical methods to obtain solutions.

It is important to note that the solution of a dynamic model will always converge to a steady-state solution. However, the behavior during the initial stages depends on free volume in the package and the initial gas concentrations. Of course, the steady-state conditions are obtained faster when free volume is smaller and when the initial gas concentrations are closer to the steady state concentrations.

Observations on designing modified atmosphere packages

The permeability ratio of films determines the range of modified atmospherses that may be obtained in MAP applications. As noted in this chapter, this ratio is within a very narrow range for all the polymers. By careful introduction of pinholes or microporous windows, one can extend the range of attainable atmospheres considerably. The atmospheres that lie outside this extended range cannot be attained by the equilibrium interaction of produce respiration and package permeability. Therefore, injecting the required atmosphere into a relatively impermeable package can be expected to maintain a favorable atmosphere for a considerable length of time. This particular concept is frequently used in the air shipment of stawberries. This problem can be analyzed by using the dynamic model as presented in this chapter.

The respiration rate and film permeability are influenced by temperature. However, the change in respiration rate is usually much greater than the change in gas permeability. This can result in anaerobic conditions inside optimally designed packages when subjected to temperature abuse. To avoid such conditions, one may rely on suboptimal designs. It may be noted that the respiration activity is only one of the factors that influences the shelf life. There is a direct correlation between the cummulative respiration activity and the shelf life during storage of some commodities.

Another important factor to consider in MAP applications is the water vapor transfer. When the vapor pressure surrounding a product is too low it causes wilting of the produce. This is often observed in packages where holes are made to allow for respiration. On the other hand, when the vapor pressure is too high then condensation of water occurs. Antifog films prevent accummulation of droplets on the film but not the spoilage effects of free water. The water vapor transfer may be controlled by selective permeation or by the use of absorbers.

In summary, the design of modified atmosphere packages requires the knowledge of mass transport properties of polymeric films and the respiration rates of fresh produce placed inside the packages. Mathematical models are useful in determining the changes in gas concentrations in a package caused by the respiring fruits and vegetables inside the package during storage.

Bibliography

1. Ben-Yehoshua, S. 1985. Individual seal-packaging of fruits and vegetables in plastic film. A new postharvest technique. HortScience 20:32-37.
2. Ben-Yehoshua, S. and A. C. Cameron. 1989. Exchange determination of water vapor, carbon dioxide, oxygen, ethylene, and other gases of fruits and vegetables. In: Modern Methods of Plant Analysis, New Series, vol. 9, Gases in Plant and Microbial Cells, eds. H.F. Linskens and J.F. Jackson, 177-193. Berlin: Springer-Verlag.
3. Blankenship, S. M., ed. 1985. Controlled atmospheres for storage and transport of perishable agricultural commodities. Proc. Fourth Natl. Controlled Atmos. Res. Conf. July 23-26, 1985, Dept. of Hortic., Univ. of North Carolina, Raleigh, NC.
4. Brecht, P. E. 1980. Use of controlled atmospheres to retard deterioration of produce. Food Technol. 34(3):45-50.
5. Brody, A. L., ed. 1989. Controlled Modified Atmosphere Vacuum Packaging of Foods. Trumbull, CT: Food & Nutrition Press.
6. Calderon, M. and R. Barkai-Golan, eds. 1990. Food Preservation by Modified Atmospheres. Boca Raton, FL: CRC Press.
7. Cameron, A. C., Boylan-Pett, W. E. and Lee, J. 1989. Design of modified atmosphere systems: Modeling oxygen concentrations within sealed packages of tomato fruits. J. Food Sci. 54(6):1413

8. Deily, K. H. and Rizvi, S. S. H. 1981. Optimization of parameters for packaging of fresh peaches in polymeric films. J. Food Process Eng. 5:23.
9. Dewey, D. H., ed. 1977. Controlled atmospheres for the storage and transport of perishable agricultural commodities (Proc. 2nd Natl. CA Res. Conf., April 1977). Mich. State Univ. Dept. Hortic. Rept. 28.
10. Dewey, D. H., R. C. Herner, and D. R. Dilley, eds. 1969. Controlled atmospheres for the storage and transport of horticultural crops (Proc. 2nd Natl. CA Res. Conf, January 1969). Mich. State Univ. Dept. Hortic. Rept. 9.
11. El-Goorani, M. A. and N. F. Sommer. 1981. Effects of modified atmospheres on postharvest pathogens of fruits and vegetables. Hortic. Rev. 3:412-61.
12. Fellman, J.K., ed. 1989. Proceedings of the ffth international controlled atmosphere research conference. June 14-16, 1989, Washington State Univ. Research and Extension Center, Wenatchee, WA). Vol. 1, 515 pp.; vol. 2, 374 pp.
13. Hardenburg, R.E., A.E. Watada, and C-Y. Wang. 1986. The commercial storage of fruits, vegetables, and florist and nursery stocks. USDA, Agric. Hb. No. 66.
14. Hayakawa, K. I., Henig, Y. S. and Gilbert, S. G. 1975. Formulae for predicting gas exchange of fresh produce in polymeric film package. J. Food Sci. 40:187.
15. Henig, Y. S. and Gilbert, S. G. 1975. Computer analysis of the variables affecting respiration and quality of produce packaged in polymeric films. J. Food Sci. 40: 1033.
16. Jurin V. and Karel. M. 1963. Studies on control of respiration of McIntosh apples by packaging methods. Food Technol. 17(782).
17. Kader, A. A. 1980. Prevention of ripening in fruits by use of controlled atmospheres. Food Technol. 34(3):51-54.
18. Kader, A. A. 1986. Biochemical and physiological basis for effects of controlled and modified atmospheres on fruits and vegetables. Food Technol. 40(5):99-100, 102-104.
19. Kader, A. A., D. Zagory, and E. L. Kerbel. 1989. Modified atmosphere packaging of fruits and vegetables. CRC Crit. Rev. Food Sci. Nutr. 28(1):1-30.
20. Lougheed, E. C. 1987. Interactions of oxygen, carbon dioxide, temperature, and ethylene that may induce injuries in vegetables. HortScience 22:791-94.
21. Mannapperuma, J. D. and Singh, R. P. 1987. A computer-aided model for gas exchange in fruits and vegetables in polymeric packages. Presented at the 1987 International Meeting of the American Society of Agricultural Engineers, Chicago, IL, December 15-18.
22. Mannapperuma, J. D. and Singh, R. P. 1994. Modeling of gas exchange in polymeric polymeric packages of fresh fruits and vegetables. In *Minimal Processing of Foods and Process Optimization — An Interface*. Eds. Singh, R. P. and F. A. R. Oliveira. pp. 437–458. CRC Press, Boca Raton, FL.
23. Mannapperuma, J. D. and R. P. Singh. 1994. Modeling of gas exchange in polymeric packages of fresh fruits and vegetables. In: Minimal Processing of Foods and Process Optimization — An Interface, eds. R. P. Singh and F. A. R. Oliverira, p. 437-458, June 14-16, 1989; CRC Press, Boca Raton, FL.
24. O'Beirne, D. 1990. Modified atmosphere packaging of fruits and vegetables. In: Chilled Foods: the State of the Art, ed. T.R. Gromley, 183-199. New York: Elsevier Science Publishers.
25. Richardson, D. G. and M. Meheriuk, eds. 1982. Controlled atmospheres for storage and transport of perishable agricultural commodities (Proc. 3rd Natl. CA Res. Conf., July 1981). Beaverton, OR: Timber Press.
26. Rizvi, S. S. H. 1981. Requirements for foods packaged in polymeric films. CRC Crit. Rev. Food Sci. Nutr. 14(2): 111.
27. Smith, S., J. Geeson, and J. Stow. 1987. Production of modified atmospheres in deciduous fruits by the use of films and coatings. HortScience 22:772-76.
28. Solomos, T. 1987. Principles of gas exchange in bulky plant tissues. HortScience 22:766-71.
29. Stannett, V. 1968. Simple gases. In: Diffusion of Cases in Polymers, ed. Crank, J. and Park, G. S., London: Academic Press.
30. Veeraju, P. and Karel, M. 1966. Controlling atmosphere in a fresh fruit package. Modern Packaging 40(2):166.

31. Wade, N. L. and Graham, D. 1987. A model to describe the modified atmospheres developed during the storage of fruit in plastic films. ASEAN Food J. 3:(3&4):105.

32. Weichmann, J. 1986. The effect of controlled atmosphere storage on the sensory and nutritional quality of fruits and vegetables. Hortic. Rev. 8: 101-27.

33. Yasuda, H. 1975. Units of gas permeability constants. J. Appl. Polymer Sci. 19:2529.

34. Zagory, D. and A. A. Kader. 1988. Modified atmospheres packaging of fresh produce. Food Technol. 42(9):70-74, 76-77.

35. Zagory, D. and A. A. Kader. 1989. Quality maintenance in fresh fruits and vegetables by controlled atmospheres. In Q~ y Factors of Fruits and Vegetables—Chemistry and Technology, ed. J. J. Jen, 174-188. Washington, D.C.: American Chemical Society.

36. Zagory, D., J. D. Mannapperuma, A. A. Kader and R. P. Singh. 1989. Use of computer model in the design of modified atmosphere packages for fresh fruits and vegetables. In: Proceedings of the Fifth International Controlled Atmosphere Research Conference, vol 1, Pome Fruits, ed. J. K. Fellman, p. 479-486, June 14-16, 1989; Wenatchee, WA.

chapter seventeen

Ambient storage

Tom C. S. Yang

Contents

0-8493-2646-X/97/$0.00+$.50
© 1998 by CRC Press LLC

Introduction

Storage of raw material is one of the integral aspects of all industrial food processing. In the fruits and vegetables industry, proper storage control is one of the key prerequisites for further processing of fresh produce and successful marketing.[1] After harvest, the metabolic activity of the living plant will continue, causing rapid deterioration of quality and loss of valuable nutrients if storage conditions are not adequate. Controlled atmosphere (CA) storage and modified atmosphere packaging (MAP) have been developed to extend the shelf life of fresh produce, but most need cool temperature as a primary requirement. Ambient storage of foods, which corresponds to an "uncontrolled" environment, often involves a great diversity of temperature, relative humidity, oxygen tension, and light intensity. For instance, many of the deteriorative changes in the nutritional quality of foods are initiated or accelerated by light.[2] Such changes may include the effect of light on initiating free radical reactions involved in fat oxidation, destruction of fat-soluble vitamins, loss of riboflavin and other water-soluble vitamins, on inducing changes in proteins and amino acids, and on inducing changes in food pigments. Most important, many foods stored at ambient temperature would be subjected to frequent changes in these environmental variables during transportation, distribution, and stocking. Almost all canned and dehydrated foods keep well at 4–5°C and in many instances at 21°C. At 38°C, however, a temperature often reached in common warehouse storage, deterioration could be rapid—in some cases occurring within 12 months. Some food products if stored for a prolonged period at a certain combination of ambient conditions, either intentionally or unintentionally, will deteriorate in quality. Consequently, it is instructive to consider the basis for the impact of environment on storageability.

Factors affecting storage stability

Bacterial survival and growth

Water activity, relative humidity, and pH

Most bacteria multiply very rapidly between 21–38°C; therefore, control of postprocess bacterial contamination is very important. Even after 4 weeks at 16°C, 22% of a wet pasta packaged under a CO_2:N_2 (20:80) mixture were contaminated with 10^3 to 10^7/g *Staphylococcus aureus* and 8% of the samples surveyed contained staphylococcal enterotoxin. These results emphasized the importance of refrigeration storage for this type of products.[3]

The survival of microorganisms in foods with water activity (A_w) levels below that required for growth is influenced by factors other than availability of water.[4] Christian and Stewart[5] studied rates of decrease of viable cells of *S. aureus* and *Salmonella newport* in cake mix, skim milk, onion soup, and gelatin-based dessert (A_w 0–0.53) over a 2-week period at 25°C. Above $A_w = 0.22$, survival of both bacteria was influenced by pH of the food, which was greatest in cake mix (pH 6.8) and least in dessert (pH 3.1). Staphylococci were particularly susceptible to the adverse effects of acidity.

The retention of viability of microbial cells during storage in "dry" foods ($A_w < 0.60$) is greatly affected by temperature. Increased rate of inactivation is generally correlerated with increased storage temperature above 0°C.

Table 1 Effect of Water Activity on Half-Lives (Days)
of Patulin and Citrinin in Grains

	A_w	Barley	Corn	Wheat
Patulin	0.7	12.7	4.4	4.4
	0.9	6.8	2.4	1.9
Citrinin	0.7	7.8	15.5	11.9
	0.9	1.8	10.4	3.0

Studies of the survival of conidiospore of *A. flavus* over a 48-week period at 21°C in lemon-flavored gelatin, flaked coconut, potato fakes, cocoa, peanut flour, wheat flour, corn flour, and cake mix adjusted to A_w values ranging from 0.32–0.78 have revealed that the rate of death is enhanced by increasing A_w and, in general, decreasing pH.[6] Rayman et al. studied the death kinetics of several microorganisms during storage of dried (12% moisture) pasta at room temperature; they reported that *D* values for vegetative cells of *Saccharomyces* and conidiospore of *Penicillium expansum* ranged from 40–54 and 130–160 days, respectively.[7]

The stability of mycotoxin as affected by A_w has not been studied extensively. In general, vegetative cells and spores of microorganisms are more tolerant to heat when the A_w of the suspending medium is reduced by additional solutes. Thus, at a given elevated temperature, a longer period of time is required to reduce viable populations in "dried" foods compared to "wet" ones. For example, the effects of A_w on the halflives of disappearance of patulin and citrinin from grains have been reported by Harwig et al.[8] and are listed in Table 1.

Solutes appear to have a protective effect against loss of bacterial viability during long-term storage. The relative humidity of the atmosphere in contact with microbial cells is known to influence the degree of lethality of gasses used for pasteurization. Tawaratani and Shibasaki showed that the lethal effects of propylene oxide increased with increasing moisture content of mold spores and the relative humidity in the exposure atmosphere.[9] Rates of inactivation of *Escherichia coli* have similarly been demonstrated to be enhanced in atmospheres with elevated relative humidity.[10]

Knowledge of the behavior of microorganisms in raw and processed foods and feeds over a range of A_w can be extremely valuable. Microbial stability of grains in silos, for example, and the shelf life of intermediate-moisture and high-moisture foods are largely dependent upon A_w. Control of A_w is critical at every step in the harvesting, storing, processing, and marketing chain if microbial deterioration is to be prevented. Staphylococci are widely distributed in meats and are known to be capable of growth under adverse conditions like those prevailing in fermented sausage. The combinations of $A_w = 0.925$/pH 5, or $A_w = 0.915$/pH 5.5, or $A_w = 0.9$/pH 6, together with the addition of sodium nitrite and nitrite, resulted in synergistic growth inhibition of *S. aureus* in salami systems.[11] With the exception of *S. aureus*, the minimum A_w levels for growth of toxin production by bacteria known to cause food borne infection or intoxication is ≥ 0.91. The minimum A_w at which molds and yeasts are capable of growing is about 0.61. Growth of molds on cereal grains and oilseeds has gained new significance since the discovery of various mycotoxin over the past 2 decades.

Preservatives
Several chemicals or procedures shown below were found to increase the storage life of a food when it was stored at 38°C and RH 75% and the mechanism for their effectiveness was described by Melpar, Inc.[12]:

Preservatives

>Cysteine
>Thiosalicylic acid
>Monothioglycerol
>Acetic anhydride
>Calcium hydride
>Lithium aluminum hydride
>Sodium hypophosphite
>Potassium borohydride
>Iso-octyl thioglycolate
>Nitrous acid

Preservation Procedures

>Treatment with glucose oxidase, catalase, dextrose with the pellet packaged in metallized aluminum foil
>Flushing with carbon dioxide and packaging under carbon dioxide
>Immersing in liquid nitrogen and packaging under nitrogen
>Displacement of the oxygen from the surface of the food with beeswax

This report also recommends preservation of the foods at 38°C and RH 75% in the following way: (1) for haddock, carrots, potato granules, and spinach—purge with CO_2 for a minimum time of 1 day but preferably for 1 week; (2) blend into the food a sulfur compound soluble in water, such as cysteine and a sulfur compound, or an iso-octyl thioglycolate; (3) place food with solid carbon dioxide in a rigid container of glass or metal, and seal the container after vaporized CO_2 replaces the air in it.

Sych et al. examined the effects of initial moisture content and storage RH on textural changes in layer cakes during storage for 42 days at 20°C.[13] Increased cake moisture content led to a decrease in initial cake firmness, but did not reduce final cake firmness. Without proper packaging, freshly baked bread will undergo a rather rapid change in texture (i.e., staling). In addition to special ingredients (various dough conditioners and emulsifiers) used to maintain the desirable texture of fresh bread, packaging it in a moisture-impermeable plastic is an effective way to preserve the quality of the final product. Retaining moisture in the bread, by preventing evaporation from the package, is an important prerequisite for retarding staling. The moisture-impermeable packaging, however, results in development of high humidity inside the package, which would enhance spoilage due to mold growth if food-approved chemical preservatives (e.g., propionic acid) are not used.[1]

Hurdle technology

Many of the aforementioned storage factors are the foundation for food preservation by combined methods—the so-called hurdle technology. Leistner stated that foods preserved by hurdle technology remain stable and safe even without refrigeration, and are high in sensory and nutritive attributes due to the minimal processing applied.[14] The original application was on shelf-stable meat items, and was later used for the preservation of many other foods. To secure the total desired quality of a food, the safety and quality hurdles should be kept within the optimal range. Leistner further linked the hurdle technology with the hazard analysis critical control points (HACCP) concept for producing shelf-stable food products with optimum quality.[15]

Chemical reactions

Some examples of chemical deterioration that occur during storage follow.

Browning reaction

This type of deterioration results in darkening or "browning" of the product, a loss in flavor, a decrease in certain nutrients, undesirable changes in taste and texture, and even the production of toxic substances. Some of the commodities that undergo this type of deterioration are dried fruits and vegetables, powdered eggs and milk, fruit juice concentrates, certain beverages, jams, jellies, canned bread, certain canned vegetables, meat items, and many other foods. To demonstrate the relationship between chemical changes and acceptability as temperature rises, a sharp increase in browning, as measured by the saline fluorescence, was found in dehydrated white potatoes stored for 2 years at 38°C. These potatoes were not acceptable for use after 8 months; similar results were found with sandwich-type cookies, for which that the acceptability was limited to 6 months at 38°C. Extreme deterioration was observed in many military rations stored at 38°C for 2 years. There was excessive browning observed in the cereal bar, giving it a distinctly mottled appearance.[16]

Fat deterioration

Fats undergo changes resulting in the production of off-odors and taste, loss of nutritive value, and perhaps the production of toxic substances. A few items so affected are biscuits, cookies, prepared mixes, ice cream powder, dried whole milk, etc. Crackers used in C-rations showed a sharp rise in peroxide value, indicating fat oxidation, and a rapid drop in acceptability at 38°C. Cheese spread, another example, showed declining acceptability tied to increased saline fluorescence, free fatty acid (indicating fat hydrolysis), and riboflavin losses at 38°C.[16] Snacks traditionally have had a limited shelf life; the normal shelf life for salted potato chips is 10 weeks, and for tortilla chips is 12 weeks. By applying natural antioxidants (e.g., natural tocopherol and oil of rosemary) with salt crystals, shelf life can be extended for additional 2–4 weeks.

Changes in texture

Several types of chemical changes could take place to cause textural changes. In dried cranberries, changes in the pectin molecule may occur to the extent that the gelling power is decreased or lost. In other cases, the product may actually toughen. Other products such as dry apricots, peaches, and pound cakes will no longer be suitable for use after approximately 18 months of storage at 38°C. A chemical change in the sucrose was found in the starch–jelly candy stored at 38°C. The acidity of the candy (pH 4.5) had caused the sucrose to invert to glucose and fructose, which causes the candy to become more sticky and gummy.[17]

Color changes

It is well known that the naturally occurring colors of many food items undergo undesirable changes during storage. The anthocyanin pigments of strawberry jam gradually disappear with storage. At the same time, browning takes place with the result that the jam becomes brown or black. Canned items such as berries, prunes, and beets undergo color changes during storage under adverse conditions. Soluble tea and coffee products take on a black, tarry appearance. This is probably caused by absorption of moisture. Bouillon powder can become so dark that it is described as similar to charcoal. Gum pellets become cream-to-tan in color and chocolate bars become dry, almost powdery, and show considerable evidence of "bloom." Fruit bars take on a mottled black-and-brown appearance. In the studies showing these results, it was quite evident that there had been a transfer of flavor from one item to another, and chemical analyses disclosed that moisture transfer had occurred.[17]

Staling reaction

In principle, staling is related to a complex chemical phenomenon of starch retrogradation, during which the hydrated, gelatinized starch granules are converted to a more permanent, hard structure. Staling is influenced by temperature (most rapid at 0°C) and by the size, shape, and concentration of starch molecules; however, retrograded amylopectin, but not amylose, can be reversed by heat.[17]

Vitamin changes

Frequently, there is a gradual loss in vitamins during storage. This is true in practically all foods that have been studied. High temperature further reduces the digestibility. In a study of combat rations stored at 26.7°C for 24 months, minimum ascorbic acid retention ranged from 48% in apricots to 84% in Alaska peas; minimum thiamin retention ranged from 46% in green asparagus to 72% in sweet peas; minimum niacin retention ranged from 71% in green asparagus to 94% in lima beans; and minimum riboflavin retention range from 27% in green beans to 77% in sweet peas. Carotene was least affected by all storage conditions.[18] In a storage study conducted on individual combat ration at 38°C for 18–24 months,[19] no destruction of essential amino acids was observed. However, rations stored at 38°C for 12 months or cycled between 21–43°C for 18 months did lead to reduced availability of amino acids as demonstrated biologically by a decrement in protein digestibility and in rat growth.

Packaging considerations

Can corrosion

Moisture deposits on products removed from cold storage and movement of vapor after placing unpacked goods into or out of cold storage are some of the major reasons for can corrosion. Water in the liquid phase is necessary for the rusting of cans, and such corrosion increases with time and temperature.[20] Table 2 shows such a relationship for wet cans of vegetables held in an atmosphere saturated with moisture. In various cartons held under unfavorable conditions, condensation formed first on the sides of cans nearest the case wall, but was heaviest on cans in the center of the cartons. Factors contributing to moisture condensation on packaged canned products were (1) entrance of vapor from without, (2) proximity to damp carton wall or similar surface, and (3) slow warming of product and slow drying of can surface. Condensation on improperly constructed cases caused carton "fatigue," weakening of the glue, wrinkling and staining of labels, and rusting of cans. Features of a protective package were (1) container wall made impervious to water vapor by aluminum foil, paint, asphalt, or other materials, (2) flaps glued in place, and (3) opening sealed with waterproof tape. Pads, dividers, absorbers, or desiccants to remove traces of moisture were beneficial.

Assembly conditions had more influence on condensation inside the cases than did conditions of either storage or warm-up. Condensate was heavier on dense products with high heat capacity and thermal conductivity, such as canned apricots, corn, or meatballs and spaghetti, than on lighter products such as soluble coffee, dry cream, or cookies. Rusting was accentuated by mechanical injury or other imperfections, including

Table 2 Time–Temperature Profile for Rust to Develop on Wet Cans of Vegetables Held in a Saturated Moisture Atmosphere

Temperature (°C)	Storage Limits
26.7–37.8	4 hours
15.6–21	1 day
10	2 days
4	4 days
0	6 days
–6.7 or below	≥ 90 days

preexisting spots of surface or pitted rust on the can surfaces, and by the presence of any type of corrosive or hygroscopic contaminant not completely removed before storage.[21]

The harmful effect of cargo damage, caused by condensation developed during transportation, can often translate into lost revenue and jeopardize credibility for manufacturers and distributors of food products. One way to overcome this problem is to use lightweight desiccant bags in between packages of cargo as they are being loaded into a container prior to shipment and then to remove them upon arrival. The quantity of bags depends on time and mode of transportation, quantity and composition of the transported items, length of the trip, and temperature and climate change experienced during transportation.

Package-flavor interactions

Packaging materials often function as a dynamic system that will affect the stability and composition of flavor and aroma components.[22] Some examples of detrimental effects include the following: (1) light penetration into packages can lead to photooxidation of food ingredients, such as lipids in milk products, resulting in oxidized flavor development; (2) exposed metal surfaces can catalyze the production of oxidative free radicals that react with flavor components; (3) metal ions can also form complexes or chelates with other ingredients, resulting in mushy, bitter, or astringent off-flavor; (4) flavor materials can migrate out of food products (e.g., *d*-limonene, benzaldehyde) or migrate into food products from the atmosphere (e.g., garlic, heptane)—some flavor ingredients can also adsorb from the food product onto packaging material surfaces; (5) packaging material, solvents, or contaminants (e.g., acrylonitrile, toluene) can migrate into food products and cause off-flavors; (6) moisture transfer into foods can accelerate enzymatic reactions resulting in off-flavor development; and (7) oxygen migrating through packaging materials can promote oxidative deterioration of flavoring materials or off-flavor generation.

Packaging for oxygen and moisture sensitivity

To inhibit oxygen induced deteriorations, there are several oxygen-absorbing materials available, some commercially marketed and others described in patents. An oxygen scavenger developed by Mitsubishi Gas Chemical Co. in Japan and based on reaction with iron compounds is an excellent example of chemically removing oxygen. It can prevent oxidation and minimize deterioration of fats and oils, discoloration, and rust formation of metal containers; it also preserves flavor, taste, and oxygen sensitive nutrients. It prevents vermin damage, thus preserving grains, beans, and their packages. Of particular significance is its ability to prevent the growth of microorganisms (mold and aerobic bacteria) and thus to preserve such foods as pizza crust, salami, sponge cake, and bread, as well as its ability to be a simple controlled atmosphere device to maintain the freshness of fruits and vegetables.

The idea of utilizing organic oxygen-absorbing materials to remove oxygen from food/beverage processing and packaging streams had been developed recently.[23] The molecular structure of the material can be engineered to control the driving force for oxygen binding and the rate of reaction, so as to irreversibly absorb oxygen or to recycle it. This technology can be applied to varieties of foods—from solid or powder to viscous or liquid—and be applied during processing, or during bulk storage, or within individual packages. In bulk shipments, recirculating fluid, membrane-based systems could continually absorb oxygen from the product during transit. It can also be coated onto the interior of closures or container sidewalls (and can serve double duty as temperature indicators that change color when exposed to oxygen) or it could be coextruded right into the container polymer.

A state of the art polymeric oxygen scavenger system was introduced at the Future-Pak '89 Ryder Conference by CMB (Carnaud Metalbox Packaging Technology plc, England).[24] It is a monolayer PET container with built-in, active oxygen inhibitor; by blending PET with an oxidizable nylon and a cobalt carboxylic acid salt (which acts as the metallic catalyst), the packaging can eliminate oxygen. The cobalt catalyzes the reaction of the oxidizable compound with any oxygen that passes through the container wall, thereby preventing the oxygen from reacting with the food product. Bottles made from this blend were successful in demonstrating the inhibition of oxygen reacting with such oxygen-sensitive products as orange juice, beer, and wine. The effectiveness of this container is a function of the structure's oxygen permeation rate and its oxygen scavenging capacity. It also depends on container wall thickness, percentage of oxidizable compound and cobalt in the blend, temperature, and relative humidity.

OXYGUARD® package (Toyo Seikan Kaisha, Ltd., Tokyo, Japan) is an oxygen-absorbing container that can absorb residual oxygen and permeated oxygen during retort sterilization and storage. It takes a small amount of moisture to activate the oxygen absorption, so is suitable for such products as cooked rice, pasta, baked foods, beverage, stew, tuna, fruit jelly, and fruits. It also has medical use as a transfusion solution package.[25]

For moisture-sensitive foods, there is a new system that incorporates desiccant crystals right into the closure, thereby removing any chance of their accidental ingestion. The system can promote desirable product aroma, flavor, and texture over extended, multiuse storage. It might restore a soggy cracker or snack chip to a crisp state. As one of the quality indicators, the crystals change color from a bright royal blue to a light blue to pink when the moisture absorption takes place. The consumer can then simply place the closure into a microwave or conventional oven to restore its original absorption capabilities and blue color.

Determinination of shelf life

Accelerated testing at 38°C is based on the principle that "reactions are reduced half by each drop of 10°C in temperature."[26] After determining the number of days it takes for flavor, color, texture, consistency, and nutritional qualities to deteriorate, this number may be doubled for storage at 28°C, and again at 18°C, 8°C, and –2°C, etc. An example is that 15 weeks at 50°C will be equivalent to 120 weeks at 20°C. Any product successfully passing the 18 months' storage at 38°C will usually be the equivalent of more than 3 years storage at 25°C. An Arrhenius-like mathematical model was developed by Ross et al. to estimate the dependence of average consumer-acceptance scores on storage time and temperature, as well as the effect of temperature on shelf life of the military ration, "Meal, Ready to Eat" (MRE), stored at 4°C, 21°C, 30°C, and 38°C for ≤ 60 months.[27]

Chepurnoi et al. analyzed various infant milk formulae, chicory extract, and honey during storage at 20°C for 2–120 months.[28] They found that the concentration of hydroxymethylfurfural increased with storage time and temperature, as well as with the length of time the product had been heated. It also varied with the source of carbohydrates (e.g., lactose and malt extract). Such data correspond directly with the influence of temperature and time on chemical reactions and reflect the effect on associated quality attributes. More information on using this type of chemical markers to track the time-temperature profile of a product either through processing or during storage is covered elsewhere in this book.

Practical applications

Bakery/cereal products

Canned pound cake should be stored for less than 18 months at 38°C to avoid discoloration and the development of off-aroma and -flavor[29]; premixed cereal remained acceptable for 2 years at 15.6°C/60% RH if packaged in grease-proof bags.[30] Army-type biscuits made with 100-hour shortening has a 12-month shelf life at 38°C when stored in sealed or breather-type cans.[31,32] Storage life of soda crackers is 12-18 months,[29] and crackers canned alone have better acceptability than those packed with cocoa or coffee, due to flavor transfer. Spaghetti in fiberboard cartons has a 9-month shelf life at 32°C with rancidity, checking, or cracking, and mustiness as signs of deterioration.[33]

Bread usually retained its fresh texture and flavor for 1 week at –9.4°C or 4 weeks at –17.8°C, but lost freshness thereafter. A bread product in the form of a miniloaf or roll and packed for individual use in a flexible pouch has been developed primarily for inclusion in the MRE pack.[34] Its development has been pursued as part of the efforts to improve the MRE, and its application can be extended to support other operational rations, such as hamburger buns for the "tray can" ration. This highly desired complement to the MRE crackers has been specially formulated to resist quality changes upon long-term storage, even at high temperatures. A combination of ingredients including sucrose esters, humectant, emulsifiers, and an antimycotic has proven to be effective in preventing microbial growth and in minimizing staling. Additional protection is afforded by a low oxygen tension in the pouch, achieved using a simple oxygen scavenger. Accelerated storage of samples at 63°C had demonstrated the stability. Briggs took the concept and further developed a shelf-stable sandwich-type individual ration (Figure 1).[35] The product contains a shelf-stable meat item enrobed in a shelf-stable bread to form a roll-type, pocket-size ration.

Stollman and Lundgren conducted a study to evaluate the textural attributes of the crust and crumb of white bread produced with four sets of time–temperature conditions and stored for 0–7 days at 21°C.[36] No significant organoleptic differences were observed in freshly baked (i.e., within 2 h) loaves prepared with different sets of conditions. During the first 2 days of storage, the crust was significantly more brittle and the soft core of the loaf was larger in the bread made on the slowest production line (i.e., longest times and lowest temperature) than in bread made on the faster production line. After 2-day storage, the time factor started to dominate textural differences, so aluminum foil as the packaging had a more marked effect on the bread crust than on the crumb. During the first 24-h storage, brittleness of the crust decreased drastically while its toughness increased, with maximum toughness after 24 h. At the end of the storage test, the crumb became tougher and less able to be rolled it into a ball, and an increase in crumbliness and dryness was observed. Maximum spring-back of the crumb occurred after 1 day of storage.

Compressibility of baked goods stored in pure carbon dioxide and processed and stored under carbon dioxide or air atmosphere was evaluated by Knorr.[37] Storage in carbon dioxide significantly reduced compressibility of French bread and white bread. Bread processed under carbon dioxide was softer than the air-processed samples, especially

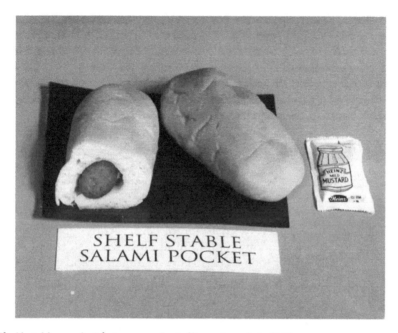

Figure 1   Shelf-stable sandwich-type product. (See color plate P–1)

when no relative humidity control was maintained during storage. Lower water activity values resulted when bread samples were stored in air than in carbon dioxide.

Ochi et al. investigated the stabilities and antioxidant effects of α- and δ-tocopherol in cookies.[38] After losing 20% of both tocopherols during baking, more loss in α-tocopherol was found when cookies were stored at 25–60°C than that in δ-tocopherol. However, adding whole milk powder or egg (1–5%) tended to retain both tocopherols. δ-tocopherol is especially effective against oxidation, with peroxide values of lipid fraction in cookies being 0 just after baking and only about 1.0 after storage at 40°C for 5 months. Pafumi and Durham used a commercially available antimold agent, which employed ethanol vapor as the preservative, to effectively inhibit mold growth in a packaged Madeira cake.[39] However, sensory deterioration was not successfully extended beyond 3 weeks of room temperature storage. Another approach that can be used is to incorporate encapsulated food-grade ethanol into sachets of paper/ethyl vinyl acetate and to pack the food with the sachet; ethanol is slowly released into the package headspace.[40] Freshly baked apple turnovers ($A_w = 0.93$) with this type of package showed a minimum of 21-day shelf life at 25°C, without any yeast growth (*Saccharomyces cerevisiae*). Ethanol absorption by the turnovers was very low.

To determine if retrogradation and storage temperature were indeed correlated, Zeleznak and Hoseney stored bread samples for 1, 3, and 5 days at 4°C, 25°C, and 40°C and analyzed the change in enthalpy values, which reflect the extent of recrystallization by differential scanning calorimetry (DSC).[41] Enthalpy values increased at higher storage temperature. However, rate of recrystallization was not linear at any temperature studied. They concluded that enthalpy values tended to converge with time, irrespective of storage temperature. Another effect of increased storage temperature was an increase in the peak temperature (T_p) of the amylopectin endotherm from which the enthalpy is derived. This indicates either that storage temperature affects the type of crystal being formed as the starch recrystallizes or that annealing of crystals occurs at higher storage temperatures. X-ray diffractometry of starch isolated from the crumb gave similar diffraction patterns, indicating no difference in crystal type among treatments. In addition, the DSC endotherm became sharper at higher temperature. Thus, increase in T_p was apparently due to annealing of crystals at higher storage temperatures.

Dehydrated fruits and vegetables

Typical dried fruits and vegetables have a 1-year shelf life at 38°C and 1 to 2 years at 21°C.[42] Darkening and loss of flavor of dried fruits are the major types of deterioration in storage. Dry apricots, for example, will show increased oxygen consumption, carbon dioxide production, disappearance of sulfur dioxide, and darkening with increasing temperature. Molds and yeasts grow slowly in dried fruits at 4°C, unless RH is low (i.e., ≤50%). The optimum conditions for storing these are 55% RH and a temperature just above freezing (–3°C to –5°C), which also reduces the sugaring effect. Before storage, dried fruits should be brought to the desired moisture content and packed in moisture-proof containers. For dried apples, browning develops gradually at temperatures of 4°C and above. To increase the stability of dried fruits and juice, one can reduce pH by adding acids or blending with more acid fruits (e.g., kiwi fruits) or juice (e.g., guava juice), increase solids using sugar (dry or liquid), pack under vacuum or in nitrogen, and add antioxidants such as ascorbic acid. By using a modified atmosphere package (MAP) system, a snorkel-type vacuumizing equipment, and a high-barrier packaging material, the shelf life of many bulk-packed, freeze-dried ingredients can be extended from 30 days to 1 year.[43] Bakar et al. examined the storage stability of a spray-dried coconut milk powder prepared with addition of skim milk and dextrin and stored for 6 months at 30°C/80% RH.[44] Lipid oxidation is the limiting factor in the stability of the powder. Its maximum shelf life is 4 months when air or vacuum packaged in aluminum laminates and 1.5 months in polypropylene.

Branch et al. studied the effect of hot water (80°C) immersion on the quality of ambient-stored peanuts.[45] By immersing peanuts in hot water for 90 s, the process disrupted the structural make-up of the surface to form a glaze and caused coalescence of subcellular bodies, as revealed by scanning electron microscopy. The hot water immersion decreases lipoxygenase activity, lowered peroxide values, and lowered free fatty acid values. Hence, it ensures better quality at the end of 8 month storage at 23–35°C.

An intermediate moisture (IM) product, such as dry or semi-dry fruits and vegetables, offers a high nutritional content, because dehydration has little effect on the mineral content, and leads to vitamin losses equal to or less than those of other preservation methods. The same weight of IM blueberry may have more than five times the mineral and stable vitamin contents than the fresh counterpart. The ready-to-eat texture and easier rehydration have made the IM berries a promising product in the snack and convenience food industry. While sun drying is one of the more primitive and inexpensive methods of drying certain fruits, it is limited to climates with hot sun and dry atmosphere.[46] An explosion-puffing process.[47–49] has been applied to dry fruits and vegetables, but many berry fruits, due to their smaller size, thinner skin, and delicate flavor, might not be suitable under the high pressure and temperature associated with this process.

Most products dried with mechanical methods are eventually rehydrated, since a bland flavor and a spongy, woody texture are found when these dry products are consumed directly. Rahman et al. proposed a thermal conditioning to plasticize the crunchy and sometimes fragile freeze-dried fruits to produce a space-saving, compressed, dehydrated product.[50] Freeze-drying can result in excellent quality.[50–52] However, a major drawback of freeze-drying is the economics of the process, which limits large-scale application. Several accelerated methods were proposed with some improvement in reducing processing cost.[51–53] Several alternative methods to freeze-drying have also been explored.

Osmotic drying, proposed by Ponting et al.[54] and Farkas and Lazar,[55] can predehydrate the fruits by removing with either dry sugar or syrup 50% of the initial weight of the fruit.

Since the cost of freeze-drying depends on the amount of moisture removed, the pretreatment with osmotic drying would reduce the processing expense. Also, the syrup remaining after osmotic drying can be recycled, as suggested by Bolin et al.,[56] or be used for other value-added products.[57,58] A raisin-type product was prepared by Yang from lowbush blueberries using a sequence of osmotic dehydration, freeze drying, and thermal plasticization.[59,60] Many similar products from cherry and cranberry were commercialized later using more economical dehydration methods. By evaluating qualities of osmotically dehydrated blueberries, residual syrup, and final products prepared with various ratios of berries and sugar, it was found that individually quick-frozen (IQF) berries were preferred to fresh berries as a starting material, and a berry to sugar ratio of 3:1 or 4:1 was appropriate. Final products had good texture, flavor, and overall acceptability, and a predicted shelf life of 16 months at 25°C. Torreggiani et al. used chemical, physicochemical, and sensory evaluations to select suitable varieties of cherry for processing into shelf-stable products by such osmotic dehydration.[61] The preserved cherries should have good flavor, texture, and especially color; they indicated that not all varieties were suitable for this type of product.

Attempts had been made to produce a direct-consumed fruit chip by Koshida et al.,[62] who first soaked diced or sliced fruit chips in a 5–30% sugar syrup and proceeded with a sequence of freeze drying, vacuum–microwave drying, and a final vacuum drying to achieve a final moisture of less than 5%. The product can be consumed as a crunchy confection-type snack. Another type of snack can be made from fruit puree that is either restructured by reshaping or extruded and combined with alginate to give a chewy texture with excellent flavor and color.

Combination drying methods were reported by Cohen et al. to tailor a centrifugal fluidized bed dryer, a material-carrying ball dryer, and a microwave-assisted freeze dryer for better-quality products with reduced processing cost.[63]

To retain quality of heat-sensitive liquid products, there are several novel technologies available to remove excess water from the products. One is a uniquely designed centrifugal-flow, thin-film vacuum evaporator (Okawara Mfg. Co., Ltd., Japan) that allows very low processing temperatures (e.g., 60°C) and extremely short holding times. The ultrathin film provides such enormous contact surface area for evaporation that the residence time is often less than a second. This technique would be ideal for heat-sensitive liquids as well as for liquids that foam or froth and could be beneficial in reducing liquid volume before freeze drying. Another technology is a pulse combustion drying system (Hosokawa Bepex Corp., MN) that applies high sound levels to enhance the drying process. The high-frequency sound wave easily separates water molecule from food particles and provides a very effective method to dry heat-sensitive liquid, and within a very short time. Recent progress in food dehydration has been reported by Cohen and Yang.[64]

Dehydrated dairy products

Estimated storage lives of some dry dairy products are given in Table 3. Nonfat dry milk is acceptable after 1 year at 18–35°C and at a relative humidity of 30–75% if sealed in jars[65] or 2 years at 15°C/60% RH in fiberboard drums.[30] Whole milk powder needs to be packed in nitrogen,[66] and such product has more than a 1-year shelf life at 21°C. With moisture under 2.5% and if packed in less than 1% oxygen, a shelf life of 2 years at 38°C can be expected for their product.[67] The major browning rate of protein–sugar reaction is largely determined by the moisture content (5% or lower) of the powder and is independent of oxygen in the container.[68,69] For successful storage at 38°C, Remaley et al. suggested preheating milk in excess of the standard for pasteurization and packaged with less than 1 ppm of copper, 4 ppm of iron, 2% moisture, and 2% oxygen.[70] Bryce and Pearse reported that dry milk with 30% butterfat was more stable than that with 26% or 28% at 27°C or lower,

Table 3 Estimated Storage Lives of Some Dry Dairy Products

	Lifetime (months)	
	21°C	32°C
Dry whole milk in can	10–12	5–6
Instant dry milk in carton	6	4
Dry malted milk in can	12–24	12

while at 38–60°C this trend is reversed.[71] A moisture content of 3% was also preferred to 2% or 5% at each temperature. Hollender suggested that dry milk be stored at or near the melting temperature of the fat to increase dispersibility, especially with surfactant added.[72] Nitrogen packed, spray-dried whole milk powder,[73] when stored 6 months at 21°C, retained its vitamin A- and B-complex contents while retaining less than 15% of the vitamin A if stored for 6 months at 38°C.

Brenner et al. reported a cocoa beverage powder to have acceptable quality and to have retained 85–100% of added ascorbic acid after 6 months storage at 38°C.[74] Rinschler further reported a 3-year shelf life at 15°C/60% RH if it is packaged in grease-proof bags and chipboard cartons.[30] A 15-month shelf life at 35°C/50% RH was found in dry coffee cream packaged under vacuum or nitrogen in cellulose acetate, aluminum foil-Pliofilm® laminated pouches.[75] Storage life of ice cream mixes[33] is 12 months at 21°C and 4 months at 32°C; above 21°C the powder became darker, lumpy, less soluble, and developed tallowy and rancid flavors. The discoloration and off-flavor can be delayed if ice cream mixes are packed in nitrogen. Margarine in cans[33] has an estimated storage life of 2 years at 21°C and 3–12 months at 32°C.

Dried egg product

While fresh eggs are sold primarily to appeal to consumers' interests in fresh food, the industrial egg-white powder is sold for its functional properties (i.e., whippability and heat coagulability). Rao and Murali studied spray-dried, foam-mat-dried, and freeze-dried whole egg powder,[76] and reported that, despite the methods of dehydration, no significant changes were observed after being stored at 38°C for 6 months, in some of the key amino acids, namely, lysine, methionine, threonine, tryptophan, and cystine, as well as in the protein efficiency ratio.

Dry fish and fishery products

The final quality of these products upon storage is critically dependent on humidity. Usually, molds start to grow as the humidity pick up rate reaches 25%; bacteria start to cause decay at 40–50%. Accordingly, storing the products at low humidity with moisture absorbents enclosed in the container is recommended. Polyethylene or moisture-proof cellophane should be used as packaging material. Oxidation of the highly unsaturated fatty acids in fish meat plus the hydrolysis caused by the lipase are responsible for quality deterioration. Storage of dried fish at room temperature and at low RH is not feasible because of increased oxidation, particularly with mackerel, whose storage stability is low in terms of lipid oxidation even at an RH of 70–90%. However, the storage of herring and anchovy at ≤90% RH is feasible in terms of stability toward lipid oxidation and microbial growth.[77] Packaging materials with low oxygen permeation rate in addition to the use of antioxidants such as BHA or ascorbic acid will be ideal for shelf life extension. Insect infestation is another major problem of dry fishery products, especially at warm climates. Storing in a tightly sealed container or at low temperature can reduce the insect problem.

There are many fumigants available, but their toxicity and safe operating procedures limit their application; the use of dry ice and the vaporized carbon dioxide is effective but the cost is high. High-temperature treatment such as with microwaves or exposure to radiation is suitable for destroying insect eggs.

Preserves

The preservation principle in this type of foods is to control water activity by adding substantial amount of sugar, thereby reducing potential bacterial spoilage. To study storage stability, O'Beirne et al. prepared such preserves from frozen black currants, gooseberries, raspberries, strawberries, and rhubarb and from fresh wild blackberries by two processes: (1) adding "Sure Set®" jam sugar (a mixture of sugar, pectin, and citric acid) and boiling for 4 min, and (2) adding granulated sugar and fresh lemon juice and boiling for times ranging from 11 min for gooseberries to 27 min for black currants.[78] The jams were stored in glass jars for 6–7 months at ambient temperature, in darkness. Results indicated that method 1 produced jams that had a significantly fruitier and more acceptable aroma and flavor than jams produced by method 2. More natural fruit color was retained by method 1. Ascorbic acid contents were largely unaffected by the process used, except in the case of black currant jam, in which case the ascorbic acid content was much higher with method 1 than with method 2. The aroma and flavor benefits of method 1 were mostly lost after storage for 6–7 months. Although the high concentration of sugar and a good seal are usually adequate to prevent fermentation or molding of preserves, cool storage is an extra safeguard and aids in the preservation of color.[79] Protection from light is also desirable to avoid fading or other discoloration.

Canned fruits and vegetables

Retention of physical quality and palatability are the most important criteria for determining the storage life of canned fruits. The order of breakdown in quality of canned fruits and juices is flavor, color, texture, and nutritive losses. The critical storage temperature for most fruits and juices is 27–29°C, and temperatures higher than this should be limited to a few weeks. Brenner et al.[74] and Brenner[80] studied several representative canned foods used in Army rations and reported that apricots had less than 12 month shelf life at 38°C, while green beans, yellow corn, peas, spinach, limas, and carrots had 18 months at 21°C, 32°C, or 38°C. Canned fruit juice will lose half of its vitamin C in 24 months at 27°C, and in 4 months if stored at 38°C. Orange juice is acceptable up to 12 months at 21°C, while tomato juice is acceptable for 18 months at 21°C and 32°C, but only for 12 months at 38°C. Usually, the deterioration is reflected in fading of color, cloud formation, gelation, staleness, off-flavor and off-odors, nontypical texture and appearance, and can corrosion. Many of these problems were caused by oxygen and can be protected by tailoring packaging material (e.g., built-in oxygen-absorbing package), by avoiding prooxidants, or by using antioxidants.

A storage stability study (i.e., 24°C for 20 weeks) was conducted on aseptically packaged orange drinks,[81] which contained 10% orange juice and various combinations of ascorbic acid, amino acids (aspartic acid, arginine, and 4-aminobutyric acid), sugars, citric acid, and potassium citrate. In the presence of oxygen, ascorbic acid was the ingredient most responsible for darkening the orange drinks. High levels (1.26%) of amino acid enhanced the rate and extent of ascorbic acid loss as well as the extent of browning pigment formation. An HPLC method was developed to rapidly and simultaneously estimate ascorbic acid and dehydroascorbic acid in fresh fruits and vegetable; it was recommended for use in determining the storage stability of this type of drink.

Plate 1 Shelf-stable sandwich-type product.

Plate 2 Ohmically heated mixed vegetables.

Plate 3 Microwave-sterilized pasta product.

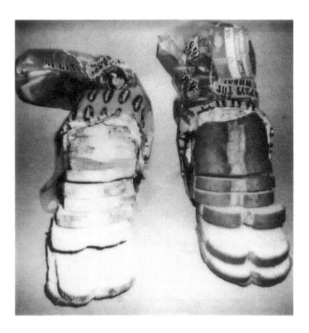

Plate 4 Bread samples with (*r*) and without (*l*) WO labels. Storage condition: 1 month at 21°C.

Table 4 Loss of Thiamin in Dehydrated Pork Stored at
Various Temperatures

Storage Temperature (°C)	Thiamin Retained (%)		
	7 days	14 days	21 days
–29	100	100	100
3	100	100	96
27	–	89	77
37	70	55	43
49	15	7	0
63	4	0	0

From Rice, E.E., et al. 1994. Food Res. 9:491. With permission.

Canned and dehydrated meat products

The most dangerous consequence of spoilage in canned low acid foods is the toxin formed by viable *Clostridium botulinum*, the spores of which survive as a result of inadequate processing. Physical characteristics of canned meat that may indicate such spoilage are (1) gas (although numerous microorganisms cause gas) and associated can swelling, (2) an odor of rancid cheese, and (3) a soft, disintegrated condition of the food.

For otherwise adequately processed military canned meats, Gardner reported a 1-year shelf life at 38°C for 31 different items, and suggested using interior-enameled cans for better appearance and flavor.[82,83]

Dry or semidry meat products are consumed either after rehydration or as is, such as jerky. Generally, the moisture content of dehydrated meat is too low to allow bacterial growth, but mold growth may occur after some weeks (e.g., *Penicillium* and *Aspergillus* spp.) if it rises above 10%. For long-term storage, the products must be compressed to exclude pockets of air or moisture and kept in an air-tight and moisture-proof container, preferably in a tin-plate can.[84] Nonoxidative changes, whether enzymatic or chemical, are pronounced at high storage temperatures. While restriction of oxygen will maintain the flavor of dry meat for 12 months or longer at 16°C, nonoxidative deterioration may develop under nitrogen at 38°C. In the absence of oxygen, the main changes during storage are caused by the Maillard reaction, so that dehydrated meat becomes unpalatable (i.e., a dark brown coloration and a bitter, burnt flavor) in 6 months when kept at high temperature[85]; moreover, the thiamin content diminishes quickly (Table 4). Concomitant losses in the water-holding capacity of the proteins cause brittleness of texture. The storage life can be considerably extended by drying to very low moisture contents. In the presence of oxygen, the storage of dehydrated meat (of high moisture content) at high temperature causes it to become pale and yellow, due to the conversion of myoglobin to bile pigments. A mealy odor develops and fat oxidation occurs giving rise to paint-like odor.[87]

Ultrahigh temperature milk

The conventional heat sterilization of liquid milk products destroys not only microorganisms, but also greater portions of some micronutrients such as vitamin B_1, B_6, or B_{12}. It also produces a strong, cooked flavor, or even a color change. Ultrahigh temperature (UHT) processing overcomes most of these deficiencies by treating the milk outside of containers with high heat for no more than 2–4 s, and then aseptically filling the shelf-stable milk into germ-free packages. The final product can be stored without refrigeration for several months. Sur and Joshi evaluated commercial UHT milk when fresh and during storage at 21 and 38°C.[88,89] Highest flavor scores were noted on day 4 for samples stored at 21°C, which then gradually decreased. A steady decline in flavor scores was observed for samples stored at 38°C. The recommended periods of storage at 21 and 38°C are 22 and 19 days, respectively. After 26 days at either temperature, a sharp increase in volatile fatty acids was found. Viscosity and pH of UHT milk showed little change and gelation was not observed after 5 months of storage at either temperature. Sediment volume increased slightly after the first 7 days of storage, and visible fat separation was formed on the surface of the milk and on milk carton walls after 40 days at 21°C and 33 days at 38°C.

Rerkrai et al. found that aseptically packaged UTH milk, when stored at 25°C for 24 weeks, showed an increase in acetaldehyde, propanol, *n*-hexanal, 2-pentanone, 2-hexanone, and 2-heptanone; these increases paralleled the stale flavor development.[90] Although the concentration of the carbonyl compounds measured were far below flavor threshold values when UHT milks were characterized as stale, the combination of acetaldehyde, acetone, 2-pentanone, and 2-heptanone could have an effect on the flavor. A temperature effect was demonstrated by increases in acid degree value (from fat lipolysis), dissolved oxygen, and titratable acidity, which also contributed to stale flavor.

Carrageenan is one of the few hydrocolloids capable of permanently suspending insoluble particles in fluids.[24] Its stabilizing effect in foods can extend their shelf life significantly even when processed under UHT, retort, or aseptic conditions. In whole milk beverages, carrageenan will interact with milk proteins to develop a three-dimensional network. A thixotropic system is formed that results in a stable suspension while simultaneously stabilizing the emulsion and preventing cream separation. However, the concentration of carrageenan should be carefully controlled to avoid overstabilization (i.e., gelation, wheying off, or mottling). Functionally, it can stabilize UHT-processed chocolate milk for up to 9 months of storage. Various milk beverages might require a combination of carrageenan and pectin or tapioca starch for smooth texture and better flavor release.

Innovative technologies

Novel foods

Enclosing foods in a capsule is an innovative method to preserve foods. (The original intention was to take advantage of human psychological repulsion to medicine capsules and thus curb the appetite of those consumers who want to be on diet.) By using a gelatin-based capsule to enclose such items as spice, candy, or healthy foods, it prevents loss of flavor, avoids direct contact with air, preserves quality, and makes it easier to digest and to store for long periods of time. The foods can be in powder, granular, or oily forms, and should be kept away from direct sunlight, low temperature, or extraneous pressure. The suitable storage condition is 20°C with 40–60% RH.

Torres and his associates reported an extensive study of microbial stabilization of intermediate moisture foods (IMF).[91–95] They had developed a refrigerated IMF product that uses lower solute concentration to control A_w and to offset sensory limitation caused by high solute concentration. However, the practicality of the type of IMF for use in uncontrolled temperature environments, such as encountered by the military, is doubtful.

Temperature changes or nonuniform freeze drying would cause a nonuniform distribution of moisture in IMF products, and thus lead to growth of microorganisms on the surface of foods. Surface treatments with edible barriers (e.g., zein) can be used to improve the surface microbial stability of IMF. The thickness of such films should be determined by considering maximum sensory acceptability. Reducing surface pH by coating IMF products with low molecular weight acids (e.g., lactic acid and sorbic acid) in a zein-based coating can also increase the surface microbial stability.

Table 5 Processing Conditions of Several Products in Split-Phase Aseptic System

Product	Processing Time (min)	F_o Value Range (min)[a]
Mixed vegetable	20	8.5–22.7
Macaroni and cheese	13	5.5–17.8
BBQ pork	12	8.0–15.3
Beef stew	15	6.5–24.4

Note: Data reported are for solids in the tank. All the sauce portions were sterilized separately.

[a] F_o value calculations were based on the temperature taken by a thermocouple at the cold spot and simultaneously converted into F_o values.

Novel thermoprocessing of particulates

Conventional thermoprocessing for aseptic packaging, which involves flow-through heat transfer devices such as a scraped surface heat exchanger, has been widely used for producing high-quality, shelf-stable fluid and semifluid food products. However, the high shear force damages food particulates, especially in viscous, stew-type products. The U.S. Army Natick RD&E Center has been exploring numerous innovative thermoprocessing techniques in place of such aseptic processing for military ration entrees.[96] These techniques include split-phase aseptic processing, ohmic heating, and microwave sterilization.

Split-phase aseptic processing

This system works by separately sterilizing the food solids batchwise and the sauce portions continuously and then aseptically packaging them together. Both solid and sauce portions receive an adequate amount of thermal treatment, thus preventing overprocessing.[97] The sauce, for example, was processed at 101°C for 1.1 min before being cooled down to 31°C to obtain a F_o value of about 4–5 in the fastest moving segment (laminar flow, $V_{max} = 2V_{average}$). The solid was sterilized by direct injection of steam in an evacuated tank with mild agitation to maintain solid integrity during the holding time and to achieve adequate F_o values (Table 5). Subsequently, evaporative cooling took place and sterilized nitrogen gas was introduced to release the vacuum in the tank.

Ohmic heating

Ohmic heating (direct-resistance heating) is a process in which food liquids and solids are heated simultaneously by passing an electric current through them. Because relatively uniform heating occurs throughout the food, process times can be shortened, resulting in greatly improved flavor and nutrient retention.[98–100] In commercial ohmic heating systems, the food is pumped between a series of electrodes, to which an alternating current is applied, which results in the food being heated up. Typically, a food remains in the ohmic heater for not more than 90 s, after which it is pumped to a holding tube for temperature equilibration and added lethality accumulation. The food, now sterile, is subsequently cooled and aseptically packaged (Figure 2). Since the food is heated from within, it can be thoroughly cooked without the need for separate cooking and external conductive heat transfer from surface to center. Thus, overheating of food can be avoided. This makes the process particularly suited to processing pumpable foods containing solid pieces, such as stews or other similar ready-to-serve meals (Table 6). Yang et al. evaluated the microbiological and sensory attributes of six ohmically heated products before and after 3 years storage at 27°C.[101] All six products were found to be commercially sterile and, in general, to have excellent retention of sensory quality, as determined by measurement of color, appearance, flavor, texture and integrity attributes and of overall quality ratings.

Figure 2 Ohmically heated mixed vegetables. (See color plate P–1)

Table 6 Processing Conditions of Several Products in Ohmic Heating System

Product	pH	Holding Time (s)	Temp.[a] (°C)	F_o[b]
Ham stew	5.8	70	132	14
Vegetable soup	5.4	70	133	18
Mushroom sauce	4.7	34	137	22
Beef stew	5.4	109	130	14
Pasta	4.6	30	133	15
Mixed vegetables	4.2	45	90	—[c]

[a] Measured at end of holding tube.

[b] Calculations based on Z value of 18.

[c] Pasteurization.

Microwave sterilization

Microwave sterilization is being explored because of its unique and rapid heating rate, flexibility, and ease of control. It can sterilize liquid, semiliquid, and chunky solids, and it can sterilize prepacked food products. The well-known overheating of edge zones and angles of food contained in trays, one of the results of poor microwave energy distribution, can be overcome by precisely dosed energy input from individual magnetrons by taking into account the dielectric parameters of the products such as dielectric constant and loss factor, as well as their density and geometry.[102] A commercial pressurized microwave sterilization system moves the products past the apertures of the various energy supply channels, which deliver the microwave energy from the magnetron into the treatment chamber. During this process, the product temperature is increased step by step by ΔT according to the product of the number of installed magnetrons times the treatment time. The construction design of the microwave output from the waveguides with a computer-controlled dosing of the power supplied by the individual magnetrons, it is possible to obtained a temperature distribution of $\Delta T \leq 5°C$ throughout the product.[102] Many commercial products are manufactured by such system (e.g., Omac Micronix®) and

Figure 3 Microwave-sterilized pasta product. (See color plate P–2)

packed in single- or dual-compartment, barrier plastic containers; many solid foods (e.g., rice or spaghetti) are also packed with modified atmosphere (MA), before micro-wave treatments, to achieve ambient shelf stability with less heat treatment. Product sterilization is tracked by Ball Datatracer® probes inserted into sample products and removed after processing for computer analyses of time, temperature, F_i and F_o values during each minute of processing: come-up, sterilization, holding, and cooling. Yang et al., working with Belgian and Japanese companies, produced excellent quality products such as sweet and sour chicken with rice, tortellini Bolognese, chili con carne, lasagna, and curry rice (Figure 3).[103] Significant retention of flavor and texture was observed even after 6-week storage at 50°C.

Antifungal/antibacterial agents

Delaquis and Mazza reviewed the antimicrobial properties of isothiocyanates in food preservation and suggested potential applications in packaged foods.[104] The Carex Corp. (Osaka, Japan) has been exploring a WasaOuro® (WO) system and Yang et al. had validated the effects and explored the potential applications of WO for military ration systems.[105] WO is a system that uses an extraction of a natural compound, allyl isothiocyanate (AIT), from horseradish, mustard seeds, or wasabi radish, to extend the shelf life of various food products, especially the ready-to-serve foods (Figure 4). WO-impregnated labels inhibit most bacterial growth and slowed down listeria growth on the surface of sliced chicken. Fruits had longer refrigerated shelf life with WO treatment, especially in those that had been minimally handled. Fresh pasta enclosed with the WO label maintained excellent quality beyond 7 weeks in the refrigerator. WO labels inhibited the growth of mold isolated from a commercial fruit and nut snack; many of these snack items were used in military Meal, Ready to Eat (MRE) components. WO gave extra protection for a shelf-stable sandwich, developed at the U.S.

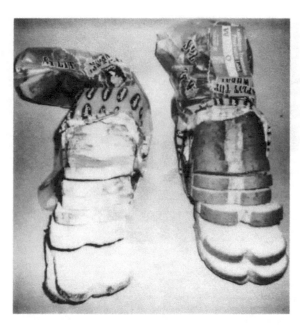

Figure 4 Bread samples with (*r*) and without (*l*) WO labels. Storage condition: 1 month at 21°C. (See color plate P–2)

Army Natick RD&E Center to have meat embedded in a roll of bread and rendered shelf-stable by a carefully designed formulation. WO pouch also slowed down hatching of *Lasioderma serricorna* and *Tribolium confusum*, two of the common bread flour beetles, for at least 2 months. An analytical method (GC/FTIR) was developed to detect WO absorbed by and distributed in foods, on packaging blends when containing WO, and to study the lifespan of WO. For many military freshly prepared ration items (A-ration) and other fresh-like field rations that are difficult to provide because of bacterial spoilage or possible yeast or mold growth, the WO systems appear to provide potentially practical solutions.

Conclusion

Ambient storage is the most economical and convenient way to preserve foods. The foods need to be microbiologically safe, organoleptically wholesome, and stable against uncontrolled environmental variations such as relative humidity, temperature fluctuation, adverse chemical reactions, and packaging deteriorations. Many of the desirable qualities could be retained at ambient temperature by applying the hurdle technology as well as the state-of-the-art novel preservation methods.

References

1. Jelen, P. 1985. Introduction to Food Processing. Reston Publishing, Reston, VA.
2. Karel, M. and Heidelbaugh, N.D. 1975. Effect of packaging on nutrients. In: Nutritional Evaluation of Food Processing. R.S. Harris and E. Karmas (Eds.), 2nd ed., p. 412-462. AVI, Westport, CT.
3. Park, C.E., Szabo, R., and Jean, A. 1988. A survey of wet pasta packaged under a $CO_2:N_2$ (20:80) mixture for staphylococci and their enterotoxins. Can. Inst. Food Sci. Technol. J. 21(1):109-111.
4. Beuchat, L.R. 1981. Microbial stability as affected by water activity. Cereal Foods World 26(7):345-349.

5. Christian, J.H.B. and Stewart, B.J. 1973. Survival of *Staphylococcus aureus* and *Salmonella newport* in dried foods, as induced by water activity and oxygen. In: The Microbiological Safety of Foods. Hobbs, B. C., and Christian. J. H. B. (Eds.). p. 107-119. Academic Press, London.

6. Beuchat, L. R. 1979. Survival of conidia of *Aspergillus flavus* in dried foods. J. Stored Prod. Res. 15:25.

7. Rayman, M. K., D'Aoust, J.-Y., Aoris, B., Maishment. C., and Wasik, R. 1979. Survival of microorganisms in stored pasta. J. Food Prot. 42:330.

8. Harwig, J., Blanchfield, B. J., and Jarvis, C. 1977. Effect of water activity on disappearance of patulin and citrinin from grains. J. Food Sci. 42:1225.

9. Tawaratani, T. and Shibasaki, I. 1972. Effect of moisture content on the microbiocidal activity of propylene oxide and the residue of foodstuffs. J. Ferment. Technol. 50:349.

10. Beuchat, L. R. 1973. *Escherichia coli* on pecans: Survival under various storage conditions and disinfection with propylene oxide. J. Food Sci. 38:1064.

11. Martinez, E. J., Bonino, N. and Alzamora, S.M. 1986. Combined effect of water activity, pH, and additives on growth of *Staphylococcus aureus* in model salami systems. Food Microbiol. 3:321-329.

12. Anonymous. 1965. Final Report-Stabilization of Food at Elevated Moisture Levels. Prepared by Melpar, Inc. October, 1965. Contract No. N 19-129-AMC-520(N).

13. Sych, J., Castaigne, F., and Lacroix, C. 1987. Effects of initial moisture content and storage relative humidity on textural changes of layer cakes during storage. J. Food Sci. 52(6):1604-1610.

14. Leistner, L. 1992. Food preservation by combined methods. Food Res. Int. 25:151-158.

15. Leistner, L. 1994. Further development in the utilization of hurdle technology for food preservation. J. Food Eng. 22:421-432.

16. Mitchell, J.H., Jr. 1955a. Stability studies on rations at the QMFCIAF. Establishing optimum conditions for storage and handling of semiperishable subsistence items. Series IV. 1., p. 7-21. Dept. of The Army, Office of the Quartermaster General, Washington, DC.

17. Mrak, E. 1955. Past and current efforts to increase the storage life of subsistence. Proceeding of Establishing Optimum Conditions for Storage and Handling of Semiperishable Subsistence Items, sponsored by the Research and Development Division, Office of the Quartermaster General, Washington, DC. 3 December 1953. Eds. H.E. Goresline, N.J. Leinen, and E.M. Mrak.

18. Guerrant, N.B., Fardig, O.B., Vavich, M.G., and Ellenberger, H.E. 1948. Nutritive value of canned foods; influence of temperature and time of storage on vitamin control. Ind. Eng. Chem. 40:2258-2263.

19. Anonymous. 1956. Biological Evaluation of Ration, Individual, Combat: the Effects of Long-Term Storage. Interim Report of Quartermaster Food and Container Institute for the Armed Forces, Nutrition Branch, Food Laboratories. Project 7-84-13-002.

20. Woodroof, J.G. and Heaton, E.K. 1955. Protective packaging of foods against moisture condensation. Food Technol. 9:510-518.

21. Heaton, E.K. and Woodroof, J.G. 1958. Moisture condensation and cold stored military rations. Food Technol. 12 (1):24-29.

22. Best, D. 1987. Engineering flavor systems. Prepared Foods 156(12):116-126.

23. Rice, J. 1988. Oxygen eliminators. Food Proc. 49(6):58.

24. Rice, J. 1990. Carrageenan. Food Proc. 51(7):44.

25. Goryoda, T. 1995. Oxygen scavenging package: its concept and application "OXYGUARD®" tray and film. Personal communication.

26. Woodroof, J.G. 1975. Storage life of canned, frozen, dehydrated, and preserved fruits. In: Commercial Fruit Processing, J.G. Woodroof and B.S. Luh, (Eds.). AVI, Westport, CT, p. 595.

27. Ross, E.W., Klicka, M.V., Kalick, J., and Branagan, M.T. 1987. A time-temperature model for sensory acceptance of military ration. J. Food Sci. 52(6):1712-1717.

28. Chepurnoi, I.P., Kunizhev, S.M., and Chebotareva, N.G. 1987. Formation of hydroymethyl-furfural during storage and processing of some foods. Voprosy Pitaniya 6:67-68.

29. Mitchell, J.H., Jr. 1955. Rate and extent of deterioration of packaged rations during storage and transportation. Proceeding of Establishing Optimum Conditions for Storage and Handling of Semiperishable Subsistence Items, sponsored by the Research and Development Division, Office of the Quartermaster General, Washington, DC. 3 December 1953. Eds. H.E. Goresline, N.J. Leinen, and E.M. Mrak.

30. Rinschler, R.A. 1954. Underground storage test. Phase I. Summary Rpt., Quartermaster Corps.

31. Horne, L.M., Stevens, H.H., and Thompson, J.B. 1946. Effect of shortening stability on commercially produced army ration biscuit. I. Initial data and results of accelerated stability tests. J. Am. Oil Chem. Soc. 25:314-318.

32. Stevens, H.H. and Thompson, J.B. 1948. Effect of shortening stability on commercially produced army ration biscuit. II. Development of oxidation during storage. J. Am. Oil Chem. Soc. 25:389-394.

33. Anonymous. 1950. Storage of Quartermaster Supplies. Dept. of The Army Tech. Manual TM 10-250. U.S. Gov't. Printing Office, Washington, DC.

34. Berkowitz, D. and Oleksyk, L.E. 1991. Levened breads with extended shelf life. U.S. Patent 5,059,432.

35. Briggs. J.L. 1992. Personal communications. Sustanability Directorate, U.S. Army Natick Reserach, Development and Engineering Center, Natick, MA.

36. Stollman, U. and Lundgren, B. 1987. Texture changes in white bread: effects of processing and storage. Cereal Chem. 64(4):230-236.

37. Knorr, D. 1987. Compressibility of baked goods after carbon dioxide atmosphere processing and storage. Cereal Chem. 64(3):150-153.

38. Ochi, T., Tsuchiya, K., Aoyama, M., Maruyama, T., and Niiya, I. 1988. Effects of tocopherols on qualitative stability of cookies and influence of powdered milk and egg. J. Jap. Soc. Food Sci. Technol. [Nippon Shokuhin Kogyo Gakkaishi] 35(4):259-264.

39. Pafumi, J. and Durham, R. 1987. Cake shelf life extension. Food Technol. Aust. 39(6):286-287.

40. Smith, J.P., Ooraikul, B., Koersen, W.J., van de Voort, F.R., Jackson, E.D., and Lawrence, R.A. 1987. Shelf life extension of a bakery product using ethanol vapor. Food Micro. 4(4):329-337.

41. Zeleznak, K.J. and Hoseney, R.C. 1987. Characterization of starch from bread aged at different temperatures. Starch/Starke 39(7):231-233.

42. King, J. 1948. Scientific problems in feeding a modern army in the field. J. Soc. Chem. Ind. 47:739-743.

43. Duxbury, D.D. 1987. Freeze-dried ingredients use MAP to extend shelf life 12-fold. Food Proc. 48(9):28.

44. Bakar, J., Hassan, M.A., and Ahmad, A. 1988. Storage stability of coconut milk powder. J. Sci. Food Agri. 43(1):95-100.

45. Branch, A.L., Worthington, R.E., Roth, I.L., Chinnan, M.S., and Nakayama, T.O.M. 1987. Effect of hot water immersion on storage stability of non-refrigerated peanuts. Peanut Sci. 14(1):26-30.

46. Somogyi, L.P. and Luh, B.S. 1975. Dehydration of fruits. In: Commercial Fruit Processing. J.G. Woodroof and B.S. Luh, (Eds.). AVI, Westport, CT.

47. Eisenhardt, N.H., Eskew, R.K., and Cording, J., Jr. 1964. Explosive puffing applied to apples and blueberries. Food Eng. 36(6):53.

48. Eisenhardt, N.H., Eskew, R.K., Cording, J. Jr., Talley, F.B., and Huhtanen, C.N. 1967. Dehydrated explosion puffed blueberries. USDA, ARS73-54.

49. Sullivan, J.F., Craig, J.C. Jr., Dekazos, E.D., Leiby, S.M., and Konstance, R.P. 1982. Dehydrated blueberries by the continuous explosion-buffing process. J. Food Sci. 47:445.

50. Rahman, A.R., Taylor, G.R., Schafer, G., and Westcott, D.E. 1970. Studies of reversible compression of freeze dried RTP cherries and blueberries. U.S. Army Natick RD&E Center. Tech. Rep. 70-52 F1.

51. Hanson, S.W.F. 1961. The Accelerated Freeze-Drying Method (AFIJ) of Food Preservation. H.M. Stationary Office, London.

52. Scharschmidt, R.K. and Kenyon, R.E. 1971. Freeze drying of blueberries. U.S. Pat. 3,467,530.

53. Vollink, W.I., Kenyon, R.E., Rarnett, S., and Rowden, H. 19#1. Freeze dried strawberries for incorporation into breakfast cereal. U.S. Pat. 3,395,022.

54. Ponting, J.D., Watters, G.G., Forrey, R.R., Jackson, R., and Stanley, W.L. 1966. Osmotic dehydration of fruits. Food Technol. 20:1365.
55. Farkas, D.F. and Lazar, M.E. 1969. Osmotic dehydration of apple pieces: effect of temperature and syrup concentration on rates. Food Technol. 23:688.
56. Bolin, H.R., Huxsoll, C.C., Jackson, R., and Ng, K.C. 1983. Effect of osmotic agents and concentration on fruit quality. J. Food Sci. 48:202.
57. Yang, T.C.S. 1984. The effects of juice extraction methods on the quality of low-calorie blueberry jellies. Maine Agric. Exp. Sta. Bull. 803.
58. Yang, T.C.S. 1984. Blueberrv roll-ups. Dept. Food Sci., Univ. of Maine, Orono,ME. Unpublished data.
59. Yang, T.C.S. 1986. Osmosis-freeze dried blueberries vs. freeze dried blueberries. Dept. Food Sci., Univ. of Maine, Orono, ME. Unpublished data.
60. Yang, A.P.P., Wills, C., and Yang, T.C.S. 1987. Use of a combination process of osmotic dehydration and freeze drying to produce a raisin-type lowbush blueberry product. J. Food Sci. 52(6):1651-1653, 1664.
61. Torreggiani, D., Giangiacomo, R., Bertolo, G., and Abbo, E. 1986. Osmotic dehydration of fruits. I. Suitability of cherry varieties. Industria Conserve 61(2):101-107.
62. Koshida, D., Sigisawa, K., Majima, J., and Hattori, R. 1982. Method for producing dry fruit chip. U.S. Patent 4,341,803.
63. Cohen, J.S., Rees, C., Hallberg, L. and Yang, T.C.S. 1994. Vegetable drying in two novel food dryers. U.S. Army Natick RD&E Center Tech. Rept. TR-95/008.
64. Cohen, J.S. and Yang, T.C.S. 1995. Progress in food dehydration. Trends Food Sci. Technol. 6:20-25.
65. Paul, P. and Plummer, J. 1949. Home storage of nonfat milk solids. J. Home Econ. 41:198-200.
66. Hetrick, J.H. and Tracy, P.H. 1945. Keeping quality of spray-dried whole milk. J. Dairy Sci. 118:687-700.
67. Coulter, J.S. 1947. The keeping quality of dry whole milk spray-dried in an atmosphere of an inert gas. J. Dairy Sci. 30:115-1002.
68. Henry, K.M., Kon, S.K., Lea, C.H., and White, J.C.D. 1948. Deterioration on storage of dried skim milk. J. Dairy Res. 15:292-363.
69. Henry, K.M., Kon, S.K., Lea, C.H., and White, J.C.D. 1949. Protein degradation in stored milk powder. 12th. Internat. Dairy Congress, Stockholm, 1948. Sec. II. Subj. 2:166-174. 1949.
70. Remaley, R.J., Stoltz, P.C., and Hening, J.C. 1949. Dairy Products. Operations Studies Number One. Subsist. Res. and Dev. Lab. QMFCI. 7:1-171.
71. Bryce, W.A. and Pearce, J.A. 1946. Dried milk powder. Can. J. Res. 24F:61-69.
72. Holloender, H.A. 1955. Method for improving the dispersibility of dry whole milk. Dry Whole Milk. Series I, 6. pp. 144-157. U.S. Army Quartermaster Corps. National Research Council, Washington, DC.
73. Sharp, P.F., Shields, J.B., and Steward, A.P., Jr. 1945. Nutritive quality of spray-dried whole milk in relation to manufacture and storage. Proc. Inst. Food Technol. 54-57.
74. Brenner, S., Wodicka, V.O., and Dunlop, S.G. 1947. Stability of ascorbic acid in various carriers. Food Res. 12:253-269.
75. Kemp, J.D., Ducker, A.J., Ballantyne, R. M., and Acheson, G.G. 1957. A study of the flexible packaging of dry cream and potato powders. Def. Res. Med. Lab. Rpt. 174-3. Toronto.
76. Rao, T.S.S. and Murali, H.S. 1987. Studies on the spray-dried, foam-mat-dried and freeze dried whole egg powders: changes in the nutritive qualities on storage. Nutr. Rep. Int. 36(6):1317-1323.
77. Mangaban, M.L. and Consolacion, F.I. 1986. Storage stability of dried fish: lipid oxidation and microbial growth at different reative humidity levels. Philippine Agri. 69(1):29-32.
78. O'Beirne, D., Egan, S., and Healy, N. 1987. Some effects of reduced boiling time on the quality of fruit preserves. Lebensmittel-Wissenschaft und -Technologie 20(5):241-244.
79. Spayd, S.E. and Morris, J.R. 1981. Influence of immature fruits on strawberry jam quality and storage stability. J. Food Sci. 46:414.
80. Brenner, S. 1947. Cooperative high temperature canned food storage study. QMFCI Proj. 7-84-12-02. Interim Rpt. No. 4. Chicago, IL.

81. Kacem, B. 1987. Storage stability of aseptically packaged single strength orange juice and orange drinks. Dissertation Abstracts Int. B47(11):4364.

82. Gardner, B.W., Jr. 1949. An organoleptic evaluation of the keeping qualities of army canned meats before and after storage. QMFCl Proj. 7-84-06-22. Interim Rpt. I.

83. Gardner, B.W., Jr. 1949. Army tests reveal how storage affects canned meat flavor. Food Ind. 21(7):881-890.

84. Sharp, J.G. 1953. Spec. Rept. Fd. Invest. Bd., Lond., No. 57.

85. Sharp, J.G. and Rolfe, E.J. 1958. Fundamental Aspects of the Dehydration of Foodstuffs. Society of Chemical Industry, London, p. 197.

86. Rice, E.E., Beuk, J.F., Kauffman, F.L., Schultz, H.W. and Robinson, H.E. 1944. Food Res. 9:491.

87. Lawrie, R.A. 1974. Meat Science. 2nd ed. Pergamon Press, New York.

88. Sur, A. and Joshi, V.K. 1989. Changes in flavour characteristics and volatile fatty acids contents of UHT milk during storage. Ind. J. Dairy Sci. 42(1):125-126.

89. Sur, A. and Joshi, V.K. 1989. Changes in viscosity, pH, oxygen content, sedimentation characteristics and fat separation in UHT milk during storage. Ind. J. Dairy Sci. 42(1):130-131.

90. Rerkrai, S., Jeon, I.J., and Bassette, R. 1987. Effects of various direct ultra-high temperature heat treatments on flavor of commercially prepared milks. J. Dairy Sci. 70:2046-2054.

91. Torres, J.A. 1987. Microbial stabilization of intermediate moisture food surfaces. Ch. 14. In: Water Activity: Theory and Applications to Food. L.B. Rockland and L.R. Beuchat, (Eds.). Marcel Dekker, New York, p. 329.

92. Torres, J.A., Bouzas, J.O., and Karel, M. 1989. Sorbic acid stability during processing and storage of an intermediate moisture cheese analog. J. Food Proc. Pres. 13:409-415.

93. Torres, J.A., Motoki, M., and Karel, M. 1985. Microbial stabilization of intermediate moisture food surfaces. I. Control of surface preservative concentration. J. Food Proc. Pres. 9:75-92.

94. Torres, J.A., Bouzas, J.O., and Karel, M. 1985. Microbial stabilization of intermediate moisture food surfaces. II. Control of surface pH. J. Food Proc. Pres. 9:93-106.

95. Torres, J.A. and Karel, M. 1985. Microbial stabilization of intermediate moisture food surfaces. III. Effects of surface preservative concentration and surface pH control on microbial stability of an intermediate moisture cheese analg. J. Food Proc. Pres. 9:107-119.

96. Yang, T.C.S., Cohen, J.S., Kluter, R.A. and Driver, M.G. 1994. Feasibility of applying ohmic heating and split-phase aseptic processing for ration entree preservation. U.S. Army Natick RD&E Center. Tech. Rept. TR-94/021.

97. Alkskog, L. And Mejvik, L.-G. 1989. Recent findings in the sterilization of food containing particles. Food Focus No. 1.

98. Palaniappan, S. and Sastry, S.K. 1991. Electrical conductivities of selected solid foods during ohmic heating. J. Food Proc. Eng. 14:221.

99. Sastry, S.K. and Palaniappan, S. 1991. Ohmic heating of liquid-particle mixtures. Food Technol. 45(12):64.

100. Parrott, D.L. 1992. Use of ohmic heating for aseptic processing of food particulates. Food Technol. 46(12):68.

101. Yang, T.C.S., Cohen, J.S., Kluter, R.A., Tempest, P., Manvell, C., Blackmore, S.J. and Adams, S. 1997. Microbiological and sensory evaluation of ohmically heated stew type foods. J. Food Quality 20(4): 303–313.

102. Koch, K. 1993. Pasteurization and sterilization of unpackaged liquid food containing solid parts in a continuous process by means of microwaves. In: Aseptic Processing of Foods. H. Reuter, (Ed.). Technomic, Lancaster, PA, p. 107-117.

103. Yang, T.C.S., Yang, P.P.A. and Taub, I.A. 1995. Microbiological and sensory evaluations of microwave sterilized meals. Unpublished data.

104. Delaquis, P.J. and Mazza, G. 1995. Antimicrobial properties of isothiocyanates in food preservation. Food Technol. 49(11):73-84.

105. Yang, T.C.S., Conca, K., Powers, E., Sikes, A. and Worfel, R. 1995. Potential applications of an antifungal/antibacterial agent (WasaOuro® systems) for military ration systems. Unpublished data.

chapter eighteen

Toxicological implications of postprocessing in stored or shipped foods

Larry D. Brown, Richard C. Worfel, and John T. Fruin

Contents

Introduction

The United States of America has one of the safest food supplies in the world. During the late 1980s, foods in the U.S. moved through 25,000 food manufacturers, 35,000 wholesalers,

250,000 food stores, 275,000 restaurants, and all households.[1] Of 32,500 food processors reported in the U.S. in 1996, around 100 accounted for 70% of all profits. An average food item is estimated to travel 1300 miles before being eaten. Americans produce approximately $275 billion (US) worth of processed food and beverage products per year. Agricultural exports amounted to $54 billion in 1995. U.S. agricultural production is evolving towards vertically integrated large-scale operations, which facilitate standardized approaches for economy, production, processing, quality assurance and safety. As such, a highly centralized and industrialized food supply has little margin for error. Mistakes under such a system have far-reaching consequences, as evidenced by the 1973 large-scale PBB catastrophe in Michigan.

Quality and safety of our food is maintained by inspections and sampling at critical control points along the supply continuum from farm to table. The U.S. Department of Agriculture Food Safety and Inspection Service (FSIS) and Animal and Plant Health Inspection Service (APHIS), Food and Drug Administration, National Oceanic and Atmospheric Administration, Centers for Disease Control and Prevention, veterinarians, food technologists and microbiologists, state and local authorities, and industry inspectors and quality assurance officials interact, fulfilling important regulatory roles. Also, to maintain safety and quality, protective or beneficial chemicals are required in the manufacture, packaging, storage, and use of foods. Since chemicals in foods may become a detriment to the health of humans or animals, their toxicological implications and potential adverse effects need to be considered. A popular quote[2] from the Middle Ages by Paracelsus (1493–1541) still rings true today: "all substances are toxic—the dose makes the poison." This translates into current day food safety practices that allow chemical additives and residues to be consumed if the total exposure (cumulative dose) is held below threshold levels for toxic manifestations. Tolerances for chemicals and drugs are currently based on animal no-adverse-effect-levels (NOAEL) with a 10- or 100-, and sometimes a 1,000-fold margin of safety built in. The guiding principle is 10-fold less for uncertainty in the dose-response curve and 10-fold less for susceptibility differences among humans. Another 10-fold may be added if a chemical crosses the placenta and is a hazard to the fetus. Bioengineered agricultural products and foods are also evolving, and their acceptance and use will expand over the next decade. BST (Posilac®, Monsanto Corp.) enhanced milk production is the most obvious example of this emerging technology.

Background concepts

Activities of toxicologists addressing safety issues fall into three main categories: descriptive, mechanistic, and regulatory. The following list of specialties indicates the many ways that toxicologists apply their expertise: (1) *system or target organ*—neuro or behavioral, cardiovascular, hepatic, immuno, reproductive, respiratory, ocular, cutaneous, renal, hematologic, biochemical, and genetic; (2) *classification of toxins*—radiation, metal, plant, natural product; (3) *economics*—legal, pharmaceutical, pesticide/chemical, and food; (4) *target population*—industrial, occupational, aquatic, environmental, forensic, veterinary, and human clinical; and (5) *approach or method of analyzing toxicity data*—epidemiological, mathematical risk analysis, historical, and pathological.

Food toxicology, one of approximately 25 subspecialties in the field of toxicology, deals with natural (endogenous/inherent) toxic constituents, introduced toxic chemicals (intentional or accidental, natural or synthetic, inorganic or organic), and toxic products from microbial or biochemical degradation of human foods. It spans a broad arena and interacts within diverse scientific fields. These include natural product carcinogens and biological toxins; chemistry and detection of toxic elements; mycology and microbiology; epidemiology and preventive medicine; human forensic and clinical medicine; toxicological risk assessment and law; animal medicine and production agriculture; biochemistry and molecular biology; pathology; toxicokinetics and biological fate; medicinal chemistry, fisheries science; botany; nutrition; restaurant management; and food and environmental sciences.

The accumulation or introduction of hazardous levels of toxic or infectious agents in foods may result from the influence of many factors. These can be cataloged as follows: (1) natural toxins preformed by the food item itself (plants: cyanide, favism); (2) toxins bioaccumulated from the environment or diet during plant or animal growth (fish: ciguateria; plants: nitrates; grazing cattle and dairy products: radionuclides—strontium 90, cesium 137, iodine 131); (3) toxins formed in response to improper storage of foods (fish: scombroid histamine poisoning); (4) hazards resulting from inadequate processing or inadvertent contamination of the food item with pathogenic microorganisms (cheeses: listeriosis; vegetables: *Escherichia coli* O157:H7, El Tor cholera; raspberries/strawberries: *Cyclospora* spp. parasite; cantaloupes: *Salmonellae*); (5) excessive growth of previously coexisting bacteria on foods (poultry and eggs: *Salmonellae*); (6) the introduction into food of toxin-elaborating bacteria (cream-filled pastry, potato salad: *SEB*; vacuum-packed fish, fermented sausage, home-canned foods: botulism) or fungi (mycotoxins); (7) intentional human treatment of foods or food animals with chemicals deemed beneficial to agriculture or food manufacturing (apples: growth regulators or pesticides; beef cattle: antibiotics, hormones, or insecticides; meats: nitrates to prevent botulinum; excessive GRAS food additives); (8) inadvertent or accidental introduction of toxic chemicals into foods (warehoused foods: pesticides; dairy products: PBBs, heptachlor; ginger extract, vegetable cooking oils (TOCP); (9) inadvertent introduction of chemicals into foods from packaging leachates or eroded containers (packaging PVC plasticizers; spangling of tin cans); (10) the consequence of processing upon food additives; (11) nonmicrobial biochemical or enzymatic degradation of natural food components into toxic products; and (12) intentional adulteration with chemicals, tampering, or sabotage.

Poisoning by consumption of chemicals is uncommon and is usually manifested by an acute onset of symptoms after poisonous food is eaten. Chronic effects have also been documented as evidenced by the poisoning of Japanese mothers after consumption of fish and shellfish contaminated with methyl mercury (MeHg). Antimony, arsenic, cadmium, tin, copper, cobalt, mercury, fluoride, lead, zinc, nitrite, cyanide, PCBs (Yusho rice oil disease), PBBs, dioxins, chlorinated hydrocarbons (DDT, benzene hexachloride, toxaphene, lindane, methoxychlor, aldrin, dieldrin, endrin, heptachlor, nemagon), organophosphates (TEPP, carbophenothion, diazinon, malathion, parathion), carbamates (sevin), warfarin rodenticide, cresyl phosphates, various poisonous plant and marine animal toxins, and several intentional food additives (monosodium glutamate, niacin [nicotinic acid], sodium chloride, calcium chloride, potassium emulsifiers [ME18]) have been implicated in food poisoning incidents.[3,4] Most of these poisoning episodes were limited in terms of the number of people affected. A few incidents at foreign locations were quite significant with as many as 50,000 individuals involved.

Hazardous levels of chemicals have been naturally, accidentally, inadvertently or intentionally introduced into foods from many varied sources.[2–7b] Example sources include (1) utensils (copperware, cadmium-plated ware, lead or antimony from cheap enameled earthenware); (2) leaking mechanical refrigerators (methyl chloride); (3) barbecuing foods on chrome or cadmium-plated refrigerator shelving (cadmium); (4) use of galvanized garbage cans (zinc) to store fruit drink at large gatherings; (5) transferring acidic fruit drinks in copper tubing leading to copper toxicity; (6) powdered white insecticides (sodium fluoride) mistakenly substituted for baking ingredients; (7) over-spray of insecticides onto foods in warehouses or via combined storage of leaking or unsealed containers of insecticides in food storage warehouses; (8) agricultural use of pesticides on grain, fruit, and vegetable crops; (9) intentional use of cobalt (1 ppm) to prevent foam in beer leading to cardiomyopathy; (10) packaging such as PCB-contaminated recycled paper used to wrap chicken carcasses; (11) accidental or intentional contamination of cooking oils by motor lubricants (TOCP, cresyl phosphates); (12) Mexican family consumption of mercury-contaminated pork derived from swine fed seed grain treated with methyl mercury fungicide or direct consumption by humans of dressed grain in bread or bakery products (Pakistan, 1963; Iraq, 1961 and 1971–72; Guatemala, 1966); (13) live steam (formerly used in cooking or in "clipper" dishwashers) containing anticorrosive chemicals (morpholines) for descaling boilers; (14) excessive use of vitamin additives or preservatives (niacin) in foods prepared in nursing homes; (15) naturally occurring toxins (favism, mushroom toxins) present at toxic levels in fungi, plants, seeds, nuts, spices, or other products (goitrogens, tung nut saponin, oxalates, nutmeg myristicin, tannic acids) consumed as part of these foods; (16) residues of toxic plants that may contaminate edible foods (bulk grain or vegetables) when harvested (pyrrolizidine alkaloids, *Lolium* sp., *Meliotus* sp., products of Euphorbiaceae and Thymelaeaceae, cyanogenic plants, other alkaloids, phenolics, glycosides, protein and amino acid toxins); (17) seafoods that naturally possess genetic-based toxic chemical components (Japanese fugu fish, tetrodotoxin), that are predisposed to developing toxins under the right set of conditions (scombroid spoilage), or that concentrate toxins or foodborne disease agents from marine algae/dinoflagellates (shellfish: PSP, DSP, ASP) or directly from seawater (ciguatera, *Vibrio vulnificus*, hepatitis A).

Radionuclides may also contaminate the food chain as evidenced by the aftermath of the Russian Chernobyl nuclear power station accident on 26 April 1986.[8-10] No acute effects have been reported outside the Soviet Union where 237 cases of acute radiation sickness, including 31 deaths, were reported.[11] In 1986, it was necessary to discard milk, meat, 20–30% of all berries, fruits with pits, and table greens in the areas most contaminated with cesium. In 1987, all products from private plots met radioactivity limits. Certain foods (some forest berries, mushrooms, fish) were discarded for long periods. By autumn 1989, potatoes, vegetables and fruits met specifications in virtually all polluted territories. Excepting milk, virtually all other categories of foods in these areas were acceptable. The milk radioactivity problem will string out for a number of years in the most contaminated regions. Because of the early radioactive iodine (^{131}I) threat, 5.4 million people underwent preventive iodine treatment, and milk was processed into cheeses and butter and then held beyond the short half-life of ^{131}I.[12]

Scope

Since most food toxicological risk events (and associated critical control points) appear to occur prior to or after processing and storage, this chapter would necessarily have been very sparse if restricted to only toxic elements or events that are formed or introduced during the postprocessing storage or distribution phases of the commercial food manufacturing cycle. One clear deviation from the "in-storage and toxicological themes" is the overview of the foodborne problem in the next section. It was included to provide a perspective on the magnitude of the toxic vs. nontoxic components of the foodborne problem in the U.S. Many concerns routinely addressed in general food science, food safety, food microbiology, and food toxicology references are not covered here.[13-30] The following food safety information is offered because it is deemed significant and has importance in the overall scheme of things.

Traveler's diarrhea, gastroenteritis, and the foodborne problem: an overview

It has been estimated that over 100 million travelers experience diarrhea each year.[31] Acute gastroenteritis has also been found to be a significant risk for U.S. military personnel.[32]

Traveler's diarrhea has multiple theories of causation, including bacterial, viral, and parasitic foodborne or waterborne pathogenic enteric agents; stress; change in diet; or exposure to chemicals or toxins in foods.[33] Benefits gained from consumption of food clearly outweigh the negative health risks associated with intake of this requirement for life. Consumption of one food meal has a relatively low health risk associated with it; however, daily cumulative risk adds up over an average lifetime (3 meals/day × 365 days × 70 years = 76,650 lifetime meals). Few people will live an average lifespan without experiencing at least one episode of foodborne gastroenteritis. Most individuals will experience a few episodes of gastroenteritis each decade. The daily risk of "foodborne illness" in the U.S. is relatively low when compared to rates outside this country. Governmental regulations, food industry "good management practices," voluntary food safety programs, high-quality food service operations (restaurants, fast-food outlets) and retail supermarkets, adequate storage facilities and transport carriers, and the general level of education of the population are a few of the factors that enable the U.S. to enjoy this safe food supply stature. Risk increases significantly for HIV immunocompromised AIDS patients, elderly patients in nursing homes, patients with liver disease, and for the very young immunoniave infants. The World Health Organization has predicted that 40 million AIDS cases will be present throughout the world by year 2000.[34] In 1991, WHO estimated that 8 to 10 million persons were infected worldwide.[35] The following data were reported at the 11th International Conference on AIDS in 1996:

Infections in the U.S.: 650,000 to 900,000 (0.4–0.5% of 270 million are HIV-positive)
Yearly infections worldwide: 3.1 million
Yearly infections, U.S.: at least 40,000
Western Europe infections: 470,000
Eastern Europe and Central Asia infections: 30,000
Caribbean infections: 270,000
North Africa and Middle East infections: 200,000
Sub-Saharan Africa infections: 14 million
East Asia and Pacific infections: 35,000
Australia and the South Pacific infections: 13,000

The WHO reported in 1990 that children less than 5 years of age from 43 African countries averaged 4.4 episodes of diarrhea per child per year for CY 1988. The total estimated diarrhea episodes for these 87 million children amounted to 382,669,400 bouts for this 1-year period.[36] Globally (excluding China), 1.362 billion diarrhea episodes were estimated for children less than 5 years old. The food production environment and processing and storage conditions in developing nations undoubtedly impinges upon the safety and wholesomeness of their end-products and the health of indigenous peoples. Besides food as a source or carrier for disease-causing agents, other etiologies such as waterborne disease, viral gastroenteritis, lack of hand washing facilities, poor sanitation, stress, lack of good medical care and malnutrition, and lower immune status also contribute to this outcome. Many epidemiological risk factors acting in unison lead to this Mount Everest of human morbidity.

Developing nations have not been as fortunate as Western society in having the same low frequency of foodborne/waterborne diarrheal diseases or to experience the less severe types of foodborne pathogens observed in developed nations (United States, Canada, Western Europe or Japan, etc.). This is evidenced by the plight of the Iraqi Kurds, Bangladeshi peoples, Africans, and South Americans (Peruvians) facing cholera, dysentery, and other serious waterborne and foodborne diseases causing severe mortality in these populations. The cholera epidemic of 1991 in the Western Hemisphere infected more than

300,000 persons. The Pan American Health Organization (PAHO) places the probable cases at 312,573 and deaths at 3288, the majority of which occurred in Peru followed next by Ecuador. Both waterborne and foodborne routes were implicated, especially undercooked seafoods.[37]

The sheer number of meals consumed per year (estimated at 295.6 billion meals per year in the U.S. for a population of 270 million individuals) provides a large denominator that extrapolates into an overall low annual foodborne risk figure for the U.S. Despite this favorable risk, unavoidable foodborne disease occurs in the U.S. each year. Between 1983 and 1987, the National Centers for Disease Control and Prevention (CDC) recorded 2397 foodborne outbreaks representing 91,678 cases.[38] According to U.S. Food and Drug Administration (FDA) estimates, between 21 million and 81 million cases of foodborne diarrheal disease occur annually in the U.S.[39,40] Most of these cases go unreported. If even the low estimate is real, this means that only 0.476% (100,000 cases reported each year out of 21 million) of foodborne illness cases are reported and some 20.9 million cases must therefore go unreported each year. This is hard to fathom, but is probably true knowing the way individuals reserve their medical visits to the persistent problems. Also, medical clinicians/services do not routinely culture patients or report diarrhea cases to the local health board or epidemiology service. A national rate of 21 million cases of foodborne illnesses out of a U.S. population of 270 million reflects an annual case rate (incidence) of 7,777 cases per 100,000 population. Assuming a duration of illness of 5 days results in a prevalence rate of 106 cases per 100,000 population. By comparison, the U.S. case rate (prevalence) of HIV/AIDS is 12.6 per 100,000.[35a] Diarrheal disease has a much shorter duration or time course (generally reflected in a lower prevalence) and U.S. mortality/fatality rates are clearly higher for AIDS (average of 25 foodborne deaths reported per year, estimated foodborne deaths are 523 to 7,041 per year, while AIDS is blamed for 10,000–15,000 human deaths per year). If the FDA estimates on foodborne disease are accurate, then foodborne disease is by far the number one medical condition in the U.S. with respect to cases per year, but not necessarily number one in importance or federal funding. The significance of foodborne disease in the U.S. ranks behind cancer (500,000 deaths per year), chronic lung/smoking disease (450,000 deaths per year), accidental death, driving accidents (50,000 deaths per year), HIV/AIDS (average 13,200 deaths per year, 1981–91), heart disease, suicide, homicide, stroke and cerebrovascular disease, pneumonia and influenza, and diabetes. The relative position of federal research funding for foodborne diseases is at about the same relative hierarchical position.

Morris[35b] and the USDA FSIS[35c] reported that although exact figures are not available, a "best estimate" is that there were 3.6 to 7.1 million people affected by illness related to specific food-associated pathogens in 1993. Of these illnesses, 2.1 to 5 million were associated with meat or poultry consumption. Between 2695 and 6587 deaths were attributed to foodborne pathogens of which 1436 to 4232 were from meat or poultry consumption. Recall that the U.S. population consumes around 300 billion meals per year; therefore, a small percentage of food safety incidents and unsafe meals can be expected. Absolute safety of the U.S. food supply is not possible. Pathogen reduction programs are designed to minimize foodborne microbial illnesses.

The estimated average costs for all food-associated illness in the U.S. are over $23 billion per year; $4.8 billion (20%) was designated for the bacterial outbreaks.[41a] In response to this significant economic aspect, Dr. Edward Menning, Executive Vice President, National Association of Federal Veterinarians (NAFV) has stated: "we're not talking here as has been said in the past, that, foodborne illnesses are a little cathartic that is periodically good for you." Most cases of foodborne disease do recover after a short bout of diarrhea and/or vomiting, but the lost productivity to the nation is enormous, not to mention the occasional loss of life, medical expenses, and other indirect costs. Mauskopf and French have estimated the value of avoiding morbidity and mortality from foodborne disease.[41b] They look at this issue from two metrics (quality-adjusted life-years and dollars). Future regulatory decisions by FDA, USDA, and U.S. Environmental Protection Agency (EPA) in setting priorities and allocating resources toward the various facets of food safety concerns will be based on innovative analytical approaches such as the quality-adjusted life-year-based method in concert with decision trees, algorithms, risk assessment, and the hazard analysis critical control point (HACCP) program. According to testimony before the Senate Committee on Agriculture, Nutrition and Forestry in 1987, three major foodborne diseases caused the following estimated losses: (1) salmonellosis, $553 million; (2) campylobacteriosis, $723 million; and (3) congenital toxoplasmosis, $215 million.[41c]

Sanford A. Miller, of the University of Texas Health Sciences Center, reported in January 1990 that a comparison of the potential statistical risk associated with naturally occurring toxic substances in food is revealing.[42a] Risk associated with pesticide residues in food averages on the order of magnitude of 10^{-6}. Risk associated with naturally occurring toxic substances in food, particularly carcinogens, is on the order of magnitude of 10^{-4} or 10^{-3}. In contrast, the actuarial calculation of risk associated with microbiological contamination of foods reveals that the risk for morbidity, that is, the number of people who become ill, is on the order of 10^{-2}, while the risk for mortality is 10^{-3}.[42a] Cothern and Marcus have classified cancer risk.[42b] U.S. population risk below 200 (10^{-6}; 1 case/1.3 million population or essentially "one in a million") cases per lifetime is considered *de minimis* while 20,000 (10^{-4}; 1 case/13,500 population) cases per lifetime would require regulatory intervention.

CDC reported that during the period 1983–87, bacterial pathogens accounted for 66% of foodborne outbreaks and 92% of cases while chemical agents (scombrotoxin, ciguatoxin, heavy metals, mushrooms, PSP shellfish toxin, monosodium glutamate, and other chemicals) accounted for 26% of all foodborne outbreaks but only 2% of cases.[43] Parasites caused 4% of outbreaks and <1% of cases, while viruses accounted for 5% of outbreaks and cases. Hepatitis A caused 71% of outbreaks due to viruses. The etiologic agent was not determined in 62% of outbreaks, which is a major shortcoming in our national epidemiological surveillance program. Table 1 lists bacteria involved in foodborne disease and Table 2 lists those disease conditions caused by these bacteria. Certain of these bacterial organisms are emerging pathogens in the foodborne disease arena. Newly emerging parasites such as *Cyclospora* sp. are also in the news.

In addition to bacteria, there are various protozoa, viruses (hepatitis A, enterically transmitted Non-A, Non-B [ET-NANB] hepatitis virus, reoviruses, rotaviruses, Norwalk agent, Norwalk-like virus, ECHO virus, coxsackie group A virus, coxsackie group B virus, adenoviruses, caliciviruses, astroviruses, coronaviruses, parvoviruses), marine dinoflagellates and algae (diatoms), chemical additives, environmental contaminants, drug and pesticide residues, and naturally occurring toxins that contribute to the foodborne problem. These etiological classes of agents may be weighed or rank-ordered from the most to least significant with respect to risk, outbreaks, cases, morbidity, mortality, and economic considerations. Salmonellosis, campylobacteriosis, and staphylococcal intoxication were reported to be the most extensive and expensive of the foodborne diseases.[47] Paige reported annual bacterial foodborne case or attack rates per 100,000 population for the "Seattle-King County, Washington Study,"[48] and Dorn for the State of Washington[49] as follows: Paige - *Campylobacter* (100 cases), *Salmonella* (40 cases); Dorn - *Campylobacter* (50 cases), *Salmonella* (21 cases), *E. coli* (8 cases), and *Shigella* (7 cases).

Between 1950 and 1989, 35 (13%) of 365 foodborne botulism outbreaks in the U.S. were associated with consumption of processed fish.[50] In 1990, an unusual outbreak of botulism occurred in three of four Hawaiian family members and was associated with fresh fish (palani).[50]

Table 1   A Partial List of Bacteria Involved/Implicated in Foodborne
Infections/Intoxications

Enterotoxin Producers

- *Clostridium perfringens*
- *Escherichia coli* (7 classes,verotoxin or VTEC or ETEC; 3,000 U.S. strains overall)
- *Staphylococcus aureus*
- *Salmonella enteritidis*, serotype typhimurium, others (over 1,700 serotypes and variants; overall 2,296 different salmonella serovars exist)
- *Shigella dysenteriae, S. flexneri, S. sonnei, S. boydei*
- *Vibrio cholerae* (7 biotypes, including El Tor)
- *Vibrio vulnificus*
- *Aeromonas hydrophila*
- *Yersinia enterocolitica* (34 serotypes and 5 biotypes)

Neurotoxin Producers

- *Clostridium botulinum*

Histamine Producers

- *Morganella morganis*
- *Klebsiella pneumonia*

Note: Sources of data: Sleisenger and Fordtran,[33] Archer and Young,[44a] Braude, Davis and Fierer,[45] Menning,[1] Madden.[46]

Listeria monocytogenes is reported to be a major concern for the food industry because the organism is common to animals and the environment. Sanitation in processing plants, adequate heat processing of foods, and reliable cold storage are food service aspects where controls must be established to reduce the prevalence of this organism in foods. Approximately 2000 human cases of listeriosis occur in the U.S. per year and the annual cost has been estimated at $480 million assuming that 1860 cases occur each year.[5] Included in the cost estimates are $35.6 million for medical care, $221.2 million for productivity losses, and at least $225.5 million for psychic (mental anguish) losses. The per-case cost of listeriosis was $137,000, as compared to $500 to $700 for salmonellosis. Listeriosis occurs over 300-fold more often in AIDS patients than in the general public.[44a] Listeria leads to a serious infection of the brain stem (meningoencephalitis), and death frequently follows.[44b,44c]

In 1985, the worst food safety emergency ever to hit the U.S. occurred in Illinois, Indiana, Iowa, and Michigan. *Salmonella typhimurium* contamination of bulk liquid 2% milk in a Melrose Park, IL dairy plant affected possibly 200,000 people patronizing a 217-store supermarket chain operating in the northern Midwest. During the period 1983–1987,[52] *Salmonellae* accounted for 57% of all foodborne bacterial disease outbreaks occurring in the U.S. Illnesses from *S. enteritidis* (SE), mostly from consumption of eggs, rose sixfold in the U.K. and sevenfold in the northeastern U.S. between 1976 and 1986.[12] During 1988–1990, SE in Grade A raw or inadequately cooked eggs caused over 6600 human cases, with 43 deaths.[1,53] In 1989 alone, 77 outbreaks involving 2394 cases and 14 deaths were recorded.[53a] During 1985–1989, state and territorial health departments reported 244 SE outbreaks, which accounted for 8607 cases of illness, 1094 hospitalizations, and 44 deaths. Of the 109 outbreaks in which a food vehicle was identified, 89 (82%) were associated with shell eggs.[54] During 1988, there were 39 SE outbreaks (8 deaths) followed by 78 outbreaks in 1989 (13 deaths) and 74 outbreaks in 1990.[55] The FDA issued an advisory in 1989 to nursing homes and hospitals to use pasteurized eggs only, as elderly and immunocompromised persons are at high risk.[56] The SE problem diminished somewhat during CY 1991 but is no longer just occurring in New England and the mid-Atlantic region. During 1991, surface contaminated cantaloupe and watermelon were associated

Table 2 A Partial List of Bacterial Foodborne Human Conditions in the United States with Estimated Cases/Deaths per Year and Infectious Dose When Known

Intoxications or Food Poisoning

Botulism, 30-40 cases, 3-5 deaths/year[c]

Bacillus cereus intoxication[c]

Staphylococcal intoxication[c]

Clostridium perfringens intoxication, 10,000 cases/yr[b]

Food Infections/Bacterial Enteritis

Salmonellosis, 1.9 million cases/2,000 deaths/yr, ID[d]100

Shigellosis[b], ID 20

Campylobacteriosis (*C. jejuni*), 2–4 million cases/48–480 deaths/yr, ID 500[a]

Yersiniosis (*Yersinia enterocolitica*), 3200 cases/yr, ID >10^6[a]

Colibacillosis (*E. coli* O157:H7 watery diarrhea, hemorrhagic colitis/HUS/TTP (classes: VTEC, ETEC, EPEC, EHEC, EIEC, AEEC), 20,000 cases/415 deaths/yr, ID >10^6 [a,b,e]

Vibriosis (*V. vulnificus* from shellfish), 45 cases/17 deaths/yr[a, b]

Vibriosis (*V. cholerae*, El Tor), ID 10^8 [b]

Vibriosis (*V. parahemolyticus*), ID 10^{5-6}

Listeriosis (*L. monocytogenes*), 1700 cases/450 deaths/yr[a, b]

Streptococcal infection, Group A and others

Brucellosis (*B. melitensis*)

Aeromonas hydrophila infection[a, b]

Plesiomonas shigelloides infection[a]

Other bacterial conditions

Note: Sources of data: Sleisenger and Fordtran,[33] Archer and Young,[44a] Braude, Davis and Fierer,[45] Menning,[1] Madden.[46]

[a] Emerging foodborne diseases.

[b] Toxin elaborated in intestine by bacteria.

[c] Toxin ingested preformed.

[d] ID = human infective dose.

[e] Hemorrhagic Uremia Syndrome (HUS) and Thrombotic Thrombocytopenic Purpura (TTP) are possible sequelae to O157:H7 infection; HUS is the leading cause of renal failure in children; TTP has a fatality rate >50% in the elderly. The U.S. has 3000, Canada 2000 and the U.K. 15,000 strains of *E. coli*; O157:H7 is one of these serotypes or strains; more than 56 strains produce verotoxin.

with *Salmonella* cases in 15 states and two Canadian provinces. Many of the victims had eaten at salad bars. Minnesota released a consumer announcement that noted 185 laboratory-confirmed U.S. cases and another 56 Canadian cases of *S. poona*.[57] U.S. congressional funding for USDA *Salmonella enteritidis* surveillance in shell eggs and *Salmonella* serotyping at the USDA National Veterinary Services Laboratory (NVSL, Ames, IA) was apparently discontinued in 1996, but this maybe a short-term or temporary oversight.

Contaminated ice was implicated as the vehicle for a large 5000 case viral gastroenteritis outbreak in Pennsylvania and Delaware during 1987.[44] Since ice is frequently used to chill foods, this case is considered noteworthy.

The U.S. dairy industry has an excellent safety record over the past 20 years. Occasional outbreaks of food poisoning due to contamination of milk products do occur. These unfortunate incidents include: (1) Illinois, Indiana, Wisconsin, Michigan, Iowa, Minnesota—1985, low-fat 2% milk, *Salmonella*, 16,000 illnesses, at least 2 deaths; (2) California—1985, Mexican-style cheese, *Listeria monocytogenes*, 150 illnesses, 84 deaths; (3) Massachusetts—1983, whole and low-fat milk, 49 illnesses, 14 deaths; (4) Tennessee, Arkansas, Mississippi—1982, *Yersinia*, milk, 172 illnesses; (5) Kentucky—1985, *Staphylococcal* enterotoxin, chocolate milk, 860 illnesses; (6) Florida, California—1985, milk, cleaning solution contamination.[58] The U.S. collective dairy herd amounted to 11.1 million dairy cows in 1983 and 9.3 million head in 1996. The dairy industry produces billions of pounds of milk and tons of cheeses annually. A recent 1996 issue of *Hoard's Dairyman* reported that 50 U.S. dairy cooperatives processed 121 billion pounds of milk per year, or 80% of the nation's supply. Total U.S. milk output for 1996 was 154.3 billion pounds. One record-holding Holstein cow in Wisconsin even produced 63,444 pounds of milk in a single year with twice-daily milking. Over that period of a year, the health of that one cow could have impacted 126,888 consumers each drinking one 8-ounce glass of milk. Public health officials would have most likely "nipped the problem in the bud" during the first few weeks limiting the magnitude of such a problem. Essentially 99.999% of all dairy products are safe and wholesome, but dairy consumers are concerned over BST, pesticide, and antibiotic residues, Salmonella, Listeria, *E. coli*, and BSE. Drug residues from udder infusions/treatment of sick dairy cows are of regulatory concern today. Veterinarians follow FDA established veterinary drug withdrawal guidelines before releasing animals or products to be sold as food. Other nations are rightfully envious of the U.S. dairy quality history and product keeping quality. Fortunately, U.S. dairy cattle have remained free of bovine spongiform encephalopathy (BSE). The system is working and there is no justified reason to change it for sake of change or to return to natural or pure organic farming.

During 1989, multiple outbreaks (>100 cases) of *Staphylococcal* enterotoxin (SEB) food poisoning occurred in the U.S. from consumption of canned mushrooms (in 100-ounce cans) imported from the People's Republic of China. Based upon 82 cases in Mississippi, Pennsylvania, and New York, the FDA recalled the mushrooms and placed an order to prohibit entry of all incoming shipments.[59a]

Todd reported in 1990 on five estimates of the number of cases and costs per year for foodborne gastroenteritis in the U.S. Estimates ranged from 6.3 million to 99 million cases and costs from 4.8 to 23 billion dollars.[47] The Council for Agricultural Science and Technology (CAST) task force reported in 1994 that foodborne illness in the U.S. affects 6.5 to 33 million people each year with approximately 9000 annual deaths.[47b] For the years 1978–1982, 14,340 cases of foodborne disease were reported in the U.S. compared with 6190 in Canada.[47] Between 1983 and 1986, the mean annual number of notifiable cases of *Salmonella* in the U.S. was 50,110, laboratory isolates 43,434, and foodborne-outbreak-associated cases 7,350. The actual number of cases of Salmonella was estimated to be much higher: between 790,000 and 3.69 million, with the average number of deaths estimated at 523 to 7,041. Actual foodborne disease fatalities (all causes) occurring during the period 1983–1987 totaled 129 deaths and was distributed by year as follows: 1983 (35), 1984 (12), 1985 (66), 1986 (11), and 1987 (5). The average over these 5 years was 25.8 deaths per year.

Vibrio vulnificus, a waterborne or shellfish-borne organism, can lead to a serious human infection (septicemia) and frequently death; it can survive in refrigerated raw oysters up to 14 days. Individuals with liver disease should not eat raw shellfish because of the hepatitis A and *V. vulnificus* risk. This bacterium is highly fatal.

Illness from amnestic shellfish poisoning (ASP) was documented for the first time in the world from mussels harvested in Prince Edward Island waters during 1987. Domoic acid, an excitatory neurotoxic amino acid produced by the algal diatom, *Nitzschia pungens*, was responsible for the brain injury (neurological deficit or long-term loss of short-term memory) and gastroenteritis produced in this condition.[47] Domoic acid is a potent analog of the excitotoxin kainic acid; both compounds are related to glutamate, an excitatory amino acid (EAA) neurotransmitter. Saxitoxin-generated paralytic shellfish poisoning (PSP) is more common and can be lethal. There are about 20 different phycotoxins in the PSP complex; the most well known being saxitoxin. PSP toxins are produced by marine algae from the genera *Alexandrium* and/or *Gonyaulax catanella* and are concentrated in seafood through the feeding practice of screening large volumes of seawater. PSP toxins block sodium channels, resulting in skeletal and nerve paralysis. Consumption of highly burdened mussels can be lethal. Diarrhetic shellfish poisoning (DSP), the third of the major shellfish poisonings, is not usually life-threatening, but people can be very sick for several days with abdominal cramps, vomiting, diarrhea, nausea, and headache. DSP toxins are dominated by okadaic acid and dinophysistoxin (DTX-1). The DSP toxins are also produced by dinoflagellates, mainly from the genus *Dinophysis*. Ciguatera poisoning is the most common but least serious of all seafood intoxications. All evidence points to the fact that the fish do not produce the poison, but acquire it through the feed chain.[47a]

Since 1982 when it was first recognized as a human pathogen due to an outbreak from ground beef, *Escherichia coli* O157:H7 has emerged as an important pathogen that produces a number of different verotoxins and is therefore classed as verotoxigenic *E. coli* (VTEC or ECO157VT). More than 50 stains of *E. coli* are known to elaborate these verotoxins. These heat-labile protein toxins are closely related by amino acid homology to Shiga toxin. ECO157 is a bacterial enteric pathogen that causes diarrhea, hemorrhagic colitis, and bloody diarrhea, and the life-threatening postdiarrhea disorders, hemolytic uremic syndrome (HUS, the leading cause of kidney failure in children) and/or thrombotic thrombocytopenic purpura (TTP). *E. coli* O157:H7 may not have always been a pathogen, but somewhere along the line the bacteria's genome changed, resulting in it becoming pathogenic from transferred genetic material (plasmids or "R" factors).[59b] Although *E. coli* O157:H7 was not reported as a human pathogen until 1982, it has become one of the most frequently recovered bacterial enteric pathogens, due in part to the widespread use of improved screening methods in clinical microbial laboratories.[59c] From November 15, 1992 through February 28, 1993, more than 500 laboratory-confirmed infections with *E. coli* O157:H7 and four associated deaths occurred in four states: Washington, Idaho, California, and Nevada.[59d] The cause of this outbreak has been linked to undercooked hamburgers from a popular fast-food restaurant. In 1993 a multistate outbreak of ECO157 diarrhea from fast-food hamburgers occurred, and in 1994 more than 25 outbreaks occurred due to home-cooked hamburger.[59e] Since 1993, many other outbreaks have occurred throughout the country, including one in California and another in Washington involving 23 cases that were caused by eating sliced dry salami. This brings into question the safety of dry salami and other fermented meat products that had previously been considered free of *E. coli* O157:H7 contamination.[59f] Although undercooked hamburger is involved in most cases, other undercooked meats, apple juice, raw milk, radishes, and certain city water supplies are known to harbor toxigenic *E. coli*. Sources include animal and human excreters or meats/vegetables contaminated by feces, and it can readily multiply on unclean surfaces of equipment in food establishments.[59g] During the 1990s, studies to determine the prevalence of genes for VT1, VT2, or both toxins, in 2100 *E. coli* strains from the feces of healthy farm animals were conducted employing new DNA–DNA colony hybridization technology. Ten of 82 milk cows, 20 of 212 beef cattle, and 5 out of 75 pigs were reported to carry *E coli* bacteria with toxic genes. Several of the serotyped isolates are pathogenic for humans (O157:H7, O82:H8, O116, O113, O126, and O91). Of 6894 dairy heifer calves sampled by USDA/APHIS across 28 states in 1068 herds, 3.6/1000 were positive for *E. coli* O157:H7. The herd prevalence was estimated to approach 5%.[59g] Prevalence in feedlot steers is between 1 and 2%.[59gg] Industry, academia, and governmental agencies are actively researching systems that will reduce the level of *E. coli* O157:H7 present in raw foods brought into food processing and food serving facilities. The emergence of this pathogen has greatly accelerated a push for mandatory hazard analysis critical control point (HACCP) programs in the meat industry, and "good dairy practices" and other on-farm "pathogen reduction programs."

During the 6-month build-up leading to the 1991 Gulf War with Iraq, U.S. forces moved $35 to $40 million worth of food per month to some 600 dining facilities in the Operations Desert Shield/Desert Storm theater in Saudi Arabia.[60a] Despite unfavorable environmental conditions (ambient desert temperatures as high as 130°F and higher interior temperatures inside metal food storage containers in conex/sealand vans) and the absence of refrigeration at some outlying locations, fewer foodborne illness cases were reported than predicted from the 0.5 million deployed U.S. troops. Diarrhea was by far the most prevalent complaint among the troops. Early in August 1990, as many as 4% of troops in some units sought medical attention for diarrhea. Shelf-stable combat Meals, Ready to Eat (MREs), Meals Ordered, Ready to Eat (MOREs), and group feeding tray rations ("B" and "T") accounted for approximately 10–25% of all foods consumed by soldiers. The government of Saudi Arabia provided the majority of food service through contracts and local procurement. Some diarrhea cases that were traced to Shigella-contaminated, locally grown lettuce disappeared after U.S. Marines banned the lettuce. A U.S. Navy survey of 2000 ground troops found that 57% of them complained of at least one diarrheal episode during the deployment. Saudi Arabia's dry climate, sophisticated infrastructure, and safe water supply were listed as factors that assisted U.S. forces in prevention of diarrheal disease.[60b] When previous deployments to the desert (Brightstar '85, '87, '90) were used as yardsticks, Operations Desert Shield/Storm in the hostile Arab desert environment resulted in comparatively lower disease nonbattle injury (DNBI) rates (4.8%/1.6/week using passive surveillance). The "Desert Storm Syndrome" (DSS) of many apparently unrelated medical conditions has now reached serious levels in the pool of 697,000 soldiers who served in this theater. Primary complaints in veterans include fatigue, memory loss, diarrhea, and insomnia. The morbidity levels are quite significant but there has been little mortality. A medical survey of afflicted veterans conducted by the U.S. Department of Defense Assistant Secretary of Defense for Health Affairs resulted in the following patterns: 36% had psychological conditions, 47% had musculoskeletal conditions, and 43% symptoms, signs, and undefined conditions. The condition(s) is also known by the term "Gulf War Syndrome" or "Gulf War Illnesses." The cause(s) is under intense study but it is highly unlikely that the etiology, if a common etiology exists, is related to food consumption during the period. According to the Presidential Committee on Gulf War Illnesses (Joyce Lashov and Philip Landrigran), battle stress/fatigue appears to be the most likely cause and the only clear common denominator at this time (January, 1997). Lashov stated that stress can lead to organic disease beyond psychiatric sequelae normally associated with mental stress. Lashov encouraged Congress to support more peer-reviewed independent research by nongovernmental parties. Congress has approved funds for medical treatment of offspring of DSS veterans born with the condition known as spinal bifida under a "presumption of cause and effect." On January 8, 1997, the *Lycos* server/engine on the Internet listed over 40,000 "hits" under the search term "Desert Storm Syndrome."

If one excludes bacterial, fungal, or dinoflagellate (including algal diatoms) elaborated toxins (botulinum, ETEC, SEB, Shiga toxin, cholera toxin, *C. perfringens* enterotoxemia, mycotoxins, psoralens, alimentary toxic aleukia (ATA), yellow rice disease, ergotism, scombrotoxin, ciguatera toxin, PSP, DSP, ASP), classical inorganic or synthetic chemical contamination of the world's food supply accounts for a relatively minor portion of overall food safety risk. Toxicology events or poisoning episodes do occasionally occur due to consumption of foods contaminated by chemicals. Industrialized nations continue to discover and manufacture new chemicals each year, and the number grows by thousands each week. According to an estimate based on the Chemical Abstracts Service (CAS) Registry, the chemical universe consists of over 5 million distinct chemical entities. Of these, 60,000 to 70,000 are available in the commercial marketplace pool to which approximately 1,000 are added each year worldwide. EPA-registered active pesticides and inert ingredients total 3,350 compounds; the Toxic Substances Control Act (TSCA) Inventory includes 48,523 chemical substances in commerce.[62,63a] Similarly, the FDA's program must cope with over 2000 food-related chemicals (900 flavors, 700 GRAS items, 350 food additives, 175 animal drugs, and 60 color additives) and an additional 12,000 indirect food additives.[63b] Complete toxicology health hazard assessment profiles were available for only 1160 (10% of pesticides, 18% of drugs, 5% of food additives) of these chemicals in 1984.[63a] A 1984 study sponsored by the NTP reported that it was remarkable how few chemicals had been subjected to systematic neurobehavioral analysis.[63c] Landrigran et al. reported that too frequently the neurotoxic or other toxic properties (such as immunotoxicological potential) of new compounds have not been recognized before their introduction to the market.[63d] The U.S. Congressional Office of Technology Assessment (OTA) has conducted assessments on neuroscience and identified the need for neurotoxicology testing of new and previously approved (GRAS) chemicals.[63e] The Executive Office of Science and Technology Policy (OSTP), the OTA, and the National Science Foundation have designated the 1990s as the "Decade of the Brain."[63f] Neurotoxicology testing is one important component of this neuroscience initiatives program. The FDA Center for Food Safety and Applied Nutrition (CFSAN) has undertaken an effort to update the 1982 "Redbook" to reflect new requirements for toxicological testing and assessment of neurotoxicity in the evaluation of food additives and chemical safety.[63g] A draft Redbook II, "Toxicological Principles for Safety Assessment of Direct Food Additives and Color Additives Used in Food," was released in March 1993.

During 1973–1974, the largest mass poisoning of farm animals documented in the U.S. to date occurred in Michigan livestock (6000 swine, 30,000 cattle, 1.5 million chickens). Farm animals were fed commercial rations accidentally contaminated by 7500 lb of Firemaster™, a flame-retardant containing polybrominated biphenyls (PBBs). This chemical had been manufactured, misbagged, and sold by a chemical company to a major animal feed mill in Michigan where it was used to prepare dairy feeds initially. Firemaster was mistakenly substituted in the warehouse for magnesium oxide marketed as Nutrimaster™.[64] The similarity of names was reported as the basis for the error in feed manufacturing practices. A warehouse laborer purportedly placed the fire retardant in brown Nutrimaster bags when he ran out of red bags for the bins, and feed mixers became contaminated. An alert engineer turned dairy farmer uncovered the problem when his persistence in searching for the cause of a decrement in his herd's milk production led to the detection of the first poisoned herd. The chemical mill managers/state inspectors should have foreseen the potential for such a mistake as many laborers either cannot read or fail to pay close attention even when they can. Segregation of chemical operations (food vs. industrial chemicals) could have prevented this disaster; the use of red and brown bags was only the first step towards segregation. The practice of segregating food/grain/food chemical operations and industrial chemicals/pesticides operations also must be applied to the transportation phase (truck tankers, ships, and railroad cars). Additionally, eggs, dry milk, butter, and livestock feed in Michigan were condemned. Estimated losses to the state of Michigan and the companies involved amounted to $71 million. In addition to livestock losses, many humans who unknowingly consumed contaminated animal products have reported a wide variety of subjective illnesses and clinical symptoms. The most severely affected were over 2000 members of farm families which had consumed their own farm-grown meat, dairy, and egg products. The U.S. PBB Exposure Registry now includes 1925 women.[64a] The "Michigan PBB Cohort" includes 4545 individuals: 2148 who lived or worked on contaminated farms, 1421 individuals who consumed contaminated farm products, 251 chemical workers, and 668 others who volunteered to be in the federal (CDC) health study. Women with serum levels of 2.0–3.0 ppb or 4.0 ppb or greater of PBBs have a higher estimated risk for breast cancer than women with less than 2.0 ppb. The estimated half-life of PBB is 10.8 years; 163 persons met the criteria of having a median PBB level of 45.5 ppb.[64b] It was estimated that it will take more than 60 years for these levels (45.5 ppb) to fall below 1 ppb. In Michigan, 95% of residents have PBB residues exceeding that allowed in meat (0.3 ppm) by FDA. Chen has written a popular account of the Michigan disaster.[64c]

Heptachlor is an highly lipid soluble organochlorine insecticide heavily used in pest control and agriculture from the 1950s through the early 1980s. It is converted by animals and stored in the body fat as heptachlor epoxide. Roughly 10% of FDA directed milk samples taken nationwide between 1978 and 1982 were found to contain heptachlor, and about 25% of samples were at levels higher than 0.5ppm (the FDA action level).[64d] During early 1986, dairy animals, swine, and beef cattle herds in five Midwestern states (primarily Arkansas and Missouri) were unknowingly fed heptachlor pesticide-contaminated feed. Spent fermentation grain mash, primarily from corn, was used as feeding supplements. These materials contained unacceptable levels of heptachlor.[65] FDA analysis of feed showed heptachlor levels over 1000 times that allowed and analysis of milk demonstrated levels 120 times allowable limits.[66] More than 50 dairies in southwest Missouri were quarantined when heptachlor residues were first detected in milk samples. The source of contamination was traced to a feed mill in Van Buren, Arkansas that had utilized by-products from an industrial ethanol (fuel-grade or gasohol) plant to manufacture livestock feeds. In 1996, the Missouri Corn Growers and the Missouri Legislature sanctioned a gasohol plant that is in final planning stages for siting near Macon, MO. The majority of the nation's corn ethanol plants are located in the cornbelt (Illinois, Indiana, and Ohio). Every new technology brings with it a different set of problems that either must be foreseen and prevented or worked out through trial and error. The industry produced 828 million gallons of corn ethanol in 1988 (without flaw) and more than 1 billion gallons in 1991.

Both of these previous scenarios (Michigan and Arkansas/Missouri) led to contamination of the food chain and necessary destruction of valuable animals and "dumping" of milk and processed milk products. In the Arkansas/Missouri case, widespread human illness was prevented by early detection and removal of contaminated products from the market. Schwabe has reviewed toxicological incidents in food animals such as the bovine hyperkeratosis epidemic in the 1940s and 1950s from chlorinated napthalenes, a 1970 episode of PCB contamination in New York chickens requiring destruction of 146,000 birds, and two smaller poultry PCB contamination incidents in 1971 (Minnesota, 50,000 turkeys destroyed; Southeast U.S. 88,000 chickens and 105,000 eggs destroyed).[67] So far, the U.S. population has been extremely fortunate to have lived without fear of widespread contamination of the food supply.

If future technology expansion occurs, world populations double, and environmental degradation continues unabated, as predicted, the number of toxicological events associated with the consumption of foods will rise, especially in developing nations. Many of these adverse situations will result from chronic low-level chemical build-up on agricultural lands or in the oceans. Today over 40 million individuals live within a 4-mile radius of a Superfund site.[67a] The U.S. federal Agency for Toxic Substances and Disease Registry (ATSDR) suggests that this puts people's health at risk. The same implications could apply to livestock and grain fields that are within proximity to these sites and where contaminated groundwater is used in irrigating fields or in watering livestock. Education, epidemiological/toxicological field surveillance, veterinary herd health/preventive medicine programs, continued regulation of chemical manufacture, transport, and use, and discovery of innovative methods for better storage and transport of all classes of foods and chemicals used in and around food processing, food service, and retail establishments will help to minimize these deleterious outcomes of our modern technological and chemically based global society. Innovation in the food and chemical industries can also build in additional safety measures.

Historically, technologies of pasteurization, refrigeration, cooking, aging, curing, drying, controlling water activity (a_w), using additives/preservatives, increasing acidity, modifying oxidation–reduction potential (Eh), and limiting keeping times and time in the danger zone (45–140°F) have all provided assurances in the past that foods would be safe. G.A. Mitchell has reported that he is not so optimistic, based on (1) the *Salmonella* scenario in 1985 where 18,000 Chicagoans were made ill from pasteurized milk; (2) 29 deaths experienced near Los Angeles in 1985 from *Listeria*-contaminated Mexican-style soft cheese; and (3) 80–100 egg-associated *Salmonella* outbreaks (over 5,000 cases) during the past 5 years.[68] Despite all our technology and resources, we have been unable to eliminate the bacterial foodborne problem. It seems to be growing, if anything. This remains an enigma or paradox for the food safety scientist. On average, 42% of raw broiler chicken carcasses in this country contain *Campylobacter* surface contamination. At times of the year it approaches 90%. Microbiological surveys for *Salmonella* on raw poultry show a contamination range from 60 to 100% of carcasses in some processing plants. In the food industry, food service arena, and in household kitchens, we face such overwhelming food contamination odds, yet seem to get by without a tremendous amount of illness or disease. Adequate cooking corrects these food hygiene shortcomings. Irradiation of raw poultry carcasses/products to control these bacterial threats is the hope for the future. Considering the sheer volume of foods processed and the number of meals prepared and consumed in the U.S. annually, one must conclude that the risk of "foodborne illness" and more specifically "toxicological foodborne illness" is negligble when compared to other deadly risk events in life (driving to work, smoking a cigarette, or a sexual encounter).

One must also keep in mind that toxicants may perturb metabolic processes or generate transient or permanent changes (mutational events) in genetic material or biological systems. These effects may be seen at nanogram or picogram concentrations, which do not always elicit outward observable clinical effects (overt illness or acute physiological changes). Chronic effects of repeated exposure to low levels of toxicants in foods must be taken into consideration; the cumulative effect could manifest itself in future generations as heritable changes. This is the big unknown for the toxicologist.

Food animal production chemicals, packaging leachates, and intentional food additives

Residues of drugs and chemicals deemed beneficial to food animal production and agriculture

Chemicals that persist as residues in foods of animal origin can be subclassified as drugs, pesticides, environmental contaminants, and naturally occurring toxicants. Natural toxicants in foods are numerous and common, which can be a problem if found in excess.[68a] Annually, the USDA Food Safety and Inspection Service (FSIS) analyzes approximately 25,000 samples for chemicals from all market classes of food-producing animals. FSIS analysis for 1990 was programmed to deal with 70,445 domestic samples and 22,184 imported samples.[69]

Drugs, unlike environmental contaminants and naturally occurring toxicants, are intentionally administered to food-producing animals. The prevalence of violative drug residues in animal products will generally be less than 1% when FDA guidelines are followed. The 1990 European ban on importation of U.S. meat led to a critical review of residues of hormones used for growth promotion. This subject received the attention of the USDA and the CODEX Committee on Residues of Veterinary Drugs in Foods.[70] Because of consumer, regulatory agency, producer, and veterinary profession emphasis, the prevalence of beta-lactam antibiotic residues in milk has declined from 13% in 1962 to 0.2–0.5% in 1991. Residues from other classes of antibiotics, such as tetracyclines and sulfamethazine, may occur at greater frequency in dairy products, albeit at very low concentrations. Adverse human reactions to chemical residues in foods are overwhelmingly allergic reactions to drugs (primarily penicillin).[65] The FDA initiated the National Drug Residue Milk Monitoring Program (NDRMMP) in 1991 to monitor raw milk for specified types and amounts of drug residues at 250 locations throughout the U.S. Initial testing targeted eight sulfas and three tetracycline drugs that have been widely misused in dairy cows. The "safe levels" or tolerances set for these drugs were chlor and oxytetracycline (30 ppb), tetracycline (80 ppb), and all eight sulfas (10 ppb).[71a] The Food Animal Residue Avoidance Data bank (FARAD), University of Florida, Gainesville has published an FDA-approved, 550 page, comprehensive compendium of drugs for use by food animal veterinarians and food safety officials.[71b] This reference details generic formulation, manufacturer, species indications, routes of administration, disease indications, directions, withdrawal, and milk discard times, covering six categories of animals. USDA FSIS has estimated that up to 400 compounds have the potential to produce residues in animal products. Rico[71c] has documented toxic effects of anabolic agents, antibiotics, and other drugs; Crawford and Franco,[71d] the human health impact of animal drug use; Riviere et al.[71e] and Sundlof et al.,[71f] residues of antibiotics in food-producing animals; Haagasma et al.,[71g] drug residues in food-producing animals; and the Moats and Medina,[71h] residues of veterinary drugs in animal-derived foods, especially milk.

James H. Steele has classified the perceived and actual risk of agricultural pesticide use and food safety among 30 causes of accidental death rank ordered from government statistics.[72] Food additives (preservatives, color additives) and pesticides were positioned on the lower end of the risk spectrum (i.e., low risk).

Studies with carbaryl residues on fresh vegetables demonstrated that simple washing reduced levels by 82–99% in tomatoes and 41–77% in broccoli. Cooking also reduces carbaryl residues in vegetables by 95–99%.[72a]

William B. Buck of the National Animal Poison Control Center (NAPCC), University of Illinois, Champaign-Urbana, has summarized principles of pesticide use and safety in food animals. These precepts include the following: (1) some insecticides used on food animals have established withdrawal times; (2) monitoring and surveillance by USDA-FSIS indicate no significant pesticide residues in foods of animal origin; (3) livestock tend to rapidly biodegrade and eliminate all insecticides recommended for control of parasitic arthropods; (4) there is 50–100-fold less contamination of meat, milk, or eggs than in grains and other components of animal rations; (5) foods of animal origin are dynamic buffers against human exposure to various toxic chemicals in crops, other vegetation, and the environment; and (6) we ingest in our diet at least 10,000 times more weight of natural pesticides than of man-made pesticides.[73]

Herbicides, the fastest growing group of pesticides, have as a class the lowest relative toxicity of the three major pesticidal groups and are selectively more toxic to plants than to other forms of life.[74] Of 70 herbicides in use, 55% have oral LD_{50} values in rats of greater than 1250 mg/kg, which places these compounds in the moderately toxic class. Only 4% were extremely toxic (LD_{50} of 0–50 mg/kg in rats). In 1986, 51 herbicides were available for use on field crops, pastures, rangeland, and nonrangeland. Between 1986 and 1991, 15 new herbicides were added and 8 deleted. Insecticides available for use amounts to 56 organophosphates and 49 others, while available fungicides amounts to only 22. Most herbicide exposures are accidental and nonfatal. Deaths resulting from herbicides are usually intentional (suicide) and complicated by the presence of alcohol and/or medications.[74] During 1989, the worldwide number of reported poisonings by pesticide category were insecticides (48,283), rodenticides (11,762), herbicides (5,531), and fungicides (1,502).[74] Most of these occurred outside the U.S.

Each year, FSIS tests more than 40,000 samples for more than 100 animal drug and pesticide products. Over the last 10 years, the violative rate has steadily declined. In 1990, only 0.3% of all samples tested showed illegal levels.[74a] FDA's sampling shows that 96–98% of tested foods (an average of 15,000 domestic and imported samples tested per year) are within legal tolerances for pesticides. Some of these FDA samples may be of fruits and vegetables, as well as foods of animal origin (milk, cheese). During 1988, FDA analyzed 18,114 samples of foods from all states, including Puerto Rico, along with products from 89 foreign countries. Of 256 detectable pesticides, 118 were actually found using a lower detection limit of 0.01 ppm. More than 96% of samples either had no detectable residues or the level was within legal tolerance.[75] A joint USDA APHIS and FDA surveillance program for pesticide residues in processed foods reported 174 residue detections in 3502 analyses. No nonpermitted pesticides were found, nor was there any overtolerance of any permitted pesticide. No residue level exceeded 15% of existing tolerances.[76] Of 2375 violative animals found by FSIS during 1989 in its 12-state Western region, 2009 of the violations (antibiotics, sulfas, others) occurred in California.[77a] FDA reported that during 1985 failure to observe withdrawal times accounted for 51% and use of unapproved drugs accounted for 17% of violative drug residues in food-producing animals.[65]

The National Advisory Committee on Chemical Toxicants in Foods, formed in 1990, chaired by FDA and cochaired by USDA, was charged to recommend criteria for chemical toxicants in foods—such as pesticide residues, animal drugs, and environmental contaminants.[76] Natural carcinogens are present in greater quantity (10,000-fold greater amounts) in foodstuffs than are accidental or intentionally introduced chemicals or drugs.[77b] Public perceptions that foodborne residues pose a major health risk are not supported by actual case reports.

The California Department of Food and Agriculture's (CDFA) marketplace surveillance program reported that less than 1% of 9000 samples of fresh fruits and vegetables taken in 1989 had illegal residues of pesticides (0.71% had illegal residues, 0.22% had residues exceeding tolerance levels, and 0.49% had residues for which no tolerance is established).[78] In 1988, of 9293 CDFA samples, only 1.16% showed illegal pesticide residues and only 0.23% had residues in excess of tolerance levels established by EPA. Since the EPA tolerances include a margin of safety, these illegal residues rarely ever present a health risk, according to scientific experts, including the WHO. The FDA reported in their 1988 market basket survey that it found no pesticide residues in more than 61% of samples compared with 56% in 1987. FDA reported that 5.6% of the 9080 vegetables sampled violated residue standards. Violation rates for other categories of foods were 2.5% of grains/grain products; 2.0% of the 1030 miscellaneous samples; 1.9% of 5256 fruit samples; 1.8% of 878 fish/seafood/other meat samples; and 0.9% of 1222 samples of milk/dairy products. Less than 1% of domestic samples violated residue tolerances. FDA analysis for 1988 involved 18,114 food samples—7639 domestic and 10,475 imported—which represented an increased sampling frequency of 17% and 31% over 1987. FDA used analytical methods capable of detecting 256 pesticides, but found only 118.[79]

James Seiber has stated that the regulatory focus on pesticides ignores chemicals that may be added to foods during processing or that may contaminate water and bioaccumulate in the food chain.[12] Seiber suggests regulatory programs should be more concerned with toxic wastes, solvents, combustion products, materials in plastics, industrial byproducts, and others. Plasticizers, such as PCBs, phthalate esters, and aryl phosphates are of particular concern. Between 1930 and 1975, 10 billion pounds of PCBs were produced, and 90% are still in the environment. Some 900 million pounds of phthalate esters are produced annually for use in buildings and home furnishings—much of these chemicals cycle back into landfills and then slowly into the environment.[12] Plasticizers are also used in food packaging and may contaminate foods by direct contact or when heated in microwave ovens. Food safety experts tend to rank the risk of chemicals in foods below that of microbial risk or nutritional considerations, while the public ranks chemical risk as the top concern.[12] Dennis Heldman, formerly of the National Food Processors Association, has stated that "chemical residues command much greater public concern than microbial issues." The Alar™ scare in apples, cyanide-contaminated Chilean grapes, hormones in rejected beef destined for Europe, drugs in milk and meat, use of preservatives and food additives, and tampering all worry consumers. Only 23% of the public is confident that their food is safe. Nevertheless, people in the U.S. are healthier than they have ever been before: they are taller and live longer, and the leading cause of death has shifted from infectious etiologies to heart disease, cancer, cerebral vascular disease, and accidents. If anything, people in the U.S. are consuming too much safe but high-calorie food.

In January 1990, the "Pesticide Residues & Food Safety Conference" was held in Pt. Clear, Alabama.[80] A set of seven recommendations was drafted targeted for four key groups—EPA, FDA, USDA, National Agriculture Chemicals Association—to guide assessment of dietary exposure risk and the communication of these risks. Zilberman et al. has reported on the economics of pesticide use and regulation in the U.S.[81] The NAS/NRC Committee on Scientific and Regulatory Issues reported on regulating pesticides in foods.[82] These publications offer insight into the issues, including that of food safety, oncogenic risks, economics, food productivity, worker safety, pesticide innovation, and alternatives. The NAS summarized federal regulations pertaining to pesticides and made estimates of potential human risk posed by pesticide residues. Under their worst case assumptions, as many as 1.46 million cases of cancer in the U.S. could be linked to residues of 28 pesticides. The FDA ranks pesticides as their fifth priority after microbial contamination, nutritional imbalance, environmental contaminants, and naturally occurring toxins and ahead of food additives, which are ranked sixth. Of the three classes of pesticides—_insecticides, herbicides, and fungicides_—the latter is thought to present the greatest risk to consumers (about 60% of the total estimated dietary oncogenic risk); nearly 90% of fungicides are oncogenic.[83]

Consumers may also be exposed to pesticides in households. The 1985 congressionally directed EPA Non-Occupational Pesticide Exposure Study (NOPES) reported that data collected in 173 households in Jacksonville, FL and 86 households in Springfield and Chicopee, MA showed that five pesticides—chlordane, chlorpyrifos, heptachlor, ortho-phenolphenol, and propoxur—were detected at very low levels (nonadverse health effect levels) in the majority of household air samples.[84] A 1978 EPA literature survey of over 600 published studies found that nonoccupationally exposed U.S. residents carried measurable residues of 94 chemical contaminants.[84a]

Chemical residues from food contact packaging

The National Science Foundation (NSF) estimated in 1973 that food packaging may contribute close to 3,000 substances to the indirect food additive category.[85] Many of these compounds are classed as "generally recognized as safe" (GRAS) or have FDA approval as indirect food additives. Leaching of packaging chemicals into foods while under storage conditions may present risk for humans; the precise magnitude has yet to be determined. Some suggest that a _de minimus_—negligible or insignificant—risk exists. Components of plastics that enter foods during microwave heating of meals are also of concern, and only certain types of plastic materials are microwave safe. Components and chemicals used in the manufacture of packaging include plasticizers, polyvinyl chloride (PVC), methyl acrylate, acrylonitrile, chloroprene, bisphenol A, polyethylene, styrene, polyurethane, vinyl acetate, polyvinyl pyrrolidone, acrolein, polyethylene terephthalate (PET); contaminants, such as TCDD, benzene and other solvents, and the heavy metals tin, zinc, and lead have been found in food packaging. The National Food Processors Association (NFPA) reported in 1991 than production of lead-soldered food containers would cease by July 1991. During 1990 more than 400 million cans were produced, most for packaging pet food, vegetable oil, and fish.[86] Canned condensed milk has historically been sold in lead-soldered cans. Acidic foods present the greatest concern for leaching of lead from containers into foods, and should not be stored in lead-soldered cans. Such foods as tomato-based products or citric juices may lead to "spangling" or degradation of the lining of tin containers. The organoleptic off-flavor produced by leached tin into food items or the cans' internal appearance usually leads to avoidance of the contents, and therefore tin poisoning cases go unreported. Modern container enamel or plastic liners preclude acid erosion of interior metal.

Bisphenol A (BPA, Dow Chemical Co.), a monomer of plastics, is used in the manufacture of many consumer products and in polymer linings of about 100 billion food cans per year. Nagel et al. have reported that this chemical has strong estrogenic activity and may be an "endocrine disruptor" (xenoestrogen) like DDT, PCBs, DES, nonylphenol, octylphenol, methoxychlor, and other pesticides.[86a]

The paper industry is participating in a voluntary program to reduce TCDD levels in all food-contact paper products to 2 parts per trillion (ppt) or less. Most (82%) of these food-contact bleached paper products met the standard as of December 31, 1990. Milk carton paper averages less (0.4 ppt TCDD).[87]

The FDA is concerned about the use of recycled plastics in food packaging; single-polymer containers are the best source for recycling.[88] Migration of chemicals such as benzene into packaged roast beef (24 ppb) has been reported to result in off-odor in the meat item.[89] Temperatures for melting of recycled plastics must be very high—so high that chemical contaminants are removed.[87]

The American Chemical Society has sponsored two conferences (1987, 1991) on stored food interactions with packaging materials.[90,90a]

Direct and indirect food additives

The Food Protection Committee of the Food and Nutrition Board of the National Academy of Sciences (NAS) defines a food additive as "a substance or mixture of substances other than a basic foodstuff which is present in a food as a result of any aspect of production, processing, storage or packaging."[85] Approximately 8600 chemicals are approved by the FDA for food additive purposes. These additives can be classified as color additives, flavor ingredients, sweeteners, vitamins and minerals, chemical preservatives (antioxidants, mold inhibitors, and bactericides), anticaking agents, emulsifiers, sequestrants, stabilizers, and miscellaneous additives. As many as 52 categories of food additives are listed in the 1985 edition of the *Kirk-Othmer Concise Encyclopedia of Chemical Technology*.[91a] A few of these additives are listed here for familiarity with classes and names of food additives: antioxidants—butylated hydroxyanisole, erythorbic acid, nitrogen gas; colorants—caramel, certified colors, beet powder; emulsifiers and surface active agents—mono- and diglycerides, propylene glycol, sorbitol, and glycerol monostearate; enzymes—papain, rennet, glucose-isomerase, pectin methylesterase; flavor enhancers—monosodium glutamate; flavors and spices—mustard, pepper, ethyl acetate, carbon dioxide, vanillin; leavening agents and baking aids—benzoyl peroxide, sodium acid pyrophosphate, various carbonates; nonnutritive sweeteners—sodium saccharin; nutrient supplements—ascorbic acid, niacin, ferrous sulfate, thiamin hydrochloride, zinc sulfate; pH control agents—acetic acid, calcium carbonate, calcium phosphates, citric acid, phosphoric acid, sodium bicarbonate, malic acid; preservatives—sodium benzoate, sulfur dioxide, sorbic acid, propionic acid; stabilizers, thickeners—modified food starch, various gums, carboxymethyl cellulose, cellulose; anticaking agents—calcium silicate, magnesium stearate; stimulants—caffeine; protein hydrolysate; and sequestrants—citric acid, disodium ethylene diaminetetraacetate.

U.S. food additive use per year in 1975 was reported to range from 1000 t (metric tons) of caffeine to 215,000 t of pH control agents.[91a] On April 26, 1983 the FDA approved the first new direct food additive in over a decade. The compound, Anoxomer(l)™, is an antioxidant and represents the first in a class of nonabsorbable, polymeric food additives.[91a] For purposes of FDA regulations, chemicals added to foods (other than pesticides and animal drugs) fall into four categories: (1) food additives; (2) generally recognized as safe (GRAS) additives; (3) prior-sanctioned substances; and (4) color additives.[91b] Direct or intentional (2800 chemicals) and indirect or nonintentional (over 3000 chemicals) food additives have been cataloged.

FDA instituted the Adverse Reaction Monitoring System (ARMS) in 1985. As of October 1988, ARMS had received 6000 passive complaints. Aspartame and sulfite preservatives topped the adverse reaction list, accounting for 95% of complaints (aspartame received 80% and sulfites 15% of complaints). Other compounds reported as inducing adverse reaction included the flavor enhancer monosodium glutamate, nitrate preservatives, the emulsifier polysorbate, vitamin and mineral supplements, and some dyes.[92a] About 1% of the population is sensitive to sulfites. According to ARMS data, 27 individuals may have died from sulfites. Cyclamates were banned from use in foods during the 1960s because of their suspected carcinogenicity. Congress and the FDA gave the artificial sweetener saccharin a special exemption from the Delaney Amendment. Saccharin is an animal epigenetic carcinogen or promoter of heritable genetic changes leading to bladder tumors in F1 generation rats; however, the predicted adverse health effects from sucrose sugars in its absence outweighed the risks of cancer.[92b,93] Nearly 200 colors were on the FDA provisional list in 1960, meaning that further investigation was needed to confirm safety. The food additive industry generates $10 billion a year in business.[91b]

Reddy and Hayes reported in 1989 that among the food additives subjected to regulatory action, only one direct food additive, i.e., cobaltous salts, has been strongly associated with an adverse effect (beer drinker's cardiomyopathy) in humans.[6] Food additives rank at the bottom of their list of foodborne hazards, behind pesticides, natural toxicants, environmental contaminants, and microbial toxins.

The USDA and FDA contracted with the NAS in 1980 to examine the effects of dietary nitrate and nitrite added as food preservatives.[91c] These direct food additives prevent outgrowth of *Clostridium botulinum* spores and protect consumers from botulism. The NAS Expert Committee's conclusions included the following: (1) nitrate is neither a carcinogen or mutagen; (2) exposure to high levels of nitrate may be associated with an increased incidence of cancer of the stomach and esophagus; (3) nitrosation of nitrate to N-nitrosoamine compounds may be involved; (4) future epidemiology studies should focus on correlation of cancer incidence with exposure to nitrate, nitrite, N-nitrosoamine compounds, nitrosatable substances, and inhibitors or enhancers of nitrosation; (5) evidence shows that nitrite is not a direct carcinogen but is mutagenic in test systems and may be a promoter or cocarcinogen; (6) most N-nitrosoamine compounds are carcinogenic in laboratory animals; (7) nitrate and nitrite can exert toxic effects such as methemoglobinemia and probably contribute significantly to the total body burden of N-nitrosoamine compounds; (8) with the exception of dry-cured products and fermented sausage products, the use of nitrate or nitrite should be discontinued in all meat and poultry products; (9) the nitrosation-inhibiting effects of ascorbate and other substances is established, and this knowledge has been put to commercial use to inhibit formation of nitrosamines in bacon; and (10) alternative curing agents should be identified for nitrite substitution. Riemann suggested that regulatory scientists should evaluate this issue based upon the following consideration: risk of botulinum poisoning in the absence of nitrate and nitrite divided by the risk of adverse effects from the use of nitrate and nitrite in foods as preservatives (i.e., the risk of *C. botulinum* Type A poisoning in the absence of additive divided by the risk of adverse health effects associated with using the additive).[91d] A ratio greater than 1 would suggest that these additives should continue to be used in foods. The NAS committee suggests that it is impossible to estimate the potential morbidity or mortality from *C. botulinum* in the absence of nitrate as a curing agent. Botulism rates from overseas countries where nitrate/nitrite food additives are not used is probably the best source of such data. The question then remains: Are residents of these countries consuming the same classes of foods as are U.S. residents, e.g.,fermented sausages, dry-cured meats, etc.?

Microbial decomposition or elaborated toxic products

Mycotoxins

Mycology is the study of eukaryotic fungi, which consist of three groups of major importance: the filamentous molds, yeasts, and mushrooms. Fungi are generally larger than bacteria, lack chlorophyll, contain a nucleus and mitochondria, and live in many diverse habitats. Over 100,000 species of fungi exist, but only approximately 175 are pathogenic for humans and other animals.[94] A large number are parasites of terrestrial plants and cause the majority of economically significant diseases of cultivated plants. Select strains of fungi have evolved the ability to elaborate toxic products as a defensive adaptation for enhanced competition with other microbes. Such mycotoxins from fungi that grow on foods or feeds are chemically diverse. They may pass up through the food chain, but do not generally bioaccumulate in vertebrates. Food animals act as dynamic buffers to metabolize and excrete much of these mycotoxins. Aflatoxin is one type of mycotoxin. Ratios of aflatoxin concentration in feed to the concentration in beef animal tissue are as high as 10,000:1, demonstrating that beef animals tend not to bioconcentrate these toxins. Swine, poultry, and fish are the farm animal species most sensitive to the effects of aflatoxins. The clinical illness caused by these natural toxins is called *mycotoxicosis* and is an intoxication rather than an infection by fungi. Systemic or dermatological fungal infection is not normally accompanied by *in situ* toxin production or overt clinical intoxication. Fungi elaborate mycotoxins best while growing on plants. By July 1978, Hesseltine reported that some 23 types of mycotoxicoses were clearly delineated, while an additional 25 diseases of animals and humans were suspected to be of mycotoxin origin.[94a] Many of these suspected diseases have now been conclusively linked to mycotoxins, and were recorded as far back as 5000 years ago in China. Ergot poisoning was one of the earliest of the mycotoxicoses to cause illness. Both Romans and Arabs have documented early medicinal use of ergot. In 1891, Sakaki reported moldy rice (yellow rice disease) in Japan to be toxic to humans, resulting in cardiac-type beriberi-like disease.

The literature on the subject of mycotoxins is substantial and covers many aspects and ramifications (history, mycology, occurrence, epidemiology, food safety, veterinary aspects, chemistry, detection, prevention and control, elimination, toxicology, pathology, detoxification, toxicokinetics, human disease, comparative animal models, carcinogenicity, acute vs. chronic exposure, metabolites, congeners and pathways, teratogenicity, mutagenicity and structure–activity relationships and the social, agricultural, and political aspects).[95-125] Well over 250 structurally characterized mycotoxins are currently known.[125a] Mossel presented a tabular listing of 70 mycotoxins and their morbid toxic effects.[126] Reddy and Hayes also published a tabular listing of mycotoxins, major producing organisms, source of fungi, and principal toxic effects.[6] Over 140 trichothecenes are recognized, although T-2, DON or vomitoxin, and DAS are the most important to veterinary medicine. DAS or diacetoxyscirpenol (anguidine), a trichothecene mycotoxin, was tested for use in cancer treatment but had deleterious toxic side-effects and is not used as an antineoplastic agent.[6a]

Aflatoxins (AFs) is a subclass of about 10 low molecular weight mycotoxins first isolated from Brazilian groundnut (peanut) meal associated with an epizootic of "turkey X disease" in England during 1960. Turkey X disease killed 100,000 turkey poults, 14,000 ducklings, and thousands of partridge and pheasants.[6b] AFs are produced by *Aspergillus flavus* or *A. parasiticus*, and have been shown to occur in corn (maize), wheat, oats, cottonseed meal, grain sorghum, rice, green coffee, cocoa beans, pistachio nuts, coconut products (copra), figs, peanuts and peanut butter, walnuts, almonds, pecans, oils from oil seed, spices, beverages, fruits and vegetables, dairy products, meat, eggs, and human milk.[127] Aflatoxins have been identified from tissues of exposed poultry, swine, and cattle, as well as in milk and other fresh dairy products. AFs, especially the aflatoxin B_1, are carcinogenic, mutagenic, teratogenic, and hepatotoxic to animals and humans. These compounds have been responsible for the deaths of large numbers of domestic animals and possibly for the high incidence of liver cancer among exposed human populations. The ultimate aflatoxin metabolite, aflatoxin 2,3-oxide, is formed in the liver by P-450 mediated oxidation of AFs at the 2,3 double bond. This product of hepatic transformation is reactive, covalently binding to macromolecular cellular components, including nucleic acids (DNA, RNA) and proteins. Interaction with DNA through alkylation or adduct formation may lead to nucleic acid damage, which if not repaired (i.e., excision of adduct and replacement of nucleoside) by polymerase enzymes could lead to permanent mutational events and potentially carcinogenesis. Aflatoxins are also reported to inhibit protein synthesis and alter immune function as an immunotoxicant or immunosuppressive agent in animals. They also will cross the placenta of humans. The animal species most sensitive to the hepatic carcinogenic effects of AFB_1 is the rainbow trout, followed by the duckling, rabbit, cat, pig, dog, guinea pig, sheep, Fischer-344 rat, macaque monkey, mouse, and then the Syrian hamster.[128] Human sensitivity is assumed to be near the range of toxicity for monkeys. Very small quantities (2.02–7.8 mg/kg per os) of aflatoxin are required to produce acute oral intoxication in primates, while chronic daily exposure to ng/kg to µg/kg levels (145–1,350 mg total dose over 5–12 years) may lead to carcinogenesis of the liver in rhesus, cynomolgus, or African green monkeys. Cancer of the kidney or colon has also been reported in albino laboratory rats. There is strong circumstantial evidence that aflatoxin also contributes to cirrhosis of the human liver in developing countries.

Mycotoxins include the aflatoxins B_1, B_2, G_1, G_2, M_1, M_2, and aflatoxicol (B_1 metabolite) as well as cyclopiazonic acid, ochratoxin A&B, citrinin, citreoviridin, penitrems, gliotoxin, cytochalasins, slaframine, fumonisin B_1 (FB1), rubratoxins, phomopsins, luteoskyrin, patulin, penicillic acid, sterigmatocystin, kojic acid, sporidesmin, ergot alkaloids, alternariol, psoralens, alternariol, ipomeanol, trichothecenes (T-2 toxin, HT-2 toxin, diacetoxyscirpenol (DAS), deoxynivalenol (DON) or vomitoxin) and zearalenone. Many of these compounds have been reported in both animal feeds and foods intended for human consumption. Hesseltine ranked the global significance of mycotoxins: most important was aflatoxin (hepatotoxin), followed by ochratoxin (nephrotoxin), trichothecenes (dermatotoxin), zearlenone (estrogen), deoxynivalenol (dermatoxin), citrinin (nephrotoxin), etc.[128a] Fumonisins are responsible for equine leucoencephalomalacia (ELEM), porcine pulmonary edema (PPE), and possibly esophageal cancer in humans. Fumonisins (1, 2, 3) are produced by *Fusarium moniliforme*, a true endophyte of corn. Fumonisins disrupt sphingolipid biosynthesis, inducing neurological disease in the horse. Over 5000 horses died in Illinois between 1934–1935 with ELEM (fumonisin intoxication from moldy grain) following a summer drought and early fall rains. Zearalenone, an estrogenic mycotoxin, is principally a problem for swine producers, as it causes reproductive failure in adult animals and evidence of premature estrus in young post-weanling females. Research on environmental estrogenic substances is currently receiving priority federal attention.[130a] Vomitoxin, another *Fusarium* toxin produced by *F. graminearum* or *F. roseum* in wheat, corn, rye, and barley in the Midwest and Canada, is also principally a problem for swine producers. Fumonisin and aflatoxin are warm-weather toxins found primarily in maize that has been damaged from drought or insects. Vomitoxin and ergot are cold- and wet-weather toxins produced principally in cereal grains during seed head development. In the winter of 1989–1990, ELEM and PPE cases in the U.S. were associated with high levels of fumonisin B_1 in corn screenings. Diener reported that the 1995 Texas corn crop was generating the worst aflatoxin problem in 10 years.[133a] In 1971, a major aflatoxin problem occurring in the U.S. corn crop led the FDA to seize some of the contaminated grains. In the early 1970s, the maize crop in North America was devastated by an epizootic of leaf blight fungus, *Helminthosporium maydis*.[133b]

Table 3 Regulatory Action Guidance—Aflatoxin

Feed Ingredient	Action Level
Corn	
Finishing beef cattle	300 ppb
Finishing swine—100 lb or more	200 ppb
Breeding beef cattle, breeding swine, or mature poultry	100 ppb
Dairy animals and immature animals	20 ppb
Cottonseed Meal	
Finishing beef cattle	300 ppb

Note: Adapted from FDA Veterinarian, Nov/Dec 1991. Reference - CPG 7126.33.

Table 4 FDA Guidance Levels for Vomitoxin (DON) in Livestock Feed

Animal Class	Feed Ingredient	Guidance Level
Feed and feedlot cattle	Grain and by-product not to exceed 50% of diet	10 ppm
Chickens	Grain and by-product not to exceed 50% of diet	10 ppm
Swine	Grain and by-product not to exceed 20% of diet	5 ppm
All other animals	Grain and by-product not to exceed 40% of diet	5 ppm

Note: Reference: Diener UL (Ed.). Mycotoxicology Newsletter. Dept. of Plant Pathology, Auburn University, AL, 7:1, (1996), p 2.

Table 5 Guidance Levels for Fumonisin B_1 In Livestock Feed

Animal Class	Feed Ingredient	Guidance Level
Horses	Nonroughage portion of diet Do not feed corn screenings	5 ppm
Swine	Total ration	10 ppm
Poultry	Total ration	50 ppm
Beef cattle	Nonroughage portion of ration	50 ppm
Dairy cattle	No recommendation	

Note: Reference: Diener UL (Ed.). Mycotoxicology Newsletter. Dept. of Plant Pathology, Auburn University, AL. 7:1, (1996), p 2.

Although aflatoxins are very stable in foods and feeds, they are susceptible to *in situ* degradation/deactivation by heat, gamma radiation, visible light, extremes of pH (<3 or >10), washing, autodegradation, oxidizing agents, or exposure to UV light in the presence of oxygen. Binders such as Novasil® (aluminosilicate at 1% of feed) or special clay materials have been incorporated into contaminated feeds. The FDA action level or regulatory upper limit for aflatoxins in cereal grains intended for human consumption is 20 ppb, while the action level in meat and milk is 20 ppb and 0.5 ppb, respectively. Feed for finishing beef cattle has a regulatory limit of 300 ppb, while feed for dairy cattle has a limit of 20 ppb. Table 3 presents FDA regulatory action guidance for aflatoxins.[129] The FDA has released guidance levels for vomitoxin in livestock feed (see Table 4) and in food intended for human consumption (1 ppm in finished product, 4 ppm in starting product). The American Association of Veterinary Laboratory Diagnosticians (AAVLD) Mycotoxin Committee has suggested guidelines for fumonisin B_1 in livestock feed (see Table 5).

Mycotoxin contamination occurs mostly under inadequate storage conditions. Controlling mycotoxin contamination in the field is growing in importance. Grain farmers cannot control the weather or conditions conducive to fungal mold growth in the grain field, so they can have little impact on minimizing this toxin during the preharvest period other than implementing progressive farming practices, such as good insect control (pesticide use) and use of modern combines that minimize grain damage. This limitation may change in the future as fungal resistance is engineered into the genetics of cultivated grains. Prevention of postharvesting mold formation and toxin elaboration on cereal grains under farm dry storage is possible. Proper harvesting, drying, storage, and handling of grain on the farm and in the bulk agricultural grain commodity system, as well as proper dry storage prior to processing and during end-product storage will tend to minimize these contaminants. Aflatoxin present in oil from oil seed can be removed during the refining process.

The many strains of mold necessary for cheese processing are generally considered safe; however, trace amounts of aflatoxin have been reported in cheese surveys conducted in West Germany during the 1970s. Moldy cheeses and certain fermentation-cured delicatessen meats (salami, etc.) should be trimmed of mold, if present and practical, prior to consumption. The mold *Penicillium roquefortii* can produce PR toxin, roquefortine, and patulin, while *P. camembertii* can produce cyclopiazonic acid.[130b]

Baking temperatures (120–250°C for 2–3 min) will destroy no more than approximately 20% of aflatoxin B_1 in bread at the concentration of 0.1 μg/g; higher temperatures (350°C, 2 min) destroy an additional 36% of this toxin. The FDA and the University of Nebraska have determined that foods heated to 175°C showed 90% loss of fumonisin B_1 after 60 min of processing, regardless of the pH levels.[130c]

Mycotoxin contamination levels in commercial animal grains may be reduced to below regulatory action levels by mixing with unaffected grain prior to feeding, by treating grain chemically with hydrated sodium calcium aluminosilicate, or by detoxifying grain chemically (bisulfite oxidation, ammoniation, and chlorination with sodium hypochlorite). Novasil® is the best known "binder" for aflatoxin; more broad spectrum clay binders (adsorbents) for zearalenone, fumonisin, and vomitoxin are under evaluation in animal trials. The structural diversity of these mycotoxins makes finding one binder that works for all very difficult. The FDA planned to allow ammoniation of aflatoxin-contaminated ingredients by May 1996. Preventive chemical grain treatments (propionic or acetic acid) are available to inhibit growth of fungi on commercial animal feed grains while stored in on-farm grain bins. Animals may be utilized as biological buffers to dilute these toxins until safe levels are reached in animal flesh. In general, detoxification of mycotoxins in biological systems occurs in the liver; for this reason, livers from animals fed contaminated grain should be discarded. Catfish or other aquaculture species, swine, sheep, and poultry may also serve in this biological detoxification role. Dairy goats or dairy cattle should not be used for this purpose, as aflatoxins are readily excreted into milk. Utilization of naturally resistant plant species or plants bioengineered or crossbred for resistance to mycotoxin-generating species of fungi will be the future approach for control of this detriment to agricultural commodity and human food production.

Since mycotoxins are found in foods and grain in concentrations ranging from nanogram to microgram quantities, their detection requires trace analytical techniques.[130d] Detection methods for mycotoxins include UV fluorescent spectrophotometry, thin-layer chromatography (TLC), high-performance thin-layer chromatography (HPTLC), gas–liquid chromatography (GLC), high-performance liquid chromatography (HPLC), mass spectrometry (MS), nuclear magnetic resonance (NMR), polarography, and enzyme-linked immunoassays (e.g., RIA, ELISA, and others).[130] HPLC is the method of choice for fumonisin; GC, HPLC, and ELISA are used for vomitoxin detection; and HPLC and ELISA are preferred methods for aflatoxin. Animal and nonvertebrate bioassays have received considerable study; however, only the 1–2-day-old Leghorn cockerel chick assay became widely accepted, but is seldom used today. The current FDA tolerance guidelines for aflatoxin in commodities is less than 20 μg of total aflatoxin/kg (20 ppb) raw product. The exceptions are raw peanuts, which may contain 25 μg/kg (25 ppb), and milk which can contain no more than 0.5 μg of AFM_1 per liter of fluid milk (0.5 ppb). Current analytical methods can detect aflatoxins at levels of 10 µg/kg in mixed animal feeds, 1.0 µg/kg in grains or other human food commodities of plant origin, and 0.1 µg/kg (0.1 ppb) of meat, milk and eggs.[131] Regulatory tolerance levels outside the U.S. range from the level of detectability up to 50 ppb.

An assessment of lifetime risk for liver cancer from consumption of aflatoxin-contaminated foods in the U.S. concluded that previous epidemiology studies conducted in sub-Saharan Africa or Southeast Asia, areas heavily infected with hepatitis B, a well-recognized risk factor for liver cancer (hepatocellular carcinoma, HCC), cannot be directly extrapolated to the U.S. population where hepatitis B rates are at least one-tenth as prevalent. It was assumed that, at most, 10% of all U.S. liver cancer was caused by aflatoxin and that the estimated excess lifetime risk was 2.17×10^{-6} per ng/kg/day of aflatoxin intake.[128,132]

Approximately 30% of the 1990 peanut crop in the Southeastern U.S. was placed in "segregation 3," indicating widespread preharvest contamination with aflatoxin. That same year, aflatoxin in corn exceeded 300 ppb in some samples analyzed by states. Georgia found aflatoxin in 28 of 53 corn samples, with 22 samples ranging from 21–100 ppb, 5 from 101–200 ppb, and 1 above 300 ppb. The FDA found aflatoxin at the maximum level of 0.8 ppb in 21 of 69 feed corn samples and in 4 of 96 milk samples; it found aflatoxin at levels below 100 ppb in 10 of 99 food corn samples, in 1 of 30 peanut samples, and in 1 of 14 pistachio samples. Scabby wheat caused by *Fusarium* fungi was present in the 1991 crop in at least eight states. Deoxynivalenol (DON) levels up to 26 ppm were also found; average levels were about 5 ppm. Barley samples averaged 7.0 ppm DON.[133] E. B. Smalley et al. of the Wisconsin Cooperative Regional Project NC-129 reported in early 1994 that, during 1992–1993, Wisconsin experienced a "moldy corn pandemic," which repeated what had happened there in 1962–1965. Smalley's team reported that, coincidentally, these pandemic years were both preceded by major volcanic eruptions and subsequent global low temperature anomalies.[133a] FDA mycotoxin analyses for FY 1995 showed that of 151 samples analyzed for aflatoxin, 24 were positive and 8 exceeded guidelines; of 17 vomitoxin samples, 9 were positive and 4 exceeded guidelines; of 46 fumonisin samples, 20 were positive and 15 exceeded guidelines; and of 29 zearalenone samples, 2 were positive (no guidance levels have been established for this mycotoxin).[133b]

Botulism

Botulism is a rare but very serious and frequently fatal intoxication caused by a high molecular weight heat-labile protein neurotoxin of the bacteria *Clostridium botulinum*.[134,135] Mortality among untreated cases can be as high as 60%. Fortunately, the introduction of a multivalent antitoxin has significantly reduced this mortality. From 1965 to 1990, the University Hospital of Poitiers (France) treated 108 cases of botulism.[85a] In 83% of patients, the food responsible was home-cured ham. The mean incubation time was 3.4 days; digestive symptoms were observed in 93% of cases, ocular symptoms in 92%, and urinary dysfunction in 22%. Six patients were classed as severe, 50 had intermediate disease, and 52 had the mild form of disease. Botulinum Type B was found in 52% of 69 blood samples and in 51% of 81 samples of suspected food. From 1965 to 1976, 44 patients were treated with guanidine hydrochloride (GH, 35 mg/kg daily) and 35 patients were given GH plus heterologous serotherapy. All 108 patients recovered without any sequelae.

Data reported to the Centers for Disease Control (CDC) indicate that in the U.S. from 1985 to 1987 only 89 foodborne botulism cases were reported.[136–138a] Frazier and Westhoff reported that in the U.S. between 1899 and 1972 there were 672 recorded outbreaks of botulism, involving 1731 cases and 963 deaths.[138b] The case fatality rate for botulism has dropped from over 60% (1899–1949) to 23% (1970–1973). Today it is even lower.

There are seven major serological types of *C. botulinum* and they are characterized as Types A through G, based on the antigenicity of distinct toxins. Botulism in humans is caused by Types A, B, E, and F and is associated with foods that have been inadequately preserved.[139] Types C and D are usually associated with animals and birds. A new Type G toxin was discovered in Argentine soil by Gimenez and Ciccarelli in 1970.[139a] At that time, Type G had not been implicated as a cause of botulism in humans. It must be noted that the etiology of infant and wound botulisms are different than foodborne botulism and will not be discussed here.[139b]

There is no doubt that botulism from many different food sources had taken its toll on humans long before an outbreak attributed to blood sausage was documented in Wildbad, Germany in 1793 and gained the medical profession's attention.[134] During the next few decades, many sausage outbreaks were documented along with outbreaks from smoked and lightly salted preserved fish. Although historically significant as a source of the intoxication, sausage is rarely implicated today. In 1987, the CDC reported an outbreak of Type E botulism associated with ungutted, salted whitefish and in 1988 reported a major outbreak of botulism, which manifested delayed recognition, from chopped garlic.[140,141]

By 1900, low-acid canned foods had become the main source of foodborne botulism in the U.S. Commercial canners took steps in the 1920s to reduce the risk of botulism from commercially canned products. During this era, K.F. Meyer of the University of California, San Francisco (UCSF) and the Hooper Foundation led U.S. medical research efforts on botulism epidemiology, microbiology, and prevention.[142] Home canned/preserved low-acid foods remain an important cause of botulism in the U.S.[135] Type E *C. botulinum* is frequently associated with fish and is capable of toxin formation even at relatively low refrigerated temperatures. Film-packed fish held without adequate refrigeration was recognized as a source of Type E botulism in the 1960s.[142] Vacuum-packaged and modified atmosphere packaged foods must include means to prevent or to signal the growth of *C. botulinum*.

The onset of the disease in humans is varied, ranging from 2 hours to as long as 8 days. Variability between botulinum toxin (BoTX) or botulin types has also been recognized. Nausea, vomiting, and gastrointestinal symptoms may occur, but may be due to substances other than the neurotoxin. Often the first neurological symptom is disturbance of vision resulting from loss of reflex to light stimulation and double vision. As the disease progresses, there is a descending muscle paralysis that can progress to respiratory failure and death. BoTX binds to the presynaptic plasma membrane of peripheral cholinergic synapses with a half-time of 12 min and is endocytosed with a half-time of 5 min. The mechanism of action of the toxin on the motor nerve terminal at the molecular level is unknown.[45]

The amount of neurotoxin required to cause illness and death in man is not precisely known, but must be quite small. Deaths have been reported from the mere organoleptic sampling of spoiled foods. Currently, it is estimated that 0.3 μg is the approximate human toxic dose for botulinum toxin.[147] Aminoglycoside antibiotics are reported to potentiate the toxicity of botulinum A toxin and have been proposed as a diagnostic aid.[143] Cardiac glycosides, 4-aminopyridine, brefeldin-A, guanidine nitrate, and guanidine hydrochloride (GH) have been reported as putative treatment drugs for botulinum poisoning based on preliminary animal trials.[144-148] GH, as mentioned above, has been used in France for treating human cases. An injectable form of sterile, purified botulinum toxin Type A, Oculinum™, has been approved by the FDA to treat two human eye disorders (strabismus and blepharospasm).[149]

The risk of botulism can be eliminated by the use of commercial products that have been properly processed (retorted) and stored.[134,135] Reheating foods just prior to consumption also destorys this protein toxin.

Scombroid fish poisoning

Scombroid fish (histamine) poisoning occurs relatively infrequently in the U.S. Data reported to the CDC indicate that, during the period 1977–1981, 68 outbreaks occurred involving 461 people. Historically, this illness was attributed to the consumption of microbiologically deteriorated fish of the families Scombridae (tuna, mackerel, bonito, Spanish mackerel) and Scomberesocidae (Atlantic saury). Many other types of nonscombroid fish are also commonly implicated in the condition including mahimahi, bluefish, jack mackerel, yellowtail, amber jack, herring, sardines, and anchovies. Even less common, cheeses (particularly swiss cheese) have been implicated.[150-152] The term *histamine poisoning* would be more appropriate, since the illness is caused by the chemical intoxication resulting from the consumption of foods that contain high levels of histamine. *Morganella morganis* and *Klebsiella pneumoniae* are the most important histamine-forming bacteria in or on fish.[150,153]

The time from consumption to onset of symptoms is short, ranging from a few minutes to a few hours, with symptoms lasting only a few hours. Symptoms come in four categories: (1) *neurological*—headache, palpitations, flushing tingle, burning, and itching; (2) *gastrointestinal*—nausea, vomiting, diarrhea, and cramping; (3) *hemodynamic* hypotension; and (4) cutaneous—rash, urticaria, edema, and localized inflammation.[150]

Relatively high levels of histamine, 150–200 mg, are required to cause illness. There is both documented and clinical evidence to indicate that other compounds contained in microbiologically deteriorated fish may potentiate the toxicity of histamine. The putrefactive amines, cadaverine and putrescine, enhance the lethality of histamine when orally administered to guinea pigs.[150] Spoilage is considered to have occurred at 20 mg histamine per 100 g of fish and the hazard threshold is 50 mg histamine per 100 g of fish flesh.

Histamine poisoning can be prevented by controlling multiplication of spoilage bacteria. Rapid cooling and then maintaining fish at a low temperature is the best way to control bacterial multiplication and histamine accumulation. Although bacteria in the normal flora of the fish are capable of histamine formation, postcatch contamination can contribute to, or be the source of, the bacteria that produce sufficient levels of histamine to be toxic. Controlling histamine formation in cheese depends on avoiding or destroying histamine-producing bacteria in milk used for cheese production. Good refrigeration, proper hygienic practices, and good quality assurance programs will prevent histamine poisoning.[150,153]

Toxins from miscellaneous spoilage, psychrotropic, and thermophilic bacteria

Biogenic amines are biochemical products of microorganisms growing in fermented foods (cheese, wine, dry sausage, sauerkraut, miso, soy sauce, kimchee, Japanese pickled vegetables, fermented fish, and fish paste).[154,155] Biogenic amines include tyramine, β-phenylethylamine, putrescine, cadaverine, and tryptamine. These amines and certain drugs (antihistamines, antimalarials, and other histamine-metabolizing enzyme inhibitors) may be potentiators of histamine toxicity in humans, leading to hypertensive crisis or dietary-induced migraine. One organism, *Lactobacillus buchneri*, may be important to the dairy industry due to its involvement in cheese-related outbreaks of histamine poisoning. Organisms containing the enzyme histidine decarboxylase are able to convert the amino acid histidine to histamine. Because histamine poisoning resembles food allergy, it is frequently misdiagnosed.

Minor toxicological concerns

Potato sprout teratogen

White cooking potatoes, *S. tuberosum*, a member of the nightshade family Solanaceae, undergo sprouting after prolonged storage under less than optimal conditions. Potatoes should be stored at 41°F and 85% relative humidity.[156] Keeler tested the teratogenicity of dried, ground potato sprout preparations in hamsters.[157] Deformities were produced with the dried sprout material, but neither peel nor tuber material from sprouted or control tubers was teratogenic. Oral doses of the sprout material in hamsters ranged from 2500 to 3500 mg/kg. The average 50 to 70 kg pregnant woman would have to consume 125 to 250 g of dry sprout material (900–1750 g of fresh weight) in a single dose to ingest comparable amounts. Sprouts are normally removed and not consumed by humans to any great extent. Consequently, trimmed sprouted potatoes are deemed perfectly safe for human consumption, and certainly could be used under nonroutine feeding scenarios such as disasters or famines. The identity of the sprout teratogen, a product of natural metabolism, is unknown.[157]

Moldy sweet potato associated lung toxicity

4-Ipomeanol (IPO) is a naturally occurring pulmonary mycotoxin first isolated and identified in 1972 by Boyd and coworkers from slice cultures of the common sweet potato (*Ipomnea batatas*) infected with the fungus *Fusarium javanicum* or *F. solani*. IPO is an oily fungal mycotoxin derived from a furanoterpenoid precursor produced by the sweet potato in response to microbial infection. Pulmonary toxic effects have been demonstrated in a mouse assay.[158,159] A dose of 1 mg intravenously in mice causes death in 5 to 8 hours from severe pulmonary interstitial edema.[160] Outbreaks of acute lethal interstitial pneumonia have occurred in feedlot cattle eating moldy sweet potatoes.[161–163] IPO is in preclinical development at the National Cancer Institute of the National Institutes of Health as an anticancer drug against bronchogenic carcinomas.[164a] It is a tissue-specific cytotoxin that requires metabolic activation to form a reactive intermediate, predominantly by enzymes in Clara cells of pulmonary bronchioles. In rodents, the IPO concentrates in Clara cells within 2 h of dosing, and the reactive metabolite binds to nucleophilic macromolecules leading to necrosis of these cells within 16 h. Loss of Clara cells ultimately affects maintenance of pulmonary bronchial and bronchiolar epithelium. Intravenous doses >15 mg/kg in rats and >24 mg/kg in dogs were lethal. Avian species are unaffected by IPO. Human-associated illness from consumption of moldy sweet potatoes has yet to be documented in the medical literature. The effect of cooking temperatures upon this pneumotoxin is unknown. It is prudent for humans to avoid consuming moldy sweet potatoes. Irradiation technology eliminates the sprouting problem.

Inadvertent chemical contamination

Several mass food poisonings of humans or animals induced by chemical or mycotoxin contamination that are of historical and modern-day interest have been reported (see Table 6). Accidental exposure to arsenic occurred in many of these incidents.[63c] The most significant arsenic event involved 40,000 individuals in France (1928) who consumed arsenious acid accidentally mixed with wine and bread. In an earlier instance (England, 1903), sugar used in beer manufacture was contaminated by arsenic and 6000 individuals were poisoned; 70 fatalities resulted.[164b]

Lead poisoning continues to haunt humankind as it did in ancient times. The FDA is attempting to educate the U.S. population on the hazards of lead from cookware (glazed pottery) and lead crystal decanters. Lead soldering of metal containers for foods has been phased out. Mahaffey has reviewed the dietary aspects of lead exposure.[164c] Oral exposure in children (pica), exposure through inhalation, and exposure through drinking water supplies are also responsible for a portion of the lead burden in humans.

Table 6 Human and Animal Food/Beverage Safety Disasters Involving Chemical Contamination or Mycotoxins

Year	Location	Chemical	Food Vehicle/ Disease/ Animals Condemned	Human Morbidity	Human Mortality	Economic Cost ($)
59	Menkes, Morocco	TOCP/cresyl phosphates, surplus jet lubricant	Intentionally resold as cooking oil	>10,000		
60	Netherlands	Novel emulsifier (food additive)	Margarine/'Margarine Disease'	>50,000		
61	Iraq	Mercury fungicide-treated seed grain	Bread, bakery products	> 1000		
63	Pakistan	Mercury fungicide-treated seed grain	Bread, bakery products	Unknown		
66	Guatemala	Mercury fungicide-treated seed grain	Bread, bakery products	Unknown		
64	Japan	Methyl mercury	Seafood	646		
68–1975	Fukuoka and Nagasaki prefectures, Japan	PCBs PCDFs PCQs	Contaminated rice oil/'Yusho Disease' millions of chickens also condemned	>15,000 had jaundice, 1665 weakness choracne darkened skin		
70	New York State	PCBs	146,000 Chickens condemned	0		
71	Minnesota	PCBs	50,000 Turkeys condemned	0		
71–1972	Iraq	Mercury fungicide	Bread, bakery products	>6000 hospitalized, 5000 severe poisonings, prenatal infant cases unrecorded, >50,000 estimated to be affected	>500 recorded, 5000 estimated	
50s– early 70s	Minamata Bay and Niigata, Japan	Methyl mercury (MeHg)	Fish, seafood/'Minamata Disease'	>1200 birth defects, neurologic sequelae		
73–1974	Michigan	PBBs in fire retardant FireMaster BP-6 FireMaster FF-1 mistakenly used as mineral supplement in animal feeds	Meat, milk, cheese, 5 million eggs, 6000 swine, 30,000 cattle, 1.5 million chickens, 1,200 sheep, 865 tons feed condemned; FDA allowable level of PBB in meat is 0.3ppm	4545 in Michigan cohort: 2148 from farms, 1421 ate PBB-containing products, 241 chemical workers, 668 others, >2000 in Human PBB Registry, 95% of Michigan residents have residues > 0.3 ppm		$71–$100 million in losses; $200 million in lawsuits were filed

Table 6 Human and Animal Food/Beverage Safety Disasters Involving Chemical Contamination or Mycotoxins (continued)

Year	Location	Chemical	Food Vehicle/ Disease/ Animals Condemned	Human Morbidity	Human Mortality	Econor Cost (
1979	YuCheng, Taiwan	PCBs/PCDF	Rice oil	>2000		
1979	Billings, Montana	PCBs accidentally introduced into rendered fat, spead to 10 states in animal feed; 20 PCB isomers/congeners action level in meat is 3.0 ppm	612,000 chickens, processed meats,eggs, grease scattered in 17 states, Canada and Japan	Unknown		
1986	Arkansas, Missouri, 3 other states	Heptachlor in gasohol spent grain mash	Milk, 50 dairies	High levels of human consumption resulted, but no adverse health effects were observed		Several million
1981–1982	Hawaiian Islands	Heptachlor metabolized *in vivo* to heptachlor epoxide, FDA action level set at 0.3 ppm; WHO action level 0.15 ppm	Contaminated pineapple leaf-chop fed to cattle resulting in residues in dairy products over most of Hawaii, public warned 57 days after error discovered	A few thousand individuals (mostly children) are in the 'Hawaiian Heptachlor Registry'		$31 Mill lawsuit been file against Monte Corp. ar Dole subsida:
1987	Argentina	Sodium arsenite, intentional introduction	Comminuted meat	817		
1994	Minnesota (General Mills Company), Fumicon, Inc. pesticide applicator subcontract	Chlorpyrifos-ethyl pesticide contaminate in 16.8 million bushels of cereal grain (oats) and 160 million boxes of breakfast cereal, all condemned	Cereal grains, cereals	0		
1928	France	Arsenious acid	Bread, wine	40,000		
1903	England	Arsenic	Beer	6000	70	
1981	Spain	Aniline denatured rapeseed oil	Toxic oil syndrome (TOS), cooking oil, many suffer severe neuropathy	>20,000	>500	
1978	St. Louis, MO (Ralston Purina Company)	PCB contaminate in fish meal	Animal feed, 400,000 chickens condemned	0		

Table 6 Human and Animal Food/Beverage Safety Disasters Involving Chemical Contamination or Mycotoxins (continued)

Year	Location	Chemical	Food Vehicle/ Disease/ Animals Condemned	Human Morbidity	Human Mortality	Economic Cost ($)
985	U.S. and Canada	Aldicarb pesticide	Melons, neuromuscular and cardiac illness	>1000		
956	Not reported	Endrin insecticide	Bakery foods convulsions	49		
987	Prince Edward Island, Canada	Domoic acid	Mussels, memory loss, seizures, 'ASP'	129	2	
913, 1932, 933–1934, 941–1945, 952, 1953, 955	Orenburg District Siberia, USSR	Mycotoxic trichothecenes, primarily T-2 toxin, *Fusarium sporotrichoides*, alimentary toxic aleukia (ATA), septic angina	Cereal grains, bakery products, wheat, barley, rye, millet, oats, buckwheat	Hundreds of thousands, 10% of human population in affected regions	Thousands; 60% of sick died in some counties	
941–1953	U.S.	Chlorinated napthalene contaminant in animal feed	>10,000 Cattle condemned, 'X' disease or bovine hyperkeratosis	Unknown		$20 million in losses
957	U.S. Southeast	Dioxin (hexa-CDD) and PCB in commercial poultry feed	Toxic fat added to feed; millions of poultry died or were killed; >1 million in NC alone; called 'chick edema disease'	0		
960	England	Aflatoxin contaminated peanut meal used in animal feed	100,000 Turkey poults, 14,000 ducklings, thousands of partridge and pheasants died	0		
955–1959	Southeast Turkey	Hexa-chlorobenzene (HCB) fungicide treated seed grain (wheat)	'Monkey Disease,' Porphyria cutanae tarda (PCT), enlarged livers, porphyria, darkened skin	3000 to 4000, mostly children	300, 10%	
937	South Africa	TOCP	Contaminated cooking oil, nerve paralysis	60		
932	California	Thallium rodenticide	Barley laced with thallium sulfate used to make tortillas, neurological symptoms	13 members of one family	6	
972–1989	China 'acute mildewed sugarcane poisoning'	Mycotoxin 3-nitropropionic acid *Arthridium* spp.	Nausea, vomiting, anorexia, headache, dizziness, somnolence, dystonia, encephalopathy, CNS symptoms prominent	Primarily children; 217 outbreaks, 884 cases	88	

Table 6   Human and Animal Food/Beverage Safety Disasters Involving Chemical Contamination or Mycotoxins (continued)

Year	Location	Chemical	Food Vehicle/ Disease/ Animals Condemned	Human Morbidity	Human Mortality	Econom Cost ($
1930s	SE U.S.	TOCP contaminated beverage	Alcoholic beverage, 'Ginger Jake'	>5000 paralysed, 20,000 to 100,000 affected	Unknown	
1941–1970	U.S. Midwest (the Ohio State Dept of Agriculture uncovered this situation)	'Cumar' silo coating = 18% PCBs 5% PCTs. Interior lining of silos on several farms contained this dangerous material	Dairy cows eating PCB-contaminated silage led to contaminated milk; problem first identified in Ohio during 1970, FDA then set PCB tolerances in dairy products	Unknown		
1977	Puerto Rico and U.S.A.	PCBs; a fire in an animal-feed warehouse led to PCB contamination of fish meal	Feed subsequently fed to poultry resulted in destruction of 400,000 chickens and millions of eggs in U.S.A.	No human illness reported	Unknown	Unknown but presumal millions
1940s–1977	Great Lakes, U.S.A.	PCBs discharged into lakes from manufacturing along lakeshores; PCB manufacture discontinued in 1977	Uknown losses, fish consumption warnings posted on Great Lakes for many years	Unknown	Unknown	Unknow
1946–1976	Hudson River and N.W. U.S.A.	Contamination of fish from discharging PCBs into river	Unknown, fish consumption warnings by local health agencies	Unknown	Unknown	Unknow
1952–1989	Foggia region, Italy (1952); Spain (1987); Turkey (1989); West Bank (1985)	Lead (Pb), heavy metal, metallic lead	Contaminated flour from local mill repaired with lead, >500 ppm Pb in flour used in baking	Plumbism, chronic and subacute; 200 symptomatic in Italy; 11 family members in Turkey; West Bank Arabs	Unknown	Unknow
1958	Sophie, Serbia	Lead	Contaminated flour used in baking	Chronic renal insufficiency, Plumbism; 37 deaths, 23 ill, 12 families affected	Unknown	Unknow
1891	Japan	Mycotoxin citreoviridin	Moldy rice 'yellow rice disease'	Thousands, acute cardiac beriberi, 'Shoshin-Kakke'	Unknown	

Various poisons and toxic chemicals are used to treat seed stock and control pests in and around food storage facilities. These pesticidal chemicals include mercury fungicides, organophosphorus insecticides, pyrethrin and pyrethroid insecticides, Phostoxin™ grain/cereal fumigants, and dicoumarol-based rodenticides, among others. The use of these compounds is controlled by governmental regulations and good manufacturing practices (GMPs), which are very specific. Except for fumigants for stored product pests, direct contact of these chemicals with foods is prohibited. However, inadvertent food usage of treated seed stock or accidental contamination of food may create potential toxicologic problems. It occurs occasionally in the food industry, but is an extremely remote event in developed societies.[64,165] Methyl mercury (MeHg) was implicated in over 1200 cases of mercury poisoning in Minamata Bay, Japan during the 1950s and 1960s. These poisonings were linked to consumption of mercury-contaminated fish and shellfish. Industrial releases of MeHg-laced effluents into a large ocean bay was determined to be the source. MeHg is biotoxic, highly bioavailable, and crosses the placenta to cause birth defects and nervous disorders. The high lipid solubility of MeHg allows this xenobiotic to cross the blood–brain barrier (BBB) and enter the CNS. The biological half-life is reported to be between 73 and 76 days. An even larger mercury outbreak occurred in Iraq in 1971–1972 where more than 6000 people were affected with 500 deaths when mercury-fungicide-treated seed grain (dyed red and placed in bags with the skull and crossbones) was mistakenly used to manufacture homemade bread.[63c,166]

Between 1968 and 1978, the U.S. congressional "watchdog" OTA surveyed the 50 states and 10 federal agencies for reports of chemical contamination of the food supply.[166a] Responses were received from 32 states. The OTA collated 243 food contamination incidents that were reported to cost $282 million (OTA admitted that the actual economic cost was much greater). The majority of state-reported food contamination incidents involved pesticides (56 of 88). Federal agencies reported 155 incidents over this 10-year period. A major PCB contamination event was reported by APHIS to FDA on August 15, 1979. PCBs from a damaged electrical transformer made its way into rendered fat in a packing plant in Billings, MT and was subsequently spread by incorporation into animal feed. Contaminated feed spread throughout 10 states, and polluted poultry, eggs, pork products, and a variety of processed foods, including strawberry cake and mayonnaise, were found scattered in 17 states. A distribution chart provided by the FDA investigators listed 612,000 chickens and an undetermined number of eggs destroyed in Franklin, ID. Top-name food companies were involved. Contaminated Swift and Co. processed chicken was found as far away from Billings as Secaucus, NJ, and contaminated grease and grease byproducts were traced to Columbia, Canada and Tokyo, Japan.

In 1978, the Ralston Purina Co. of St. Louis negotiated compensation for 400,000 chickens that had to be destroyed because of PCB-contamination of fish meal sold to poultry producers.[166b]

Higuchi reported on a large Japanese outbreak of human chloracne-like poisoning in Fukuoka, Japan between 1968 and the mid-1970s.[166c] Patients reported ocular swelling, skin eruptions and enhanced pigmentation (skin darkening), headache, memory loss, numbness, hypoesthesia, and neuralgia of the limbs. Epidemiological studies revealed that this condition was the result of poisoning by PCB-contaminated rice oil called yusho. This condition then became known as yusho rice-oil disease. By September 1973, 1200 poisoned patients had been identified. In 1979, 11 years after the Japanese yusho incident, more than 2000 people in YuCheng, Taiwan were exposed to PBBs and polychlorinated dibenzofurans (PCDFs) by eating contaminated rice oil. This condition was also labeled "yusho disease." Clinical manifestations included chloracne, pigmentation of skin, and hepatic necrosis. Immunological irregularities were also noted.[166d] More than 200 isomers or congeners exist within the class of PCBs. The FDA allowable level for PCBs is 3.0 ppm; chemical manufacture of PCBs has been discontinued.

On 30 October 1985, 32 school children in Berlin, WI, experienced illness (burning or irritated lips, mouth, and throat) after consuming milk accidentally contaminated with an ammonia refrigerant at a local dairy bottling plant.[167]

Food manufacturing misadventures have led to some fairly large human medical disasters. A foodborne disaster was reported in Spain during 1981.[167a] Over 20,000 individuals reported allergic and autoimmune health complaints after consumption of foods cooked in rapeseed cooking oil that had been "denatured" by chemicals. The disease was labeled "toxic oil syndrome" (TOS). Non-necrotizing vasculitis was seen during the acute phase, while the chronic state involved severe debilitating neuromuscular disease with scleroderma-like symptoms. Scleroderma is an immunological illness seen after exposure to a "superantigen." Multiple deaths (350 occurred overall) were the result of respiratory insufficiency. Aniline denatured cooking oil sold in 5-liter unlabeled containers was the causative agent. The illicit rapeseed oil had been purchased in France and diverted from industrial use, refined in Spain to remove the aniline denaturant, mixed with various other oils, and then sold by traveling salesmen, often in mercadillos (little shops). A symposium on TOS was held at the EUROTOX '96 Congress.[167b] A larger poisoning event had occurred 20 years earlier in the Netherlands where approximately 50,000 consumers developed food-associated toxic erythema called "margarine disease."[166d] This inadvertent condition was ascribed to a novel emulsifier in margarine.

The U.S. veterinary profession (AVMA PAC) and the American Academy of Comparative and Veterinary Toxicology (AAC&VT) are working with the U.S. Congress for mandatory addition of bitter taste avoidance chemicals to pesticide- or fungicide-treated seed grains to protect the health of farm animals, pets, and wild birds.[168] Such technology could also assist developing nations in the prevention of accidental poisoning in human populations. Such a situation appeared to be surfacing in India during August 1996. Contaminated grain used in bread making was reported to be the cause of over 50 deaths and many cases of human illness. The exact contaminant is as yet unreported, but the outbreak was described as India's "worst food poisoning incident" in many years.

Intentional chemical adulteration/sabotage

One of the most egregious examples of intentional mass food poisoning occurred in 1959 in Meknes, Morocco where more than 10,000 humans were poisoned by cheap cooking oil deliberately adulterated with surplus turbojet lubricant containing cresyl phosphates.[168a] These lubricants had been purchased at a military depot in Casablanca and were clearly identified as nonconsumable industrial chemicals.

Chilean grapes imported into the U.S. for consumption in 1989 were found to contain cyanide, purportedly placed there by terrorists. One bunch of grapes out of 2000 bunches examined at the Philadelphia port contained trace amounts of cyanide—enough that one grape could be lethal.[169] Based upon this potential immediate threat to the health of U.S. residents, the FDA condemned millions of pounds of Chilean grapes. The indirect additive procymidone (a fungicide) has also been reported in table grapes and in 19 of 203 sampled imported wines from France and Italy.[170] The wise adage of "when in doubt, throw it out" still applies, even today.

Mass food service presents one potential avenue for sabotage of organizations, military units, or groups of people. Military medical experts have described the foodborne biological risk as *de minimis,* but there are no guarantees. Military veterinary food inspection programs at U.S. commissaries in Europe during the 1970s identified elemental mercury in sabotaged oranges procured from Israel. The poisoned Chilean grape incident serves as a reminder of this threat and demonstrates the economic impact a "food terrorist" can theoretically generate with only three cyanide-contaminated grapes.[171-173]

Argentina has had some fairly recent experience with food terrorism (1987) when a sodium arsenite insecticidal acaricide was intentionally added to comminuted meats and sold unknowingly to consumers at a meat market. An acute massive epidemic of meatborne illness ensued involving 718 subjects.[174]

Defense medical food procurement and food service sanitation inspection activities act as a deterrent to sabotage. Veterinary inspections are normally conducted on each shipment of high-cost or highly perishable subsistence, such as carcass beef or dairy products, and randomly on other deliveries at the commissary—at least one veterinarian technician is normally assigned to each defense commissary full-time. During 1987, the U.S. Army Veterinary Corps, the Executive Agency for triservice veterinary support, inspected 13.1 billion pounds of food for all services, performed 3993 sanitary inspections of civilian food establishments and 3763 inspections of military food establishments and warehouses, inspected 8398 food vans/carriers, and performed 68,883,380 ante- and post-mortem inspections of food animals (primarily overseas).[175] The military food safety program is productive, economically justified, and tailored proportionally to the need. No system in the world, however, not even the military system, is specifically designed to detect biological agent or biotoxin sabotage.

The U.S. civilian food marketing and food service industry is more vulnerable than is the military system. Because of inspector shortages, the FDA has had difficulty sending inspectors to civilian food manufacturing plants, food service operations, or food marketing units more frequently than once every few years.[176] Moreover, civilian facilities are vulnerable to disgruntled employee retaliation. The possibility also exists for clandestine entry of a terrorist into a food processing or serving establishment with the intention of sabotaging the food supply. The size of the establishment and the daily meals served limits the numbers of casualties one could potentially inflict—McCormick Center in Chicago serves thousands daily and some catering services have catered parties for corporations serving as many as 15,000 people at one setting. Fast food outlet chains collectively serve many millions of meals daily. Centralized preparation of foods (e.g., soft-serve ice cream mix) in one large facility, with simultaneous distribution of a given lot or batch to hundreds of chain outlets in the region, offers the potential for sabotage and for generating thousands of immediate biotoxin casualties and even more delayed agent casualties. The daily risk of intentional foodborne intoxication or sabotage is less than driving a car (50,000 casualties per year), smoking cigarettes (450,000 deaths per year), or getting AIDS from unprotected sex. This situation might change overnight if a terrorist goes to work in a major retail food or restaurant chain's East or West Coast distribution plant or a knowledgeable disgruntled employee decides to strike back. The food industry, federal agencies, and state and local authorities must remain vigilant in their efforts to ensure that our civilian food supply is safe.

Conclusions

A variety of toxic developments or contaminants in stored foods are potentially possible, but rarely occur. These include botulinum, scombroid (histamine) poisoning, residues of chemicals or drugs used in producing foods of animal origin, packaging leachates, food additive degradation products or excessive use of approved additives, enzymatic decomposition products, mycotoxins, toxins from spoilage organisms, ipomeanol in moldy sweet potatoes, potato sprout teratogen, inadvertent industrial chemical contamination, and intentional chemical adulteration or tampering. However, toxicological contamination of the food chain occurs very infrequently. The relative risk or concern to humans for these events is insignificant, negligible or essentially *de minimis*. Risk for these events during storage or in transport is less than risk occurring during the preharvest on-farm period, processing, or food service (postharvest) phases of the food cycle.

Although the threat of toxins or chemicals delivered to humans through food vehicles or in spoiled foods is minor, the food industry needs to remain vigilant and uphold strict standards. Moreover, food processing and food service industry standards and governmental regulatory controls can be credited with preventing potentially serious chemical poisoning events in human populations. Although concerns or probability predictions for such events are not high, one major mistake on the part of warehouse workers, quality control personnel, or truckers could lead to a catastrophic poisoning scenario. Large-scale vertical integration of our food supply system leaves little room for error, and mistakes can play out into far-reaching, widespread food safety events. Only 7500 lb of PBBs were needed to contaminate the state of Michigan's dairy industry. The amount or volume of chemical required for highly toxic agents is not great. Chemical contamination of a stainless steel truck tanker or railroad car quantity of cooking oil, syrup, dairy products, or liquid foods with a highly toxic delayed neurotoxin or immunotoxic agent could lead to significant morbidity in thousands of individuals. It might be totally consumed before the hazard is realized. Early warning signs are not always a given condition.

Fortunately, accidental mass poisoning via food occurs so infrequently that reports of occurrences are not easily accessible. Foodborne botulism due to oral ingestion of preformed toxin is rare in humans but not in animals, especially cattle and domestic fowl. Pesticide tolerances in foods reflect a very conservative margin of safety; these are normally set at 1% or 0.1% of the "no adverse effect" levels (NOAEL) in test animals. FDA's sampling shows that 96–98% of tested foods are within legal tolerances for pesticides. CDC reports of chemically associated or linked foodborne human illnesses are very rare. However, emerging problems in food animals and foods around the world are raising new questions of food safety that must be resolved through research and epidemiology. These new challenges have included "spiking mortality of turkeys (SMT) syndrome" in young poults in North Carolina; BSE's potential relationship to human spongiform encephalopathies (CJD) and the safety of cattle-derived foods in the U.K.; *Cyclospora* parasites in imported raspberries from Guatamala causing over 850 human cases of gastroenteritis illness in the U.S. and Canada during May–June 1996; the ECO157 outbreak in more than 9500 Japanese (Osaka suburb of Sakai) citizens, primarily school children, eating contaminated school lunches and became ill—11 died in the summer of 1996 with virulent *E. coli* O157:H7, probably from eating contaminated radish sprouts; and the Hudson Foods beef condemnation (ECO157 outbreak/25million pounds condemned).

The threat changes periodically and programs must shift accordingly. The food safety scientist must now also attempt to protect subpopulations of individuals who are immunodeficient from HIV exposure. All of these issues notwithstanding, the U.S. food supply is affordable, bountiful, high-quality, wholesome, and safe, reflecting the achievements in agricultural, food science, and nutritional research, in developing applicable harvesting, processing, and storage technologies, and in implementing prevention strategies involving food safety training, consumer education, and voluntary and mandatory regulatory inspection programs.

These food safety facts are presented to inform scientists and officials—not to worry or alarm consumers. One should attempt to keep a reasonable perspective, but not let down one's guard. Chauncey M. DePue once delivered a poem with an emphasis on food as an after-dinner speech.[177] It remains prophetic even today.

> In these days of indigestion,
>   It is often times a question
>   As to what to eat and what to leave alone,
>   For each microbe and bacillus
>   Has a different way to kill us
> And in time they always claim us for their own.
>   There are germs of every kind
>   In most foods that you find,
>   In the market and upon the bill of fare.
>   Drinking water is just as risky
>   As the so-called deadly whiskey
> And its often times a mistake to breathe the air.

The inviting green cucumber
 Gets most every-body's number,
While the green corn has a system of its own,
 Though a radish seems nutritious,
Its behavior is quite vicious
 And a doctor will be coming to your home.
Eating lobster, cooked or plain
 Is only flirting with ptomaine
While an oyster sometimes has a lot to say
 But the clams we eat in chowder
Makes the angels chant the louder,
 For they know that we'll be with them right away.

Take a slice of nice fried onion
 And you're fit for Dr. Munyon;
Apple dumplings kill you quicker than a train,
 Chew a cheesy midnight "rabbit"
And a grave you'll soon inhabit
 Ah—to eat at all is such a foolish game!
Eating huckleberry pie
 Is a pleasant way to die,
While sauerkraut brings on softening of the brain.
 Every undertaker titters
And the casket makers nearly go insane.

When cold storage vaults I visit
 I can only say what it is
Makes poor mortals fill their systems with such "stuff"
 How for breakfast prunes are dandy,
If a stomach pump is handy
 And your doctor can be found quite soon enough,
Eat a plate of fine pigs knuckles
 And the headstone cutter chuckles
While the grave digger makes a note upon his cuff.

 Eat that lovely red bologna
And you'll wear a wooden kimona
 As your relatives start scrapping 'bout your stuff.
All those crazy foods they mix
 Will float us 'cross the River Styx
Or they'll start us climbing up the Milky Way
 And the meals we eat in courses
Mean a hearse and two black horses
 So—before a meal some people always pray
Luscious grapes breed pendicitis
 So—there's only death to greet us either way.
Friends will soon ride slow behind you
 And the papers then will have nice things to say—
"Some little bug will get you some day."

Appendix A

A List of Relevant Food Safety Topics/Issues

Initial or additional discussions of these topics are available in general food science, food safety, food microbiology, and food toxicology library references.[13-30]

1. Ciguateria fish poisoning
2. Ethylene dibromide (EDB) use as soil fumigant on agricultural soils
3. Ethylenebisdithiocarbamates (EBDC) or ethylenethiourea (ETU) residues in foods
4. Artificial sweetner (aspartame, saccharin or cyclamate) safety considerations
5. hlorofluorocarbon propellants in foods
6. Potato solanine from sun scald in the field
7. Mushroom (amanita toxin, muscarine, gyromitrin, ibotenic acid, psilocybin) poisoning
8. Mercury seafood contamination (Minamata Bay, Japan)
9. Cyanide-contaminated Chilean grapes
10. Intentional fungicide (procymidone) or ethylene glycol wine sweetener "additives"
11. L-Tryptophan-associated eosinophilic myalgia syndrome (EMS); 39 deaths / 1500 severely ill, 1989
12. Puffer or fugu fish (tetrodotoxin) poisoning
13. Saxitoxin/paralytic shellfish poisoning (PSP) or domoic acid/amnesic shellfish poisoning(ASP)
14. Food-borne bacterial infections due to contamination on the farm, during processing, or during preparation in the food service outlet (yersiniosis, salmonellosis, campylobacteriosis, listeriosis)
15. Food intoxications (staphylococcal food poisoning/SEB; *C. perfringens*) due to contamination or excessive growth prior to post-processing storage or growth on the establishment serving line
16. MSG and Chinese restaurant syndrome
17. Naturally occurring biogenic amines in foods
18. Poisonous plant toxins
19. Hypervitaminosis A or D
20. PCBs/PBBs/heptachlor in dairy products
21. Alar pesticide use on apples
22. Aldicarb pesticide contamination of California watermelons
23. Pesticide residuals in fruits or vegetables accumulated while in the field or under cultivation
24. Nitrate toxicity from accumulation in growing plants
25. Nitrate/nitrite additives in meats and thermally generated nitrosamine/carcinogen formation
26. Metal (cadmium, lead, thallium, copper, mercury, arsenic, zinc) or inorganic trace element (selenium) accumulation in foods prior to harvesting
27. Sulfite antioxidants/preservatives for freshness and asthmatics
28. Oilspills and hydrocarbon (PAHs) residues in seafood
29. Heptachlor-tainted grain/fuel-grade ethanol (gasohol) plant mash and introduction into dairy cattle/products
30. Neurotoxic substances (TOCP) introduced into cooking oils or ginger extracts
31. *Salmonella enteritidis* outbreaks and Grade A shell eggs
32. ETEC or *E. coli* O157:H7 (hemorrhagic enteritis/colitis food poisoning
33. Listeriosis and ethnic soft cheeses

34. Cholera, shigellosis, giardiasis
35. Food irradiation or fallout radionuclide accumulation in foods
36. Dioxin levels in foods
37. Campylobacteriosis and foods
38. Shell eggs containing excess levels of iodine
39. Foodborne brucellosis and Q-fever (dairy products)
40. Hepatitis A risk associated with raw oysters/clams.

References

1. Menning E. Emerging Challenges and Concerns An Enigma of Advancing Technology, In: *Proceedings Ninety-Third Annual Meeting of the United States Animal Health Association.* Richmond, VA: USAHA, 1989, p 300.
2. Klaassen CD. Principles of Toxicology, Ch. 2, In: Klaassen CD, Amdur MO, Doull J (Eds). *Casarett and Doull's Toxicology: The Basic Science of Poisons,* 3rd ed. New York: MacMillan Press, 1986, p 11.
3. Frazier WC, Westhoff DC. Food-Borne Poisonings, Infections, and Intoxications: Nonbacterial, Ch. 24, In: *Food Microbiology.* New York: McGraw Hill, 1978, p 470.
4. Bryan FL. *Diseases Transmitted By Foods.* HEW Publication No. (CDC) 80-8237, Atlanta, GA: USDHEW/PHS, CDC, Bureau of Training, December 1979, 63p.
5. Winter CK, Seiber JN, Nuckton CF (Eds). *Chemicals In The Human Food Chain.* Florence, KY: Van Nostrand Reinold, 1991.
6. Reddy CS, Hayes AW. Food-Borne Toxicants, Ch 3, In: Hayes AW (Ed). *Principles and Methods of Toxicology,* 2nd ed. New York: Raven Press, 1989, p 67-110.
6a. Rousseaux CG. Trichothecene Mycotoxins: Real or Imaginary Toxins? *Comments on Toxicology,* 2:1 (1988), p 37-49.
6b. ApSimon JW. The Biosynthetic Diversity of Secondary Metabolites, Ch. 1, In: Miller JD, Trenholm HL (Eds), *Mycotoxins In Grain: Compounds Other Than Aflatoxin.* St. Paul, MN: Eagan Press, 1994, p 3.
7a. Rizk AM. *Poisonous Plant Contamination of Edible Plants.* Boca Raton, FL: CRC Press, 1990.
7b. Cheeke PR. *Toxicants of Plant Origin.* Vols. 1–4: Alkaloids (1), Glycosides (2), Proteins and Amino Acids (3), Phenolics (4). Boca Raton, FL: CRC Press, 1989.
8. U.S. Department of Energy. *Report on the Accident at the Chernobyl Nuclear Power Station.* Washington, DC: U.S. GPO, U.S. Nuclear Regulatory Commission, 1987.
9. Mould RF. *Chernobyl The Real Story: The History of the World's Worst Civil Nuclear Disaster and Its Aftermath.* New York: Pergamon Press, 1988.
10. Carter MW (Ed). *Radionuclides In The Food Chain.* New York: Springer-Verlag, 1988.
11. Anspaugh LR, Catlin RJ, Goldman M. The Global Impact of the Chernobyl Reactor Accident. *Science,* 242, 16 Dec (1988), p 1513-1519.
12. Murdock BS (Ed). Food Safety and Toxicology. *Health & Environment Digest,* 4:7, (1990), p 1-11.
13. Graham HD. *The Safety of Foods.* 2nd ed. Westport, CT: AVI Publishing, 1980.
14. Borgstrom G. *Principles of Food Science.* Vol. I, Food Technology. London: Macmillan, 1969.
15. Ayres JC, Kirshman JC (Eds). *Impact of Toxicology on Food Processing.* Westport, CT: AVI Publishing, 1981.
16. Boyd E (Ed). *Toxicity of Pure Foods.* Cleveland, OH: CRC Press, 1973.
17. Furia TE (Ed). *Handbook of Food Additives.* 2nd ed. 2 Vols. Cleveland, OH: CRC Press, 1972, 1980.
18. Leonard BJ. *Toxicological Aspects of Food Safety.* Archives of Toxicology (Suppl.1, Vol.1). Berlin: Springer-Verlag, 1978.
19. Rechcigl M Jr. (Ed). *CRC Handbook of Naturally Occurring Food Toxicants.* CRC Series in Nutrition and Food. Boca Raton, FL: CRC Press, 1983.
20. U.S. Congress, Office of Technology Assessment. *Environmental Contaminants in Food.* Washington, DC: Office of Technology Assessment, 1979.

21. National Research Council. *Toxicants Occurring Naturally in Foods*. 2nd ed. Washington, DC: National Academy of Sciences, 1973.
22. Frazier WC, Westhoff DC. *Food Microbiology*. New York: McGraw-Hill, 1978.
23. Friedman M (Ed). *Nutritional and Toxicological Aspects of Food Safety*. New York: Plenum Press, 1984.
24. Hathcock JN (Ed). *Nutritional Toxicology*, Vols 1–3. San Diego, CA: Academic Press, 1982, 1987, 1989.
25. Riemann H, Bryan FL (Eds). *Foodborne Infections and Intoxications*. 2nd ed. New York: Academic Press, 1979.
26. Liener IE (Ed). *Toxic Constituents of Plant Foodstuffs*. 2nd ed. New York: Academic Press, 1980.
27. Reilly C. *Metal Contamination of Food*. London: Applied Science Publishers, 1991.
28. Doyle MP (Ed). *Foodborne Bacterial Pathogens*. New York: Marcel Dekker, 1989.
29. Gilchrist A. *Foodborne Disease and Food Safety*. Monroe, WI: American Medical Association, 1981.
30. FASEB. *Emerging Issues In Food Safety and Quality for the Next Decade*. Rockville, MD: Federation of American Societies for Experimental Biology (FASEB), Life Science Research Office (LSRO), 1991.
31. Haberberger RL, et al. Traveler's Diarrhea Among United States Military Personnel during Joint American–Egyptian Armed Forces Exercises in Cairo, Egypt. *Military Med.*, 156:1, (1991), p 27.
32. Echeverria P, Ramirez G, Blacklow NR, et al. Traveler's diarrhea among U.S. army troops in South Korea. *J. Infect Dis.*, 139:215-219, 1979.
33. Gorbach SL. Infectious Diarrhea, Ch. 63, In: Sleisenger MH, Fordtran JS (Eds). *Gastrointestinal Disease*, Vol. 2, Pathophysiology, Diagnosis, Management. 4th ed. Philadelphia: W. B. Saunders, 1989, p 1215.
34. ASTMH. *Tropical Medicine and Hygiene News*. Vol. 40, No. 5, Washington, DC: American Society of Tropical Medicine and Hygiene, November 1991, p 142.
35a. Blattner WA. HIV Epidemiology: Past, Present, and Future. *FASEB J.*, 5:10, (1991), p 2340-2348.
35b. Morris JG. Current Trends in Human Diseases Associated with Foods of Animal Origin. *JAVMA*, 209:12, (1996), p 2045-2047.
35c. Food Safety and Inspection Service, USDA. Pathogen Reduction; Hazard Analysis and Critical Control Point (HACCP) Systems. *Fed. Reg.* Feb 3, 1995; 60: 6774-6889.
36. WHO. *Programme for Control of Diarrhoeal Diseases, Seventh Programme Report*. Geneva: World Health Organization/CDD/90.34, Annex 1, 1988-1989, p 96-98.
37. Anonymous. Cholera Epidemic Continues Spread, But Risks In U.S. Considered Minimal. *U.S. Med.*, 27:21 & 22, (1991), p 6.
38. Bean N, et al. Foodborne Disease Outbreaks, 5-Year Summary, 1983-1987. *MMWR*, Atlanta, GA: USDHEW, PHS/CDC, 39:SS-l, (1990), p 15.
39. American Association of Food Hygiene Veterinarians. CDC Says Foodborne Illness Underreported. *News-O-Gram*, 14:3, (1990), p 14.
40. Young FE. Safety First: Protecting America's Food Supply Part 1. *FDA Consumer*, 22:6, (1988), p 6-7.
41a. Roberts T. Foodborne Illnesses Cost $4.8 Billion in 1987. *Conference of Public Health Veterinarians-Newsletter*, CPHV, East Lansing, MI, Dec 1989, p 13.
41b. Mauskopf JA, French MT. Estimating the Value of Avoiding Morbidity and Mortality from Foodborne Illnesses. *Risk Analysis*, 11:4, (1991), p 619-631.
41c. Craigmill AL. Food Safety Issues. *Environmental Toxicology Newsletter*, 9:1, (1989), p 9.
42a. Miller SA. Food Safety: The Case for an Integrated View. *Vet. Hum. Toxicol.*, 32:3, (1990), p 262.
42b. Cothern RC, Marcus WL. Estimating Risk for Carcinogenic Environmental Contaminants and Its Impact on Regulatory Decision Making. *Regulatory Toxicol. Pharmacol.*, 4 (1984), p 265-274.
43. CDC. Foodborne Disease Outbreaks, 5-Year Summary, 1983-1987. Atlanta, GA: *USPHS, CDC MMWR*, 39:SS-l, CDC Surveillance Summaries, Mar 1990, p 15.
44a. Archer DL, Young FE. Contemporary Issues: Diseases with a Food Vector. *Clin. Microbiol. Rev.*, 1:4, (1988), p 377-398.
44b. Kampelmacher EH. *Foodborne Listeriosis*. Verlag, Germany: Behr's 1989.

44c. Miller AJ, Smith J, Somkuti GA (Eds). *Foodborne Listeriosis.* Amsterdam, Netherlands: Elsevier Science Publishers, Society of Industrial Microbiology, 1990.

45. van Heyningen WE. Bacterial Exotoxins, Ch. 5, In: Braude AI, Davis CE, Fierer J (Eds). *Infectious Diseases and Medical Microbiology.* Philadelphia: W.B. Saunders, 1986, p 45.

46. Madden JM. Concerns Involving Emerging Bacterial Pathogens. In: *AVMA. Veterinary Perspectives on Safety of Foods of Animal Origin. Proceedings of a Symposium,* September 19-20, 1989, at the U.S. Department of Agriculture, Washington, DC. Schaumburg, IL: American Veterinary Medical Association, p 17.

47. Todd E. Epidemiology of Foodborne Illness: North America. *Lancet,* 336, Sept 29 (1990), p 788-93.

47a. Schantz EJ. Seafood Toxicants, Ch. 19, In: *Toxicants Occurring Naturally In Foods.* 2nd ed. Washington, DC: National Academy of Sciences, Committee on Food Protection, 1973, p 424-447.

47b. Foegeding PM, Roberts T, et al. *Foodborne Pathogens: Risks and Consequences.* CAST Task Force No. 122, September 1994; Ames, IA: Council for Agricultural Science and Technology, 1994.

48. Paige JC. A Summary of the Seattle-King County Study: Surveillance of the Flow of Salmonella and Campylobacter In a Community. Presentation before NAFV Academy of Veterinary Preventive Medicine Seminar: Meat/Poultry Microbiology-Disease Transmission. Conducted At: The USAHA Annual Meeting, October 29, 1985, Milwaukee, WI.

49. Dorn CR. Serotype 0157:H7 and Other Enterohemorrhagic *Escherichia coli.* Lecture and Handout at 1990 AVMA Annual Meeting, Orlando, FL, Dept of Epidemiology and Public Health, College of Veterinary Medicine, Ohio State University, Columbus, OH, July 1990.

50. AAFHV. Fish Botulism. *News-O-Gram,* American Assn of Food Hygiene Veterinarians, 15:4, (1991), p 14.

51. AAFHV. Annual Costs of Listeriosis Put At $480 Million. *News-O-Gram.* American Assn of Food Hygiene Veterinarians, 14:3, (1990), p 8.

52. Lecos C. Of Microbes and Milk: Probing America's Worst Salmonella Outbreak. *FDA Consumer,* 20:1, (1986), p 18-21.

53. AAFHV. Food Animal Contamination Up, Consumer Food Safety Knowledge Down. *News-O-Gram.* 13:6, (1989).

53a. USDHHS. *Healthy People 2000: National Health Promotion and Disease Prevention Objectives.*Ch. 12. Food and Drug Safety. Washington, DC: U.S. PO, DHHS Publication No. (PHS) 91-50212, 1991, p 342.

54. FDA. Update on Salmonella Enteritidis In Eggs. *FDA Veterinarian.* 6:2, (1991), p 10.

55. King L. *Salmonella enteriditis* (SE) Control. Presentation at The Food Safety and Inspection Service (FSIS) and The American Association of Food Hygiene Veterinarians (AAFHV) Symposium on Food Safety, Washington, DC, October 23, 1990.

56. AVMA. FDA Urges Institutions to Use Pasteurized Eggs. *JAVMA,* 195:7, (1989), p 877.

57. AAFHV. Salmonella Outbreak Linked To Cantaloupe. *News-O-Gram,* American Association of Food Hygiene Veterinarians, 15:4, (1991), p 6.

58. Lecos CW. A Closer Look at Dairy Safety. *FDA Consumer,* 20:3, (1986), p 14-17.

59a. CDC. Multiple Outbreaks of Staphylococcal Food Poisoning Caused by Canned Mushrooms. *MMWR Epidemiological Notes and Reports,* 38:24, (1989), p 417-18.

59b. Bjerklie, S and Nunes, K. 1994. The Glass House. *Meat Poultry,* USDA FSIS, May 1994, p 40-44.

59c. Centers for Disease Control. Emerging Infectious Diseases. *Morbidity Mortality Weekly Report,* 42, (1993), p 257-260.

59d. CDC (Centers for Disease Control). Update: Multistate Outbreak of *Escherichia coli* 0157:H7 Infections from Hamburgers—Western United States, 1992–1993. *Morbidity Mortality Weekly Report,* 42, (1993), 258-263.

59e. Berkelman RL. Emerging infectious diseases in the United States, 1993. *J. Infect. Dis.,* 170, (1994), p 272.

59f. Anon. New *E. coli* Outbreak Hits Dry Salami. *Meat Poultry,* UDSA FSIS, January (1995), p 6.

59g. Bryan, F. Public Health Problems of Foodborne Diseases and Their Prevention. In: Tu Anthony (Ed). *Food Poisoning.* New York: Marcel Dekker, 1992, p 7.

59gg. AAFHV. Epidemiology of *E. coli* 0157 in Feedlot Cattle. *AAFHV News-o-Gram* Vol. 21(2): 4, April–June 1977.

59h. Cullor J. Human Pathogens on the Farm. *Proceedings of 80th Annual Meeting of the Livestock Conservation Institute (LCI)*. Colorado Springs, CO, April 10-12, 1996, p 142-155.

60a. NAFV. Army Veterinarian Saves Kuwaiti Cattle. *Federal Veterinarian*, National Association of Federal Veterinarians (NAFV), 48:12, (1991), p 11.

60b. Daly CB. Diseases From the Gulf War: American Troops Suffer the Fewest Illnesses of Any Army in History. *Washington Post Health*, December 10, 1991, p 1.

61. Recent CENTCOM disease experience came from a military report authored by LCDR K. Hanson, NAVCENT Preventive Medicine Officer. "Summary Disease Surveillance Information From Operation Desert Storm," Memorandum To: Capt J. Crim, I MEF Surgeon,12 Mar 1991,3p with enclosures. LCDR Hanson concluded that, "disease had a relatively minor impact on the operation." MAJ T. Sanchez, Division of Preventive Medicine, WRAIR, provided a copy of this Memorandum on 3 May 1991 as well as two WRAIR PM Briefing Charts On the Impact of Diseases on Military Operations. Desert Shield/Storm Disease Non-Battle Injury (DNBI) rates (%/week) were apparently lower—anywhere from 10–30% of the rates experienced in Brightstar Exercises during years 1985, 1987, and 1990. These lower rates may reflect the passive (Desert Shield/Storm) vs. active (Brightstar Exercises) epidemiological surveillance employed during these events and not a real difference in disease levels. A falsely diminished DNBI rate could be reflected by differences in collection and reporting of cases. Hospital admission rates for DNBIs during the first 5 weeks of Desert Shield/Storm were apparently one half of the rate in WWII, one sixth the Korean War, twice the Vietnam War, and one tenth the Civil War. Overall, for the 6-month Gulf campaign, the DNBI rates were obviously very low, resulting in little impact on military operations. This information contained no classification restrictions.

62. Newill VA. Significance of Risk Assessment in the Management of Environmental Exposures to Chemical Mixtures. In: Mehlman MA, Lutkenhoff SD (Eds). *Risk Assessment and Risk Management*. Princeton, NJ: Princeton Scientific Publishing, Toxicology and Industrial Health, Vol 5, No 5, 1989, p 635.

63a. NRC. *Toxicity Testing: Strategies to Determine Needs and Priorities*. Washington, DC: National Academy Press, Board on Toxicology, Commission on Life Sciences, National Research Council, 1984, p 3-5.

63b. NAS/CLS/IOM. Introduction. In: *Risk Assessment in the Federal Government: Managing the Process*. Washington, DC: National Academy Press, 1983, p 12.

63c. Hartman DE. *Neuropsychological Toxicology: Identification and Assessment of Human Neurotoxic Syndromes*. New York: Pergamon Press, 1988, p 7, 62, 96.

63d. Landrigran PJ, et al. Clinical Epidemiology of Occupational Neurotoxic Disease. *Neurobehavioral Toxicol.*, 2, (1980), p 43-48.

63e. OTA. *Neurotoxicity: Identifying and Controlling Poisons of the Nervous System*. OTA-BA-436 and 37 (Summary), 1990. Washington, DC: Publications Office, Office of Technology Assessment, U.S. Congress, 1990.

63f. Beall JR. OSTP Brain and Behavior Subcommittee. *NCAC-SOT Newsletter*, Rockville, MD: National Capitol Area Chapter of the Society of Toxicology, 1:2, (1991), p 3.

63g. NCAC-SOT. Neurotoxicity Screening In the Toxicological Evaluation of Food Chemicals. *NCAC-SOT Newsletter*, Rockville, MD: National Capitol Area Chapter of the Society of Toxicology, 1:2, (1991), p 2.

64. Sundlof SF. Drug and Chemical Residues in Livestock, In: G.E. Burrows (Ed.), *Clinical Toxicology, The Veterinary Clinics of North America-Food Animal Practice*, Vol. 5. No 2, Philadelphia: W.B. Saunders, (1989), p 421.

64a. Henderson AK et al. Breast cancer among women exposed to polybrominated biphenyls. *Epidemiology*, 6:5, (1995), p 544-546.

64b. Rosen DH et al. Half-Life of Polybrominated Biphenyl in Human Sera. *Environ. Health Perspectives*, 103:3, (1995), p 272-274.

64c. Chen E. *PBB, An American Tragedy*. Englewood Cliffs, NJ: Prentice-Hall, 1979.

64d. Smith JR. Hawaiian Milk Contamination Creates Alarm. *Science*, 217, 9 July (1982), p 137-140.

65. Sundlof SF. Drug and Chemical Residues in Livestock, In: G.E. Burrows (Ed.), *Clinical Toxicology. The Veterinary Clinics of North America-Food Animal Practice*, Vol. 5, No 2, Philadelphia: W.B. Saunders, 1989, p 411-449.

66. Farley D. Setting Safe Limits on Pesticide Residues. *FDA Consumer*, Rockville, MD: FDA Public Affairs Staff, 22:8, (1988), p 9.

67. Schwabe CW. Food Safety. Ch. 21, In: *Veterinary Medicine and Human Health*. 3rd ed. Baltimore: Williams and Wilkins, 1984, p 539-561.

67a. NRC. *Environmental Epidemiology: Public Health and Hazardous Wastes*. Washington, DC: National Academy Press, 1991.

68. AAFHV. FDA Fails To Find *S. Enteritidis* In Rendered Products. *News-O-Gram*. American Association of Food Hygiene Veterinarians, 14:1, (1990), p 2.

68a. NAS. *Toxicants Occurring Naturally In Foods*. 2nd ed. Washington, DC: National Academy of Sciences, Committee on Food Protection, 1973.

69. AAFHV. End Notes. *News-O-Gram*. American Association of Food Hygiene Veterinarians, 14:4, (1990), p 16.

70. AAFHV. Codex Proper Forum for Hormone Discussion. *News-O-Gram*. American Association of Food Hygiene Veterinarians, 14:3, (1990), p 4.

71a. FDA. Levels Set For Milk Monitoring Program. *FDA Veterinarian*, 6:3, (1991), p 14.

7lb. AVMA. New Publications from FARAD. Announcement in *JAVMA*, Vol 199, No 12, Dec 15, 1991, pl681.

71c. Rico AG (Ed). *Drug Residues in Animals*. Orlando, FL: Academic Press, 1986, 233p.

71d. Crawford LM, Franco DA (Eds). *Animal Drugs and Human Health*. Lancaster, PA: Technomic Publishing Co., Inc., 1994.

71e. Riviere JE, Craigmill AL, Sundlof SF. *CRC Handbook of Comparative Pharmacokinetics and Residues of Veterinary Antimicrobials*. Boca Raton, FL: CRC Press, 1990, 575p.

71f. Sundlof SS, Riviere JE, Craigmill AL. *FARAD: The Food Animal Residue Avoidance Databank-Tradename File-A Comprehensive Compendium of Food Animal Drugs*. Gainesville, FL: University of Florida, 1988, 597p.

71g. Haagsman N, Ruiter A, and Czedik-Eysenbert PB, Eds. *Proceedings Conference on Residues of Veterinary Drugs in Food* (2 volumes). Veldhoven, Netherlands: University of Utrecht, 3–5 May 1993, 719 p.

71h. Moats WA and Medina MB, Eds. *Veterinary Drug Residues — Food Safety*. ACS Symp. Series 636, Washington, DC: ACS, 1996, 192 p.

72. Steele JH. Pesticides and Food Safety: Perception vs. Reality. In: *Proceedings of the Ninety-Third Annual Meeting of the U.S. Animal Health Association (USAHA)*. Richmond, VA: USAHA, 1989, p 275-293.

72a. Cranmer MF, Jr. Carbaryl: A Toxicological Review and Risk Analysis. *Neurotoxicology*, 7:1, (19xx), p 1-78.

73. Buck WB, et al. Overview of Insecticide Use. Lecture Handout at 1991 AVMA Annual Meeting, Seattle, WA, National Animal Poison Control Center, College of Veterinary Medicine, University of Illinois, Urbana, IL, July 1991.

74. Smith EA, Oehme FW. A Review of Selected Herbicides and Their Toxicities. *Vet Hum Toxicol*, 33:6, (1991), p 596-608.

74a. AHI. Food Safety and Animal Drugs Brochure. Alexandria, VA: Animal Health Institute, 1994.

75. FDA. Updates: Pesticide Report. *FDA Consumer*, 24:2, (1990), p 4.

76. AAFHV. Committee On Chemical Toxicants To Begin Work Soon. *News-O-Gram*, American Association of Food Hygiene Veterinarians, 14:2, (1990), p 15.

77a. AAFHV. CA Residue Law To Fight High Violation Rate. *News-O-Gram*. American Association of Food Hygiene Veterinarians, 14:4, (1990), p 7.

77b. Ames BN. Food Constituents as a Source of Mutagens, Carcinogens, and Anticarcinogens. *Prog. Clin. Biol. Res.*, 206, (1986), p 3-32.

78. AFPMB. *Defense Pest Management Information Analysis Center Technical Information Bulletin (TIB)*. Washington, DC: Armed Forces Pest Management Board (AFPMB), September-October 1990, p 3.

79. Anon. California and FDA Report Food Monitoring Results. *Vet. Hum. Toxicol.*, 32:2, (1990), p 105.

80. Ainsworth SJ. Pesticide Conference Drafts Food Safety Guidelines for Key Groups. *C&EN*, 68:7, (1990), p 18-19.

81. Zilberman D, et al. The Economics of Pesticide Use and Regulation. *Science*, 253, 2 August (1991), p 518-522.

82. NAS/NRC. *Regulating Pesticides In Food: The Delaney Paradox.* Washington, DC: National Academy Press, 1987.

83. Kurz PJ, Desk R, Harrington RM. Pesticides. Ch. 5. In: *Principles and Methods of Toxicology.* 2nd ed. Hayes AW (Ed.). New York: Raven Press, 1989, p 156.

84. EPA. Non-Occupational Pesticide Exposure Study. *Vet. Hum. Toxicol.*, 32:5, (1990), p 511.

84a. OTA. *Environmental Contaminants in Food.* Washington, DC: U.S. Congress Office of Technology Assessment, Dec 1979, p 19.

85. Hayes JR, Campbell TC. Food Additives and Contaminants. Ch. 24. In: *Casarett and Doull's Toxicology: The Basic Science of Poisons*, 3rd ed. Klaassen CD, Amdur MO, Doull J (Eds.). New York: MacMillan Press, 1986, p 771-800.

85a. Roblot P et al. Retrospective Study of 108 Cases of Botulism in Poiters, France. *J. Med. Microbiol.*, 40:6, (1994), p 379-84.

86. AAFHV. *News-O-Gram.* American Association of Food Hygiene Veterinarians, 15:2, (1991), p 14.

86a. Nagel SC, Vom Saal FS, Thayer KA, Dhar MG, Boechler M, and Welshons WV. Relative binding affinity-serum modified access (RBA-SMA) assay predicts the relative *in vivo* bioactivity of the xenoestrogens bisphenol A and octylphenol. *Environmental Health Perspectives* 105(1), 1997, 70–76.

87. FDA. Recycled Plastics. *FDA Consumer*, 25:9, (1991), p 11.

88. AAFHV. Concerns About Recycling Materials For Food-Contact Use. *News-O-Gram.* American Association of Food Hygiene Veterinarians, 14:5, (1990), p 11.

89. AAFHV. Health Risk Small From Benzene In Plastic Packaging. *News-O-Gram.* American Association of Food Hygiene Veterinarians, 15:2, (1991), p 9.

90. Hotchkiss JH. *Chemical Interactions Between Foods and Food Packaging.* SS 365. ACS Symposium Proceedings-193rd Denver, CO, Spring 1987, ACS Advances In Chemistry Series Symposium Series (SS), Washington, DC: ACS, 1987.

90a. Risch SJ, Hotchkiss JH (Eds.). *Food and Packaging Interactions II.* American Chemical Society Symposium Series 473, Washington, DC: ACS, 1991.

91a. Grayson M (Ed.). *Kirk-Othmer Concise Encyclopedia of Chemical Technology.* New York: Wiley-Interscience, 1985, p 521-522.

91b. FDA. A Primer on Food Additives. *FDA Consumer*, 22:8, (1988), p 13.

91c. NAS. *The Health Effects of Nitrate, Nitrite, and N-Nitroso Compounds.* Part 1 of a 2-Part Study by the Committee on Nitrate and Alternative Curing Agents in Foods. Washington, DC: National Academy Press, 1981, p 1-11.

91d. Riemann H. Lecture Handout. Food Hygiene and Safety Course. Masters in Preventive Veterinary Medicine (MPVM) Program, University of California, Davis, 1981.

92a. FDA. Reporting Reactions to Additives. *FDA Consumer*, 22:8, (1988), p 16.

92b. Furia TE (Ed.). *Handbook of Food Additives.* Vols 1-2, Boca Raton, FL: CRC Press, 1972, 1980.

93. Klaassen CD. Principles of Toxicology. Ch. 2. In: *Casarett and Doull's Toxicology: The Basic Science of Poisons*, 3rd ed. Klaassen CD, Amdur MO, Doull J (Eds.). New York: MacMillan Press, 1986, p 103,123,133.

94. Rippon JW. *Medical Mycology.* 2nd ed. Philadelphia: W.B. Saunders, 1982, p 3.

94a. Hesseltine CW. Introduction, Definition, and History of Mycotoxins of Importance to Animal Production. In: *Interactions of Mycotoxins in Animal Production.* Washington, DC: National Academy of Sciences, 1979, p 3-18.

95. Christensen CM, Nelson GH. Mycotoxins and Mycotoxicosis. *Modern Veterinary Practice*, May 1976, p 367-371.

95a. Miller JD, Trenholm HL (Eds.). *Mycotoxins In Grain: Compounds Other Than Aflatoxin.* St. Paul, MN: Eagan Press, 1994.

95b. Smith JE, Henderson RS (Eds.). *Mycotoxins and Animal Foods.* Boca Raton, FL: CRC Press, 1991.

95c. Sharma RP, Salunkhe DK. *Mycotoxins and Phytoalexins.* Boca Raton, FL: CRC Press, 1991.

95d. Betina V (Ed.). *Mycotoxins: Production, Isolation, Separation and Purification.* Developments in Food Science, Vol. 8. New York: Elsevier, 1984.

95e. Richard JL, Thurston JR (Eds.). *Diagnosis of Mycotoxicoses.* Boston, MA: Martinus Nijhoff, 1986.

95f. Eaton DL, Groopman JD. *The Toxicology of Aflatoxins: Human Health, Veterinary, and Agricultural Significance.* Orlando, FL: Academic Press, 1993.

95g. Chu FS. Mycotoxin Analysis: General Considerations and Immunological Techniques. Ch. 27. In: *Foodborne Disease Handbook: Diseases Caused By Viruses, Parasites and Fungi.* Vol. 2. Hui YH, Gorham JR, Murrell KD, Cliver DO (Eds.). New York: Marcel Dekker, 1994.

95h. Jackson LS, DeVries JW, Bullerman LB (Eds.). *Fumonisins in Food.* Advances In Experimental Medicine and Biology, Vol. 392. New York: Plenum Press, 1996.

95i. Wyllie TD, Morehouse LG (Eds.). *Mycotoxic Fungi, Mycotoxins, Mycotoxicoses: An Encyclopedic Handbook.* Vol 2. New York: Marcel Dekker, 1978.

96. Steyn PS, Vleggaar R (Eds.). *Mycotoxins and Phycotoxins, Bioactive Molecules.* Vol. 1. New York: Elsevier Science, 1986.

97. Stahr HM (Ed.). *Mycotoxins. Supplement to Analytical Toxicology Manual.* Ames, IA: Iowa State University Press, 1970.

98. Shank RC (Ed.). *Mycotoxins and N-Nitroso Compounds: Environmental Risks.* Vols. 1-2. Boca Raton, FL: CRC Press, 1981.

99. Gorelick NJ. Risk Assessment for Aflatoxin: I. Metabolism of Aflatoxin B_1 By Different Species. *Risk Analysis,* 10:4, (1990), p 539-559.

100. Kuiper-Goodman T. Risk Assessment to Humans of Mycotoxins in Animal-Derived Food Products. *Vet. Hum. Toxicol.,* 33:4, (1991), p 325-333.

101. Rodricks JV, Hesseltine CW, Mehlman MA (Eds.). *Mycotoxins in Human and Animal Health.* Park Forest South, IL: Pathotox, 1977.

102. Palti J. *Toxigenic Fusaria: Their Distribution and Significance as Causes of Disease in Animals and Man.* Berlin: Verlag Paul Parey, 1978.

103. Osweiler GD et al. (Eds.). Mycotoxicoses. In: *Clinical and Diagnostic Toxicology.* 3rd ed. Dubuque, Iowa: Kendall/Hunt, 1985, p 409-442.

104. Moreau C. *Moulds, Toxins and Food.* Chichester, U.K.: John Wiley & Sons, 1979.

105. Marasas WFP, Nelson PE, Toussoun TA (Eds). *Toxigenic Fusarium Species: Identity and Mycotoxicology.* University Park, PA: Pennsylvania State University Press, 1984.

106. Lacey J (Ed.). *Trichothecenes and other Mycotoxins.* New York: John Wiley and Sons, 1985.

107. Kurata H, Uneo Y (Eds.). *Trichothecenes: Chemical, Biological, and Toxicological Aspects. Developments in Food Science.* Vol 4. New York: Elsevier (Kodansha LTD), 1983.

107a. Kurata H, Uneo Y (Eds.). *Toxigenic Fungi: Their Toxins and Health Hazards.* Developments in Food Science. Vol 7. Tokyo: Kodansha, 1984.

108. Krogh P. *Mycotoxins in Food.* New York: Academic Press, 1987.

109. Keeler RF, Tu AT. *Handbook of Natural Toxins.* Vol. 1. Plant and Fungal Toxins. New York: Marcel Dekker, 1983.

110. Joffe AZ. *Fusarium Species: Their Biology and Toxicology.* New York: John Wiley & Sons, 1986.

111. Edds GT. Aflatoxins. In: *Conference on Mycotoxins in Animal Feeds and Grains Related to Animal Health.* Shimoda W (Ed.). FDA Bulletin BVM-80/132. Washington, DC: U.S. Government Printing Office, 1980; or Springfield, VA: National Technical Information Service Bulletin, 1979; p 80-164.

112. Conning, L. *Toxic Hazards in Food.* New York: Raven Press, 1983.

113. Cole RJ, Cox RH . *Handbook of Toxic Fungal Metabolites.* Orlando FL: Academic Press, 1981.

114. Cheeke PR, Shull LR. *Natural Toxicants in Feeds and Poisonous Plants.* Avitvan Nostrand Reinhold, New York, 1987.

115. Ceigler A, Burmeister HR, Vesonder RF. Poisonous Fungi: Mycotoxins and Mycotoxicoses. In: *Fungi Pathogenic for Humans and Animals-Part B. Pathogenicity and Detection.* Howard D (Ed.). New York: Marcel Dekker, 1983.

116. Beasley V. *Trichothecene Mycotoxicosis: Pathophysiologic Effects*, Vols 1-2. Boca Raton, FL: CRC Press, 1989.

117. Van Rensburg SJ. Role of Epidemiology in the Elucidation of Mycotoxin Health Risks. In: *Mvcotoxins in Human and Animal Health*. Rodricks JV, et al. (Eds.). Park Forest South, IL: Pathotox Publishers, 1977, p 699-712.

118. Busby WF, Wogan GN. Aflatoxins. In: *Chemical Carcinogens*. Vol. 2. ACS Monograph 182, C.E. Searle (Ed.). Washington, DC: American Chemical Society, 1984, p 945-1136.

119. Council for Agricultural Science and Technology (CAST). *Mycotoxins: Economic and Health Risks*, Task Force Report No. 116. Ames, Iowa: CAST, 1989.

120. National Research Council (NRC). *Interactions of Mycotoxins in Animal Production*. Proceedings of a Symposium, Jut 13, 1978, Michigan State University. Washington, DC: National Academy of Sciences, 1979.

121. National Research Council (NRC). *Protection Against Trichothecene Mycotoxins*. Washington, DC: National Academy of Sciences, 1983.

122. Uneo Y (Ed.). *Trichothecenes, Chemical, Biological and Toxicological Aspects*. Developments in Food Science Series, Vol. 4. New York: Elsevier (Kodansha Ltd), 1983.

123. Hayes AW. *Mycotoxin Teratogenicity and Mutagenicity*. Boca Raton, FL: CRC Press, 1981.

124. Matossian MK. *Poisons of the Past-Molds, Epidemics, and History*. New Haven, CT: Yale University Press, 1989.

125. Brook TD, Madigan MT. *Biology of Microorganisms*. 6th ed. Englewood Cliffs, NJ: Prentice Hall, 1991, p 817-819.

125a. Osweiler GD. Mycotoxins Ch. 29. In: *Toxicology*. Osweiler GD (Ed.). The National Veterinary Medical Series. Philadelphia, PA: Williams and Wilkins, 1996, p 409.

126. Mossel DAA. *Microbiology of Foods: The Ecological Essentials of Assurance and Assessment of Safety and Quality*. 3rd ed. Utrecht: The University of Utrecht, Faculty of Veterinary Medicine, 1982, Table 6, p 21.

127. Uraguchi K, Yamazaki M (Eds.). *Toxicology. Biochemistry, and Pathology of Mycotoxins*. New York: John Wiley/Halstead Communications, 1978.

128. Busby WF, Wogan GN. Aflatoxins. In: *Chemical Carcinogens*, Vol. 2. ACS Monograph 182, C.E. Searle (Ed.). Washington, DC: American Chemical Society, 1984, p 945-1136.

128a. Hesseltine CW. Global Significance of Mycotoxins. In: *Mycotoxins and Phycotoxins*. Steyn PS,Vleggaar R (Eds.). Amsterdam: Elsevier, 1986, p 1.

129. FDA. Bacteria, Mycotoxins and Environmental Contamination of Feed. *FDA Veterinarian*, 6:6, (1991), p 7-8.

130. Keller WC, Beasley VR, Robens JF. Proceedings of the Symposium on the Public Health Significance of Natural Toxicants in Animal Feeds, Feb 6-7, 1989. *Veterinary Human Toxicol.*, 32, Supplement (1990).

130a. McLachlan JA (Ed.). *Estrogens in the Environment: Influences on Development*. New York: Elsevier, 1985.

130b. Smith JE, Henderson RS (Eds.). *Mycotoxins and Animal Foods*. Boca Raton, FL: CRC Press, 1991.

130c. Diener UL, Ed. Mycotoxicology Newsletter, Dept. of Plant Pathology, Auburn University, AL. 7: 1, 1996. Data submitted to *Food Chem. News*, Nov. j20, 1995.

130d. Steyn PS, Thiel PG, Trinder DW. Detection and Quantitation of Mycotoxins by Chemical Analysis. Ch. 9. In: *Mycotoxins and Animal Foods*. Smith JE, Henderson RS (Eds.). Boca Raton, FL: CRC Press, 1991, p 165-221.

131. Dvorackova I. *Aflatoxins and Human Health*. Boca Raton, FL: CRC Press, 1989.

132. Bruce RD. Risk Assessment for Aflatoxin: II. Implications of Human Epidemiology Data. *Risk Analysis*, 10:4, (1990), p 561-569.

133. Anon. Mycotoxin Occurrence In USA In 1990-91. *Vet. Hum. Toxicol.*, 33:5, (1991), p 440.

133a. Smalley EB et al., Wisconsin Cooperative Regional Project NC-129 Annual Progress Report, 1994 (for Jan. 1, 1993–Dec. 31, 1993).

133b. Diener UL (Ed.). *Mycotoxicology Newsletter*. Dept. of Plant Pathology, Auburn University, AL. 7:1, (1996), p 2.

133c. Lisker N, Lillehoj EB. Prevention of Mycotoxin Contamination (Principally Aflatoxins and Fusarium Toxins) at the Preharvest Stage. Ch. 29. In: *Mycotoxins and Animal Foods*. Smith JE, Henderson RS (Eds.). Boca Raton, FL: CRC Press, 1991, p 689-719.

134. Smith LD. *Botulism: The Organism, Its Toxins, the Disease.* Springfield, IL: Charles C.Thomas, 1977.

135. DHEW. *Botulism in the United States.* Washington, DC: U.S. Department of Health, Education, and Welfare, 1967.

136. CDC. *Morbidity and Mortality Weekly Report (MMWR).* Atlanta, GA: Centers for Disease Control, 34:54, (1985), p 3.

137. CDC. *Morbidity and Mortality Weekly Report (MMWR).* Atlanta, GA: Centers for Disease Control, 34:55, (1986), p 3.

138a. CDC. *Morbidity and Mortality Weekly Report (MMWR).* Atlanta, GA: Centers for Disease Control, 36:54, (1987), p 3.

138b. Frazier WC, Westhoff DC. Food-Borne Infections and Intoxications: Bacterial. Ch. 23. In: *Food Microbiology.* New York: McGraw Hill, 1978, p 423.

139. CDC. *Morbidity and Mortality Weekly Report (MMWR).* Atlanta, GA: Centers for Disease Control, 40:24, (1991), p 412-414.

139a. Sugiyama H. *Clostridium botulinum* Neurotoxin. *Microbiol. Rev.* 44:3, (1980), p 419-448.

139b. Miller DA. Nonfoodborne Clostridial Diseases. In: *Handbook of Zoonoses,* 2nd ed. GW Beran, JH Steele (Eds.). Boca Raton, FL: CRC Press, 1994, p 127-147.

140. CDC. International Outbreak of Type E Botulism Associated with Ungutted, Salted Whitefish. *MMWR* 36, (1987), p 812-813.

141. St. Louis ME, Peck SHS, Bowering D, et al. Botulism from Chopped Garlic: Delayed Recognition of a Major Outbreak. *Ann. Intern. Med.,* 1087, (1988), p 363-368.

142. Meyer KF, Eddie B. *Sixty-Five Years of Human Botulism in the United States and Canada.* Berkeley, CA: The University of California, 1965.

143. Wang Y, Burr DH, Korthals GJ, Sugiyama H. Acute Toxicity of Aminoglycoside Antibiotics as an Aid in Detecting Botulism. *Appl. Environ. Microbiol.,* 48:5, (1984), p 951-955.

144. Siegel LS, Johnson-Winegar AD, Sellin LC. Effect of 3,4-Diaminopyridine on the Survival of Mice Injected with Botulinum Neurotoxin Type A B E or F. *Toxicol. Appl. Pharmacol.,* 84:2, (1986), p 255-263.

145. Lewis GE, Wood RM. Effects of 3,4-Diaminopyridine in Cynomolgus Monkeys Poisoned with Type A Botulinum Toxin. *Adv. Biosci. (England),* 35, (1982), p 313-323.

146. Simpson LL. A Preclinical Evaluation of Aminopyridines as Putative Therapeutic Agents in the Treatment of Botulism. *Infect. Immun.,* 52:3, (1986), p 858-862.

147. Tyner CF. WRAIR Research: Ricin and Botulinum Toxins. Washington, DC: Walter Reed Army Institute of Research, *WRAIR Annual Report,* January 1990, p 28.

148. Brown L et al. *Acute Oral Toxicity of Guanidine Nitrate In Mice.* LAIR Institute Report No. 276, San Francisco, CA: Letterman Army Institute of Research, Presidio of San Francisco, CA, July 1988, p 13.

149. FDA. Updates: Drug from Botulism Toxin. *FDA Consumer,* 24:2, (1990), p 2.

150. Taylor SL. Marine Toxins of Microbial Origin. *Food Technol.* 42:3, (1988), p 94-98.

151. Taylor SL. Histamine Food Poisoning: Toxicology and Clinical Aspects. *CRC Crit. Rev. Toxicol.* 17:2, (1986), p 91-128.

152. Taylor SL, Bush RK. Allergy by Ingestion of Seafoods. Ch. 7. In: *Marine Toxins and Venoms.* Vol. 3 of Handbook of Natural Toxins. Tu A. (Ed.). New York: Marcel Dekker, 1988, p 149-183.

153. Taylor SL, Stratton JE, Nordlee JA. Histamine Poisoning (Scombroid Fish Poisoning): Allergy-Like Intoxication. *J. Toxicol. Clin. Toxicol.,* 27:4, (1989), p 225-240.

154. AAFHV. Biogenic Amines. *News-O-Gram.* American Association of Food Hygiene Veterinarians, 15:5, (1991), p 17.

155. AAFHV. Biogenic Amines In Foods. *News-O-Gram.* American Association of Food Hygiene Veterinarians, 15:4, (1991), p 17.

156. Desrosier NW. Preserving of Food With Ionizing Radiation. Ch. 10. In: *The Technology of Food Preservation.* Westport, CT: AVI Publishing, 1963, p 349.

157. Keeler RF. Naturally Occurring Teratogens From Plants. Ch. 5. In: *Plant and Fungal Toxins.* Vol. 1 of Handbook of Natural Toxins. RF Keeler, AT Tu (Eds.). New York: Marcel Dekker, 1983, p 189.

158. Boyd MR et al. Lung-Toxic Furanoterpenoids Produced by Sweet Potatoes following Microbial Infection. *Biochim. Biophys. Acta,* 337, (1974), p 184-195.

159. Boyd MR, Wilson BJ. Isolation and Characterization of 4-Ipomeanol, a Lung-Toxic Furanoterpenoid, Produced by Sweet Potatoes (*Ipomoea batatas*). *J. Agric. Food Chem.,* 20, (1972), p 428-430.

160. Rippon JW. *Medical Mycology.* 2nd ed. Philadelphia: W.B. Saunders, 1982, p 712.

161. Wilson BJ.Toxicity of Mold Damaged Sweet Potatoes. *Nutr. Rev.,* 31, (1973), p 73-78.

162. Peckham JC et al. Atypical Interstitial Pneumonia in Cattle Fed Moldy Sweet Potatoes. *JAVMA,* 160, (1972), p 169-172.

163. Doster AR et al. Effects of 4-Ipomeanol, a Product from Mold Damaged Sweet Potatoes on the Bovine Lung. *Vet. Pathol.,* 15, (1978), p 367-375.

164a. Christian MC et al. 4-Ipomeanol: A Novel Investigational New Drug for Lung Cancer. *JNCI,* 81:15, (1989), p 1133-1143.

164b. Katz GV. Metals and Metalloids Other Than Mercury and Lead. In: *Neurotoxicity of Industrial and Commercial Chemicals.* Vol 1. O'Donoghue JL (Ed.). Boca Raton, FL: CRC Press, 1985, p 171-191.

164c. Mahaffey KR. *Dietary and Environmental Lead: Human Health Effects.* Vol. 7, Topics In Environmental Health. New York: Elsevier, 1985.

165. Reddy CS, Hayes AW. Food-Borne Toxicants. Ch. 3. In: *Principles and Methods of Toxicology,* 2nd ed. Hayes AW (Ed.). New York: Raven Press, 1989, p 101.

166. Goyer RA. Toxic Effects Of Metals. Ch. 19. In: *Casarett and Doull's Toxicology: The Basic Science of Poisons,* 3rd ed. Klaassen CD, Amdur MO, Doull J (Eds.). New York: MacMillan Press, 1986, p 608.

166a. OTA. *Environmental Contaminants in Food.* Washington, DC: U.S. Congress Office of Technology Assessment, 1979.

166b. Jaroslovsky R. Contamination of Fish-Meal with PCBs Sparks FDA Study of Mishap at Ralston Purina Plant. *Wall Street J.,* July 12, 1978, p 46.

166c. Higuchi K (Ed.). *PCB Poisoning and Pollution.* New York: Academic Press, (Kodansha Ltd) 1976.

166d. Schuurman H-J, Krajnc-Franken MAM, Kuper CF, van Loveren H, Vos JG. Immune System. Ch. 17. In: *Handbook of Toxicologic Pathology.* Haschek WM, Rousseaux CG (Eds.). San Diego, CA: Academic Press, 1991, p 453.

167. FDA. School Children Injured by Contaminated Milk, *FDA Consumer,* 20:6, (1986), p 35-36.

167a. Black C, Welsh K. Occupationally and Environmentally Induced Scleroderma-Like Illness: Etiology, Pathogenesis, Diagnosis, and Treatment. *Internal Med.,* 9:6, (1988), p 135-154.

167b. EUROTOX. Symposium 1. Toxicologists versus Toxicological Disasters: Toxic Oil Syndrome (TOS). Abstracts of the 35th European Congress of Toxicology-EUROTOX '96, Alicante, Spain, 22-25 September 1996, *Toxicol. Lett.,* 88, Suppl. 1, (1996), p 1-2.

168. Beasley V. *NAPINeT Report.* Urbana, IL: National Animal Poison Information Network (NAPINeT), 1991.

168a. Dolev E. Medical Aspects of Terrorist Activities. *Military Med.,* 153:5, (1988), p 243-44.

169. Young FE. Health Talk With Frank Young: Weighing Food Safety Risks. *FDA Consumer.* 23:7, (1989), p 12.

170. Anon. Fungicide in Imported Wines. *FDA Consumer,* 24:6, (1990), p 3.

171. Young. Weighing Food Safety Risks, *FDA Consumer,* 23:7, (1989), p 8-13.

172. Anon. Is Anything Safe. *Time,* Mar. 27, 1989.

173. Anon. How Safe Is Your Food. *NEWSWEEK,* Mar. 27, 1989.

174. Roses OE et al. Mass Poisoning By Sodium Arsenite. *Clin. Toxicol.,* 29:2, (1991), p 209-213.

175. Finnegan N, Brown L, Smith W. Report To US Congress On Special Pays for Veterinary Professionals, In: *Report to U.S. Congress on Special Pays for AMEDD Professionals.* Falls Church, VA: Department of the Army, Office of The Surgeon General, Feb. 1989.

176. Russell L, Jr. The Future Role of the Veterinarian in Food Safety and His Preparation for It. *JAVMA,* 163:9, (1973), p 1071-1074.

177. Townsend JF. Professor, Pathology and Anatomical Science. University of Missouri School of Medicine, Columbia, MO, Pathology 200 Coursenotes, Fall 1996 Semester.

chapter nineteen

Consumer attitudes and perceptions

Christine M. Bruhn

Contents

Introduction: food is more than mere subsistence

Meal patterns have changed markedly in the last decades. The typical meal of the 1950s that began with bacon and eggs, prepared by the wife/mother full-time homemaker is no longer the pattern of today. Now the day begins with most family members taking care of themselves, probably with a bowl of cereal or a microwaveable meal.[1]

Food habits have changed due to a variety of factors, including changing taste preferences, lifestyle and convenience factors, attitudes and perceptions about diet and health, advances in technology that influence food quality and availability, and economic factors. Taste, nutrition, and convenience are the driving forces in today's market. Safety, wholesomeness, and even social issues also influence today's consumers. Shaping these perceptions are the overlaying changing demographics and lifestyle of the population.

Demographic influences on food selection

One of the major social–economic trends in the last few decades is the increased participation of women in the labor force.[1] The overall labor force participation rate for women went from 34.8% in 1960 to 56.6% in 1988.[2] This high participation rate is even found among women whose youngest child is less than 6 years old. Dual-earning households means rising family incomes and increased expenditures on more expensive items, including food. It also means a changing lifestyle in which people are increasingly willing to pay for convenience. People are exchanging money for time-saving restaurant dining, take-out meals, and convenient home preparation.

Household size

Household size has declined from 3.3 in 1960 to 2.6 in 1989.[2] Singles living alone, on the average both the young and the old, account for about a quarter of all U.S. households. Over half of all households are make up of one or two persons, which translates into an increased demand for food products in smaller sizes.

Income

Household income affects food expenditure. Higher-income houses spend 66% more per person on weekly food expenditures.[3] Despite growing affluence, over 16% of the population receive food assistance in the form of food stamps, vouchers, and food packages.[4]

Ethnicity

The ethnic mix in the U.S. population is changing. The fastest-growing ethnic groups are Latinos and Asian-Americans. Latinos in the U.S. are projected to be 30 million by 2000.[5]

Race also influences food expenditure. Households headed by African-Americans spend less per person than those headed by Caucasians or other races. In 1986, Caucasian households spent $19.98 per week per person, African-American $12.96 per person, a 54% difference. Native Americans, Eskimos, and Asian-Americans spent 10% less than Caucasians, but 39% more than African-Americans.[3] Some food choices are culturally or regionally based. Preference and the ability to pay both influence food choice.

Age

The number of people 65 years of age and older is projected to more than double over the next 50 years, going from 30 million now to 68 million in 2040.[6] As the age of household members increases, per person food spending tends to rise until age 65, primarily due to higher incomes.[3] Both the increased size of this group and its greater buying power will impact market demand. Older Americans, more concerned about nutrition and health, will find the health aspects of low-sodium, low-fat, high fiber foods appealing.

Region of the Country

About half of the population growth from now to the year 2000 will occur in Sunbelt states, California, Texas, and Florida.[7] When people move, they often leave behind some of their old food habits and acquire tastes of the new region. The increased consumption of Mexican and Southern-style food may be partly explained by regional migration.[8] Household spending also reflects regional differences; however some of these differences are the result of regional variation in food prices, household incomes, and food types.[3]

Facets of food quality

Quality is an illusive, ever-changing concept. It includes sensory properties of flavor and texture, as well as psychological and other factors that the food contributes. People select a food because they anticipate it will provide value, in terms of sensory satisfaction, contributions to health, and compatibility with lifestyle conditions.

Although good taste is essential, there is more behind perceptions of quality than taste. The traditional factors associated with food choice are cultural and regional, and many food preferences are formed during childhood. A product can be perceived as high quality if it offers a nostalgic taste and flavor experiences comparable to that enjoyed in the past. Some foods convey status in their use; although these have changed over time, status foods are generally expensive, difficult or time-consuming to prepare, and often associated with animal protein. During the great depression of the 1930s, chicken was a status food, associated with the "chicken in every pot" political promise. Today, chicken is seen as a routine healthy food, while high-status foods are lobster, filet mignon, or specialty items prepared from high-quality fresh ingredients. Foods that signify affection, security, and comfort are reward foods. Mom's cookies, homemade bread, and premium ice cream are examples. Foods also create moods that can be conveyed to the consumer. Champagne and sparkling juices, for example, are celebration foods appropriate for festive occasions. Other foods convey body images, such as grapefruit for slimming and steak for athletic power. Some are thought to carry "magical" properties; oysters, for example, are considered aphrodisiacs. Special properties of food differ by culture; because of their pinkish and reddish color, peaches signify good luck and longevity in Japan.

Today, "fresh" is a highly desirable characteristic. A quality processed product should provide flavor and texture comparable to that of a freshly prepared counterpart. Convenience is also of paramount concern to many households. A quality product, then, should require minimal preparation and/or cooking time. Health aspects are of importance to an increasingly large segment of the population. A quality processed product should provide the nutrients comparable to the fresh counterpart without hazardous compounds added or formed during processing.

Attitudinal influences that impact perceptions of quality

In the late 1980s, the Food Marketing Institute identified 10 trends in consumer attitudes, consistent with other observations, that are likely to continue for several years and to affect quality perception.[1]

Neotraditionalism

This trend is associated with a desire for premium-quality products and a willingness to pay more to obtain them. It is reflected in the increased demand for gourmet foods, such as gourmet coffee, specialty cheeses, and premium-quality fruits and vegetables. The demand for quality is reflected in increased demand for fresh and frozen produce, but a drop in consumption of canned produce. It also includes the desire for goods with minimal environmental impact. Consumers want foods, packaging, and processing that do not threaten the environment. Food should protect the long-term health and wellbeing of consumers. Food should taste good and be in good taste (i.e., it should add to the value of life).

Price is important, especially in hard economic times, but for the average consumer, it ranks as less important than other factors. The Food Marketing Institute found that coupon use peaked in 1992 and that consumers use of coupons and other time-intensive techniques peaked in 1992.[9] In 1994 about 20% compare prices in different supermarkets, down from 25% in 1992, and 30% stock up on items when they are a bargain.

Adventure

The trend for adventure expresses itself in a desire for variety, new tastes, and new foods. With increased income, consumers seek more variety in the goods they purchase. Small households and those with more adult women are more likely to display diversity in food purchases.

A population that is wealthier and better traveled also desires variety. This desire is enhanced by an increasingly diverse population with a multitude of ethnic and regional groups. As ethnic groups move to new locations, they bring their distinctive food choices. The large Asian and Hispanic population in the West accounts for much of the consumption of exotic vegetables.[1]

Individualism

An individualistic lifestyle allows people to make statements about themselves and their beliefs through their product choices. The microwave oven has fostered individualism by allowing each family member to easily prepare a snack or meal independent of others. At least two thirds of the children prepared at least one meal a week without supervision in 1990, and even preschoolers used microwave ovens to prepare food for themselves.

Indulgence

The trend toward indulgent self-gratification is consistent with the trend toward neotraditionalism, adventure, and individualism. It is driven by a large number of working adults who have little leisure time and who feel they have earned the right to indulge. People indulge in treats that enhance their pleasure or status. With more income, these treats will be more frequent and more conspicuous.

While buying behavior reflects nutritional awareness, a counter trend is also apparent. Candy consumption rose 11%, cola drinks 13.5%, and snack foods 23%.[3] Many lower-fat and lower-calorie foods entered and left the market because they failed to deliver the taste consumers sought.

Cocooning

This term represents the tendency for people to stay home. Cocooning reflects people's desire to protect themselves from the hassles of public appearances and social occasions. It can been seen by the number of people who eat while doing something else, such as watching television, reading, working, or doing some other activity.

The working population is likely to continue cocooning, while elderly and single-person households will seek opportunities for social interaction. People are expected to spend more time working and playing at home. The tendency for cocooning suggests an increase in foods prepared away from home (take-home) and in foods prepared in a simple manner while at home.

Grazing

The tendency to nibble throughout the day and eat fewer full, sit-down meals is consistent with the need for adventure, variety, and cocooning. Take-out food, finger food, and microwave ovens facilitate this tendency.

Wellness

Physical fitness and health are increasingly important. This trend is consistent with the generally aging and better informed public. People will make the effort to modify dietary habits if they believe it will enhance their likelihood of living to a healthy, independent old age.

Nutritional awareness is at an all-time high.[9-11] About 70% of men and women believe their diet could be at least somewhat healthier. Perceptions of working women are closely aligned to that of men, with more unemployed women believing their diet is already healthy enough. The belief that diet could be healthier is found among all income levels, although it is lower among those under $15,000 per year and at all age groups except those over 65. Presumably, older Americans believe that they are already eating as healthy as they can.

People tell us they are increasing their consumption of produce and decreasing their intake of foods high in saturated fat and cholesterol.[9-12] Sales records confirm that consumers are eating more grains and poultry, less red meat.[4,13] Consumer concern about dietary issues is greatest for fat and cholesterol, which has resulted in a proliferation of reduced-fat or fat-free new products. Despite annual statements that they are eating more fruits and vegetables, people have not significantly increased produce purchases. Vegetable consumption (including potatoes) increased 1.5 lb per capita between 1989 and 1992, while fruit consumption peaked in 1989 and has decreased 6.5 lb by 1992.[14]

A survey among 1000 health-conscious Americans provides a closer look at those who buy for health.[1] The major motivations for buying healthy foods were identified as for medical reasons, to promote daily stamina, to lose weight, or for value-related "spiritual" reasons. "Investors" (45%) eat well to ensure future good health; they are likely to be college-educated men and they work longer hours than any other group. "Managers" (36%) choose healthy foods to gain an edge in short-term performance; they are likely to be college-educated managers or administrators. "Healers" (9%) are mostly older, low-income women. "Strugglers" (7%) are low-income parents torn between their love for high-fat, high-sugar food and their health awareness. "Disciples" (3%) are animal-rights activists, Seventh-Day Adventists, and others who follow a dietary regimen for philosophical reasons. Information about these categories is used to position products and prepare advertising, package copy, and nutritional information to appeal to investors and managers.

Controlling time (through convenience)

An increased need to manage time will lead to a great demand for convenience in foods and other goods and services. The need to balance time is in direct conflict with other demands, for example, for variety, quality, and health. Achieving all these ends will continue to be a challenge.

Convenience is one of the most important attributes of food products. Fast-paced, two-income life-styles have reduced the amount of time available for preparing food at home and increased the demand for quick, easy-to-prepare meals. There is increased willingness to pay more to buy convenience and service.[1] As the adult population ages, its tendency to use more convenient foods, especially those that are microwaveable, increases.[3] Even in dual-employment households, women still do over 90% of meal preparation.[15] As incomes rise and pressure on available time increases, many consumers are feeling dollar-rich and time-poor.[1]

The demand for convenience impacts shopping as well as meal preparation. Only 30% of consumers like to shop. Increasingly, people would prefer to order from a catalog. People especially hate standing in check-out lines and waiting for home deliveries. Between 1981 and 1994, the average number of stores visited per week dropped from 2.6 to 2.1.[9,16]

Teens are doing an increased amount of shopping, both for themselves and for the family.[17] Both boys and girls do family shopping, but girls do a bit more, with teens spending $47.7 billion in 1988. In 70% of the homes where both or single parents work, teenagers do much of the grocery shopping. Teens say they consider price and taste most important, and brand name is a distant third. Nevertheless, 70% say they often chose a national brand they had used and liked. Fast and easy-to-prepare foods were important. Teens were quite familiar with microwaves, and 77% said they use microwaves daily.

Consumer demand for convenience food implies continued introduction of new products. The microwave oven has had an important impact on at-home food preparation. An increase in the percent of households owning microwave ovens has contributed to the growth of shelf-stable foods. These account for $250 million in sales in 1988, and could reach $696 million in 1993.[4] The value of food prepared for the microwave is expected to grow from $5 billion in 1988 to $7 billion in 1993. Now some microwaved food products are developed specifically for children and include activity kits.

Supermarkets have added convenience by cleaning and precutting fresh vegetables, packaging vegetables and meat together for stir frying, and precutting melons and pineapples. The food service sector of the supermarket has experienced significant growth.[18] Two thirds of all supermarkets now have service delis, half offer salad and soup bars, 42% hot pizza, and 42% hot pasta. Many stores have in-store microwave ovens. Catering services are now offered in 40% of supermarkets. Customer conveniences in the future will include drive-up windows and separate entrances for deli.

People are eating out more as incomes rise and as more women enter the work force. Since 1965, the away-from-home food expenditures have increased nearly ninefold, more than double the at-home rate.[3] It was found that 61% of personal food expenditures went for food to be used at home, and about one third for food eaten away from home.

Single consumers and those under age 25 spend nearly 50% of their food budgets on food away from home. Oldest households—75 years and over—spent the lowest share of their food dollar in restaurants (32%).[4,19] Large households spent less money eating out than smaller households.[3]

Selectivity

Consumers are increasingly intolerant of unsafe, poorly constructed, and overpriced products. Many manufacturers have initiated toll-free numbers to handle complaints; however, if consumers are still dissatisfied, they might switch brands, unless the product offers unique features.

Ethics

There is an increased desire to be socially responsible with respect to environmental impact, to be tolerant of diversity, to improve health care, and to respond to the needs of the poor and homeless.

Packaging can have a major impact on the consumer's perception of ethical products. In the past, packaging has focused on product protection and marketing appeal. Newer package designs also include ease of use, with reuseable and recylable containers and microwaveable cooking containers. About one third of consumers are concerned about "over-packaging," and the ability to recycle products, particularly those with multilayer packaging materials.[9] A leading food processor notes that their *total quality management* program addresses environmental issues. They are using resources efficiently and developing creative solutions for waste management.[20]

In addition to environmental impact, about 40% of consumers say they are curtailing the purchase of products that may involve what they consider unethical treatment of animals.[9] Cosmetics are advertised as produced without animal testing, and some consumers are boycotting veal because of concern over farm management practices. Some consumer objection to the use of bovine somatotrophin (BST) for enhancing milk production relates to the concern for animal health.[21]

The promise of new technologies: how people respond

New technologies can help the food industry exploit production methods and provide characteristic products that consumers find desirable. Consumer attitudes toward the possible benefits of, and potential risks posed by, new technologies determine the rapidity with which potential commercialization is realized. Two of these technologies, food irradiation and biotechnology, have faced similar hurdles; however, their potential for market adoption varies.

Irradiation

Irradiation processing involves the exposure of food to high-energy radiation from either radioactive isotopes such as cobalt or electron- or X-ray-producing machines.[22] The effect of this treatment varies according to the total dose of energy absorbed. A high-dose treatment sterilizes spices and can substitute for the fumigant, ethylene oxide. Pathogens in raw poultry and meat can be reduced by as much as 99.9% by a lower "pasteurization" dose of radiation. Grain and produce can be disinfested of insects and larvae by still lower doses, and the natural decay of some fruits and vegetables can be retarded. Irradiation is an effective quarantine treatment for the Mediterranean fruit fly and other pests, and can substitute for methyl bromide, used in pest control.

In all cases, irradiation is a cold treatment. The food maintains a raw, fresh-like character. Irradiation is considered safe by the World Health Organization, the American Medical Association, U.S. Food and Drug Administration, and various health and safety authorities in over 30 countries.

Opposition to irradiation can be related to the use of nuclear energy, preference for unprocessed food, skepticism regarding food safety, perception that the technology is not needed, and concern for worker and environmental safety. Special interest groups have systematically distorted information regarding irradiation to the public. Congress specified in the 1950s that irradiation should be treated as a "food additive" and FDA has the responsibility for processing conditions and labeling. Extremists threaten to boycott brands using irradiated ingredients and to picket markets carrying irradiated food. Some food processors stated they do not use irradiated ingredients, and most major supermarkets did not offer irradiated food.

Research on consumer attitudes conducted in the mid-1980s found that consumers were cautious regarding food irradiation. People knew little about the technology and wanted more information before deciding on its use. Subsequent consumer research consistently demonstrated that most consumer and health professionals respond positively toward food irradiation when given factual information about the technology.[23-34] People respond more positively when the reasons for irradiation were provided and when irradiated products were available for sampling.[30,33,34] The most recent nationwide consumer survey in the U.S. indicates that concern about irradiation is less than other food related concerns.[9] Additionally the percentage expressing significant concern has decreased in recent years while those having no concerns has increased.

After limited marketing tests in the late 1980s, a specialty market in Florida offered irradiated strawberries in 1992. This initial commercial offering was rapidly followed by a specialty market in Glenview, IL selling strawberries and a host of other produce items. Fewer than ten activists picketed the Florida market.[31,35] Carrot Top, the specialty grocery store in Glenview, IL indicated that irradiated produce outsold conventional produce by 10 to 1 in the first year of sales and reached 20 to 1 thereafter. Between 1995 and 1997, more than 100,000 pounds of irradiated tropical fruit were sold in independent supermarket chains in the midwest.[36] No major processor irradiates poultry; however, irradiated poultry sold well in limited marketing.[37]

The key to successful marketing was a quality product, consumer information, and a retailer prepared to face potential critics. Consumers could experience the quality and longer shelf life of irradiated strawberries, tomatoes, mushrooms, onions, and other produce items. If accompanied by educational material and presented by a retailer the consumer trusts, irradiated food would be acceptable to the majority of consumers.

Biotechnology

Biotechnology is a general term for several techniques that use living organisms to make or modify products for a specific purpose. The techniques of biotechnology offer opportunities to address consumer issues of food quality and environmental safety. Biotechnology can be used to make fruit more flavorful, to improve the nutritional quality of fruits, vegetables, grains, and muscle foods, to grow foods in a wider climatic zone, and to grow foods in a more environmentally benign fashion. In the U.S., consumer attitudes toward biotechnology are generally positive; however, the public has questions, some of which are highlighted by special interest groups.

A recent nationwide study found that 78% of U.S. consumers believe they will benefit from biotechnology in the next 5 years.[38] People were most aware of the potential positive effect on farm economics, food quality, and nutrition.[39]

Consumers will evaluate the individual applications of biotechnology based upon their perceptions of the benefits. Generally, people are more accepting of applications relating to plant products than to animal products. People express the concern that modifications by biotechnology disrupt nature and could lead to unforeseen consequences; however, they hold this same concern about traditional cross pollination and breeding.[39] Some question the ethics and safety of taking genetic material from one source and inserting it into another. People want to be assured that biotechnology is being used to address areas they consider important. They want to be sure the benefits are equitably distributed and available to all. They want to know that product safety has been evaluated, potential risks have been fully explored, and measures have been taken to reduce risk until it is virtually nonexistent.

Surveys indicate people are interested in receiving information about biotechnology, particularly information about potential risks and benefits.[39-40] Interest was expressed by people in all income and education levels and even by some with little or no interest in science.

As with irradiation, there are special interest groups opposed to biotechnology, and some of them distort information given to the public. Opponents of biotechnology are skeptical about product safety, believe specific applications are not needed, prefer traditional "unchanged" food, charge that animals are not treated humanely, and describe the science as interfering with God or nature. They call for mandatory labeling so consumers can choose. Extremist groups threaten brand boycotting and supermarket picketing, and they call for supermarkets, processors, and chefs to pledge not to use products modified by biotechnology. Some well-known chefs followed the guideline set by activist groups; many of these after learning about potential applications of biotechnology and the safety and regulatory system disassociated themselves from the movement.

Although an enzyme used in cheese manufacture was the first biotechnology-modified product used in food, public interest and controversy centered around cows supplemented with recombinant BST to increase milk production and the tomato modified to delay softening, which enables the fruit to stay on the vine longer and develop more flavor. Despite efforts by activists, both products appear to be accepted. Controversy was reduced regarding use of the biotechnologically modified tomato, since the product is clearly labeled and the consumer can immediately determine any flavor benefit. Although marketed on a small scale due to limited product availability, the tomato has sold well. Use of recombinant BST is much more complex. Most products are not labeled, the product is consumed by children—who people believe to be vulnerable and in need of extra protection—and the modification involves animals. Although consumers have had questions about milk safety, endorsements by health professionals appears to have eased most people's minds.[40] Activists call for labeling and choice in the marketplace, but their actions indicate the want to stop use of DNA technology, even when the product is clearly labeled. At least one processor offering both a regular line and milk labeled from nonsupplemented cows has been threatened with a nationwide boycott. The activist group does not want the processor to accept any milk from supplemented cows.

Endorsements by numerous respected health and medical groups likely enhances consumer perceptions of safety. Consumer surveys indicate that the credibility of activist groups has decreased significantly in recent years.[40-41] Food industry resistance to activist threats may have also been interpreted by the public as a safety endorsement. These experiences suggest that the public is not as likely to be swayed by activists comments today as in previous years.

Market predictions

The following are some predictions regarding market trends based on our understanding of consumer attitudes and perceptions:

- Growth in convenience foods, especially lightly processed items with flavor similar to freshly prepared foods.
- Increased emphasis on food safety and increased intolerance of foodborne disease. Gradual introduction of irradiation to enhance food safety and replace less safe chemical fumigants.
- Development of nutrafoods, foods that emphasize disease protection. With increased knowledge of diet and health, interest has increased in vitamins E, C, and A and other food constituents associated with health-enhancing qualities. Consum-

ers currently choose foods such as broccoli, carrots, cabbage, and apples for cancer prevention.[42] Advanced breeding technology could produce fruits or vegetables with high levels of health-enhancing vitamins and oils with a fatty acid profile least likely to raise blood cholesterol.

- Gradual nationwide blending of ethnic specialties. Chefs are preparing innovative combinations of classic cuisine and unique local ingredients. Use of specialty items such as flavorful salad greens, lemon grass, nopales (cactus leaves), and chevre (goat cheese) seen in upscale restaurants will gradually spread across the country and become incorporated into home cuisine.
- Increased environmental labeling based upon clearly stated guidelines.

Summary: the opportunity to develop market niches

Today's fast-paced life allows for increased opportunities for people to pursue individual preferences. There will continue to be expansion of lower calorie, lower fat foods with the use of new ingredients and/or technology for improved flavor. Microwave meals, for example, can be individually chosen for each family member. Convenience foods also enhance the opportunity for grazing, or eating fewer sit-down meals and snacking throughout the day.

References

1. Senauer, B., Asp, E., Kinsey, J. Food Trends and the Changing Consumer, 1991, Ch. 1, 2, 3.
2. U.S. Department of Commerce, Bureau of the Census, 1990, p 378.
3. Blaylock, J. and Elitzak, H. Food Expenditures National Food Review. July-September, 1990, 17-25.
4. Schwenk, N. E. Food trends. Family Economics Review 1991, 4(3), 16-23.
5. Corchado. Campbell Soup is seeking to be numero uno where Goya reigns. Wall Street Journal, March 28, 1988, p 20.
6. Wall Street Journal, People patterns. Feb. 7, 1989, p B1.
7. Wall Street Journal, People patterns. Nov. 17, 1988, p B1.
8. Kiplinger Agriculture Letter, January 30, 1987.
9. Opinion Research, Trends 1994. Consumer attitudes and the supermarket. Food Marketing Institute, Washington, DC, 1994.
10. Gallup Organization. How are Americans making food choices? American Dietetics Association and the International Food Information Council. Chicago, and Washington, DC, 1994.
11. Packwood Research Corporation. Shopping for health, eating in America: Perception and reality. Food Marketing Institute and Prevention Magazine, Washington, DC and Emmaus, PA, 1994.
12. Needham, D. D. B. The Changing American Lifestyle Trends 1975-91. DDB Needham Worldwide, Chicago, IL, 1991.
13. Gordon, E. et al. National Restaurant Association, Foodservice Industry Forecast, National Restaurant Association, Washington, DC, 1990.
14. Putnam, J. J. Food Supply and Use. Agricultural Outlook, Economic Research Service, USDA, 1994.
15. Burros. Women: Out of the house but not out of the kitchen. New York Times, February 24, 1988.
16. Opinion Research: Trends 1989. Consumer attitudes and the supermarket. Food Marketing Institute, Washington, DC, 1989.
17. Blumenthal, D. When teens take over the shopping cart. FDA Consumer, March 1990, 131-132.
18. Putnam, J. J. Food Consumption, Prices, and Expenditures 1967-1988. USDA Economics Research Service, 1990, Statistical Bulletin No. 804.

19. U.S. Department of Labor - Bureau of Labor Statistics. Consumer Expenditures in 1989. News USDL, 1990, No. 90-616.
20. Western Grocery News, 1992.
21. McCuirk, A. M. and Kaiser, H. M. BST & milk: Benefit or bane? Choices (First Quarter), 1991, p 20.
22. Diehl, J. F. Safety of Irradiated Foods. Marcel Dekker, NY, 1990.
23. Bruhn, C. M. and J. W. Noel. Consumer in-store response to irradiated papayas. Food Technol. 1987, 41(9), 83-85.
24. Bruhn, C. M., Schutz, H. G., Sommer, R. Attitude change toward food irradiation among conventional and alternative consumers. Food Technol. 1986, 40(1), 86-91.
25. Bruhn, C. M., Sommer, R., Schutz, H. G. Effect of an educational pamphlet and posters on attitude toward food irradiation. J. Ind. Irradiat. Technol. 1986, 4(1), 1-21.
26. Bruhn, C. M., Schultz, H. G., Sommer, R. Food irradiation and consumer values. Ecol. Food Nutr. 1987, 21, 219-235.
27. Bruhn, C. M. and Schutz, H. G. Consumer awareness and outlook for acceptance of food irradiation. Food Technol. 1989, 43(7), 93-94, 97.
28. Gallup Organization. Irradiation: Consumers' attitudes. Prepared for the American Meat Institute, NJ, 1993.
29. Johnson, F. C. S. Knowledge and attitudes of selected home economists toward irradiation in food preservation. Home Econ. Res. J. 1990, 19(2), 170-183.
30. Pohlman, A. J. Influence of education and food samples on consumer acceptance of food irradiation. Masters of Science Thesis, Purdue University, W. Lafayette, IN, 1993.
31. Pszczola, D. E. Irradiated produce reaches midwest market. Food Technol. 1992, 46(5), 89-92.
32. Pszczola, D. E. Irradiated poultry makes U.S. debut in midwest and Florida markets. Food Technol. 1993, 47(11), 89, 92, 94, 96.
33. Resurreccion, A. V. A. F., Galvez, C. F., Fletcher, S. M., Misra, S. K. Consumer attitudes toward irradiated food, results of a new study. Presented at 1993 Annual Meeting of the Institute of Food Technology, Chicago, IL, 1993.
34. Schutz, H. G., Bruhn, C. M., Diaz-Knauf, K. V. Consumer attitudes toward irradiated foods: Effects of labeling and benefits information. Food Technol. 1989, 43(10), 80-86.
35. Marcotte, M. Irradiated strawberries enter the U.S. market. Food Technol. 1992, 46(5), 80, 82, 84, 86.
36. Wong, L. Personal communication. Hawaiian Department of Agriculture, Honolulu, HI, 1997.
37. Anonymous. The irradiation issue. Food Safety Consortium 5(3), 1, 5, 1995.
38. Wirthlin Group. Consumer survey on biotechnology. International Food Information Council, 1997.
39. Hoban, T. J., IV and Kendall P. A. Consumer attitudes about the use of biotechnology in agriculture and food production. North Carolina State University, Raleigh, NC, 1992.
40. Hoban, T. J. IV. Consumer awareness and acceptance of bovine somatotropin (BST). Conducted by North Carolina State University for the Grocery Manufacturers of America, Raleigh, NC, 1994.
41. McNutt, K. W. Consumer acceptance of irradiated foods. Presented to R&DA Irradiated Food Products Committee, Boston, MA, 1985.
42. The Packer. Fresh Trends, a profile of fresh produce consumers, Lincolnshire, IL, 1994.

Index

T